St. Olaf College

OCT 0 4 1991

Science Library

D1786861

Nonlinear Ordinary Differential Equations and Their Applications

PURE AND APPLIED MATHEMATICS

A Program of Monographs, Textbooks, and Lecture Notes

EXECUTIVE EDITORS

Earl J. Taft
Rutgers University
New Brunswick, New Jersey

Zuhair Nashed
University of Delaware
Newark, Delaware

CHAIRMEN OF THE EDITORIAL BOARD

S. Kobayashi
University of California, Berkeley
Berkeley, California

Edwin Hewitt
University of Washington
Seattle, Washington

EDITORIAL BOARD

M. S. Baouendi
University of California, San Diego

Donald Passman
University of Wisconsin-Madison

Jack K. Hale
Georgia Institute of Technology

Fred S. Roberts
Rutgers University

Marvin Marcus
University of California, Santa Barbara

Gian-Carlo Rota
Massachusetts Institute of Technology

W. S. Massey
Yale University

David L. Russell
Virginia Polytechnic Institute and State University

Leopoldo Nachbin
Centro Brasileiro de Pesquisas Físicas and University of Rochester

Jane Cronin Scanlon
Rutgers University

Anil Nerode
Cornell University

Walter Schempp
Universität Siegen

Mark Teply
University of Wisconsin-Milwaukee

MONOGRAPHS AND TEXTBOOKS IN PURE AND APPLIED MATHEMATICS

1. *K. Yano*, Integral Formulas in Riemannian Geometry (1970) *(out of print)*
2. *S. Kobayashi*, Hyperbolic Manifolds and Holomorphic Mappings (1970) *(out of print)*
3. *V. S. Vladimirov*, Equations of Mathematical Physics (A. Jeffrey, editor; A. Littlewood, translator) (1970) *(out of print)*
4. *B. N. Pshenichnyi*, Necessary Conditions for an Extremum (L. Neustadt, translation editor; K. Makowski, translator) (1971)
5. *L. Narici, E. Beckenstein, and G. Bachman*, Functional Analysis and Valuation Theory (1971)
6. *S. S. Passman*, Infinite Group Rings (1971)
7. *L. Dornhoff*, Group Representation Theory (in two parts). Part A: Ordinary Representation Theory. Part B: Modular Representation Theory (1971, 1972)
8. *W. Boothby and G. L. Weiss (eds.)*, Symmetric Spaces: Short Courses Presented at Washington University (1972)
9. *Y. Matsushima*, Differentiable Manifolds (E. T. Kobayashi, translator) (1972)
10. *L. E. Ward, Jr.*, Topology: An Outline for a First Course (1972) *(out of print)*
11. *A. Babakhanian*, Cohomological Methods in Group Theory (1972)
12. *R. Gilmer*, Multiplicative Ideal Theory (1972)
13. *J. Yeh*, Stochastic Processes and the Wiener Integral (1973) *(out of print)*
14. *J. Barros-Neto*, Introduction to the Theory of Distributions (1973) *(out of print)*
15. *R. Larsen*, Functional Analysis: An Introduction (1973) *(out of print)*
16. *K. Yano and S. Ishihara*, Tangent and Cotangent Bundles: Differential Geometry (1973) *(out of print)*
17. *C. Procesi*, Rings with Polynomial Identities (1973)
18. *R. Hermann*, Geometry, Physics, and Systems (1973)
19. *N. R. Wallach*, Harmonic Analysis on Homogeneous Spaces (1973) *(out of print)*
20. *J. Dieudonné*, Introduction to the Theory of Formal Groups (1973)
21. *I. Vaisman*, Cohomology and Differential Forms (1973)
22. *B.-Y. Chen*, Geometry of Submanifolds (1973)
23. *M. Marcus*, Finite Dimensional Multilinear Algebra (in two parts) (1973, 1975)
24. *R. Larsen*, Banach Algebras: An Introduction (1973)
25. *R. O. Kujala and A. L. Vitter (eds.)*, Value Distribution Theory: Part A; Part B: Deficit and Bezout Estimates by Wilhelm Stoll (1973)
26. *K. B. Stolarsky*, Algebraic Numbers and Diophantine Approximation (1974)
27. *A. R. Magid*, The Separable Galois Theory of Commutative Rings (1974)
28. *B. R. McDonald*, Finite Rings with Identity (1974)
29. *J. Satake*, Linear Algebra (S. Koh, T. A. Akiba, and S. Ihara, translators) (1975)

30. *J. S. Golan*, Localization of Noncommutative Rings (1975)
31. *G. Klambauer*, Mathematical Analysis (1975)
32. *M. K. Agoston*, Algebraic Topology: A First Course (1976)
33. *K. R. Goodearl*, Ring Theory: Nonsingular Rings and Modules (1976)
34. *L. E. Mansfield*, Linear Algebra with Geometric Applications: Selected Topics (1976)
35. *N. J. Pullman*, Matrix Theory and Its Applications (1976)
36. *B. R. McDonald*, Geometric Algebra Over Local Rings (1976)
37. *C. W. Groetsch*, Generalized Inverses of Linear Operators: Representation and Approximation (1977)
38. *J. E. Kuczkowski and J. L. Gersting*, Abstract Algebra: A First Look (1977)
39. *C. O. Christenson and W. L. Voxman*, Aspects of Topology (1977)
40. *M. Nagata*, Field Theory (1977)
41. *R. L. Long*, Algebraic Number Theory (1977)
42. *W. F. Pfeffer*, Integrals and Measures (1977)
43. *R. L. Wheeden and A. Zygmund*, Measure and Integral: An Introduction to Real Analysis (1977)
44. *J. H. Curtiss*, Introduction to Functions of a Complex Variable (1978)
45. *K. Hrbacek and T. Jech*, Introduction to Set Theory (1978)
46. *W. S. Massey*, Homology and Cohomology Theory (1978)
47. *M. Marcus*, Introduction to Modern Algebra (1978)
48. *E. C. Young*, Vector and Tensor Analysis (1978)
49. *S. B. Nadler, Jr.*, Hyperspaces of Sets (1978)
50. *S. K. Segal*, Topics in Group Rings (1978)
51. *A. C. M. van Rooij*, Non-Archimedean Functional Analysis (1978)
52. *L. Corwin and R. Szczarba*, Calculus in Vector Spaces (1979)
53. *C. Sadosky*, Interpolation of Operators and Singular Integrals: An Introduction to Harmonic Analysis (1979)
54. *J. Cronin*, Differential Equations: Introduction and Quantitative Theory (1980)
55. *C. W. Groetsch*, Elements of Applicable Functional Analysis (1980)
56. *I. Vaisman*, Foundations of Three-Dimensional Euclidean Geometry (1980)
57. *H. I. Freedman*, Deterministic Mathematical Models in Population Ecology (1980)
58. *S. B. Chae*, Lebesgue Integration (1980)
59. *C. S. Rees, S. M. Shah, and C. V. Stanojević*, Theory and Applications of Fourier Analysis (1981)
60. *L. Nachbin*, Introduction to Functional Analysis: Banach Spaces and Differential Calculus (R. M. Aron, translator) (1981)
61. *G. Orzech and M. Orzech*, Plane Algebraic Curves: An Introduction Via Valuations (1981)
62. *R. Johnsonbaugh and W. E. Pfaffenberger*, Foundations of Mathematical Analysis (1981)
63. *W. L. Voxman and R. H. Goetschel*, Advanced Calculus: An Introduction to Modern Analysis (1981)
64. *L. J. Corwin and R. H. Szcarba*, Multivariable Calculus (1982)
65. *V. I. Istrătescu*, Introduction to Linear Operator Theory (1981)
66. *R. D. Järvinen*, Finite and Infinite Dimensional Linear Spaces: A Comparative Study in Algebraic and Analytic Settings (1981)

67. *J. K. Beem and P. E. Ehrlich,* Global Lorentzian Geometry (1981)
68. *D. L. Armacost,* The Structure of Locally Compact Abelian Groups (1981)
69. *J. W. Brewer and M. K. Smith, eds.,* Emmy Noether: A Tribute to Her Life and Work (1981)
70. *K. H. Kim,* Boolean Matrix Theory and Applications (1982)
71. *T. W. Wieting,* The Mathematical Theory of Chromatic Plane Ornaments (1982)
72. *D. B. Gauld,* Differential Topology: An Introduction (1982)
73. *R. L. Faber,* Foundations of Euclidean and Non-Euclidean Geometry (1983)
74. *M. Carmeli,* Statistical Theory and Random Matrices (1983)
75. *J. H. Carruth, J. A. Hildebrant, and R. J. Koch,* The Theory of Topological Semigroups (1983)
76. *R. L. Faber,* Differential Geometry and Relativity Theory: An Introduction (1983)
77. *S. Barnett,* Polynomials and Linear Control Systems (1983)
78. *G. Karpilovsky,* Commutative Group Algebras (1983)
79. *F. Van Oystaeyen and A. Verschoren,* Relative Invariants of Rings: The Commutative Theory (1983)
80. *I. Vaisman,* A First Course in Differential Geometry (1984)
81. *G. W. Swan,* Applications of Optimal Control Theory in Biomedicine (1984)
82. *T. Petrie and J. D. Randall,* Transformation Groups on Manifolds (1984)
83. *K. Goebel and S. Reich,* Uniform Convexity, Hyperbolic Geometry, and Nonexpansive Mappings (1984)
84. *T. Albu and C. Năstăsescu,* Relative Finiteness in Module Theory (1984)
85. *K. Hrbacek and T. Jech,* Introduction to Set Theory, Second Edition, Revised and Expanded (1984)
86. *F. Van Oystaeyen and A. Verschoren,* Relative Invariants of Rings: The Noncommutative Theory (1984)
87. *B. R. McDonald,* Linear Algebra Over Commutative Rings (1984)
88. *M. Namba,* Geometry of Projective Algebraic Curves (1984)
89. *G. F. Webb,* Theory of Nonlinear Age-Dependent Population Dynamics (1985)
90. *M. R. Bremner, R. V. Moody, and J. Patera,* Tables of Dominant Weight Multiplicities for Representations of Simple Lie Algebras (1985)
91. *A. E. Fekete,* Real Linear Algebra (1985)
92. *S. B. Chae,* Holomorphy and Calculus in Normed Spaces (1985)
93. *A. J. Jerri,* Introduction to Integral Equations with Applications (1985)
94. *G. Karpilovsky,* Projective Representations of Finite Groups (1985)
95. *L. Narici and E. Beckenstein,* Topological Vector Spaces (1985)
96. *J. Weeks,* The Shape of Space: How to Visualize Surfaces and Three-Dimensional Manifolds (1985)
97. *P. R. Gribik and K. O. Kortanek,* Extremal Methods of Operations Research (1985)
98. *J.-A. Chao and W. A. Woyczynski, eds.,* Probability Theory and Harmonic Analysis (1986)
99. *G. D. Crown, M. H. Fenrick, and R. J. Valenza,* Abstract Algebra (1986)
100. *J. H. Carruth, J. A. Hildebrant, and R. J. Koch,* The Theory of Topological Semigroups, Volume 2 (1986)

101. *R. S. Doran and V. A. Belfi*, Characterizations of C*-Algebras: The Gelfand-Naimark Theorems (1986)
102. *M. W. Jeter* Mathematical Programming: An Introduction to Optimization (1986)
103. *M. Altman*, A Unified Theory of Nonlinear Operator and Evolution Equations with Applications: A New Approach to Nonlinear Partial Differential Equations (1986)
104. *A. Verschoren*, Relative Invariants of Sheaves (1987)
105. *R. A. Usmani*, Applied Linear Algebra (1987)
106. *P. Blass and J. Lang*, Zariski Surfaces and Differential Equations in Characteristic $p > 0$ (1987)
107. *J. A. Reneke, R. E. Fennell, and R. B. Minton*. Structured Hereditary Systems (1987)
108. *H. Busemann and B. B. Phadke*, Spaces with Distinguished Geodesics (1987)
109. *R. Harte*, Invertibility and Singularity for Bounded Linear Operators (1988).
110. *G. S. Ladde, V. Lakshmikantham, and B. G. Zhang*, Oscillation Theory of Differential Equations with Deviating Arguments (1987)
111. *L. Dudkin, I. Rabinovich, and I. Vakhutinsky*, Iterative Aggregation Theory: Mathematical Methods of Coordinating Detailed and Aggregate Problems in Large Control Systems (1987)
112. *T. Okubo*, Differential Geometry (1987)
113. *D. L. Stancl and M. L. Stancl*, Real Analysis with Point-Set Topology (1987)
114. *T. C. Gard*, Introduction to Stochastic Differential Equations (1988)
115. *S. S. Abhyankar*, Enumerative Combinatorics of Young Tableaux (1988)
116. *H. Strade and R. Farnsteiner*, Modular Lie Algebras and Their Representations (1988)
117. *J. A. Huckaba*, Commutative Rings with Zero Divisors (1988)
118. *W. D. Wallis*, Combinatorial Designs (1988)
119. *W. Więsław*, Topological Fields (1988)
120. *G. Karpilovsky*, Field Theory: Classical Foundations and Multiplicative Groups (1988)
121. *S. Caenepeel and F. Van Oystaeyen*, Brauer Groups and the Cohomology of Graded Rings (1989)
122. *W. Kozlowski*, Modular Function Spaces (1988)
123. *E. Lowen-Colebunders*, Function Classes of Cauchy Continuous Maps (1989)
124. *M. Pavel*, Fundamentals of Pattern Recognition (1989)
125. *V. Lakshmikantham, S. Leela, and A. A. Martynyuk*, Stability Analysis of Nonlinear Systems (1989)
126. *R. Sivaramakrishnan*, The Classical Theory of Arithmetic Functions (1989)
127. *N. A. Watson*, Parabolic Equations on an Infinite Strip (1989)
128. *K. J. Hastings*, Introduction to the Mathematics of Operations Research (1989)
129. *B. Fine*, Algebraic Theory of the Bianchi Groups (1989)
130. *D. N. Dikranjan, I. R. Prodanov, and L. N. Stoyanov*, Topological Groups: Characters, Dualities, and Minimal Group Topologies (1989)

131. *J. C. Morgan II,* Point Set Theory (1990)
132. *P. Biler and A. Witkowski,* Problems in Mathematical Analysis (1990)
133. *H. J. Sussmann,* Nonlinear Controllability and Optimal Control (1990)
134. *J.-P. Florens, M. Mouchart, and J. M. Rolin,* Elements of Bayesian Statistics (1990)
135. *N. Shell,* Topological Fields and Near Valuations (1990)
136. *B. F. Doolin and C. F. Martin,* Introduction to Differential Geometry for Engineers (1990)
137. *S. S. Holland, Jr.,* Applied Analysis by the Hilbert Space Method (1990)
138. *J. Okniński,* Semigroup Algebras (1990)
139. *K. Zhu,* Operator Theory in Function Spaces (1990)
140. *G. B. Price,* An Introduction to Multicomplex Spaces and Functions (1991)
141. *R. B. Darst,* Introduction to Linear Programming: Applications and Extensions (1991)
142. *P. L. Sachdev,* Nonlinear Ordinary Differential Equations and Their Applications (1991)
143. *T. Husain,* Orthogonal Schauder Bases (1991)

Other Volumes in Preparation

Nonlinear Ordinary Differential Equations and Their Applications

P. L. SACHDEV
Indian Institute of Science
Bangalore, India

MARCEL DEKKER, INC.　　　New York • Basel • Hong Kong

Library of Congress Cataloging-in-Publication Data

Sachdev, P. L.
　　Nonlinear ordinary differential equations and their applications /
P.L. Sachdev.
　　　　p.　cm. -- (Monographs and textbooks in pure and applied
mathematics : 142)
　　Includes bibliographical references and index.
　　ISBN 0-8247-8364-6
　　1. Differential equations.　I. Title.　II. Series.
QA372.S147　1991
5.5 .352--dc20
　　　　　　　　　　　　　　　　　　　　　　　　　　　　90-46627
　　　　　　　　　　　　　　　　　　　　　　　　　　　　CIP

This book is printed on acid-free paper.

Copyright © 1991 by MARCEL DEKKER, INC.　All Rights Reserved

Neither this book nor any part may be reproduced or transmitted in any form or by any means, electronic or mechanical, including photocopying, microfilming, and recording, or by any information storage and retrieval system, without permission in writing from the publisher.

MARCEL DEKKER, INC.
270 Madison Avenue, New York, New York　10016

Current printing (last digit):
10　9　8　7　6　5　4　3　2　1

PRINTED IN THE UNITED STATES OF AMERICA

Dedicated to the memory of my parents

Maya Devi and Munshi Ram

Preface

The purpose of this book is to present simple analytic methods which are readily accessible for use in applications. These include transformations, phase plane analysis, integral equation formulation, shooting arguments, local and asymptotic analysis, singular point analysis, and others. Our approach is entirely constructive, from which we may also obtain some existence results.

The applications, reflecting the research interests of the author, are in fluid mechanics, nonlinear diffusion, nonlinear waves, and mathematical models of biology. Other nonlinear DE, such as Lane-Emden, Fermi-Thomas, and their generalizations, have always been useful for illustrating nonlinear methods. Each method will be illustrated by simple examples before the theory and application to realistic physical models are given.

Chapters 1 to 4 are elementary. Chapter 1 reviews the theory of linear ODE. Second-order DE are given special attention. The method of variation of parameters and Green's function are emphasized.

Chapter 2 deals with the many ways of transforming nonlinear ODE in order to solve them analytically. These include exact linearization, conversion to autonomous form or to a simpler nonlinear form, and reduction to a simultaneous system that can be treated in the Lie or phase plane. The chapter ends with some nontrivial examples of DE which, though amenable to several neat transformations, can still elude solution in simple explicit form. The Riccati, Abel, and Langmuir equations are discussed.

Chapter 3 concerns the solution of initial and boundary value problems via series expansions. For the latter, the nature of the singularities that nonlinear DE may possess is first discussed in great detail in the complex plane. The Briot-Bouquet equation, owing to its important role, occupies a prominent position. Other examples include the initial value problems for equations of boundary layer theory. The convergence of the series in each case is carefully analyzed. The results of numerical solutions are also provided for the sake of comparison.

Chapter 4 is in some sense a continuation of Chapter 3. It encompasses both local and global analyses. Again the series solutions are found for local analysis. The Painlevé property deciding the nature of the singularities of DE is introduced and elaborated with the help of examples. Local analysis includes the large-time (space) behavior of solutions, as dealt with by Levinson (1970), for example. The global solution of boundary value problems by integral equation formulation and subsequent iteration is illustrated with the help of the second Painlevé equation and a nonlinear Bessel equation. Connection problems for a class of second-order nonlinear DE arising in the theory of generalized Burgers equations are explained and solved analytically. Numerical results are given along with analytic ones to show the accuracy of the latter.

Chapters 5 to 9 include recent developments in the field. Chapter 5 discusses the shooting technique for boundary value problems. The analysis is simple, rigorous, and constructive. Boundary value problems solved include those related to flow in a porous medium and to some equations of electromagnetic theory and plasma physics. The second Painlevé equation is treated as a special case.

Chapter 6 concerns the phase plane and (three-dimensional) phase space analysis. Our main concern is with quadratic and three-dimensional systems. The examples include Fisher's equation, a nonlinear diffusion equation, gas-dynamic equations describing the collapse of a spherical cavity in a perfect gas, and the equations describing the dynamics of a compressible isothermal atmosphere.

Chapter 7 briefly describes one aspect of the phenomenon of chaos, namely, the relationship between the analytic structure of the singularities of the nonlinear DE and the possibility of the occurrence of chaos. The equations dealt with include the celebrated Lorenz model, the Henon-Heiles Hamiltonian system, and the Kuramoto model. This chapter extends some of the ideas developed in Chapter 4.

Chapter 8 details the analytic theory of the Painlevé equations I-VI, which is not readily available in the literature. This continues and extends the related work reported by Ince (1956) in the light of more recent investigations.

Chapter 9 deals with systems of nonlinear PDE drawn mostly from fluid mechanics. The solution of initial boundary problems is sought in terms of an infinite series in which the similarity solution of a simpler system is the first term. The higher-order terms yield the effects due to the presence of lower-order terms in the PDE. The problem reduces to solving an infinite system of ODE with suitable initial/boundary conditions. The problems investigated include N-wave solutions of nonplanar Burgers equation, expansion of a gas sphere (cylinder) into a vacuum, a converging shock wave from a spherical or cylindrical piston, the shallow water equations describing gravity current release, and the problem of the climb of a bore over a sloping beach.

As background for this book, some basic knowledge of real and complex analysis

Preface

and elementary linear ordinary differential equations is needed. The book proceeds at a leisurely pace, illustrating the ideas with numerous examples. This accounts for its relatively large size. Many of the examples have been drawn from the research literature and have been duly referenced. The choice of material reflects the research interests of the author; there is no claim to comprehensiveness. However, the topics included should help a large class of scientists and engineers understand and solve nonlinear ODE which they may encounter in their investigations.

The book follows the activist school of mathematics (cf. Bender and Orszag 1978, for example), which analyzes and organizes the ideas. There is no claim to a high level of rigor, although there are portions, such as in Chapter 5, where existence results are proved quite precisely. In any case, this book is not written to "build a logically impeccable mathematical structure in which definitions, theorems, and rigorous proofs are welded together into a formidable barrier which the reader is challenged to penetrate." Instead, the reader is invited to share the joy of discovering the richness and beauty that nonlinear ODE manifestly unfold.

It was during my visit to the Courant Institute, New York University, as a Fulbright scholar in 1983 that I conceived of this book. I discussed this project with Professors Martin Kruskal, W. F. Ames, and H. Segur, who were very enthusiastic. When returning to India, I stopped briefly at the Institute of Mathematics and Its Applications, Oxford as a Visiting Merril Fellow. I benefitted from discussions with Dr. Allen Tayler and Dr. John Ockendon, who offered critical and constructive suggestions. As I wrote different chapters, the size of the book grew beyond my earlier plans. I was fortunate to receive help in this challenging task from several colleagues. Dr. Varughese Philip and Professor M. R. Raghavachar checked through several chapters. Professors N. Rudraiah and V. G. Tikekar were helpful in various other ways. Ms. Neelam Gupta assisted in the proof correction. Several students provided generous assistance. My wife, Rita, braced me up in my low moments, as I spent long hours in completing this work.

Financial assistance was provided by the Curriculum Development Cell, established at the Indian Institute of Science by the Ministry of Education and Culture, Government of India.

P. L. Sachdev

Contents

PREFACE   v

INTRODUCTION   ix

1. REVIEW OF LINEAR ORDINARY DIFFERENTIAL EQUATIONS

    1.1   Introduction   1
    1.2   Linear and Nonlinear ODE   3
    1.3   Initial and Boundary Value Problems   6
    1.4   Elements of Second-Order Linear DE   9
    1.5   Method of Variation of Parameters   13
    1.6   Green's Function   16

2. TRANSFORMATIONS OF NONLINEAR ORDINARY DIFFERENTIAL EQUATIONS   23

    2.1   Introduction   23
    2.2   Simple Symmetries—Equidimensionality and Scale Invariance   24

	2.3	First-Order Nonlinear Equations and Their Linearization—The Riccati Equation	28
	2.4	Abel's Equation	30
	2.5	Relation Between Third-Order Linear Equations and Second-Order Nonlinear Equations	33
	2.6	Relationship Betwen Linear and Nonlinear Second-Order Equations	37
	2.7	Exact Linearization of Nonlinear Autonomous Second-Order Equations via Factoring	42
	2.8	Transformation of Nonlinear Equations to Nonlinear Integrable Forms	48
	2.9	Reducing Nonlinear Nonautonomous Second-Order Equations to First-Order Equations	52
	2.10	Transforming Nonlinear Nonautonomous Second-Order Equations to the Lie Plane	59
	2.11	Langmuir's Equation—When Transformations Do Not Work	63
	2.12	Example of an Equation Which Is "Solved" Exactly, but Has a Solution too Implicit to Be of Practical Use	66
3.	SERIES SOLUTIONS OF NONLINEAR DIFFERENTIAL EQUATIONS		69
	3.1	Introduction	69
	3.2	Classification of Singular Points of Homogeneous Linear Differential Equations	70
	3.3	The Point at Infinity	79
	3.4	Singularities of Nonlinear Differential Equations	81
	3.5	First-Order Binomial Equations	87
	3.6	Singularities of the Briot-Bouquet Equation	89
	3.7	Solutions of the Briot-Bouquet Equation	94
	3.8	Power Series Solutions of Nonlinear DE	97
	3.9	Method of Majorants	109
	3.10	Construction of Majorant Equations	112
	3.11	Series Solution of Boundary Value Problems	119
	3.12	Asymptotic Series Solution of Boundary Value Problems	123
	3.13	Computation of Radius of Convergence of a Series	129

Contents

4. **LOCAL AND GLOBAL ANALYSIS OF NONLINEAR DIFFERENTIAL EQUATIONS** — 133

   4.1 Introduction — 133
   4.2 Local Analysis About Singular Points—The Painlevé Property — 140
   4.3 Large-Time (Asymptotic) Behavior — 152
   4.4 Embedding of Asymptotic Solutions in a Larger Family—The Thomas-Fermi Equation — 165
   4.5 Global Solutions of Nonlinear Boundary Value Problems by Iteration — 176
   4.6 The Connection Problem for Euler-Painlevé Equations — 187
   4.7 Other Nonlinear DE Related to Euler-Painlevé Equations — 201

5. **EXISTENCE THEORY FOR BOUNDARY VALUE PROBLEMS VIA SHOOTING TECHNIQUES** — 213

   5.1 Introduction — 213
   5.2 Elementary Discussion of Boundary Value Problems — 214
   5.3 Lagerstrom Model for Flow at Low Reynolds Numbers — 220
   5.4 Similarity Solutions of the Porous Media Equation — 224
   5.5 A Differential Equation Arising in Electromagnetic Theory — 231
   5.6 A Boundary Value Problem Associated with an Equation of Plasma Physics — 240
   5.7 Some Other Boundary Value Problems — 243

6. **PHASE SPACE STUDY OF AUTONOMOUS SYSTEMS** — 245

   6.1 Introduction — 245
   6.2 One-Dimensional Phase Space or the Phase Line — 247
   6.3 Nature of Singular Points in the Phase Plane — 249
   6.4 Examples of Simple Critical Points — 261
   6.5 Singular Points at Infinity — 268
   6.6 Nonsimple Singular Points — 273
   6.7 Quadratic Systems — 284
   6.8 Normal Quadratic Systems — 287

6.9	Nonlinear Diffusion Equation of Population Growth—Fisher's Equation	303
6.10	Boundary Value Problems for a Nonlinear Diffusion Equation	308
6.11	Collapse of a Spherical Cavity in a Perfect Gas	316
6.12	Nonlinear Traveling Waves in an Isothermal Atmosphere	324
6.13	Singular Points of a System of Three Linear Differential Equations	330

7. SINGULARITY STRUCTURE AND CHAOTIC BEHAVIOR OF NONLINEAR ORDINARY DIFFERENTIAL EQUATIONS — 351

7.1	Introduction	351
7.2	The Lorenz System	365
7.3	Henon-Heiles Hamiltonian System	394
7.4	Some Other Hamiltonian Systems	409
7.5	The Kuramoto Model	413
7.6	Painlevé Property for Some Other Systems	419

8. PAINLEVÉ TRANSCENDENTS — 421

8.1	Introduction	421
8.2	Special Solutions of Painlevé Equations	436
8.3	Transformations and Special Solutions of Painlevé Equations	442
8.4	Solutions of Second-Order Nonlinear DE with Irrational Right Sides	452
8.5	The First Painlevé Equation	460
8.6	The Second Painlevé Equation	473
8.7	The Fourth Painlevé Equation	482
8.8	The Fifth Painlevé Equation	493
8.9	The Third Painlevé Equation	501

9. APPLICATIONS OF THE THEORY OF NONLINEAR ORDINARY DIFFERENTIAL EQUATIONS TO SOLUTIONS OF PARTIAL DIFFERENTIAL EQUATIONS—SOME PHYSICAL PROBLEMS — 512

9.1	Introduction	512
9.2	N-Wave Solutions of Nonplanar Burgers Equation	513

Contents

9.3	Collapse of a Spherical Cavity in a Perfect Gas—The Global Solution	520
9.4	The Converging Shock Wave from a Spherical or Cylindrical Piston	529
9.5	Solutions of the Shallow-Water Equations Representing Gravity-Current Releases	538
9.6	Generalized Similarity Solution for Climb of a Bore over a Sloping Beach	545

REFERENCES 559

INDEX 573

Introduction

Most natural phenomena are nonlinear. Therefore, they are described, in general, by a system of nonlinear partial differential equations (PDE), together with suitable initial/boundary conditions that are derived from a specific physical context. These equations are difficult to solve analytically, chiefly because the principle of linear superposition of solutions does not hold. Thus, most mathematical tools developed for linear systems cannot be used. Even a clear classification of nonlinear PDE has not yet been achieved.

However, in many cases, it is possible to reduce these PDE exactly to nonlinear ordinary differential equations (ODE) by using group-theoretic methods or dimensional analysis. This happens, for example, when the original system has two independent variables, while the reduced system depends upon only one. While the ODE represent a certain degenerate form of the original system, they give extremely important information. The solutions of these systems are often referred to as self-similar or traveling waves. These nonlinear ODE have a very rich structure and have solutions which, in the Soviet literature, are referred to as intermediate asymptotics (Barenblatt 1979). These solutions are not isolated mathematical curiosities, but highly representative exact solutions of the original systems of PDE to which a much larger class of solutions of initial/boundary value problems approaches asymptotically with large time or distance. This is particularly true for self-similar solutions of the second kind, which cannot be found by dimensional analysis alone, but have to be obtained, in general, by solving an eigenvalue problem (Zeldovich and Rizer 1967). In certain physical situations, the original physical system may itself be described by a nonlinear ODE, as in nonlinear mechanics.

While the physical (and mechanical) problem may be simplified by utilizing the invariance of the system of PDE with respect to certain transformations, and hence reduction to ODE (Ames 1968), the task is still formidable. Indeed, it is rather ambitious to try to take a broad view of all the nonlinear ODE and classify them, this merely reflecting the fact of extreme diversity of natural phenomena. Nevertheless it would appear that nature tends to favor second-order nonlinear DE (their linear counterparts being, in general, approximations). Thus, we are faced with the problem of solving nonlinear ODE with suitable initial/boundary conditions. There are no general rules or methods to suit all problems. However, some of the techniques are quite flexible and cover a fair amount of ground.

In the first instance, one may attempt to exactly linearize the nonlinear ODE so that the apparatus for solving linear problems can be used. There are several examples, the best known being the Riccati equation, which, at the cost of its order being raised, is exactly linearizable. There are many other examples of nonlinear ODE that can be linearized exactly. Not all such nonlinear ODE are physically meaningful.

Another class that can be handled analytically is second-order nonlinear autonomous equations. These can be treated in the phase plane. It is remarkable that there is a large class of nonlinear nonautonomous second-order equations, which, by a change of both dependent and independent variables, or by Lie group invariance, can be changed into autonomous equations (and some subsidiary equations) and analyzed in the phase or Lie plane. Therefore, the technique of local singular point analysis merits special attention. The singularities in the phase plane will, in general, not be elementary. However, reduction to the phase plane is a major simplification and permits local analysis and construction of solutions, thereby proving their existence.

The local analysis of nonlinear DE often proves useful. If the equation has a nice nonsingular structure, one may sometimes obtain a series solution giving definite information about its radius of convergence. Asymptotic solutions for large time may be found, which facilitate the analysis of the solutions and avoid expensive computation. Of course, one may also wish to prove that the solutions are asymptotic (Levinson 1970). An integral equation formulation proves very useful in this kind of analysis. It may even help to find a good approximation to the global solution by iteration, in which, for example, the linear solution is used as the first iterate (Miles 1978). This formulation may sometimes even prove the existence of the solution as well (Hastings and Mcleod 1980). Simple classical analysis throws much light on the qualitative nature of the solution (see, for example, Gilding and Peletier 1976 and Murray 1976). This may involve solving some differential inequalities.

Another class of nonlinear ODE was studied by Fuchs, Painlevé and their coworkers in the late nineteenth and early twentieth centuries (see Ince 1956, Hille 1976, and Ablowitz and Segur 1981). These mathematicians would never have thought that the equations they classified purely from their singularity structure point of view would reappear in such a unifying physical way in the theory of

Introduction xvii

nonlinear dispersive waves. It is well known that the singularities of a linear equation are entirely determined by its coefficients and are located at the singularities of the latter. These are called fixed singularities: they do not depend on the constants of integration or on the initial/boundary conditions. In contrast, for nonlinear equations the singularities may depend on the initial/boundary conditions. These are called movable or spontaneous singularities. Fuchs showed that out of all first-order equations of the form:

$$w' = F(w,z),$$

where F is rational in w and locally analytic in z, the only equations without moving critical points (by which we mean singularities such as branch points and essential singularities, but not poles) are generalized Riccati equations

$$\frac{dw}{dz} = P_0(z) + P_1(z)w + P_2(z)w^2.$$

A similar result for second-order equations

$$w'' = F(w, w', z),$$

where F is rational in w, w' and locally analytic in z, was proven by Painlevé and his group. They identified 50 canonical equations whose only movable singularities are poles. This class sometimes is said to enjoy the Painlevé property (or to be of P-type). It has been conjectured by Ablowitz et al. (1980) that a nonlinear PDE is solvable by an inverse scattering transform if every nonlinear ODE obtained from it by exact similarity reduction (and possibly by some other transformation) is of P-type. This conjecture neatly illustrates how mathematics and physics sometimes conspire to create a beautiful structure.

Of the 50 equations enumerated by Painlevé, only 6 do not admit their general solutions in terms of well-known special or elementary functions. The solutions to these six equations are referred to as Painlevé transcendents. However, for special values of the parameters in the equations, these transcendents can be related to solutions of Riccati or hypergeometric equations. Conversely, the theory of inverse scattering for nonlinear dispersive equations can be used to exactly linearize the Painlevé transcendents to obtain a single-parameter family of solutions. The Soviet contribution to Painlevé transcendents has been summarized by Erugin (1976a,b) and Gromak and Lukashevich (1982). The study of this class of equations is fascinating.

As detailed in the preface, we present in this book a variety of methods that help treat nonlinear ODE either exactly or by some approximate means that highlight their structure and facilitate their understanding, even as the precise information is obtained computationally. A special feature of this book is a profusion of examples drawn from research literature on DE dealing with applications. The final chapter is devoted to solution of systems of nonlinear PDE with appropriate initial/boundary conditions, describing time-dependent fluid dynamic problems. The methods for treating these problems are based generally on the material developed in the previous chapters. The book therefore prepares a fair ground for handling differential equations that present themselves in the study of natural phenomena.

Nonlinear Ordinary Differential Equations and Their Applications

1
Review of Linear Ordinary Differential Equations

1.1 INTRODUCTION

Change is the law of nature. Most things evolve with time. They are also diverse and nonuniform in space. In mathematics, differential equations (DE) describe the manner of this change in a quantitative way by modeling each separate phenomenon. The DE that model these phenomena are generally partial and nonlinear. They may state conservation laws of mass, momentum, and energy, or they may describe other balances, such as in heat flow. It would be a great step forward if all the partial differential equations (PDE) could be categorized in some mathematical way, but that seems to be impossible. In practice, the theory of PDE for a given physical phenomenon is developed as a distinct study. Even for ordinary differential equations (ODE), which may arise directly as descriptors of physical phenomena or from reduction of PDE Either as steady-state models or through similarity transformations, a general classification is extremely difficult. This is as it should be, since nature is so diverse in its manifestations and DE (with suitable initial and/or boundary conditions) are merely mathematical representations of physical phenomena. Most natural phenomena are nonlinear. Linear models of such phenomena are approximations of reality. Occasionally, these linear models are reasonably accurate representations. However, even linear equations with variable coefficients cannot always be solved in closed form. Still, generally speaking, the structure of linear equations is relatively simpler and well understood. Nonlinear equations have a much richer structure, and solutions in closed form are generally not available. For this reason nonlinear DE, even ordinary ones, are challenging and fascinating. They shall be our concern for most of the book.

We first discuss some linear equations to illustrate simpler models. Often, linear methods will be used to solve nonlinear problems—for example, in approximate or iterative techniques. The linearized form of the full nonlinear system contains much information and usually allows the investigator to make a good initial guess at the actual solution. Before proceeding to ODE, we give the canonical form of second-order PDE with u as the dependent variable, x and t as independent variables for time dependent problems, and x and y as independent variables for steady-state problems:

$$u_{tt} - u_{xx} = 0 \quad \text{(hyperbolic)} \tag{1}$$

$$u_t - u_{xx} = 0 \quad \text{(parabolic)} \tag{2}$$

$$u_{xx} + u_{yy} = 0 \quad \text{(elliptic).} \tag{3}$$

Equations (1) and (2) describe evolving systems, namely one-dimensional wave propagation and heat conduction, respectively. Equation (3) models steady-state phenomena such as in potential fluid flows. To complete the formulation, each equation must be supplemented by suitable initial and/or boundary conditions motivated by the physical situation that the equation represents. Each problem must then be shown to be well-posed—that is, the problem has a solution, and the solution is unique and stable. Generally, equations (1)–(3) describe highly idealized forms of the basic conservation equations and are derived by linearizing them. For example, (1) follows from the one-dimensional gas dynamic equations under the assumption that the acoustic perturbations on an otherwise quiescent uniform gas, which is assumed to be adiabatic, are small.

The relevant gas dynamic equations are

$$u_t + uu_x + \frac{1}{\rho}p_x = 0, \tag{4}$$

$$\rho_t + u\rho_x + \rho u_x = 0, \tag{5}$$

$$(p\rho^{-\gamma})_t + u(p\rho^{-\gamma})_x = 0. \tag{6}$$

Here u, ρ, and p denote particle velocity, density, and pressure, respectively. Equations (4)–(6) constitute a nonlinear system of PDE with u, ρ, and p as the dependent variables, and x and t as independent variables. Equation (1) follows from assuming that the perturbations in density, pressure, and velocity on a constant quiescent state are small, so that they are governed (in a nondimensional form) by (1) (see Courant and Friedrichs, 1948).

Review of Linear Ordinary Differential Equations

1.2 LINEAR AND NONLINEAR ODE

Now we turn to ODE. A general nth-order DE has the form

$$y^{(n)} = F(x,y,y',\ldots,y^{(n-1)}), \tag{1}$$

which is obtained by solving for $y^{(n)}$ the equation

$$\phi(x,y,y',\ldots,y^{(n)}) = 0. \tag{2}$$

Here y is the dependent variable, and x is the independent variable. Equations (1) and (2) involve algebraic and differential operators only, the latter being denoted by superscripts.

The *order* of the DE is defined as the order of the highest derivative appearing in it. The *degree* (of nonlinearity) of the DE is the degree of the highest-order derivative with respect to the independent variable.

The most general linear DE of order n can be written as

$$Ly \equiv \left[\frac{d^n}{dx^n} + p_{n-1}(x)\frac{d^{n-1}}{dx^{n-1}} + \cdots + p_0(x)\right]y = f(x) \tag{3}$$

where the coefficient of $d^n y/dx^n$ has been divided out (assuming such a division is valid). Equation (3) is homogeneous if $f(x) \equiv 0$, and is inhomogeneous otherwise.

It is possible for an equation containing derivatives to not be a DE. We give some examples.

i. $y'(x) - y = 0$

   This is a linear homogeneous DE of order 1 with constant coefficients.

ii. $y'(x) = y(x+1) = y(x+1) - y(x) + y(x)$

   This is a difference-differential equation, not a DE. $y(x+1) - y(x)$ is the difference part.

iii. $y' = a_0(x) + a_1(x)y + a_2(x)y^2$

   This is a nonlinear DE of order 1, quadratic in y. It is called the Riccati equation.

iv. $y'' = 6x + y^2$

   This is a second-order nonlinear DE.

v. $y' = \int_0^x [1 + y^2(s)]^{1/2}\,ds$

   This is an integrodifferential equation, where the unknown function $y(x)$ also appears under the integral sign. However, if we differentiate this equation, we get the DE

$$y'' = [1 + y^2(x)]^{1/2}.$$

This DE should, however, be supplemented by the condition $y'(0) = 0$ to be equivalent to the integrodifferential equation.

vi. $y = \int_0^1 [y'^2(s) + y^2(x)]\,ds$

This is an integral equation not reducible to a differential equation. The limits of integration are definite numbers.

We now give solutions of a few linear and nonlinear DE to illustrate the distinct nature of these equations.

EXAMPLE 1 $\quad y' - (x + 1)y = (x + 1)$

This is a linear inhomogeneous first-order equation, and is easily solved by separation of variables

$$\frac{dy}{dx} = (x + 1)(y + 1)$$

so that

$$\frac{dy}{y + 1} = (x + 1)\,dx.$$

Integrating, we have

$$\ln(y + 1) = \frac{(x + 1)^2}{2} + \ln c$$

or

$$y + 1 = c e^{(x+1)^2/2}.$$

Here c is the constant of integration and appears in the solution in linear and explicit form; $y = -1$ is the "particular" solution, and $e^{(x+1)^2/2}$ is the solution of the homogeneous equation.

EXAMPLE 2 $\quad y' + y^2 = A^2/x^4$

This is a nonlinear, inhomogeneous first-order equation with a quadratic nonlinearity. It is not easy to find its solution; however, it can be easily checked that the equation is satisfied by

$$y(x) = \frac{1}{x} + \frac{A}{x^2} \frac{c - e^{2A/x}}{c + e^{2A/x}}.$$

Review of Linear Ordinary Differential Equations

In contrast to Example 1, the constant of integration c appears in a complicated way in the solution.

EXAMPLE 3 $xy'' = yy'$

This is a nonlinear second-order differential equation which is nonautonomous since x appears explicitly in it. A DE is *autonomous* if it involves only the dependent variable and its derivatives. The above DE enjoys a certain invariance with respect to x; namely, if we write $x = AX$, the equation is the same in the new independent variable X. This property is possessed by the (linear) Euler–Cauchy equation

$$x^2 y'' + a_0 xy' + a_1 y = 0.$$

Therefore, we try the same change of variables as for the latter. Put $x = e^t$, so that the given equation written as

$$x^2 y'' - xyy' = 0$$

becomes

$$\left(\frac{d^2}{dt^2} - \frac{d}{dt} \right) y - y \frac{dy}{dt} = 0$$

or

$$\frac{d^2 y}{dt^2} = (1+y) \frac{dy}{dt}.$$

Integrating gives

$$\frac{dy}{dt} = \frac{(1+y)^2}{2} + c, \tag{1}$$

where c is the constant of integration.

Separating variables and integrating gives

$$\frac{2}{(2c)^{1/2}} \tan^{-1} \frac{y+1}{(2c)^{1/2}} = t + c_0$$

or

$$y = (2c)^{1/2} \tan(c/2)^{1/2}(t + c_0) - 1$$
$$= (2c)^{1/2} \tan(c/2)^{1/2}(\ln x + c_0) - 1. \tag{2}$$

Here, c_0 is another constant of integration. Equation (2) represents the required solution, a two-parameter family, the parameters being c_0 and c. They appear in a complicated manner in the solution.

We note that $y = c_2$, a nonzero constant, is another solution of this DE. This special solution is not contained in the two-parameter family (2).

Moreover, if we choose $c = 0$ in the first integral (1), we obtain the solution

$$y = -1 - \frac{2}{c_3 + t} = -1 - \frac{2}{c_3 + \ln x},$$

where c_3 is a different constant of integration. This special solution is also not contained in the two-parameter family (2).

Example 3 shows that a nonlinear DE of order 2 has a solution involving two constants, but this solution need not exhaust all possible solutions. That is, it does not make up the general solution. There may be other solutions (in general, singular) which are not obtained by giving particular values to the two arbitrary constants. All the solutions (the two-parameter family and the singular solutions) together constitute the general solution of the nonlinear problem. Another point that emerges from this example is that the solution is in general not a linear and explicit function of the constants of integration, in contrast to linear equations, so that fitting initial or boundary conditions may require solving some implicit transcendental equation(s).

The general DE of nonlinear type may be written as

$$L(y) = f(x), \tag{3}$$

where $L(y)$ is a nonlinear differential operator. Similarly to the homogeneous part of the linear DE, we may define the null nonlinear DE as

$$L(y) = 0. \tag{4}$$

We can no longer use the solutions of the homogeneous equation to find the particular solution of the inhomogeneous equation chiefly because the principle of linear superposition of solutions does not hold. However, one particular known solution may sometimes facilitate finding other solutions (e.g., for the Riccati equation), as we shall see in Chapter 2.

1.3 INITIAL AND BOUNDARY VALUE PROBLEMS

To discuss the conditions which an nth-order DE describing a (model) physical problem may be required to satisfy, it is convenient to write it as a system of n first-order DE. That is, if in

$$y^{(n)} = F(x, y(x), \ldots, y^{(n-1)}), \tag{1}$$

Review of Linear Ordinary Differential Equations

we let $y(x) = y_0, y'(x) = y_1, y''(x) = y_2, y^{(k)}(x) = y_k, \ldots,$ and $y^{(n-1)} = y_{n-1}$, we have a system of n equations in $y_0, y_1, \ldots, y_{n-1}$:

$$\frac{d}{dx} y_k(x) = y_{k+1}, \quad k = 0, 1, 2, \ldots, n-2,$$

$$\frac{d}{dx} y_{n-1}(x) = F(x, y_0, y_1, \ldots, y_{n-1}). \tag{2}$$

Conversely, given any solution of the preceding first-order system, the first component y_0 will have other components $y_1, y_2, \ldots, y_{n-1}$ as its derivatives of order 1, ..., $n - 1$, respectively. Hence, substituting suitably for $y_1, y_2, \ldots, y_{n-1}$ in (2), we recover the nth-order DE (1) for $y = y_0$.

Whether the differential equation is linear or nonlinear, its general solution admits an infinite number of solutions, which are obtained by varying the n constants of integration. These solutions are rendered definite by the imposition of n conditions. In general, a good mathematical model with "reasonable" conditions derived from physical considerations will admit a unique solution. However, this statement needs a mathematical proof. In fact, a problem is called *well-set* or *well-posed* in the sense of Hadamard if it satisfies the following conditions:

i. The DE has at least one solution (existence).
ii. The solution is unique (uniqueness).
iii. The solution varies continuously with the initial conditions; that is, if the initial conditions are changed only slightly, the solution changes only slightly (stability).

The last requirement arises from practical considerations: since measurements cannot be made with infinite accuracy, the mathematical model should continue to represent reality when subject to small inaccuracies in the experiment. It turns out that the existence of the solution can be proved under very general conditions. It is the uniqueness (and stability) requirement which limits the generality of the conditions.

The conditions may be imposed at one point, say, at $x = x_0, y_0 = \hat{y}_0, y_1 = \hat{y}_1, y_2 = \hat{y}_2, \ldots, y_{n-1} = \hat{y}_{n-1}$. This is referred to as an initial value problem (IVP), the initial value referring to time as the independent variable. Or the conditions may be imposed at two or more points in the interval of interest; for example, if $n = 3$ and we have a third-order DE, we may set $y = \hat{y}_0, y' = \hat{y}_1$ at $x = x_0$, the left end of the interval $[x_0, x_1]$, and $y = \hat{y}_2$ at $x = x_1$, the right end of the interval. As we shall see, initial value problems are much easier to handle by local analysis about the initial point, such as through a Taylor series. The solution may be written in the form of an infinite series, and the radius of convergence of the series can be computed.

Boundary value problems, however, require a global analysis since the boundary conditions are imposed at both end points (possibly in addition to those at an internal point). Boundary value problems for nonlinear DE may admit no solution, a finite number of solutions, or even an infinite number of solutions. The global analysis is more difficult. Under special circumstances, even local analysis (a convergent series solution, for example) may give a solution in the entire interval of interest; a Taylor series solution holding in one domain may be analytically continued to cover a larger domain (see Davis, 1962). For the initial value problem, $y^{(n)} = F(x,y,y^{(1)},\ldots,y^{(n-1)})$, $y = \hat{y}_0, y^{(1)} = \hat{y}_1, \ldots, y^{(n-1)} = \hat{y}_{n-1}$ at $x = x_0$ to be well-posed in a domain D about $x = x_0$, $y = \hat{y}_0, \ldots, y^{(n-1)} = \hat{y}_{n-1}$, it is sufficient that F, $\partial F/\partial y$, $\partial F/\partial y^{(1)}, \partial F/\partial y^{(2)}, \ldots, \partial F/\partial y^{(n-1)}$ be continuous. The continuity of F with respect to its argument implies the existence of at least one solution through every point; it does not necessarily imply the uniqueness of the solution.

EXAMPLE 1 $y' = 3y^{2/3}$, $y(x_0) = c$

The right side of this DE is a continuous function near $x = 0$, $y = 0$, but $\partial F/\partial y = 2y^{-1/3}$ is not continuous on the line $y = 0$. Through every point (x_0, c), $c \neq 0$, of the (x,y) plane, there passes just one curve $y = (x - C)^3$, with $C = x_0 - c^{1/3}$. The constant C depends continuously on (x_0, c). Hence the initial value problem for the DE $y' = 3y^{2/3}$ has one and only one solution of the form $y = (x - C)^3$ passing through a point (x_0, y_0), where $y_0 \neq 0$.

However, by inspection, $y = 0$ is another solution, the envelope of the curves $y = (x - C)^3$, obtained by eliminating C from $\partial y/\partial C = -3(x - C)^2 = 0$ and $y = (x - C)^3$. There is another solution defined by

$$y = \begin{cases} (x - \alpha)^3, & x < \alpha, \\ 0, & \alpha \leq x \leq \beta, \\ (x - \beta)^3, & x > \beta. \end{cases}$$

Hence the DE has a third two-parameter family solution, the parameters being α and β. This example shows again that nonlinear DE display a great variety of solutions.

EXAMPLE 2 $y' = x^{-1/2}$, $y(0) = 1$

The right side is not a continuous function of x at $x = 0$, yet the solution of this initial value problem is unique: $y = 2x^{1/2} + 1$. The requirements of the uniqueness theorem are sufficient but not necessary. Besides, the linearity of the equation makes the solution simpler.

Review of Linear Ordinary Differential Equations

EXAMPLE 3 $y' = (\tan x)y + 1$, $y(0) = 1$

The right side of this DE is a continuous function of x and y, and $\partial F/\partial y = \tan x$, a continuous function of x and y at all points except $x = n\pi/2$, where n is an integer. Again, sufficient conditions for uniqueness are satisfied except at $x = n\pi/2$. The actual solution is obtained as follows:

$$y' - (\tan x)y = 1.$$

The integrating factor is

$$\exp\left(-\int \frac{\sin x}{\cos x}\, dx\right) = \cos x.$$

Therefore,

$$(\cos x)y' - (\sin x)y = \cos x.$$

Integrating, we have

$$(\cos x)y = \sin x + c.$$

The initial condition $y(0) = 1$ gives $c = 1$, so the solution of the initial value problem is

$$y = \frac{1 + \sin x}{\cos x}.$$

1.4 ELEMENTS OF SECOND-ORDER LINEAR DE

We shall review some important results of the theory of linear ordinary second-order DE although a similar discussion is possible for nth-order equations. Second-order DE occur most frequently in applications and are simpler to treat.

The most general second-order linear DE has the form

$$p_0(x)y'' + p_1(x)y' + p_2(x)y = p_3(x) \tag{1}$$

where the functions $p_i(x)$ ($i = 0, 1, 2, 3$) are assumed to be continuous and real-valued on a given interval $[a, b]$ of the real line, whether finite or infinite. The points on the interval of interest for which $p_0(x)$ vanishes are called *singular points*. Assuming $p_0(x) \neq 0$ at any point in the interval, we may rewrite (1) as

$$y'' + p(x)y' + q(x)y = r(x), \tag{2}$$

where

$$p(x) = \frac{p_1(x)}{p_0(x)}, \quad q(x) = \frac{p_2(x)}{p_0(x)}, \quad r(x) = \frac{p_3(x)}{p_0(x)}.$$

For the homogeneous equation

$$y'' + p(x)y' + q(x)y = 0, \tag{3}$$

the general solution can always be written as

$$y = c_1 y_1(x) + c_2 y_2(x), \tag{4}$$

where $y_1(x)$ and $y_2(x)$ are any two linearly independent solutions of (3), and c_1 and c_2 are arbitrary constants. With $p(x)$ and $q(x)$ continuous, equation (3) has exactly two linearly independent solutions. Two functions $y_1(x)$ and $y_2(x)$ are said to be *linearly independent* over an interval $a \le x \le b$ if no nontrivial constants c_1 and c_2 can be found such that

$$c_1 y_1(x) + c_2 y_2(x) = 0. \tag{5}$$

Conversely, if it is possible to find nonzero c_1 and c_2 such that (5) holds over the interval $a \le x \le b$, then $y_1(x)$ and $y_2(x)$ are *linearly dependent*. It is obvious that (3) always has the trivial solution $y = 0$, so we refer to nontrivial $y_i(x)(i = 1, 2)$ in (4) and (5). Again, it is evident that if $y_1(x)$ and $y_2(x)$ are solutions of (3), then so is the linear combination

$$y = c_1 y_1(x) + c_2 y_2(x).$$

Relations (4) and (5) are required to hold in the entire interval $a \le x \le b$; that is, linear dependence or independence is necessarily a global concept. If the equation is nonlinear, the principle of linear superposition of solutions does not hold. This loss accounts for the analytic complexity of nonlinear equations; much of the elegant theory developed for linear DE cannot be used. Note that the general solution of the inhomogeneous equation (2) is obtained by adding a particular integral of this equation to the general solution of the corresponding homogeneous equation (3). As we shall see in the next section, the particular solution of the inhomogeneous equation can be obtained with the help of the two (linearly independent) solutions of the corresponding homogeneous equation. Therefore, we must be able to obtain two linearly independent solutions of the latter. A simple test of linear dependence of the two solutions is whether one of the solutions is a constant multiple of the other. A stricter test is that if the Wronskian of the solution,

$$\begin{aligned} W(x) &= W[y_1(x), y_2(x)] \\ &= \det \begin{bmatrix} y_1(x) & y_2(x) \\ y_1'(x) & y_2'(x) \end{bmatrix} \\ &= y_1(x) y_2'(x) - y_2(x) y_1'(x), \end{aligned}$$

does not vanish identically over the given interval then the two solutions $[y_1(x), y_2(x)]$ are linearly independent.

Review of Linear Ordinary Differential Equations

We now derive the DE for the Wronskian. Since y_1 and y_2 satisfy (3), we have

$$y_1'' + p(x)y_1' + q(x)y_1 = 0,$$
$$y_2'' + p(x)y_2' + q(x)y_2 = 0.$$

Multiplying the first equations by y_2 and the second by y_1 and subtracting, we get

$$(y_1'y_2 - y_2'y_1)' + p(x)(y_1'y_2 - y_2'y_1) = 0,$$

or

$$W'(x) = -p(x)W(x). \tag{7}$$

After integrating, we get Abel's formula

$$W(x) = C\exp\left[-\int_0^x p(t)\,dt\right], \tag{8}$$

where C is the constant of integration. Thus, we can compute the Wronskian from the (coefficients of the) DE itself, without knowing the solutions. Equation (8) shows that (i) the Wronskian is obtained up to an arbitrary multiplicative constant, and (ii) because of the exponential form of W, the Wronskian of any two solutions of the homogeneous linear DE (3) is identically positive, identically negative, or identically zero.

The Wronskian has many uses. It appears naturally when we try to find a particular solution of an inhomogeneous equation, as we shall see in the next section. Its vanishing may mean that the initial value problem is not well-posed. We point out that the vanishing of the Wronskian indicates the linear dependence of solutions only for analytic solutions. For nonanalytic solutions, this may not be so. For example, the functions x^3 and $|x|^3$ are linearly independent in $-1 < x < 1$, but their Wronskian

$$\begin{vmatrix} x^3 & x^3 \\ 3x^2 & 3x^2 \end{vmatrix} \quad \text{or} \quad \begin{vmatrix} x^3 & -x^3 \\ 3x^2 & -3x^2 \end{vmatrix}$$

vanishes identically in $0 < x < 1$ and $-1 < x < 0$, respectively.

EXAMPLE 1 $(1-x)y'' + xy' - y = 0$, $y(0) = 1$, $y'(0) = -1$

By inspection the two solutions are e^x and x. The Wronskian is

$$W(x) = \begin{vmatrix} e^x & x \\ e^x & 1 \end{vmatrix}$$
$$= e^x(1-x) \neq 0 \quad \text{except at } x = 1.$$

EXAMPLE 2 $xy'' - (1+x)y' + y = 0$

i. $y(1) = 1$, $y'(1) = 2$,
ii. $y(0) = 1$, $y'(0) = 2$,
iii. $y(0) = 1$, $y'(0) = 1$.

Again by inspection, the two solutions are $1+x$ and e^x. The Wronskian is

$$W = \begin{vmatrix} e^x & 1+x \\ e^x & 1 \end{vmatrix} = -xe^x,$$

which for finite x vanishes only at $x = 0$, where

$$\frac{d}{dx}(1+x) = \frac{d}{dx}e^x;$$

that is, the two solutions have a common tangent. The two solutions are linearly independent. The general solution is

$$y(x) = c_1 e^x + c_2(1+x).$$

We now discuss cases (i)–(iii).

Case (i).

$$c_1 e + 2c_2 = 1$$
$$c_1 e + c_2 = 2$$

Thus $c_2 = -1$, $c_1 = 3/e$, and the solution is

$$y(x) = \frac{3}{e}e^x - (1+x).$$

Case (ii).

$$c_1 + c_2 = 1$$
$$c_1 + c_2 = 2$$

These two equations are not compatible; therefore, no solution of the IVP exists. It is an ill-posed problem; here $W = 0$ at $x = 0$.

Case (iii).

$$c_1 + c_2 = 1$$
$$c_1 + c_2 = 1$$

Therefore, $c_2 = 1 - c_1$, and the solution is

$$y = c_1(e^x - 1 - x) + 1 + x.$$

Here, again, $W = 0$ at $x = 0$. We have an infinite number of solutions of the IVP.

Review of Linear Ordinary Differential Equations

EXAMPLE 3 Consider the boundary value problem

$$y'' + y = 0, \quad y'(0) = 0, \quad y\left(\frac{\pi}{2}\right) = 1.$$

The general solution is

$$y(x) = c_1 \cos x + c_2 \sin x$$

and

$$y'(x) = -c_1 \sin x + c_2 \cos x.$$

The Wronskian is

$$W = \begin{vmatrix} \cos x & \sin x \\ -\sin x & \cos x \end{vmatrix} = 1.$$

The boundary conditions require $c_2 = 0$ and $c_2 = 1$. Therefore, no solution of the problem exists. Here we have an example for which coefficients in the DE are continuous, $W(x) = 1$, and yet the solution does not exist. Boundary value problems are more difficult than initial value problems.

Solutions of linear DE become more complex as the equations become more general, in the following order: homogeneous equations with constant coefficients, inhomogeneous equations with constant coefficients, homogeneous equations with variable coefficients, and inhomogeneous equations with variable coefficients. There are no general methods for the latter two types. The series solution (Frobenius method) could be used to define possibly new special functions for new equations, in addition to the existing ones. However, there is a powerful method, variation of parameters, which helps us to find a second solution if one is known or to determine a particular solution of the inhomogeneous equation if two linearly independent solutions of the homogeneous equation are known. We shall describe this method in some detail since it influences some techniques for nonlinear DE—for example, the integral equation formulation of a nonlinear DE. We shall omit other methods for solving linear DE such as the method of undetermined coefficients. In Chapter 3, we discuss the series solution method for initial value problems for nonlinear DE and compare it with the series solution method for the linear DE.

1.5 METHOD OF VARIATION OF PARAMETERS

As we noted earlier, there is no simple way of finding two linearly independent solutions of a general second-order linear DE. However, if we know one solution, say $y_1(x)$, of the equation

$$y'' + p(x)y' + q(x)y = 0, \tag{1}$$

we can find a second (linearly independent) solution. Obviously, $cy_1(x)$, where c is a constant, is also a solution of (1), but the solutions y_1 and cy_1 are not linearly independent.

Now suppose c varies with x so that $c = v(x)$. We want to find $v(x)$ so that $y_2 = vy_1$ is the second solution of (1). This is one aspect of the method of variation of parameters (or constants). Since, by assumption, v is not a constant, the second solution y_2 will be linearly independent.

Substituting $y_2 = vy_1$ in (1), we have

$$v(y_1'' + py_1' + qy_1) + v''y_1 + v'(2y_1' + py_1) = 0. \tag{2}$$

Since y_1 is a solution of (1), (2) reduces to

$$v''y_1 + v'(2y_1' + py_1) = 0$$

or

$$\frac{v''}{v'} = -\frac{2y_1'}{y_1} - p.$$

Integrating gives

$$v' = \frac{1}{y_1^2} e^{-\int p\, dx} \tag{3}$$

and

$$v = \int \frac{1}{y_1^2} e^{-\int p\, dx}\, dx. \tag{4}$$

Thus, we have found $v(x)$ in terms of quadratures and, hence, $y_2(x) = vy_1$.

Of course, it may not always be possible to evaluate (4) explicitly. This approach may also be interpreted as reducing the order of the given equation (1) by 1 (see (3)). It is not difficult to verify that y_1 and $y_2 = vy_1$ are linearly independent. The Wronskian is

$$W(y_1, vy_1) = \begin{vmatrix} y_1 & vy_1 \\ y_1' & vy_1' + v'y_1 \end{vmatrix} = v'y_1^2 \neq 0,$$

since v is not a constant and y_1 is a nontrivial solution of (1).

Now we adapt this procedure to obtain a particular solution of the inhomogeneous equation

$$y'' + p(x)y' + q(x)y = r(x). \tag{5}$$

Review of Linear Ordinary Differential Equations

We assume that two linearly independent solutions $y_1(x)$ and $y_2(x)$ of the homogeneous equation (1) have been found somehow. The general solution of (1) is

$$y = c_1 y_1(x) + c_2 y_2(x), \qquad (6)$$

where c_1 and c_2 are arbitrary constants. We let $c_1 = v_1(x)$ and $c_2 = v_2(x)$ so that they vary with x, and we assume that a particular solution of (5) is

$$y = v_1 y_1 + v_2 y_2. \qquad (7)$$

Differentiating (7), we have

$$y' = v_1 y_1' + v_2 y_2' + v_1' y_1 + v_2' y_2. \qquad (8)$$

The functions v_1 and v_2 are unknown, and we need two equations to determine them. If we differentiate (8) a second time and substitute in (5), we would have v_1'' and v_2'' appearing in this equation and v_1 and v_2 will not be determined by a single quadrature. We therefore put

$$v_1' y_1 + v_2' y_2 = 0. \qquad (9)$$

This provides one equation for v_1 and v_2. Differentiating (8) again (after taking (9) into account) and substituting for y' and y'' in (5), we get

$$v_1(y_1'' + py_1' + qy_1) + v_2(y_2'' + py_2' + qy_2) + v_1' y_1' + v_2' y_2' = r(x). \qquad (10)$$

Since y_1 and y_2 satisfy (1), (10) reduces to

$$v_1' y_1' + v_2' y_2' = r(x). \qquad (11)$$

Solving (9) and (11) for v_1' and v_2', we have

$$v_1' = -\frac{y_2 r(x)}{W(y_1, y_2)}, \qquad v_2' = \frac{y_1 r(x)}{W(y_1, y_2)}. \qquad (12)$$

Since y_1 and y_2 are linearly independent, $W(y_1, y_2) \neq 0$ and (12) is legitimate. Integrating the system (12) and substituting in (7), we get

$$y = y_1(x) \int^x (-1) \frac{y_2(t) r(t)}{W(y_1, y_2)} \, dt + y_2(x) \int^x \frac{y_1(t) r(t)}{W(y_1, y_2)} \, dt \qquad (13)$$

as the particular solution of (5). Formula (13) shows that the method is worthwhile only when two explicit solutions $y_1(x)$ and $y_2(x)$ of the homogeneous solution (1) are known, and the integrals in (13) can be evaluated explicitly.

EXAMPLE 1 $\quad x^2 y'' + xy' - y = 0 \qquad (14)$

Here we see that $y_1 = x$ is a solution by inspection or by noting that equation (14) is of Cauchy–Euler (equidimensional) type. Equation (4) with

$p = 1/x$ gives

$$v = \int \frac{1}{x^2} e^{-\int (1/x)\,dx}\, dx = -\frac{x^{-2}}{2}.$$

This yields the second solution

$$y_2 = x\left(-\frac{1}{2}x^{-2}\right) = -\frac{1}{2x}.$$

The general solution is

$$y = c_1 x + \frac{c_2}{x}.$$

EXAMPLE 2 $y'' + y = \csc x$ \hfill (15)

The two particular solutions of the homogeneous form are $y_1 = \sin x$ and $y_2 = \cos x$. Moreover,

$$W(y_1, y_2) = y_1 y_2' - y_2 y_1' = -1.$$

Therefore

$$v_1 = \int -\frac{\cos x \csc x}{-1}\, dx = \log \sin x,$$

$$v_2 = \int \frac{\sin x \csc x}{-1}\, dx = -x.$$

Hence $y = \sin x \log \sin x - x \cos x$ is a particular solution of (15). The general solution of (15) is therefore

$$y = c_1 \cos x + c_2 \sin x + \sin x \log \sin x - x \cos x.$$

1.6 GREEN'S FUNCTION

There is an alternative approach to the method of variation of parameters which is particularly useful for solving boundary value problems (BVP). If we write a BVP for the general second-order DE as

$$y'' + p_1(x) y' + p_2(x) y = \phi(x), \qquad x_0 \leq x \leq x_1, \tag{1}$$

$$y(x_0) = y_0, \qquad y(x_1) = y_1, \tag{2}$$

then a (linear) change of dependent variable

$$z = y - \frac{y_1 - y_0}{x_1 - x_0}(x - x_0) - y_0$$

reduces the boundary conditions (2) to homogeneous conditions:

$$z(x_0) = 0, \qquad z(x_1) = 0.$$

In addition, if we multiply (1) by $e^{\int p_1(x)dx}$, we can rewrite it in the form

$$L(y) = \frac{d}{dx}(p(x)y') + q(x)y = r(x), \tag{3}$$

where

$$p(x) = e^{\int p_1(x)dx}, \qquad q(x) = p_2(x)e^{\int p_1(x)dx},$$

$$r(x) = \phi(x)e^{\int p_1(x)dx}.$$

Therefore, we may solve (3) instead of (1), subject to homogeneous boundary conditions

$$y(x_0) = 0, \qquad y(x_1) = 0. \tag{4}$$

We make a few remarks about the BVP (3)–(4). If y_1 and y_2 are solutions of (3)–(4) with $r = r_1(x)$ and $r = r_2(x)$, respectively, then $c_1 y_1 + c_2 y_2$ is the solution of (3)–(4) with $r = c_1 r_1 + c_2 r_2$. This is easily checked by noting that L is a linear operator, so

$$L(c_1 y_1 + c_2 y_2) = c_1 L y_1 + c_2 L y_2 = c_1 r_1 + c_2 r_2.$$

The homogeneous boundary conditions are also satisfied by the solution $c_1 y_1 + c_2 y_2$. We shall use this linear superposition in the following. Any solutions y_1 and y_2 of (3)–(4) differ by a solution of the completely homogeneous system

$$Lu = 0, \qquad x_0 < x < x_1, \quad u(x_0) = 0, \quad u(x_1) = 0. \tag{5}$$

Since (5) has only the trivial solution, the uniqueness of the solution to (3)–(4) is immediately proved (see Birkhoff and Rota, 1978, p. 34).

We first attempt to solve the BVP with a delta function on the right side:

$$\frac{d}{dx}(p(x)y') + q(x)y = \delta(x-s) \tag{6a}$$

$$y(x_0) = 0, \qquad y(x_1) = 0. \tag{6b}$$

The delta function can be defined in many ways as a limit of a sequence of functions:

$$\delta(x-s) = \lim_{\epsilon \to 0+} \frac{\epsilon}{\pi[(x-s)^2 + \epsilon^2]}$$

or

$$\delta(x-s) = \lim_{\epsilon \to 0+} (\pi\epsilon)^{-1/2} e^{-(x-s)^2/\epsilon}.$$

For now we denote by $\delta(x)$ the limit of the continuous and nonnegative density functions $\rho(x)$, concentrated in a narrow interval $(-\epsilon,\epsilon)$ near $x = 0$, with total mass $\int_{-\epsilon}^{\epsilon}\rho(x)\,dx = 1$, as $\epsilon \to 0$. By mere translation, $\delta(x - \xi)$ represents the same limit about the point $x = \xi$. Thus, $\delta(x - s)$ is a mathematical idealization of a unit impulse. It may be conceived of as an infinitely thin spiky function centered at $x = s$ and having unit area under it.

Let $f \in C[a,b]$, $a < 0 < b$, and let ρ be the density function with support $(-\epsilon,\epsilon) \subset [a,b]$. Then, by the second mean value theorem of integral calculus,

$$\int_a^b f(x)\rho(x)\,dx = \int_{-\epsilon}^{\epsilon} f(x)\rho(x)\,dx = f(x_1)\int_{-\epsilon}^{\epsilon}\rho(x)\,dx = f(x_1),$$

where $-\epsilon < x_1 < \epsilon$. Now, let ϵ tend to zero; we get, in this limit,

$$\int_a^b f(x)\delta(x)\,dx = f(0). \tag{7}$$

Using the translated delta function, we similarly have

$$\int_a^b f(x)\delta(x - \xi)\,dx = f(\xi). \tag{7'}$$

In particular, if $f(x)$ is chosen to be 1, we get

$$\int_a^b \delta(x - \xi)\,dx = \begin{cases} 1 & \text{if } \xi \in (a,b), \\ 0 & \text{if } \xi \notin (a,b). \end{cases}$$

If we denote by $G_\epsilon(x,s)$ the continuous solution of the BVP (6) (where ϵ is the same term as in the definition of the delta function) and pass to the limit $\epsilon \to 0$, then the function

$$G(x,s) = \lim_{\epsilon \to 0} G_\epsilon(x,s) \tag{8}$$

is called the *Green's function* of the BVP (6).

The solution of the BVP (3)–(4), with $r(x)$ on the right side of (3), may be regarded as a linear superposition of the solutions of the BVP that correspond to delta functions, localized in points s_i, with momenta $r(s_i)\Delta s$, where the points s_i divide the interval $[x_0,x_1]$ into m equal parts and $\Delta s = (x_1 - x_0)/m$. In other words, an approximate solution of the linear BVP (3)–(4) is the integral sum

$$\sum_{i=1}^m G(x,s_i)r(s_i)\Delta s,$$

Review of Linear Ordinary Differential Equations

which, in the limit $m \to \infty$, becomes

$$y(x) = \int_{x_0}^{x_1} G(x,s) r(s) \, ds. \tag{9}$$

The function $y(x)$ in (9) solves (3)–(4), as we shall verify.

The physical interpretation of the Green's function (or influence function) is as follows. We may regard $y(x)$ in (3) as the displacement of some system under the influence of some force $r(x)$ continuously distributed over the interval $[x_0, x_1]$ (hence, the term *momenta* for $r(s_i)\Delta s$). This may, for example, be the deflection of a string from its equilibrium position under the effect of a distributed load with density $r(x)$. In this sense, the function $G(x,s)$ gives the displacement caused by a unit concentrated force applied at the point $x = s$, and the solution (9) may be regarded as the limit of the sum of solutions corresponding to the concentrated forces at the points s_i defined earlier.

In view of the definition (8), etc., the Green's function has the following properties:

i. $G(x,s)$, for fixed s, is a continuous function of x in $x_0 \leq x \leq x_1$, where $x_0 < s < x_1$.
ii. $G(x,s)$ is a solution of the homogeneous equation

$$\frac{d}{dx}(p(x)y') + q(x)y = 0$$

over $[x_0, x_1]$ except at the point $x = s$.
iii. $G(x,s)$ satisfies the homogeneous boundary conditions $G(x_0, s) = G(x_1, s) = 0$.
iv. At the point $x = s$, $G_x(x,s)$ has a finite discontinuity.

To demonstrate (iv), we note that the most singular term on the left side of (6a) is $\partial^2 G/\partial x^2$ because differentiation is an "unsmoothing" operation (see Chapter 3 for a discussion of singularities of a DE). This term must balance $\delta(x-s)$ on the right. The difference $d^2G/dx^2 - \delta(x-s)$ must be less singular than a delta function. Therefore, dG/dx, which is the (next) most singular term, must balance $d^2G/dx^2 - \delta(x-s)$ and have a finite discontinuity at $x = s$. To see this, we integrate the identity

$$\frac{d}{dx}(p(x)G'(x,s)) + q(x)G(x,s) = \delta(x-s)$$

from $x = s - \epsilon$ to $x = s + \epsilon$, use (6), and take the limit as $\epsilon \to 0$. We get

$$p(x)G'(x,s) \Big|_{s-\epsilon}^{s+\epsilon} + \int_{s-\epsilon}^{s+\epsilon} q(x)G(x,s)\, dx = 1$$

or

$$[G'(s+0,s) - G'(s-0,s)] = \frac{1}{p(s)} \tag{10}$$

in the limit as $\epsilon \to 0$. The second integral vanishes in this limit since $q(x)$ and $G(x,s)$ are both continuous. Thus the Green's function may be defined as the solution of BVP (6), which satisfies conditions (i)–(iv). Boundary conditions (6b) are satisfied by virtue of (iii). The uniqueness of the Green's function solution of (6a)–(6b) follows from (5) and the statement following it regarding the problem (3)–(4).

Now we give the actual method of construction of the Green's function, which will also make explicit the conditions for its existence. We find special (nontrivial) solutions $y_1(x)$ and $y_2(x)$ of the homogeneous equation

$$\frac{d}{dx}(p(x)y'(x)) + q(x)y = 0, \tag{11}$$

which satisfy the conditions $y_1(x_0) = 0$, $y_1'(x_0) \neq 0$, $y_1(x_1) \neq 0$, and $y_2(x_1) = 0$, $y_2'(x_1) \neq 0$, respectively.

Obviously, $c_1 y_1$ and $c_2 y_2$, with c_1 and c_2 nonzero arbitrary constants, also satisfy (11) and the relevant homogeneous boundary conditions. Now we seek the Green's function in the form

$$G(x,s) = \begin{cases} c_1 y_1(x) & \text{for } x_0 \leq x \leq s, \\ c_2 y_2(x) & \text{for } s < x \leq x_1. \end{cases} \tag{12}$$

We choose c_1 and c_2 in such a way that conditions (i) and (iii) in the definition of the Green's function are satisfied. The continuity of $G(x,s)$ with respect to x everywhere in $[x_0, x_1]$, particularly at $x = s$, requires that

$$c_1 y_1(s) = c_2 y_2(s), \tag{13}$$

while the jump condition on the derivative of G with respect to x at $x = s$ yields

$$c_2 y_2'(s) - c_1 y_1'(s) = \frac{1}{p(s)} \tag{14}$$

(see equation (10)). Since, by assumption, $y_1(x_1) \neq 0$ while $y_2(x_1) = 0$, the solutions $c_1 y_1(x)$ and $c_2 y_2(x)$ are linearly independent. Therefore, the Wronskian $W(y_1, y_2) = (y_1 y_2' - y_2 y_1') \neq 0$ anywhere in $[x_0, x_1]$ including $x = s$.

Equations (13) and (14) can be solved simultaneously for c_1 and c_2:

$$c_1 = \frac{y_2(s)}{W(s)p(s)}, \qquad c_2 = \frac{y_1(s)}{W(s)p(s)}. \tag{15}$$

Review of Linear Ordinary Differential Equations

Substituting (15) into (12), we get

$$G(x,s) = \begin{cases} \dfrac{y_2(s)y_1(x)}{W(s)p(s)} & \text{for } x_0 \leq x \leq s, \\ \dfrac{y_1(s)y_2(x)}{W(s)p(s)} & \text{for } s < x \leq x_1. \end{cases} \qquad (16)$$

To verify that the solution of the BVP (3)–(4) is given by (9), we note that $G(x,s)$ satisfies (6a), so

$$\frac{d}{dx}[p(x)G'(x,s)] + q(x)G(x,s) = \delta(x-s), \qquad (17)$$

while $y(x)$ satisfies (3). Both $G(x,s)$ and $y(x)$ vanish at $x = x_0$ and $x = x_1$. Multiplying (17) by $y(x)$ and (3) by $G(x,s)$ and subtracting, we get

$$[p(yG' - Gy')]' = \delta(x-s)y(x) - r(x)G(x,s). \qquad (18)$$

Integrating (18) from x_0 to x_1 and using relevant homogeneous boundary conditions on y and G, we have

$$y(s) = \int_{x_0}^{x_1} r(x)G(x,s)\,dx, \qquad (19)$$

wherein we have used the localizing property (7') of the delta function.

Now if we interchange the role of s and x in (13) and (14), etc., we have

$$G(s,x) = \begin{cases} \dfrac{y_2(x)y_1(s)}{W(x)p(x)} & \text{for } x_0 \leq s \leq x, \\ \dfrac{y_2(s)y_1(x)}{W(x)p(x)} & \text{for } x < s \leq x_1 \end{cases} \qquad (20)$$

$$= G(x,s).$$

For a given x, $G(s,x)$ satisfies all the requirements of the Green's function as a function of s. Therefore, (19) can be written as

$$y(x) = \int_{x_0}^{x_1} r(s)G(x,s)\,ds$$

This completes the proof that the solution of (3)–(4) is given by (9).

The construction of the Green's function for boundary conditions more general than (2), namely,

$$\alpha_{11}y(x_0) + \alpha_{12}y'(x_0) + \beta_{11}y(x_1) + \beta_{12}y'(x_1) = 0,$$
$$\alpha_{21}y(x_0) + \alpha_{22}y'(x_0) + \beta_{21}y(x_1) + \beta_{22}y'(x_1) = 0,$$

where α_{ij}, β_{ij} are given real numbers, is accomplished in the same manner (see Stakgold, 1967).

EXAMPLE 1 $y''(x) + y(x) = f(x)$, $y(0) = 0$, $y(\pi/2) = 0$

The solutions of the corresponding homogeneous equation which satisfy $y(0) = 0$ and $y(\pi/2) = 0$, respectively, are $y_1 = \sin x$ and $y_2 = \cos x$. Substituting these in (16), with $W = -1$ and $p(x) = 1$, we have

$$G(x,s) = \begin{cases} -\cos s \sin x, & 0 \le x \le s, \\ -\sin s \cos x, & s < x \le \pi/2. \end{cases}$$

Hence the solution is

$$y(x) = \int_0^{\pi/2} G(x,s) f(s) \, ds.$$

EXAMPLE 2 $y'' - y = f(x), [y(\pm\infty) = 0]$

The solutions of the homogeneous equation are e^x and e^{-x}. The solution satisfying $y(-\infty) = 0$ is $y_1 = e^x$, and that satisfying $y(+\infty) = 0$ is $y_2 = e^{-x}$. The Wronskian is $W(e^x, e^{-x}) = -2$, and $p(x) = 1$. Therefore, from (16), we have

$$G(x,s) = \begin{cases} -\dfrac{1}{2} e^{x-s} & \text{for } -\infty < x \le s, \\ -\dfrac{1}{2} e^{s-x} & \text{for } s < x < \infty. \end{cases}$$

The solution can be written as

$$y(x) = -\frac{1}{2} \int_{-\infty}^{\infty} e^{-|x-s|} f(s) \, ds.$$

2
Transformations of Nonlinear Ordinary Differential Equations

2.1 INTRODUCTION

Since the principle of linear superposition is not applicable to nonlinear DE, the elegant theory developed for linear equations cannot be used. Nevertheless, one attempts to analyze the equations before rushing to the computer for a numerical solution to the problem. One of the following approaches may be used:

1. Look for some hidden symmetries of the equation such as scale-invariance or equidimensionality with respect to independent or dependent variable, which may be discovered by inspection and used to reduce the order or render the equation autonomous.
2. Seek exact linearization to DE with constant or variable coefficients by a change of independent and dependent variables.
3. Transform difficult nonlinear DE to (simpler) solvable nonlinear DE.
4. Generate some classes of nonlinear DE from given linear DE by suitable nonlinear transformations.
5. Try to solve nonlinear first-order equations, such as Riccati's or Abel's, or general second-order nonlinear equations from second- or third-order linear equations with constant or variable coefficients.
6. Reduce second-order nonlinear equations to first-order equations, or change second-order nonautonomous equations to second-order

autonomous equations so that phase plane analysis (see Chapter 6) can be used.

We shall explain each of these approaches with the help of examples in an attempt to mitigate the difficulties of nonlinear problems and render them more amenable to analysis. In conclusion, we shall show, with the help of the Langmuir equation, that in spite of many symmetries and special features that a nonlinear DE may manifest, it may elude solution in a closed form. The discussion in this chapter is carried out without reference to initial or boundary conditions.

2.2 SIMPLE SYMMETRIES—EQUIDIMENSIONALITY AND SCALE INVARIANCE

An equation is *equidimensional* in the independent variable x if a change of scale $x \to ax$ leaves it invariant. Similarly, the equation is said to be equidimensional in the dependent variable y if a change of scale $y \to ay$ leaves it invariant. These invariances help by either making the equation autonomous or reducing its order.

We observed in Chapter 1 that the (linear) Euler–Cauchy equation

$$x^2 y'' + a_0 xy' + a_1 y = 0 \tag{1}$$

(sometimes referred to as homogeneous) does not change under the scale change $x \to ax$. Through the transformation

$$x = e^t, \tag{2}$$

(1) reduces to

$$\left(\frac{d^2 y}{dt^2} - \frac{dy}{dt} \right) + a_0 \frac{dy}{dt} + a_1 y = 0, \tag{3}$$

a linear equation with constant coefficients. The important point is that this equidimensionality with respect to x removes x-dependence via (2) even in a nonlinear equation and thereby renders it autonomous and amenable to phase plane analysis.

EXAMPLE 1 $y'' + (y + 1)y'/x + y/x^2 = 0$

Changing the variable according to (2), we have

$$\left(\frac{d^2 y}{dt^2} - \frac{dy}{dt} \right) + (y + 1) \frac{dy}{dt} + y = 0,$$

Transformations of Nonlinear ODE

or

$$\frac{d^2y}{dt^2} + y\frac{dy}{dt} + y = 0. \tag{4}$$

Equation (4) is autonomous, since t does not appear explicitly in the coefficients.

Writing $p = dy/dt$, we have

$$p\frac{dp}{dy} + yp + y = 0,$$

or

$$\frac{p}{p+1}dp + y\,dy = 0.$$

This equation can be integrated by separation of variables:

$$p - \ln(p+1) + \frac{y^2}{2} = c.$$

Unfortunately, it is not possible to integrate this equation explicitly. However, equation (4) may be analyzed in the $(y, dy/dt)$ phase plane.

EXAMPLE 2 $xy'' = yy'$

This equation may be written as $x^2y'' = xyy'$ so that it is equidimensional with respect to x. Making the substitution (2), we have

$$\frac{d^2y}{dt^2} - \frac{dy}{dt} = y\frac{dy}{dt},$$

or

$$\frac{d^2y}{dt^2} = (y+1)\frac{dy}{dt},$$

an autonomous equation. The substitution $dy/dt = p$, $d^2y/dt^2 = p(dp/dy)$ changes this DE into $p(dp/dy) = p(y+1)$. This gives either $p = 0$ or $dp/dy = y + 1$. The first case yields $y = $ const, while the second integrates to

$$\frac{dy}{dt} = p(y) = \frac{y^2}{2} + y + c_0 = \frac{1}{2}(y+1)^2 + c_0 - \frac{1}{2}.$$

This equation is of separable form. After integrating and substituting $t = \ln x$, we get

$$y(x) = 2c_1 \tan(c_1 \ln x + c_2) - 1,$$

where $c_1 = \frac{1}{2}\sqrt{2c_0 - 1}$, and c_2 is another constant. This demonstrates the use of the equidimensionality property for nonlinear DE.

EXAMPLE 3 $yy'' + y'^2 - yy'/(1+x) = 0$

We let $1 + x = z$ and write the equation as

$$z^2 yy'' + z^2 y'^2 - zyy' = 0,$$

where the prime now denotes differentiation with respect to z. This equation is equidimensional with respect to z. If we put $z = e^t$, the equation becomes autonomous:

$$y \frac{d^2 y}{dt^2} - 2y \frac{dy}{dt} + \left(\frac{dy}{dt}\right)^2 = 0.$$

Writing $dy/dt = p$, we get two possibilities: (i) $p = 0$; i.e., $y =$ const, and (ii) $y(dp/dy) = 2y - p$. This equation can be integrated to give $py = y^2 + c$, where c is a constant. Therefore,

$$\frac{dy}{dt} = p = \frac{y^2 + c}{y}.$$

This equation integrates to give

$$y^2 + c = c_1 e^{2t} = c_1 z^2,$$

or

$$y^2 = c_1(1 + x)^2 - c.$$

Here c_1 is another constant.

EXAMPLE 4 $x^2 y'' + 3xy' + 2y = x^{-4} y^{-3}$

This equation is scale invariant under $x \to aX, y \to a^{-1}Y$. Therefore, $xy = XY$. This suggests the substitution $u = xy$. The transformed equation is

$$x^2 u'' + xu' + u = u^{-3}.$$

This equation is equidimensional with respect to x. The substitution $x = e^t$ changes it to

$$\frac{d^2 u}{dt^2} + u = u^{-3}.$$

This is an autonomous second-order equation, and the substitution $du/dt = p$ transforms it to

$$p \frac{dp}{du} = \frac{1 - u^4}{u^3}.$$

Transformations of Nonlinear ODE

After one integration, this equation becomes

$$\left(\frac{du}{dt}\right)^2 = p^2 = \frac{1}{u^2}(2c_0 u^2 - u^4 - 1).$$

A second integration finally gives the solution in terms of the original variables as

$$y = \pm \frac{1}{x}\{\cosh c_1 + (\sinh c_1)\sin(2\ln x + c_2)\}^{1/2},$$

where $c_1 = \cosh^{-1} c_0$ and c_2 are two arbitrary constants.

EXAMPLE 5 $w'' + x^n e^w = nx^{-2}$, $x > 0$

where n is any number (SIAM Rev. 27, (1985), 83). The homogeneous equation $w'' + x^n e^w = 0$ cannot be solved, but we will show that the full inhomogeneous equation can be. We make a "plausible" substitution $v = -x^n e^w$. Then the equation in v becomes

$$vv'' = v'^2 + v^3.$$

Writing $v' = p$, $v'' = p(dp/dv)$, and integrating, we obtain the first-order equation

$$v'^2 = 2v^3 + kv^2,$$

where k is an arbitrary constant. This equation can be easily integrated to give

$$e^w = \begin{cases} \dfrac{k}{2x^n}\operatorname{sech}^2\left(\dfrac{k^{1/2}}{2}x + K\right), & k \neq 0, \\ \dfrac{-4}{x^n(K - \sqrt{2}x)^2}, & k = 0, \end{cases}$$

where K is another constant. This method of solution can be generalized to solve

$$w'' + f(x)e^w = g(x).$$

This equation can be reduced to the form $v'^2 = 2v^3 + kv^2$, under the transformation $v(x) = -f(x)e^w$ if and only if

$$f^2 g = f'^2 - ff''.$$

2.3 FIRST-ORDER NONLINEAR EQUATIONS AND THEIR LINEARIZATION—THE RICCATI EQUATION

Of all first-order nonlinear DE, the Riccati equation

$$y' = f_0(x) + f_1(x)y + f_2(x)y^2 \qquad (1)$$

occupies perhaps the most important place. There are three reasons: (i) it is closely related to the general linear (homogeneous) second-order equation; (ii) its only movable singularities are poles (all other singularities being fixed), as we shall discuss in Chapter 4; and (iii) it appears in numerous physical applications. We discuss it briefly, since a good treatment is readily accessible in Davis (1962) and Hille (1969).

We note two special cases of (1). First, when $f_2(x) = 0$, it is a linear equation of first order with variable coefficients and may be integrated by a quadrature. The second case arises when $f_0(x) = 0$ and (1) becomes a special case of Bernoulli's equation

$$y' = f_1(x)y + f_2(x)y^p \qquad (2)$$

with $p = 2$. Equation (2) is easily integrated since the substitution $u(x) = y^{1-p}$ changes it to a linear equation for u,

$$u' = (1-p)f_1(x)u + (1-p)f_2(x). \qquad (3)$$

For the general case (1), we substitute

$$y(x) = -\frac{w'(x)}{f_2(x)w(x)} \qquad (4)$$

and obtain

$$w''(x) - \left[\frac{f_2'(x)}{f_2(x)} + f_1(x)\right]w'(x) + f_0(x)f_2(x)w(x)$$
$$\equiv w''(x) + p(x)w'(x) + q(x)w = 0. \qquad (5)$$

Equation (5) is linear and second order with variable coefficients. If we know two linearly independent solutions of (5), say $w_1(x)$ and $w_2(x)$, then the solution $y(x)$ of (1) can be written in terms of the general solution

$$w(z) = c_1 w_1(x) + c_2 w_2(x)$$

of (5), namely,

$$y(x) = -\frac{1}{f_2(x)} \frac{c_1 w_1'(x) + c_2 w_2'(x)}{c_1 w_1(x) + c_2 w_2(x)}$$
$$= -\frac{1}{f_2(x)} \frac{w_1'(x) + c w_2'(x)}{w_1(x) + c w_2(x)}, \qquad (6)$$

Transformations of Nonlinear ODE

where $c = c_2/c_1$.

The solution (6) of (1), therefore, involves only one arbitrary constant c and is a fractional linear function of this constant. Thus, if we can solve the second-order equation (5), the solution of Riccati equation (1) is (6). Since there is no general method of explicitly solving all equations of type (5), not all Riccati equations can be solved in closed form. However, if we can find one nontrivial solution $y_1(x)$, however simple, of the Riccati equation, we can easily find the second solution. The required substitution [corresponding to the multiplicative substitution $y(x) = y_1(x)v(x)$ for linear second-order equations (see Section 1.3)]

$$y = y_1(x) + u(x)$$

is now additive. Equation (1) is then transformed to the Bernoulli equation

$$u'(x) = [f_1(x) + 2f_0(x)y_1(x)]u(x) + f_2(x)u^2(x),$$

which is solvable (see equation (2)).

EXAMPLE 1 $y' = \dfrac{y}{x} + x^3 y^2 - x^5$

$y = x$ is obviously a solution of this equation. Writing $y = x + u(x)$, we have

$$u' - \left(\frac{1}{x} + 2x^4\right)u = x^3 u^2.$$

Putting $v = 1/u$, we get

$$v' + \left(\frac{1}{x} + 2x^4\right)v = -x^3.$$

This equation has the solution

$$v = -\frac{1}{2x} + \frac{c}{x} e^{-2x^5/5}.$$

Substituting this expression for v into $y = x + u = x + 1/v$, we finally have

$$y = x\left(\frac{c + \tfrac{1}{2} e^{2x^5/5}}{c - \tfrac{1}{2} e^{2x^5/5}}\right)$$

as the solution.

2.4 ABEL'S EQUATION

A natural generalization of Riccati's equation is Abel's equation of the first kind,

$$\frac{dy}{dx} = f_0(x) + f_1(x)y + f_2(x)y^2 + f_3(x)y^3, \tag{1}$$

with a cubic nonlinearity (see Murphy, 1960 for special integrable cases of (1)). We shall see how special classes of Abel's equation may be related to a linear equation of third order. As expected, the relationship is much more complicated than that of the Riccati equation to the second-order linear equation (Vein, 1967). For simplicity we consider the special third-order equation

$$\frac{d^3\phi}{ds^3} - \phi = 0. \tag{2}$$

Let ϕ_1, ϕ_2, and ϕ_3 be any three linearly independent solutions of (2). Then it is evident (by simple differentiation, etc.) that ϕ_1, ϕ_2, and ϕ_3 are related as follows:

$$\begin{aligned}
\phi_1' &= \phi_3, & \phi_1'' &= \phi_2, \\
\phi_2' &= \phi_1, & \phi_2'' &= \phi_3, \\
\phi_3' &= \phi_2, & \phi_3'' &= \phi_1, & ' &= \frac{d}{ds}.
\end{aligned} \tag{3}$$

It can be verified that the functions ϕ_1, ϕ_2, ϕ_3 have the representations

$$\begin{aligned}
\phi_1(x) &= \frac{1}{3}\left\{e^x + 2e^{-x/2}\cos\frac{\sqrt{3}x}{2}\right\}, \\
\phi_2(x) &= \frac{1}{3}\left\{e^x - 2e^{-x/2}\cos\left(\frac{\sqrt{3}x}{2} + \frac{\pi}{3}\right)\right\}, \\
\phi_3(x) &= \frac{1}{3}\left\{e^x - 2e^{-x/2}\cos\left(\frac{\sqrt{3}x}{2} - \frac{\pi}{3}\right)\right\},
\end{aligned} \tag{4}$$

and are related by

$$\phi_1 + \phi_2 + \phi_3 = e^x \tag{5}$$

and

$$\phi_1^3 + \phi_2^3 + \phi_3^3 - 3\phi_1\phi_2\phi_3 = 1.$$

Transformations of Nonlinear ODE

In view of (2) and (4), ϕ_1, ϕ_2, and ϕ_3, are referred to as third-order hyperbolic functions.

Now writing a solution of (2) as

$$t = c\phi_1(s) + \phi_2(s), \tag{6}$$

where c is a constant, we have, in view of (3),

$$\begin{aligned} t'(s) &= c\phi_3(s) + \phi_1(s), \\ t''(s) &= c\phi_2(s) + \phi_3(s), \\ t''' &= t. \end{aligned} \tag{7}$$

We define a transformation

$$\begin{aligned} x &= a\left(\frac{t'}{t}\right) + b\left(\frac{t''}{t}\right) \\ &= \frac{a\phi_1(s) + bc\phi_2(s) + (ac+b)\phi_3(s)}{c\phi_1(s) + \phi_2(s)} \\ &\equiv \Phi(s), \quad \text{say,} \end{aligned} \tag{8}$$

a generalization of (4) of Section 2.3 for the Riccati equation, which defines s implicitly as a function of x. In (8), a and b are arbitrary constants. Assuming that the inverse of (8) exists, we may write it as

$$s = \Phi^{-1}(x). \tag{9}$$

We shall show that ds/dx satisfies an Abel equation of the first kind. Differentiating (8) with respect to s and substituting t for t''' from (7), we have

$$\frac{dx}{ds} = a\left[\frac{t''}{t} - \left(\frac{t'}{t}\right)^2\right] + b\left[1 - \frac{t'}{t}\left(\frac{t''}{t}\right)\right]. \tag{10}$$

Now, substituting for t''/t from (8), we get

$$b\frac{dx}{ds} = (ax + b^2) - (bx + a^2)\frac{t'(s)}{t(s)}. \tag{11}$$

Differentiating (11) with respect to x, we have

$$-\frac{b(d^2s/dx^2)}{(ds/dx)^2} = a - b\left(\frac{t'}{t}\right) - (bx + a^2)\left[\frac{t''}{t} - \left(\frac{t'}{t}\right)^2\right]\frac{ds}{dx}. \tag{12}$$

Eliminating t'/t and t''/t from (12) with the help of (8) and (11) (and after some reduction), we get

$$(bx + a^2)\frac{d^2s}{dx^2} = -2b\frac{ds}{dx} + 3(ax + b^2)\left(\frac{ds}{dx}\right)^2 \qquad (13)$$
$$+ (x^3 - 3abx - a^3 - b^3)\left(\frac{ds}{dx}\right)^3$$

or, with

$$y = \frac{ds}{dx}, \qquad (14)$$

we obtain an Abel equation of the first kind:

$$(bx + a^2)\frac{dy}{dx} = -2by + 3(ax + b^2)y^2 + (x^3 - 3abx - a^3 - b^3)y^3. \qquad (15)$$

Equation (11) gives the solution of (15) as

$$\frac{b}{y} = (ax + b^2) - (bx + a^2)\left[\frac{c\phi_3(s) + \phi_1(s)}{c\phi_1(s) + \phi_2(s)}\right]$$
$$= (ax + b^2) - (bx + a^2)\left[\frac{c\phi_3(\Phi^{-1}(x)) + \phi_1(\Phi^{-1}(x))}{c\phi_1(\Phi^{-1}(x)) + \phi_2(\Phi^{-1}(x))}\right], \qquad (16)$$

where (7) and (9) have been used.

We can obtain two other forms of the solution by writing $t = c\phi_2(s) + \phi_3(s)$ and $t = c\phi_3(s) + \phi_1(s)$ instead of (6). Inversion of (8) to give (9) is far from trivial, so the solution (16) is not quite explicit. For example, we consider the equation

$$a^2 y' = 3axy^2 + (x^3 - a^3)y^3 \qquad (17)$$

by putting $b = 0$ in (15). Here we cannot use (16), but an alternative form is easily derived. We eliminate t'/t from (11) with the help of (8) and then substitute t''/t directly from (7). We find that

$$\frac{a}{y} = (ab - x^2) + (bx + a^2)\left[\frac{c\phi_2(\Phi^{-1}(x)) + \phi_3(\Phi^{-1}(x))}{c\phi_1(\Phi^{-1}(x)) + \phi_2(\Phi^{-1}(x))}\right]. \qquad (18)$$

To find the solutions of (17), we put $b = 0$ in (18) so that

$$\frac{a}{y} = a^2\left[\frac{c\phi_2(\Phi^{-1}(x)) + \phi_3(\Phi^{-1}(x))}{c\phi_1(\Phi^{-1}(x)) + \phi_2(\Phi^{-1}(x))}\right] - x^2. \qquad (19)$$

This form of the solution is more "explicit" (provided $\Phi^{-1}(x)$ can be found) than that obtained by using a set of transformations, suggested in Murphy (1960, p. 231).

Transformations of Nonlinear ODE

Vein (1967) has indicated how the above analysis may be generalized by starting with the more general third-order linear equation

$$\psi''' = \alpha\psi + \beta\psi' + \gamma\psi''$$

instead of (2), where α, β, and γ are arbitrary constants.

The rest of this chapter is concerned with second-order nonlinear differential equations and their transformations.

2.5 RELATION BETWEEN THIRD-ORDER LINEAR EQUATIONS AND SECOND-ORDER NONLINEAR EQUATIONS

The first-order Riccati equation is related in a simple manner to second-order linear DE with variable coefficients, as was shown in Section 2.3. It is natural to expect that second-order nonlinear DE may be related to third-order linear equations with variable coefficients. This is indeed the case, as was first shown by Dasarathy and Srinivasan (1969). The transformations that accomplish this change both the dependent and independent variables. The nonlinear DE in this section will have x as the dependent variable and t as the independent variable.

We start with third-order linear DE

$$v''' + p(s)v'' + q(s)v' + r(s)v = 0, \tag{1}$$

where the prime denotes differentiation with respect to s, and p, q, and r are known functions of s. The first few steps generalize Vein's technique (see Section 2.4). Let v_1, v_2, and v_3 be three linearly independent solutions of (1). We construct a function

$$u(s) = c_1 v_1(s) + c_2 v_2(s), \tag{2}$$

which, with c_1 and c_2 as arbitrary constants, is also a solution of (1). Now we define a new dependent variable x by the relation

$$x = a(s)\frac{u'}{u} + b(s)\frac{u''}{u} \equiv V(s), \tag{3}$$

say, where $a(s)$ and $b(s)$ are arbitrary functions of s. This transformation using constants was introduced in Section 2.4 (see equation (8)). We write the inverse of (3) as

$$s = V^{-1}(x). \tag{4}$$

In this case, a (complicated) change of independent variable is also introduced:

$$\frac{ds}{dx} = f(x)\left(\frac{dx}{dt}\right)^n \equiv f(x)\dot{x}^n, \tag{5}$$

where $f(x)$ is an arbitrary function and a dot denotes differentiation with respect to t, the new independent variable; the case $n = 0$ reduces to Vein's transformation, and the case $n \neq 0$ leads to second-order DE.

Now the procedure is quite similar to that in Section 2.4. We differentiate (3) with respect to s, so that

$$\frac{dx}{ds} = V'(s). \tag{6}$$

Differentiation of (6) with respect to x gives

$$\frac{-d^2s/dx^2}{(ds/dx)^2} = V''(s)\frac{ds}{dx} \tag{7}$$

while that of (5) with respect to x leads to

$$\frac{d^2s}{dx^2} = f^*(x)\dot{x}^n + nf(x)\dot{x}^{n-2}\ddot{x}, \tag{8}$$

where $f^* = df(x)/dx$. Now, eliminating ds/dx and d^2s/dx^2 with the help of (5), (7), and (8), we get

$$n\ddot{x} + \frac{f^*(x)}{f(x)}\dot{x}^2 + V''(s)f^2(x)\dot{x}^{2n+2} = 0. \tag{9}$$

The function $V''(s)$ has to be found so that, when substituted into (9), it would change the latter to a second-order (autonomous) nonlinear DE in x alone. For this purpose, we differentiate (3) with respect to s:

$$V'(s) = a\left[\frac{u''}{u} - \left(\frac{u'}{u}\right)^2\right] + b\left[\frac{u'''}{u} - \frac{u'}{u}\frac{u''}{u}\right] + a'\frac{u'}{u} + b'\frac{u''}{u}. \tag{10}$$

The function u satisfies (1), so u'''/u can be expressed in terms of u''/u and u'/u. In turn, u''/u can be expressed in terms of u'/u and V by way of (3). Thus, (10) can be written as

$$V'(s) = G(s)V(s) - b(s)r(s) + [H(s) - V(s)]\frac{u'}{u}, \tag{11}$$

where

$$G(s) = \frac{a + b' - bp}{b}, \quad H(s) = a' - bq - aG.$$

Transformations of Nonlinear ODE

Equation (11) is now differentiated with respect to s; u''/u and u'/u are then eliminated with the help of (3) and (10). We thus obtain

$$V''(s) + A(s) + B(s)V' + C(s)V'^2 = 0, \qquad (12)$$

where

$$\begin{aligned}A(s) &= (br)' - G'V + \frac{br - GV}{H - V}(br - GV - H') \\ &\quad - \frac{(H - V)V}{b} + \frac{a(br - GV)}{b}, \\ B(s) &= \frac{a}{b} + \frac{3(br - GV) - H'}{H - V} - G, \\ C(s) &= \frac{2}{H - V}.\end{aligned} \qquad (13)$$

Combining (5) and (6), we get

$$V'(s) = \left[f(x)\left(\frac{dx}{dt}\right)^n \right]^{-1}. \qquad (14)$$

We now obtain $V''(s)$ by substituting (14) into (12). The resulting expression for V'' is inserted into (9) to get a second-order nonlinear autonomous equation in x, with t as the independent variable:

$$n\ddot{x} + \left[\frac{f^*(x)}{f(x)} - C[V^{-1}(x)]\right]\dot{x}^2 - A[V^{-1}(x)]f^2(x)\dot{x}^{2n+2}$$
$$- B[V^{-1}(x)]f(x)\dot{x}^{n+2} = 0. \qquad (15)$$

Here we express $C(s), A(s)$, and $B(s)$, etc., as functions of x, via $s = V^{-1}(x)$ (see equation (4)). A special case of (15) with $n = -1$ is of particular interest in applications:

$$\ddot{x} + P(x)\dot{x}^2 + Q(x)\dot{x} + R(x) = 0, \qquad (16)$$

where

$$\begin{aligned}P(x) &= C[V^{-1}(x)] - \frac{f^*(x)}{f(x)}, \\ Q(x) &= B[V^{-1}(x)]f(x), \\ R(x) &= A[V^{-1}(x)]f^2(x).\end{aligned}$$

Thus, the procedure to generate the nonlinear DE (15) or (16) is as follows. Choose the coefficients $p(s)$, $q(s)$, and $r(s)$ for which equation (1) is explicitly solvable. Again, choose the coefficients c_1 and c_2 and the functions $a(s)$ and $b(s)$ so that relation (3) may be explicitly inverted in the form (4), giving s as a function of x. Equation (5) then provides the connection between the old independent variable and the new dependent and independent variables. The function $f(x)$ is arbitrary and may be suitably chosen.

EXAMPLE 1 Consider the equation $v''' + k_1^2 v' = 0$, which has three solutions $v_1 = v_{10}$, a constant, $v_2 = \sin k_1 s$, and $v_3 = \cos k_1 s$. The coefficients in equation (1) are $p(s) = 0$, $q(s) = k_1^2$, and $r(s) = 0$. In this case, we choose the solution $u = c_1 v_1 + c_2 v_2(s)$, and $a(s) = 1/k_2$, a constant, $b(s) = 0$, $n = -1$, and $f(x) = 1$; thus (5) becomes

$$\frac{ds}{dx} = \frac{dt}{dx} \tag{17}$$

so that $s = t + \theta$, where θ is a constant. The transformation (3) is then

$$x = \frac{1}{k_2} \frac{u'(s)}{u(s)} = V(s). \tag{18}$$

We directly obtain

$$V'(s) = \frac{1}{k_2} \left[\frac{u''}{u} - \left(\frac{u'}{u}\right)^2 \right] = \frac{1}{k_2} \frac{u''}{u} - k_2 V^2,$$

$$V''(s) = \frac{1}{k_2} \left[\frac{u'''}{u} - \frac{u'}{u} \frac{u''}{u} \right] - 2k_2 V V' \tag{19}$$

$$= -k_1^2 V - k_2^2 V^3 - 3k_2 V V',$$

where we have used the equation

$$u''' = -k_1^2 u'.$$

Here, we have

$$V'(s) = \frac{dx}{ds} = \frac{dx}{dt} = \dot{x}.$$

Substituting $V(s) = x$ and $V'(s) = \dot{x}$ in equation (19), we get

$$V''(s) = -k_1^2 x - k_2^2 x^3 - 3k_2 x \dot{x}.$$

Equation (9) now assumes the form

$$\ddot{x} + 3k_2 x \dot{x} + k_1^2 x + k_2^2 x^3 = 0. \tag{20}$$

Transformations of Nonlinear ODE

Its solution is

$$x = \frac{1}{k_2}\frac{u'(s)}{u(s)} = \frac{1}{k_2}\frac{u'(t+\theta)}{u(t+\theta)} = \frac{1}{k_2}\frac{k_1 c_2 \cos k_1(t+\theta)}{c_1 v_{10} + c_2 \sin k_1(t+\theta)}$$

$$= \frac{\cos k_1(t+\theta)}{k_3 + k_4 \sin k_1(t+\theta)}, \qquad (21)$$

where k_3 and k_4 are some other constants.

2.6 RELATIONSHIP BETWEEN LINEAR AND NONLINEAR SECOND-ORDER EQUATIONS

So far we have related some nonlinear DE with linear equations of higher order, motivated basically by the Riccati equation. Now we connect other nonlinear second-order DE with linear second-order DE. Specifically, we ask what is the class of nonlinear second-order DE whose solutions are expressible as functions of the two linearly independent solutions of a linear second-order equation. Pinney (1950) showed that the solution of the second-order nonlinear DE

$$y'' + a_0(x)y = k_0 y^{-3}, \qquad ' = \frac{d}{dx}, \qquad (1)$$

where $a_0(x)$ is continuous and k is a constant, can be written as

$$y = (y_1^2 + k_0 w^{-2} y_2^2)^{1/2}. \qquad (2)$$

Here y_1 and y_2 form the fundamental system (linearly independent solutions) of the linear equation

$$y'' + a_0(x)y = 0 \qquad (3)$$

and $W(y_1(x), y_2(x)) = w$, a constant, is the Wronskian.

Berkovich and Rozov (1972) attribute this result to Ermakov (1880). Before we discuss this result and its generalizations, we consider an important context in which (1) appears as the equation governing one of the transformation functions which change a given linear second-order equation to another, preassigned one. Specifically, we seek functions $u(x)$ and $v(x)$ such that the substitutions

$$y = v(x)z, \qquad dt = u(x)dx \qquad (4)$$

transforms (3) to

$$\ddot{z} + b_0(t)z = 0, \qquad (5)$$

where $b_0(t)$ is a given function and the dot denotes differentiation with respect to t. Now, according to (4),

$$\frac{dy}{dx} = v'(x)z + v(x)u(x)\dot{z},$$
$$\frac{d^2y}{dx^2} = v''(x)z + 2v'(x)u(x)\dot{z} + v(x)[u'(x)\dot{z} + u^2(x)\ddot{z}]$$
(6)

Substituting (6) into (3), we get

$$vu^2\ddot{z} + (vu' + 2v'u)\dot{z} + (v'' + a_0(x)v)z = 0. \tag{7}$$

This is the same as (5) if

$$\frac{v'}{v} + \frac{1}{2}\frac{u'}{u} = 0 \tag{8}$$

and

$$(vu^2)^{-1}[v'' + a_0(x)v] = b_0(t(x)). \tag{9}$$

Equation (8) integrates to give

$$vu^{1/2} = c_0. \tag{10}$$

Choosing $c_0 = 1$ and eliminating u from (9) with the help of (10), we get

$$v'' + a_0(x)v = b_0(t(x))v^{-3}. \tag{11}$$

Eliminating v from (9) and (10), we have

$$\frac{1}{2}\frac{u''}{u} - \frac{3}{4}\left(\frac{u'}{u}\right)^2 + b_0(t(x))u^2 = a_0(x). \tag{12}$$

In particular, equation (3) is reduced to an equation with a constant coefficient,

$$\ddot{u} + k_0 u = 0, \tag{13}$$

if $b_0(t(x)) = k_0$, the function v is governed by (1), and the function u is correspondingly given by (10).

Takayama (1986) has shown that it is possible to solve analytically the DE

$$\ddot{x} + [q_1(t) + \lambda q_2(t)]x = 0$$

if the solution of the "initial" equation

$$\ddot{x} + q_1(t)x = 0$$

Transformations of Nonlinear ODE

is known in an analytic form and the function $q_2(t) = \beta^{-2}$, where β is governed by the nonlinear DE

$$\tfrac{1}{2}\beta\ddot{\beta} - \tfrac{1}{4}\dot{\beta}^2 + q_1(t)\beta^2 = 1.$$

Several examples illustrating how to obtain solutions of more general linear DE with variable coefficients from simpler ones which are explicitly solvable were also given by Takayama.

Before we proceed to generalize (1)–(2), we look for transformations (4) which change a nonlinear DE

$$y'' + a_0(x)y = \phi(x)y^\alpha \tag{14}$$

into an autonomous nonlinear form

$$\ddot{z} + k_0 z = p z^\alpha \tag{15}$$

where k_0, p, and α are constants. In the present case, (14) changes to (7) with a right side of $\phi(x)v^\alpha z^\alpha$, so the former becomes (15), provided

$$\phi(x)v^\alpha v^{-1} u^{-2} = p,$$

or

$$pu^2 = \phi(x)v^{\alpha-1}. \tag{16}$$

Since (15) is an autonomous second-order DE, it can be reduced to first-order by the substitution $dz/dt = q$, say, and hence solved. Since $dt/dx = v^{-2}$, according to (4) and (10) with $c_0 = 1$, we may write the solution of (14) in the form

$$y = v(x)z(t + c_1, c_2) = v(x)z\left[\int \frac{dx}{v^2(x)} + c_1, c_2\right],$$

where $z(t + c_1, c_2)$ is the general solution of the autonomous equation (15). In particular, if $z = \sigma$ is the constant solution of (15) so that $k_0 \sigma = p\sigma^\alpha$ (that is, $\sigma = (k_0/p)^{1/(\alpha-1)}$), then

$$y = \sigma v(x) \tag{17}$$

is a solution of (14). We recall that, in the transformation (4) used here, $v(x)$ is a solution of (1). In particular, if we choose this solution to be (2), then it follows that equation (14) with

$$\phi(x) = p(y_1^2 + k_0 w^{-2} y_2^2)^{-(\alpha+3)/2}, \qquad p = \text{constant},$$

has the solution

$$y = \sigma(y_1^2 + k_0 w^{-2} y_2^2)^{1/2}, \qquad \sigma = (k_0/p)^{1/(\alpha-1)}.$$

Now we turn to the generalization of Pinney's (and several other investigators') work by Reid (1973). Reid proved that the homogeneous function

$$y = (ay_1^m + mby_1^j y_2^n + cy_2^m)^{k/m}, \quad m = j + n, \tag{18}$$

where a, b, and c are arbitrary constants, satisfies the nonlinear DE

$$y'' + r(x)y' + kq(x)y = (1 - l)y'^2 y^{-1} + kQW^2 y^{1-2ml}, \tag{19}$$

where $kl = 1$ and

$$Q = by_1^{j-2} y_2^{n-2}[(m - j - 1)nay_1^m + (m - n - 1)jcy_2^m - bnjy_1^j y_2^n]$$
$$+ (m - 1)ac(y_1 y_2)^{m-2}.$$

Here, $y_1(x)$ and $y_2(x)$ are two linearly independent solutions of the linear second-order equation

$$y'' + r(x)y' + q(x)y = 0, \tag{20}$$

and $W(y_1 y_2)$ is the (in general nonconstant) Wronskian of the set of solutions (y_1, y_2). Pinney's case corresponds to $r(x) = 0$, $q(x) = a_0(x)$, $l = 1$ ($k = 1$), $m = 2$, $a = 1$, $b = 0$, $c = k_0 w^{-2}$, and $W = w$. To simplify the calculations, we derive a special case of (18), although the procedure is exactly the same for the general case. We ask what nonlinear equation has solutions of the form

$$y = [ay_1^m + cy_2^m]^{k/m}, \tag{21}$$

where y_1 and y_2 are linearly independent solutions of (20). Let this equation in particular be of the form

$$y'' + r(x)y' + kq(x)y = R(x, y, y'), \tag{22}$$

where $R(x, y, y')$ represents the nonlinear terms to be determined. The first and second derivatives of y are

$$y' = ky^{1-ml} A, \tag{23}$$
$$y'' = k(k - m)y^{1-2ml} A^2 + ky^{1-ml} A', \tag{24}$$

respectively, where $kl = 1$ and

$$A = ay_1^{m-1} y_1' + cy_2^{m-1} y_2'. \tag{25}$$

Substituting (23) and (24) into (22), we have an equation in terms of A and A':

$$k(k - m)y^{1-2ml} A^2 + ky^{1-ml}[A' + r(x)A + q(x)y^{ml}] = R(x, y, y'). \tag{26}$$

Transformations of Nonlinear ODE

Multiplying (26) by y^{2ml-1} and substituting for A and A' from (25), we get

$$k(k-m)A^2 + ky^{ml}[(m-1)(ay_1^{m-2}y_1'^2 + cy_2^{m-2}y_2'^2)$$
$$+ (y_1'' + r(x)y_1' + q(x)y_1)ay_1^{m-1}$$
$$+ (y_2'' + r(x)y_2' + q(x)y_2)cy_2^{m-1}] = y^{2ml-1}R. \quad (27)$$

Since $y_1(x)$ and $y_2(x)$ are solutions of (20), the coefficients of ay_1^{m-1} and cy_2^{m-1} in brackets vanish. So, (27) can conveniently be written as

$$k^2A^2 - mkA^2 + k(m-1)y^{ml}(ay_1^{m-2}y_1'^2 + cy_2^{m-2}y_2'^2) = y^{2ml-1}R. \quad (28)$$

We express y_1' and y_2' in terms of y_1 and y_2 by solving the equations $W = y_1y_2' - y_2y_1'$ and (25) simultaneously. Thus, (28) reduces, after some simplification, to

$$k(k-1)A^2 + k(m-1)ac(y_1y_2)^{m-2}W^2 = y^{2ml-1}R. \quad (29)$$

Now, writing $A = ly'y^{ml-1}$ from (23), we finally obtain the expression for the nonlinear term R:

$$R = (1-l)y'^2y^{-1} + k(m-1)ac(y_1y_2)^{m-2}W^2y^{1-2ml}. \quad (30)$$

With this expression for R on the right side of (22), the latter represents the nonlinear equation whose solution is give by (21) in terms of two linearly independent solutions of the linear equation (20). Note that starting with (2) rather than (3) (which has no first derivative term) adds some generality to the class of nonlinear equations (22). The Wronskian

$$W = \exp\left(-\int^x r(t)\,dt\right)$$

in this case is always a function of x and not a constant.

As an example of the class of equations (19), we refer to the Duffing-type equations

$$y'' \pm w^2 y \mp \beta^2 e^{-6wx} y^3 = 0, \quad (31)$$

where w and β are constants. For the upper signs, $y_1 = e^{+iwx}$, $y_2 = e^{-iwx}$ and the solution, by comparing (31) and (19),

$$y = \sqrt{10}\left(\frac{w}{\beta}\right)e^{3wx}, \quad (32)$$

follows from the values $l = 1$, $m = -1$, $j = (3i-1)/2$, $n = -(3i+1)/2$, where $i^2 = -1$, and $a = c = 0$, $b^2 = \beta^2/10w^2$, and $W^2 = -4w^2$. From the lower signs in (31), the solution is

$$y = \frac{e^{3wx}}{ae^{2wx} - i\beta/2\sqrt{2}w}$$

for $m = -1, j = -2, n = 1$ and for $c = 0$, $b^2 = \beta^2/8w^2$, and $W^2 = 4w^2$.

2.7 EXACT LINEARIZATION OF NONLINEAR AUTONOMOUS SECOND-ORDER EQUATIONS VIA FACTORING

Before considering the factoring of a nonlinear second-order differential operator, we discuss the same for a linear second-order one. We show that the conditions required for factoring are quite stringent. We consider the linear operator such that

$$Ly \equiv \left[p_0(x) + p_1(x) \frac{d}{dx} + \frac{d^2}{dx^2} \right] y. \tag{1}$$

Suppose we can write

$$\begin{aligned} Ly &= \left[\frac{d}{dx} + a_1(x) \right] \left[\frac{d}{dx} + a_2(x) \right] y \\ &= y'' + (a_1 + a_2)y' + (a_2' + a_1 a_2)y. \end{aligned} \tag{2}$$

Comparing (2) and (1), we require that $a_1 + a_2 = p_1$ and $a_2' + a_1 a_2 = p_0$; eliminating a_1 from these equations, we get

$$a_2' = a_2^2 - p_1 a_2 + p_0. \tag{3}$$

This is a Riccati equation whose solution cannot always be found explicitly (see Section 2.3). Indeed, solving (3) is equivalent to solving a second-order linear equation; the latter is tantamount to the original problem $Ly = 0$ that we started with.

Now we consider a second-order autonomous nonlinear DE defined by the operator $N(x)$ (Berkovich, 1979):

$$N(x) \equiv \ddot{x} + f(x)\dot{x}^2 + \phi(x)\dot{x} + \psi(x) = 0, \quad \cdot = \frac{d}{dt}. \tag{4}$$

We construct the class of equations of type (4) which depend on two arbitrary functions and whose solutions may be found in terms of quadratures alone. This is done by transforming both dependent and independent variables,

$$X = \frac{x}{v(x)}, \quad d\tau = u(x)dt, \quad u(x(t))v(x(t)) \neq 0, \tag{5}$$

Transformations of Nonlinear ODE

defined in an interval $a \leq t \leq b$ (cf. (4) of Section 2.6), such that (4) becomes an inhomogeneous linear second-order equation with constant coefficients,

$$X'' + b_1 X' + b_0 X + c = 0, \qquad ' = \frac{d}{d\tau} \tag{6}$$

where b_1, b_0, and c are arbitrary constants. Equation (6) can always be rewritten as

$$\left(\frac{d}{d\tau} - r_2\right)\left(\frac{d}{d\tau} - r_1\right) X + c = 0, \tag{7}$$

where r_1 and r_2 are the roots of the characteristic equation

$$r^2 + b_1 r + b_0 = 0. \tag{8}$$

Berkovich has proved that for (4) to be linearized via (5) in the form (6), it is necessary and sufficient that it may be "factored" as

$$\left(\frac{d}{dt} - \frac{\dot{v}}{v} - r_2 u - \frac{\dot{u}}{u}\right)\left(\frac{d}{dt} - \frac{\dot{v}}{v} - r_1 u\right) x + c u^2 v = 0, \tag{9}$$

or, equivalently,

$$\left(u^{-1}\frac{d}{dt} - u^{-1}v^{-1}\dot{v} - r_2\right)\left(u^{-1}\frac{d}{dt} - u^{-1}v^{-1}\dot{v} - r_1\right) x + cv = 0, \tag{10}$$

where

$$v^* = \frac{dv}{dx}, \quad \dot{v} \equiv v^* \dot{x}, \quad \text{and} \quad \dot{u} = \frac{du}{dx} \dot{x} \equiv u^* \dot{x}.$$

The operators in (9) are noncommutative, and those in (10) are commutative. We may rewrite (7) as

$$(D_\tau - r_2)(D_\tau - r_1) X + c = 0, \qquad D_\tau = \frac{d}{d\tau}, \tag{11}$$

By straightforward calculation we can prove the operator identity

$$\left(u^{-1}\frac{d}{dt} - r_k\right) u^{1-k} v^{-1} = u^{-k} v^{-1} L_k, \tag{12}$$

where

$$L_k = \frac{d}{dt} - \frac{\dot{v}}{v} - r_k u - (k-1)\frac{\dot{u}}{u}, \qquad k = 1, 2.$$

We multiply (11) by u^2v, take note of (5), and use (12) successively for $k = 1, 2$; we easily verify that

$$\left(u^{-1}\frac{d}{dt} - r_1\right)v^{-1}x = u^{-1}v^{-1}\left(\frac{d}{dt} - \frac{\dot{v}}{v} - r_1 u\right)x \equiv u^{-1}v^{-1}L_1 x$$
$$u^2 v\left(u^{-1}\frac{d}{dt} - r_2\right)u^{-1}v^{-1}L_1 = L_2 L_1. \qquad (13)$$

Multiplying (11) by u^2v and using (13) lead to (9). The equivalence of (9) and (10) can be proved by direct calculation.

To prove the sufficiency of the decomposition (9), we note that the operator identity

$$L_k u^{k-1}v = u^k v(D_\tau - r_k), \qquad k - 1, 2,$$

changes (9) to

$$u^2 v(D_\tau - r_2)(D_\tau - r_1)X + cu^2 v = 0. \qquad (14)$$

Dividing (14) by u^2v gives (11). Thus, writing out the operators in (9), where

$$\dot{u} = \frac{du}{dx}\dot{x} \quad \text{and} \quad \dot{v} = \frac{dv}{dx}\dot{x},$$

we find that (4) can be exactly linearized by (5) if the former has the form

$$EN(x) \equiv E\ddot{x} - F\dot{x}^2 + G\dot{x} + H = 0, \qquad (15)$$

where

$$E = 1 - v^{-1}v^*x, \qquad F = (2v^{-1}v^* + u^{-1}u^*)E + xv^{-1}v^{**},$$
$$G = b_1 uE, \qquad H = b_0 u^2 x + cvu^2, \qquad v \neq ax. \qquad (16)$$

Equation (15) is autonomous. It may be written in the simpler form

$$\ddot{x} + f\dot{x}^2 + b_1\phi\dot{x} + \Lambda(\phi,x) = 0, \qquad (17)$$

if we denote $-F/E$ by f, G/E by $b_1\phi$, and H/E by $\Lambda(\phi,x)$. The last function will be written out explicitly. First we note that $f = -F/E$ implies that

$$v^{**} - 2v^{-1}v^{*2} + (2x^{-1} - u^{-1}u^* - f)v^* + x^{-1}(u^{-1}u^* + f)v = 0. \qquad (18)$$

Two cases arise:
(i) $v \neq ax + b$, where a and b are constant. Writing $v = 1/V$, we have

$$V^{**} + \left(\frac{2}{x} - \frac{u^*}{u} - f\right)V^* - \frac{1}{x}\left(\frac{u^*}{u} + f\right)V = 0. \qquad (19)$$

Transformations of Nonlinear ODE

This equation may be factored as

$$\left(\frac{d}{dx} + \frac{1}{x} - \frac{1}{u}\frac{du}{dx} - f\right)\left(\frac{d}{dx} + \frac{1}{x}\right)V = 0. \qquad (20)$$

If we let $(d/dx + 1/x)V = W$, (20) becomes

$$\frac{dW}{dx} + \left(\frac{1}{x} - \frac{1}{u}\frac{du}{dx} - f\right)W = 0 \qquad (21)$$

which on integration gives

$$\frac{Wx}{u}\exp\left(\int -f(x)\,dx\right) = \beta,$$

where β is a constant. The equation $(d/dx + 1/x)V = W$ on integration yields

$$v^{-1} = V = \frac{1}{x}\left[\alpha + \beta \int u \exp\left(\int f\,dx\right)dx\right], \qquad (22)$$

where α is another constant.

The relation

$$\frac{G}{E} = b_1 u \qquad (23)$$

shows that the function $\phi(x)$ in (17) is simply $u(x)$. Equations (22)–(23) fix the transformation (5):

$$X = \frac{x}{v(x)} = \beta \int \phi \exp\left(\int f\,dx\right)dx = X_1, \quad \text{say,} \qquad (24)$$

$$d\tau = \phi(x)\,dt,$$

if we choose α in (22) to be zero. Substituting these in the expression for H in (16), we get

$$\Lambda = \frac{H}{E} = \frac{b_0 u^2 x + cvu^2}{1 - v^{-1}v^*x} = u^2 \frac{b_0 x/v + c}{v^{-1} - v^{-2}xv^*}$$

$$= \phi \exp\left(-\int f\,dx\right)\left[b_0 \int \phi \exp\left(\int f\,dx\right)dx + \frac{c}{\beta}\right]$$

$$\equiv \Lambda_1, \quad \text{say.}$$

Equation (17), therefore, becomes

$$\ddot{x} + f\dot{x}^2 + b_1\phi\dot{x} + \Lambda_1(\phi,x) = 0, \qquad (17a)$$

involving two arbitrary functions f and ϕ.

(ii) $v = ax + b$. In this case,

$$f = -\frac{F}{E} = -\left(\frac{2a}{ax+b} + \frac{\phi^*}{\phi}\right),$$

$$X = \frac{x}{ax+b} \equiv X_2,$$

$$d\tau = \phi(x)dt,$$

and

$$\Lambda = \phi^2(ax+b)b^{-1}[b_0 x + c(ax+b)] \equiv \Lambda_2.$$

Equation (17) becomes

$$\ddot{x} - \left(\frac{2a}{ax+b} + \frac{\phi^*}{\phi}\right)\dot{x}^2 + b_1\phi\dot{x} + \Lambda_2(\phi,x) = 0. \tag{17b}$$

Therefore, we restate the linearizing theorem for equation (4). For this equation to be exactly linearized by (5), it is necessary and sufficient that it be represented in the form (17a) or (17b). Each form involves two arbitrary functions. The roots r_1, r_2 of equation (8) determine the solution $\{X_1(\tau), X_2(\tau)\}$ of linear DE (7). The following cases arise:

a. $r_1 \neq r_2 \neq 0$, $X_i = C_1 \exp(r_1\tau) + C_2 \exp(r_2\tau) - c/b_0$.

b. $r_1 = r_2 = -b_1/2 \neq 0$, $X_i = (C_1\tau + C_2)\exp(-b_1\tau/2) - c/b_0$.

c. $r_1 = 0$, $r_2 \neq 0$, $X_i = C_1 + C_2\exp(-b_1\tau) - cb_1^{-1}\tau$. (25)

d. $r_1 = r_2 = 0$, $X_i = C_1 + C_2\tau - (c/2)\tau^2$.

e. $b_1 = 0$, $b_0 > 0$, $X_i = A\sin(\sqrt{b_0}\,\tau + B) - c/b_0$.

f. $b_1 = 0$, $b_0 < 0$, $X_i = A\sinh(\sqrt{-b_0}\,\tau + B) - c/b_0$.

Here $i = 1, 2$, $t = \int d\tau/\phi(x(\tau))$, and C_1, C_2, A, and B are arbitrary constants.

When $c = 0$ in (7), we have a special situation. Here, again, we consider cases (i) and (ii) separately.

(i) Differentiating $X = x/v$ logarithmically with respect to τ and using the solution X of (7) with distinct characteristics, we have

$$\frac{1}{X}\frac{dX}{d\tau} = r_k$$

or

$$\frac{\dot{x}}{x}(1 - xv^{-1}v^*) = r_k u.$$

Transformations of Nonlinear ODE

Noting that $u = \phi$ and using (22) with $\alpha = 0$ to eliminate v, we obtain

$$\dot{x} - r_k \exp\left(-\int f\,dx\right) \int \phi \exp\left(\int f\,dx\right) dx$$

$$\equiv \dot{x} - r_k \exp\left(-\int f\,dx\right)\frac{X}{\beta} = 0, \quad (26)$$

see equation (24). This can be integrated with respect to t:

$$r_k t + C_k^{(1)} = I_1, \quad I_1 = \int \frac{\exp \lambda(x)}{\beta^{-1} X(x)}\,dx, \quad \lambda(x) = \int f\,dx, \quad (27)$$

where $C_k^{(1)}$ are arbitrary constants; k assumes values 1 and 2 in (26) and (27). Thus, knowing $f(x)$ and $\phi(x)$ in (17a), we can find $v(x)$ from (22) and hence $X = xv^{-1}$. Equation (27) relates x and t. The relation between t and τ is

$$t = \int \frac{d\tau}{\phi(x(\tau))},$$

where $\phi(x) = u(x)$, and x and τ are connected by (25) and $X = xv^{-1}$.

(ii) As in (i), we have

$$\dot{x} - r_k b^{-1} x(ax + b)\phi(x) = 0 \quad (28)$$

with the integral

$$r_k t + C_k^{(2)} = I_2, \quad I_2 = b\int \frac{dx}{x(ax+b)\phi(x)}. \quad (29)$$

The explicit form is obtained in the same way as for case (i).

For (i) and (ii) we have assumed that $r_1 \neq r_2 \neq 0$; other cases arise when this is not true. Further, the solution has a different form when $b_1 = 0$ in (8). We refer the reader to the original paper of Berkovich for details and several examples from dynamical systems. We conclude this section with one illustration.

EXAMPLE 1 $\quad \ddot{x} + 3x\dot{x} + x^3 = 0 \quad (30)$

Comparing (30) with (17a), we have $f(x) = 0$, $\phi(x) = x$, $b_1 = 3$, $c = 0$, $\beta = 2$, and $b_0 = 2$. Equation (30) admits the factorization

$$\left(\frac{d}{dt} - r_2 x\right)\left(\frac{d}{dt} + \frac{\dot{x}}{x} - r_1 x\right) x \equiv 2(\ddot{x} + 3x\dot{x} + x^3) = 0.$$

This follows easily from (15) and (16), where $v = x^{-1}$ and $u = x$. Here we find that $r_1 = -2$ and $r_2 = -1$. Thus there are two one-parameter families of solutions of (30) (see equations (27) and (5)):

$$x = \frac{1}{t+c_1} \quad \text{and} \quad x = \frac{2}{t+c_2}.$$

We also note that the substitution $x^2 = X$, $d\tau = x\,dt$ (see Section 2.6) reduces (30) to the linear form

$$X'' + 3X' + 2X = 0, \quad ' = \frac{d}{dt}.$$

This gives the solution

$$x = \pm[C_1 \exp(-2\tau) + C_2 \exp(-\tau)]^{1/2}, \quad t = \int \frac{d\tau}{x}, \tag{31}$$

where C_1 and C_2 are arbitrary constants. If $C_2 \neq 0$, we have

$$x = \pm \exp(-\tau)[C_1 + C_2 \exp \tau]^{1/2}, \quad t = \pm 2 C_2^{-1}(C_1 + C_2 \exp \tau)^{1/2} - k,$$

where k is an arbitrary constant. If we eliminate the parameter τ and let $2k = b$ and $k^2 - 4C_1/C_2^2 = c$, we obtain a two-parameter family of solutions $x = (2t+b)/(t^2+bt+c)$. The earlier special case arises when $C_2 = 0$ in (31).

The factorization of coupled second-order systems has been considered by Humi (1986).

2.8 TRANSFORMATION OF NONLINEAR EQUATIONS TO NONLINEAR INTEGRABLE FORMS

In the previous sections we have described techniques to exactly linearize nonlinear DE. Now we relate certain classes of nonlinear DE to other nonlinear DE which may be integrated by suitable transformations (Braude, 1967). We illustrate this by transforming

$$y'' + f(x)y' + \phi(x)y + \theta(x)y^n = 0, \quad ' = \frac{d}{dx}, \tag{1}$$

to

$$\ddot{z} + F(t)\dot{z} + \Phi(t)z + \psi(t)z^n = 0, \quad \cdot = \frac{d}{dt} \tag{2}$$

via

$$x = \alpha(t), \quad y = \beta(t)z(t). \tag{3}$$

Transformations of Nonlinear ODE

Here and in the sequel, n and m are integers. If we choose $F(t)$, $\Phi(t)$, and $\psi(t)$ such that (2) is exactly integrable, then we can find $f(x)$, $\phi(x)$, and $\theta(x)$ via (3) such that (1) is also integrable. From (3), we have

$$\frac{dy}{dx} = (\dot{\beta}z + \beta\dot{z})(\dot{\alpha})^{-1}, \qquad \frac{dx}{dt} = \dot{\alpha},$$

$$\frac{d^2y}{dx^2} = (\ddot{\beta}z + 2\dot{\beta}\dot{z} + \beta\ddot{z})(\dot{\alpha})^{-2} - (\dot{\alpha})^{-3}\ddot{\alpha}(\dot{\beta}z + \beta\dot{z}).$$
(4)

Substituting (4) into (1), we get

$$\ddot{z} + \frac{\dot{\alpha}^2}{\beta}\left[\frac{2\dot{\beta}}{\dot{\alpha}^2} - \beta\frac{\ddot{\alpha}}{\dot{\alpha}^3} + \frac{\beta}{\dot{\alpha}}f(x)\right]\dot{z}$$

$$+ \left[\frac{\ddot{\beta}}{\dot{\alpha}^2} - \frac{\ddot{\alpha}\dot{\beta}}{\dot{\alpha}^3} + f(x)\frac{\dot{\beta}}{\dot{\alpha}} + \phi(x)\beta(t)\right]\frac{\dot{\alpha}^2}{\beta}z$$
(5)

$$+ \theta(x)\beta^{n-1}\dot{\alpha}^2 z^n = 0.$$

Equation (5) is identical to (2) provided that

$$F(t) = \frac{\dot{\alpha}^2}{\beta}\left[\frac{2\dot{\beta}}{\dot{\alpha}^2} - \frac{\ddot{\alpha}}{\dot{\alpha}^3}\beta + \frac{\beta}{\dot{\alpha}}f(x)\right],$$
(6)

$$\Phi(t) = \frac{\dot{\alpha}^2}{\beta}\left[\frac{\ddot{\beta}}{\dot{\alpha}^2} - \frac{\ddot{\alpha}\dot{\beta}}{\dot{\alpha}^3} + f(x)\frac{\dot{\beta}}{\dot{\alpha}} + \phi(x)\beta\right],$$
(7)

and

$$\psi(t) = \beta^{n-1}\dot{\alpha}^2\theta(x).$$
(8)

If we choose $(2\dot{\beta}/\dot{\alpha}^2) - (\ddot{\alpha}/\dot{\alpha}^3)\beta = 0$—that is, $\beta^2 = \dot{\alpha}$—then (6) simplifies considerably, and we have

$$\beta^2 = \dot{\alpha} = \frac{F(t)}{f(x)}.$$
(9)

Substituting (9) into (7) and (8) gives

$$\frac{1}{f^2(x)}\left[\phi(x) - \frac{uf(x)}{2} - \frac{u'}{2} + \frac{u^2}{4}\right]$$

$$= \frac{1}{F^2(t)}\left[\Phi(t) - v\frac{F(t)}{2} - \frac{\dot{v}}{2} + \frac{v^2}{4}\right]$$
(10)

and

$$\frac{\theta(x)}{[f(x)]^{(n+3)/2}} = \frac{\psi(t)}{[F(t)]^{(n+3)/2}},$$
(11)

where

$$u = \frac{f'(x)}{f(x)}, \quad v = \frac{\dot{F}(t)}{F(t)}. \tag{12}$$

Quite evidently, only a restricted class of equations of type (1) can be integrated. In the first instance, we choose functions $F(t)$, $\psi(t)$, and $\Phi(t)$ such that (2) can be solved by a quadrature alone. Then substituting these functions into (10) and (11), we obtain two relations in three unknown functions, $f(x)$, $\phi(x)$, and $\theta(x)$. Thus, one of the functions may be arbitrarily chosen. It is convenient to choose $f(x)$; then (11) and (10) give $\theta(x)$ and $\phi(x)$, respectively, in an elementary way. The transformation (3) is then obtained from (9) via a quadrature. For such a choice of $f(x)$, $\theta(x)$, and $\phi(x)$, a solution of (1) follows from (3) and (9):

$$\int f(x)\,dx = \int F(t)\,dt, \tag{13}$$

$$y = \left[\frac{F(t)}{f(x)}\right]^{1/2} z_1(t, C_{11}, C_{12}),$$

where $z_1(t, C_{11}, C_{21})$ is the "general" solution of (2), and C_{11} and C_{21} are arbitrary constants.

Braude has also related two other sets of nonlinear DE. The equation

$$y'' - my' + f(x)y + \phi(x)y^n = \theta(x), \tag{14}$$

by the substitution (3), becomes

$$\ddot{z} - m\dot{z} + F(t)z + \Phi(t)z^n = \Psi(t), \tag{15}$$

provided

$$[\phi(x)]^{-4/(n+3)}\left[f(x) + \left(\frac{u+2m}{n+3}\right)^2 - \frac{u' + m(u+2m)}{n+3}\right]e^{2m[(n-1)/(n+3)]x}$$

$$= [\Phi(t)]^{-4/(n+3)}\left[F(t) + \left(\frac{v+2m}{n+3}\right)^2 - \frac{\dot{v} + m(v+2m)}{n+3}\right]e^{2m[(n-1)/(n+3)]t},$$

$$[\phi(x)]^{-3/(n+3)}\theta(x)e^{2mnx/(n+3)} = [\Phi(t)]^{-3/(n+3)}\Psi(t)e^{2mnt/(n+3)},$$

$$u = \frac{\phi'(x)}{\phi(x)}, \quad v = \frac{\dot{\Phi}(t)}{\Phi(t)}, \tag{16}$$

$$\int [\phi(x)e^{-m(n+1)x}]^{2/(n+3)}\,dx = \int [\Phi(t)e^{-m(n+1)t}]^{2/(n+3)}\,dt.$$

Transformations of Nonlinear ODE

The solution of (14) is

$$y = \left[\frac{\Phi(t)}{\phi(x)}\right]^{1/(n+3)} e^{-2m(x-t)/(n+3)} z_2(t, C_{21}, C_{22}), \tag{17}$$

where $z_2(t, C_{21}, C_{22})$ is the general integral of (15), and C_{21} and C_{22} are arbitrary constants.

Similarly, the equation

$$y^{n-2}[yy'' + (m-1)y'^2 + f(x)yy'] + \phi(x)y^n = \theta(x) \tag{18}$$

is related to

$$z^{n-2}[z\ddot{z} + (m-1)\dot{z}^2 + \Psi(t)z\dot{z}] + F(t)z^n = \Phi(t) \tag{19}$$

via (3) and

$$[f(x)]^{-2}\left[\phi(x) + \frac{u^2}{4m} - \frac{u'}{2m} - \frac{uf(x)}{2m}\right]$$

$$= [\Psi(t)]^{-2}\left[F(t) + \frac{v^2}{4m} - \frac{\dot{v}}{2m} - \frac{v\Psi(t)}{2m}\right],$$

$$\theta(x)[f(x)]^{(n-4m)/2m} = \Phi(t)[\psi(t)]^{(n-4m)/2m}, \tag{20}$$

$$u = \frac{f'(x)}{f(x)}, \quad v = \frac{\dot{\Psi}(t)}{\Psi(t)}.$$

The solution of (18) is given by

$$\int f(x)\,dx = \int \Psi(t)\,dt,$$

$$y = \left[\frac{\Psi(t)}{f(x)}\right]^{1/2m} z_3(t, C_{31}, C_{32}), \tag{21}$$

where $z_3(t, C_{31}, C_{32})$ is the general solution of (19) with two arbitrary constants C_{31} and C_{32}.

As an example, we take the special case of (18) with $\theta(x) = 0$. Equation (20b) then implies that $\Phi(t) = 0$. The resulting equation for z is

$$z\ddot{z} + (m-1)\dot{z}^2 + \Psi(t)z\dot{z} + F(t)z^2 = 0. \tag{22}$$

This equation is equidimensional in z. Substituting $z = e^\tau$ and $\dot{\tau} = \xi$, we get a first-order equation in ξ:

$$\dot{\xi} + m\xi^2 + \Psi(t)\xi + F(t) = 0. \tag{23}$$

For $F(t) = a$, $\Psi(t) = b$, where a and b are constants, (23) can be integrated to yield

$$C_1 - t = \int \frac{d\xi}{m\xi^2 + b\xi + a}. \qquad (24)$$

The integral in (24) is elementary and can be explicitly evaluated. It has different forms, depending on the discriminant $\Delta = 4ma - b^2$. When $\Delta = 0$, (24) has the explicit form

$$\xi = \frac{1}{2m}\left(\frac{2}{t - C_1} - b\right) \qquad (25)$$

or

$$\dot{\tau} = \frac{1}{2m}\left(\frac{2}{t - C_1} - b\right).$$

This can be integrated and the result expressed in terms of the original variable z:

$$z = C_2(t - C_1)^{1/m} e^{-bt/2m}. \qquad (26)$$

Using the first of (20), and then (21), we obtain the relation between $f(x)$ and $\phi(x)$, and the solution:

$$\int f(x)\, dx = bt, \qquad (27)$$

$$y = \left[\frac{b}{f(x)}\right]^{1/2m} C_2(t - C_1)^{1/m} e^{-bt/2m}, \qquad (28)$$

$$\frac{1}{f^2(x)}\left[\phi(x) - \frac{u'}{2m} + \frac{u^2}{4m} - \frac{uf(x)}{2m}\right] = \frac{a}{b^2}, \qquad (29)$$

$$u = \frac{f'(x)}{f(x)}. \qquad (30)$$

Choosing $f(x)$, we can find x in terms of t from (27) and $\phi(x)$ from (29). Hence with this choice of $f(x)$, equations (18), which are transformed to (19) and hence integrated, can be identified.

2.9 REDUCING NONLINEAR NONAUTONOMOUS SECOND-ORDER EQUATIONS TO FIRST-ORDER EQUATIONS

Now we describe another simplification of nonlinear nonautonomous second-order equations, namely, their reduction to a first-order equation,

Transformations of Nonlinear ODE

which may be solved either by quadrature or in the phase plane, and a secondary equation which relates the solution of the former to that of the original equation. It will be shown that a remarkably large number of DE with specific physical applications is included in the class of equations which is thus transformed (Abdelkader, 1968).

This class of equations is represented by

$$y'' + \frac{a}{y}y'^2 + \frac{b}{x}y' + \frac{c}{x^2}y + dx^r y^s = 0, \quad ' = \frac{d}{dx}, \tag{1}$$

where a, b, c, d, r, and s are arbitrary constants ($r \neq -2$, $s \neq 1$.) Two variants of this equation are easily obtained. For $y = e^\psi$, (1) transforms to

$$\psi'' + (a+1)\psi'^2 + \frac{b}{x}\psi' + \frac{c}{x^2} + dx^r e^{(s-1)\psi} = 0. \tag{2}$$

Alternatively, changing the dependent and independent variables by the relations

$$x = f\frac{dx}{df}, \quad y = \frac{df}{d\xi}\left(\frac{x}{f}\right)^{(2+r)/(1-s)}, \tag{3}$$

we get

$$f_3 + \frac{a-1}{f_1}f_2^2 + \frac{b}{f}f_1 f_2 + \frac{c}{f^2}f_1^3 + df^r f_1^{s+2} = 0, \tag{4}$$

where

$$f_n = \frac{d^n f}{d\xi^n}, \quad n = 1, 2, 3.$$

Equation (4) is a third-order version of (1) with f as the dependent variable and ξ as the independent variable.

Now we reduce (1) to a system of first-order equations by the transformation

$$y = zx^{(2+r)/(1-s)}, \quad w(z) = xz^a\frac{dz}{dx}, \quad r \neq -2, \quad s \neq 1. \tag{5}$$

According to (5),

$$\frac{dy}{dx} = x^{(2+r)/(1-s)-1}\left[z^{-a}w(z) + \frac{2+r}{1-s}z\right],$$

$$\frac{d^2y}{dx^2} = x^{(2+r)/(1-s)-2}\left(-az^{-2a-1}w + z^{-2a}w' + \frac{2+r}{1-s}z^{-a}\right)w(z) \tag{6}$$

$$+ \left(\frac{2+r}{1-s} - 1\right)x^{(2+r)/(1-s)-2}\left[z^{-a}w(z) + \frac{2+r}{1-s}z\right].$$

Substituting (6) into (1) and factoring out $x^{(2+r)/(1-s)-2}w$, we get

$$\frac{dw}{dz} = F(z) + \frac{G(z)}{w}, \tag{7}$$

where

$$F(z) = \left[1 - b - 2(a+1)\frac{2+r}{1-s}\right]z^a,$$

$$G(z) = \left[(1-b)\frac{2+r}{1-s} - (a+1)\left[\frac{2+r}{1-s}\right]^2 - c\right]z^{2a+1} - dz^{2a+s}. \tag{8}$$

Setting

$$A = 1 - b - 2(a+1)\frac{2+r}{1-s} \neq 0$$

and assuming $d \neq 0$, we can write (7) in terms of the rescaled variables

$$Z = (A^2/d)^{1/(1-s)}z,$$
$$W = d^{(a+1)/(s-1)}A^{-1-2(a+1)/(s-1)}w, \tag{9}$$

as

$$\frac{dW}{dZ} = Z^a + \frac{1}{W}(KZ^{2a+1} - Z^{2a+s}), \tag{10}$$

where K is a function of a, b, c, r, and s. It may be written in two alternative forms as a quadratic rational function of either b or r, for fixed values of a, c, and s:

$$K(b,r) = \frac{\{(1-s)(2+r) - (a+1)(2+r)^2 - c(1-s)^2\} - (1-s)(2+r)b}{[\{1-s-2(a+1)(2+r)\} - (1-s)b]^2}$$

$$= \frac{\{2(1-s)(1-b) - 4(a+1) - c(1-s)^2\} + \{(1-s)(1-b) - 4(a+1)\}r - (a+1)r^2}{[(1-s)(1-b) - 4(a+1) - 2(a+1)r]^2}. \tag{11}$$

Equation (10) is of first-order. The variable Z (see (9a)) is related to x via (5b); that is,

$$\frac{dz}{dx} = x^{-1}z^{-a}w(z).$$

We now give special cases of (1), (2), and (4) which occur in applications.

1. The Emden–Lane–Fowler equation

$$y'' + \frac{2}{x}y' + y^\mu = 0$$

Transformations of Nonlinear ODE

is a special case of (1) with $a = 0$, $b = 2$, $c = 0$, $d = 1$, $r = 0$, and $s = \mu$.

2. Bellman's equation

$$y'' = kx^r y^s$$

is a simplified form of (1) with $a = 0$, $b = 0$, $c = 0$, and $d = -k$. The Thomas–Fermi equation

$$y'' = y^{3/2} x^{-1/2}$$

is a special case of Bellman's equation. The equation

$$\xi \frac{d}{d\xi} \left[\frac{1}{\xi} \frac{d}{d\xi} (\xi^2 \gamma) \right] = \frac{(\lambda^2 - \xi^2)^2}{8\xi^2 \gamma^2},$$

which occurs in the analysis of the large deflection of an annular membrane, changes via the transformation

$$y = \xi^2 \gamma, \qquad x = \xi^2 - \lambda^2$$

to

$$32 y^2 y'' = x^2$$

which is a special case of Bellman's equation.

3. The equation

$$y^{1/2} y'' = e^x,$$

a space-charge equation in one dimension, changes via $t = e^x$ to the Langmuir–Blodgett space-charge equation for cylinders:

$$\frac{d^2 y}{dt^2} + \frac{1}{t} \frac{dy}{dt} - t^{-1} y^{-1/2} = 0.$$

This is a special case of (1) with $a = 0$, $b = 1$, $c = 0$, $d = -1$, $r = -1$, and $s = -\frac{1}{2}$.

4. The Langmuir–Boguslavski equation

$$\frac{d}{dx} \left(x^n \frac{d\phi}{dx} \right) = \frac{1}{\phi^{1/2}}$$

or

$$\phi'' + nx^{-1} \phi' - x^{-n} \phi^{-1/2} = 0$$

is a special case of (1) with $a = 0$, $b = n$, $c = 0$, $d = -1$, $r = -n$ and $s = -1/2$.

5. Ivey's equation

$$y'' - \frac{y'^2}{y} + \frac{2}{x} y' + ky^2 = 0,$$

occurring in space-charge theory, is a special case of (1) with $a = -1$, $b = 2$, $c = 0$, $d = k$, $r = 0$, and $s = 2$.

6. The generalized Duffing equation

$$\ddot{z}_1 + \alpha \dot{z}_1 + \beta z_1 + \gamma z_1^m = 0, \qquad \cdot = \frac{d}{dt},$$

via the transformation

$$z_1 = ye^t = xy, \qquad x = e^t,$$

becomes

$$y'' + (\alpha + 3)\frac{y'}{x} + (\alpha + \beta + 1)\frac{y}{x^2} + \gamma x^{m-3} y^m = 0,$$

which is a special case of (1) with $a = 0$, $b = \alpha + 3$, $c = \alpha + \beta + 1$, $d = \gamma$, $r = m - 3$, and $s = m$.

7. The equation

$$y'' + \frac{\alpha}{y} y'^2 + \beta = \gamma \frac{x}{y},$$

which appears in the theory of internal ballistics of guns, is a special case of (1) with $a = \alpha$, $b = 0$, $c = 0$, $d = -\gamma$, $r = 1$, and $s = -1$, provided $\beta = 0$.

8. The equation

$$\frac{d}{dz}\left(z \frac{d}{dz} F^2\right) + 2zF = 0$$

occurs in the analysis of plasma diffusion in a magnetic field. The substitution

$$x = z^2, \qquad y = F^2$$

changes it to

$$y'' + \frac{1}{x} y' + \frac{y^{1/2}}{2x} = 0,$$

a special case of (1) with $a = 0$, $b = 1$, $c = 0$, $d = \frac{1}{2}$, $r = -1$, and $s = \frac{1}{2}$.

An example of a special case of (2) is

$$\psi'' + \frac{\psi'}{x} + dx^r e^\psi = 0,$$

with $a = -1$, $b = 1$, $c = 0$, and $s = 2$. This equation occurs in the flow of viscous fluids with viscosity depending exponentially on temperature.

Transformations of Nonlinear ODE

An example of the special case of (4) is the Blasius equation of laminar boundary layer theory,

$$\frac{d^3\xi}{d\eta^3} + \xi\frac{d^2\xi}{d\eta^2} = 0,$$

which, after the substitution $f(\xi) = d\xi/d\eta$, becomes

$$\xi + \frac{f}{f_1}f_2 + f_1 = 0, \qquad (12)$$

where again $f_n = d^n f/d\xi^n$. Differentiating (12) with respect to ξ and then eliminating ξ, we get

$$f_3 - \frac{1}{f_1}f_2^2 + \frac{2}{f}f_1 f_2 + \frac{1}{f}f_1 = 0,$$

which is a special case of (4) with $a = 0$, $b = 2$, $c = 0$, $d = 1$, $r = -1$, and $s = -1$.

Before concluding this section, we give four cases for which either (1) or (7) is integrable by quadrature alone.

CASE 1 If a, b, r, and s satisfy the relation

$$1 - b - 2(a+1)\left(\frac{2+r}{1-s}\right) = 0,$$

the coefficient of z^a in expression (8) for $F(z)$ is zero. The equation becomes separable and hence can be integrated by quadrature.

CASE 2 If $d = 0$ in (1), we have

$$y'' + \frac{a}{y}y'^2 + \frac{b}{x}y' + \frac{c}{x^2}y = 0;$$

multiplying by y^a, we get

$$\frac{d}{dx}y^a y' + \frac{b}{x}y^a y' + \frac{c}{x^2}y^{a+1} = 0.$$

Putting $y^{a+1} = w$, we get

$$\frac{d^2 w}{dx^2} + \frac{b}{x}\frac{dw}{dx} + c(a+1)\frac{w}{x^2} = 0,$$

a linear equation equidimensional with respect to x (the Euler–Cauchy equation). The solution may easily be obtained as

$$y^{a+1} = Cx^{r_1} + Dx^{r_2} \qquad (a+1 \neq 0),$$

where τ_1 and τ_2 are the roots of

$$\tau^2 + (b-1)\tau + c(a+1) = 0,$$

and C and D are arbitrary constants.

CASE 3 If $b = 0$ and $s = -a \neq 1$, equation (1) becomes

$$y'' + \frac{a}{y}y'^2 + \frac{c}{x^2}y + dx^r y^{-a} = 0.$$

As in case 2, the substitution $w = y^{a+1}$ is suggested, and the solution then can easily be written as

$$x^{-\sigma}y^{a+1} = C + D\int x^{-2\sigma}\,dx - d(a+1)\int x^{-2\sigma}\left\{\int x^{\sigma+r}\,dx\right\}dx,$$

where

$$\sigma^2 - \sigma + c(a+1) = 0,$$

and C and D are arbitrary constants.

CASE 4 If in equation (11) we assume that

$$K = -2\frac{2a+s+1}{(4a+s+3)^2} \neq 0 \text{ or } \infty,$$

and make the substitutions

$$Z^{s-1} = KU^2,$$

$$v = \frac{1}{U} + kWU^{(2a+s+1)/(1-s)},$$

and

$$k = -\frac{1}{2}(4a+s+3)K^{(1+a)/(1-s)}$$

in (10), we get a linear DE for $U = U(v)$:

$$\frac{dU}{dv} = \frac{1-s}{2a+s+1}\frac{Uv-1}{v^2-1}$$

or

$$\frac{dU}{dv} - \frac{1}{2}\frac{1-s}{2a+s+1}\frac{2v}{v^2-1}U = \frac{s-1}{2a+s+1}\frac{1}{v^2-1}.$$

The general solution of this equation is

$$(v^2-1)^\delta U = \frac{s-1}{2a+s+1}\int (v^2-1)^{-\theta}\,dv + C,$$

where

$$\delta = \frac{1}{2}\frac{s-1}{2a+s+1}, \qquad \theta = \frac{1}{2}\frac{4a+s+3}{2a+s+1},$$

and C is an arbitrary constant. One may retrace the steps to get the solution of the original equation.

Abdelkader (1968) has shown that, in view of the relations (11), for any given differential equation (1) with fixed values of a, c, and s, two finite or infinite sequences of differential equations with related solutions and having the same values of a, c, and s but different values of b and r can be generated. The coefficient $d \neq 0$ may be varied by a change of scale of x or y. As a consequence of the above relationship, the knowledge of the solution of any equation belonging to these sequences leads to the solution of all the others.

2.10 TRANSFORMING NONLINEAR NONAUTONOMOUS SECOND-ORDER EQUATIONS TO THE LIE PLANE

It is well known that a second-order nonlinear autonomous DE can be reduced to a first-order nonlinear DE by introducing the first derivative as the new dependent variable. Then, even though the equation may not be explicitly solvable, its phase plane analysis provides sufficient information regarding the solutions of the original equation: whether they are oscillatory or monotone, whether they are stable or unstable, whether they have any asymptotic limits, or whether they have zeros or singularities. We consider phase plane (and phase space) analysis in detail in Chapter 6. Here, we show how even nonautonomous DE (for which phase plane analysis is not applicable) may, through certain transformations of the dependent variable, its derivative, and the independent variable, be rendered autonomous and treated in a plane which we shall refer to as the Lie plane (similar to phase plane). Indeed, these transformations are based on a theorem of Lie (see Cohen, 1931; Ames, 1968), involving the invariance of the differential equations under a one-parameter group of transformations of the variables. The theorem states that if an invariant u and a first differential invariant v of the group are substituted in the original second-order equation, the resulting equation in u and v is of first order. We may add that an invariant u of the group is a function of x, the independent variable, and y, the dependent variable, which is invariant under the transformations of the group, while a first differential invariant v is an invariant function of x, y, and $y' = dy/dx$. There is a systematic method of finding these invariants, which may be found in Cohen (1931) and Ames (1968). We shall show, with the help of several examples, that these may often be found by inspection (Dresner, 1971).

1. Bessel equation of order zero,

$$x^2 y'' + xy' + x^2 y = 0, \qquad ' = \frac{d}{dx}, \tag{1}$$

as indeed all linear equations, is invariant under the one-parameter group

$$\bar{y} = \lambda y, \qquad \bar{x} = x, \tag{2}$$

$0 < \lambda < \infty$. An obvious invariant function not involving the derivative is $u = x$, while the first differential invariant is $v = y'/y = \bar{y}'/\bar{y}$. We find that

$$du = dx, \qquad dv = \frac{y''}{y} dx - \frac{y'^2}{y^2} dx,$$

so, on using (1), we have

$$\frac{dv}{du} = \frac{y''}{y} - \frac{y'^2}{y^2} = -\frac{1}{y}\left(\frac{y'}{x} + y\right) - \frac{y'^2}{y^2}$$

$$= -\frac{v}{u} - 1 - v^2. \tag{3}$$

This is the required equation in the Lie plane, which is a (nonlinear first-order) Riccati equation that is related to the linear second-order equation (1) by $v = y'/y$ and $u = x$ (see Section 2.3).

2. The Emden–Fowler equation

$$y'' + \frac{2}{x} y' + y^n = 0, \qquad x \geq 0, \quad n \geq 0, \tag{4}$$

is invariant under the transformation

$$\bar{x} = \lambda x, \qquad \bar{y} = \lambda^{2/(1-n)} y, \tag{5}$$

except when $n = 1$. In the latter case, the DE (1) is linear. Here

$$\frac{d\bar{y}}{d\bar{x}} = \lambda^{(1+n)/(1-n)} \frac{dy}{dx}.$$

Therefore, the invariants are $u = x^{-2/(1-n)} y$ and $v = x^{(n+1)/(n-1)} y'$. As in Example 1, we find that

$$\frac{dv}{du} = \frac{(1-n)u^n + (3-n)v}{2u - (1-n)v}. \tag{6}$$

Transformations of Nonlinear ODE

(Again, equation (4) has been used to eliminate y'' to arrive at (6).) Equation (6) is the representation of (4) in the Lie plane.

3. The Thomas–Fermi equation

$$y'' = x^{-1/2} y^{3/2} \tag{7}$$

is invariant under the transformation

$$\bar{y} = \lambda^{-3} y, \qquad \bar{x} = \lambda x, \tag{8}$$

whose invariant u and first differential invariant v are

$$u = x^3 y, \qquad v = x^4 y'. \tag{9}$$

The representation of (7) in the Lie plane is

$$\frac{dv}{du} = \frac{u^{3/2} + 4v}{v + 3u}. \tag{10}$$

4. The one-dimensional Poisson–Boltzmann equation

$$y'' + \frac{\alpha}{x} y' = e^y, \tag{11}$$

where $\alpha = 0, 1, 2$ for plane, cylindrical, and spherical symmetry, respectively, is invariant under the one-parameter group of transformations

$$\bar{x} = \lambda x, \qquad \bar{y} = y - 2 \ln \lambda. \tag{12}$$

(To see this, substitute $\bar{x} = \lambda x$, $\bar{y} = y + f(\lambda)$ in (11); the invariance of this equation requires $f(\lambda) = -2 \ln \lambda$.) Now it is not difficult to check that the functions u and v are given by

$$u = x^2 e^y, \qquad v = xy', \tag{13}$$

and equation (11) transforms to

$$\frac{dv}{du} = \frac{u + 1(1-\alpha)v}{u(v+2)}. \tag{14}$$

Finally we quote two examples from Jones (1935); here a different set of transformations changes the original system to a special first-order equation in the Lie plane.

5. Considering again the Emden–Fowler equation

$$y'' + \frac{2}{x} y' + y^n = 0, \tag{15}$$

we introduce the transformation

$$X = \frac{xy'}{y}, \qquad Y = \frac{xy^n}{y'} \tag{16}$$

(which leaves (15) invariant; cf. (5)). We now have

$$\frac{dX}{dx} = \frac{y'}{y} - \frac{xy'^2}{y^2} + \frac{xy''}{y}$$

$$= \frac{1}{x}\left[X - X^2 + \frac{x^2}{y}\left(-\frac{2}{x}y' - y''\right)\right]$$

$$= \frac{1}{x}\{-X - X^2 - XY\},$$

$$\frac{dY}{dx} = \frac{y''}{y'} + nxy^{n-1} + \frac{xy''}{y'^2}\left(\frac{2}{x}y' + y''\right) \tag{17}$$

$$= \frac{1}{x}(3Y + nXY + Y^2),$$

which with $\ln|x| = t$, become

$$\begin{aligned}\frac{dX}{dt} &= -X(1 + X + Y), \\ \frac{dY}{dt} &= Y(3 + nX + Y).\end{aligned} \tag{18}$$

The system (18) is of the type studied by Poincaré (see Section 6.8) and may be replaced by the single equation

$$\frac{dY}{dX} = -\frac{Y}{X}\frac{nX + Y + 3}{X + Y + 1}. \tag{19}$$

6. Jones (1953) also discusses the equation

$$yy'' + ay'^2 + by = cx, \tag{20}$$

which describes the motion of a piston in an expansion chamber. It is easy to check that (20) cannot be converted to a first-order DE by scaling. However, it may be transformed by

$$X = \frac{xy'}{y}, \qquad Y = \frac{x^2}{yy'}, \qquad Z = \frac{x}{y'} \tag{21}$$

into a coupled system

$$\begin{aligned}\frac{dY}{dX} &= \frac{Y}{X}\frac{2 + (a-1)X - cY + bZ}{1 - (a+1)X + cY - bZ}, \\ \frac{dZ}{dX} &= \frac{Z}{X}\frac{1 + aX - cY + bZ}{1 - (a+1)X + cY - bZ}.\end{aligned} \tag{22}$$

Transformations of Nonlinear ODE

For the special case $b = 0$, we have an autonomous equation for (22a),

$$\frac{dY}{dX} = \frac{Y}{X}\frac{2 + (a-1)X - cY}{1 - (a+1)X + cY}, \tag{23}$$

which is of the Poincaré type. We discuss this system in Section 6.8. It is clear from this example that the second-order nonautonomous systems which transform into the Lie plane are rather special.

2.11 LANGMUIR'S EQUATION—WHEN TRANSFORMATIONS DO NOT WORK

We now discuss Langmuir's equation

$$3y\frac{d^2y}{dx^2} + \left(\frac{dy}{dx}\right)^2 + 4y\frac{dy}{dx} + y^2 = 1 \tag{1}$$

to illustrate that symmetries of the equations do not necessarily lead to exact solutions. But they may deliver a form which facilitates (local or global) approximate analysis (Inselberg, 1969). We note that (1) is autonomous and its left side is of degree 2 in y and its derivatives. Indeed, if the first two terms on the left side of (1) had the same coefficients, say a, then it would be possible to put it in the form

$$\bar{L}(y^2) = 1, \tag{2}$$

where

$$\bar{L} = \frac{a}{2}\frac{d^2}{dx^2} + 2\frac{d}{dx} + 1 \tag{3}$$

is a linear operator with constant coefficients, so that (2) could be easily solved for y^2. However, the coefficients of the first two terms in (1) are 3 and 1, respectively, so unfortunately, the above argument does not apply. We nevertheless attempt the following form of (1):

$$\Lambda(y) \equiv 3y\frac{d^2y}{dx^2} + \left(\frac{dy}{dx}\right)^2 + 4y\frac{dy}{dx} + y^2 = h(y)L[f(y)], \tag{4}$$

where $h(y)$ and $f(y)$ are functions to be determined such that (4) holds. The operator L in (4) is defined as

$$L = a_2(x)\frac{d^2}{dx^2} + a_1(x)\frac{d}{dx} + a_0(x).$$

If we expand the right side of (4), we have

$$h(y)L(f(y)) = h(y)\left\{a_2\left[\frac{d^2f}{dy^2}\left(\frac{dy}{dx}\right)^2 + \frac{df}{dy}\frac{d^2y}{dx^2}\right]\right. \tag{5}$$
$$\left. + a_1\frac{df}{dy}\frac{dy}{dx} + a_0 f\right\}.$$

Matching the coefficients of d^2y/dx^2, $(dy/dx)^2$, etc., in (4) results in the following relations:

$$a_2(x)h(y)\frac{d^2f}{dy^2} = 1, \tag{6}$$

$$a_2(x)h(y)\frac{df}{dy} = 3y, \tag{7}$$

$$a_1(x)h(y)\frac{df}{dy} = 4y, \tag{8}$$

$$a_0(x)h(y)f = y^2. \tag{9}$$

It is clear from equations (6)–(9) that $a_i(x)$ ($i = 0, 1, 2$) must be constant since f and h are functions of y alone. We write

$$a_i(x) = k_i, \quad k = 0, 1, 2. \tag{10}$$

Equations (7) and (8) imply that

$$k_2 = \frac{3}{4}k_1. \tag{11}$$

Dividing (6) by (7), we have

$$\frac{df'}{f'} = \frac{dy}{3y}, \quad f' = \frac{df}{dy}$$

which integrates to give

$$f'(y) = cy^{1/3}$$

and hence we obtain, by another integration,

$$f(y) = y^{4/3}, \tag{12}$$

if for convenience we set the constant of integration c equal to $\frac{4}{3}$. The second constant of integration is put equal to zero.

Now, equations (9) and (12) yield

$$h(y) = y^{2/3}/k_0. \tag{13}$$

Transformations of Nonlinear ODE

By substituting (12) and (13) into (8), we find that

$$k_1 = 3k_0. \tag{14}$$

Relations (11) and (14) express k_2 and k_1 in terms of k_0. Choosing the latter to be 1, we have $k_1 = 3$ and $k_2 = \frac{9}{4}$. Now Langmuir's equation (1) can be written in a neat form:

$$\Lambda(y) = y^{2/3} L(y^{4/3}) = 1 \tag{15}$$

with

$$L(y) = \frac{9}{4} \frac{d^2 y}{dx^2} + 3 \frac{dy}{dx} + y$$

$$= \left(\frac{3}{2} D + 1\right)^2 y, \qquad D = \frac{d}{dx},$$

or

$$\Lambda(y) = \frac{9}{4} y^{2/3} \left(D + \frac{2}{3}\right)^2 y^{4/3} = 1. \tag{16}$$

In spite of the symmetry between the exponents of y and the translation coefficient $\frac{2}{3}$, we are nowhere near the solution. Recalling the elementary translation identity

$$(D + r)^k u = e^{-rx} D^k (e^{rx} u),$$

applying it to (16), and writing $w = e^{x/3} y^{2/3}$, we have

$$y^{2/3} e^{-(2/3)x} \frac{d^2}{dx^2} (e^{x/3} y^{2/3})^2 = \frac{4}{9}$$

or

$$w \frac{d^2 w^2}{dx^2} = \frac{4}{9} e^x. \tag{17}$$

This equation is superior to (1) in that it has only one nonlinear term on the left side but is equally recalcitrant.

We conclude with the reduction of order of (16), which is autonomous, by one of the obvious substitutions

$$v = y^{4/3}, \qquad p = \frac{dv}{dx}, \qquad \frac{d^2 y^{4/3}}{dx^2} = p \frac{dp}{dv}.$$

We arrive at the equation

$$p \frac{dp}{dv} + \frac{4}{3} p = \frac{4}{9} (v^{-1/2} - v). \tag{18}$$

This is a second-order Abel equation (Murphy, 1960) and can be put into a more compact form by letting

$$p + \frac{4}{3}v = q.$$

Thus we have

$$\left(q - \frac{4}{3}v\right)\frac{dq}{dv} = \frac{4}{9}(v^{-1/2} - v). \tag{19}$$

To remove the fractional power of v, we write $z = v^{1/2}$ so that (19) assumes the form

$$\left(q - \frac{4}{3}z^2\right)\frac{dq}{dz} = \frac{8}{9}(1 - z^3). \tag{20}$$

This is again Abel's equation, not solvable in a closed form.

2.12 EXAMPLE OF AN EQUATION WHICH IS "SOLVED" EXACTLY, BUT THE SOLUTION IS TOO IMPLICIT TO BE OF PRACTICAL USE

We give an example due to Bender and Orszag (1978) which illustrates the futility of an exact solution when its form is too complicated. We consider

$$y^2 y''' = -\frac{1}{3}. \tag{1}$$

This equation is autonomous, so its order can be reduced by introducing the derivative

$$A = A(y) = \frac{dy}{dx} \tag{2}$$

as the new dependent variable and y as the independent variable. Equation (1) can be written as

$$y^2 A^2 \frac{d^2 A}{dy^2} + A\left(y\frac{dA}{dy}\right)^2 = -\frac{1}{3}. \tag{3}$$

This equation is equidimensional in y since it is invariant under the scale change $y \to ay$. We therefore write

$$y = e^t, \qquad A(y) = B(t) \tag{4}$$

Transformations of Nonlinear ODE

so that (3) becomes

$$B^2\left(\frac{d^2B}{dt^2} - \frac{dB}{dt}\right) + B\left(\frac{dB}{dt}\right)^2 = -\frac{1}{3}. \tag{5}$$

This equation is again autonomous, so its order can be reduced by 1. By writing

$$C = C(B) = \frac{dB}{dt}, \tag{6}$$

we change (5) to

$$B^2\left(C\frac{dC}{dB} - C\right) + BC^2 = -\frac{1}{3}. \tag{7}$$

This does not appear to be much simpler. However, if we substitute

$$C(B) = \frac{D(B)}{B} \tag{8}$$

into (7), we get

$$D\frac{dD}{dB} - DB = -\frac{1}{3} \tag{9}$$

If now we employ the substitution

$$D(B) = \frac{1}{2}B^2 + E(B),$$

so that B is the dependent variable and E the independent variable, we obtain the familiar Riccati equation

$$\frac{dB}{dE} + \frac{3}{2}B^2 + 3E = 0. \tag{10}$$

This equation can be linearized by the transformation (see Section 2.3)

$$B(E) = \frac{2}{3}\frac{1}{F(E)}\frac{dF(E)}{dE}; \tag{11}$$

we thus have

$$\frac{d^2F}{dE^2} + \frac{9}{2}EF = 0. \tag{12}$$

Equation (12) is an Airy equation whose general solution is a linear combination of

$$Ai[-(\tfrac{9}{2})^{1/2}E] \quad \text{and} \quad Bi[-(\tfrac{9}{2})^{1/3}E].$$

Equation (11) then gives

$$B(E) = -\left(\frac{4}{3}\right)^{1/3} \frac{c_1 A'i\left[-\left(\frac{9}{2}\right)^{1/3} E\right] + B'i\left[-\left(\frac{9}{2}\right)^{1/3} E\right]}{c_1 Ai\left[-\left(\frac{9}{2}\right)^{1/3} E\right] + Bi\left[-\left(\frac{9}{2}\right)^{1/3} E\right]}. \tag{13}$$

Recovering y from all the preceding transformations by tracing the steps backward is a hopeless task!

3
Series Solutions of Nonlinear Differential Equations

3.1 INTRODUCTION

We observed in Chapter 2 that it is not always possible to solve even linear DE with variable coefficients in a closed form or in terms of special or elementary functions. An elementary function must be carefully defined, since even trigonometric functions with which one gets familiar in an elementary course may be defined by power series. In any case, the situation for nonlinear DE is still more difficult in that very few of them may be explicitly solved in terms of elementary functions via transformations, etc. One turns, therefore, to "approximate" methods such as a series solution in the neighborhood of a given point. This is, in its very nature, a local analysis but it may be possible to patch together local behavior in the neighborhoods of several points to get a global picture. Indeed, the same series may have a large enough radius of convergence to allow a global treatment of a problem and may even solve a boundary value problem over an infinite interval. We shall deal with some examples wherein boundary value problems over an infinite interval may be solved by a series. In the process, we may be faced with a difficult computational problem. In any case, a series solution spells some progress in tackling the problem. In this chapter, we shall study a class of initial and boundary value problems, arising from applications, via series solution. The domain of validity of a series solution, that is, its radius of convergence, is delimited by the presence of singularities. The types of singularities that encumber a DE in some sense characterize it, so we shall devote some attention to the categorization of these singularities.

Again, nonlinear DE display a much more varied and richer singular structure than linear ones. One may do well to consider these equations in the complex plane to discover the presence and nature of singularities which may otherwise go undetected. This is true even for linear equations. We shall commence our discussion with a brief review of singularities of linear homogeneous DE and find series solutions about them, and demonstrate the need for analysis with a complex independent variable. We shall assume that the functions that occur in the equations are defined for both real and complex values of their arguments.

3.2 CLASSIFICATION OF SINGULAR POINTS OF HOMOGENEOUS LINEAR DIFFERENTIAL EQUATIONS

Before we classify linear DE with regard to their singularities, we recollect some properties and definitions of analytic functions. A function $w = f(z)$ of a complex variable $z = x + iy$ is analytic at a point if it has a single-valued derivative there. This is equivalent to the definition that any complex analytic function can be expanded in a convergent power series—a Taylor series—in the neighborhood of the given point. An analytic DE is one in which the coefficient functions are all analytic; it solution then is necessarily analytic. For example, a polynomial, the exponential, and sine and cosine functions are all analytic at any (finite) point z.

Ordinary Point

The point $x = x_0 \neq \infty$ is an *ordinary point* of the DE (the point $x_0 = \infty$ would require a separate treatment)

$$y^{(n)}(x) + p_{n-1}(x)y^{(n-1)}(x) + \cdots + p_0(x)y(x) = 0 \qquad (1)$$

if the coefficients $p_{n-1}(x)$, $p_{n-2}(x)$, $\cdots$, $p_0(x)$ are analytic at $x = x_0$. We have divided by the coefficient $p_n(x)$ of $y^{(n)}(x)$ in the most general form of the equation to write (1), assuming that the former is analytic and does not vanish at $x = x_0$.

EXAMPLE 1

i. For $y'' = e^x y$, every point $x \neq \infty$ is ordinary since the coefficient function e^x is analytic.

Series Solutions

ii. For $x^3 y^{iv} = y$, every point, except $x = 0$ and (possibly) $x = \infty$, is ordinary.
iii. For $y'' = |x|y$, there is no ordinary point of the equation since the function $|x|$ is nowhere analytic.

EXAMPLE 2 We solve the equation

$$(1+x^2)y'' + 2xy' - 2y = 0 \tag{2}$$

in the neighborhood of $x_0 = 0$.

We first observe that the coefficients $1+x^2$, $2x$, -2 are all analytic about $x_0 = 0$, and the coefficient of y'' is not zero at $x_0 = 0$. We look for a solution of the form

$$y(x) = c_0 + c_1 x + c_2 x^2 + c_3 x^3 + c_4 x^4 + 5c_5 x^5 + \cdots. \tag{3}$$

Then,

$$y'(x) = c_1 + 2c_2 x + 3c_3 x^2 + 4c_4 x^3 + 5c_5 x^4 \cdots,$$
$$y''(x) = 2c_2 + 6c_3 x + 12c_4 x^2 + 20c_5 x^3 + 30c_6 x^4 \cdots. \tag{4}$$

Substituting (3) and (4) into (2), we have

$$(1+x^2)(2c_2 + 6c_3 x + 12c_4 x^2 + 20c_5 x^3 + 30c_6 x^4 + \cdots)$$
$$+ 2x(c_1 + 2c_2 x + 3c_3 x^2 + 4c_4 x^3 + 5c_5 x^4 + 6c_6 x^5 + \cdots)$$
$$- 2(c_0 + c_1 x + c_2 x^2 + c_3 x^3 + c_4 x^4 + c_5 x^5 + c_6 x^6 + \cdots)$$
$$= 0.$$

This must hold for all x in some neighborhood of $x_0 = 0$. Equating coefficients of powers of x^0, x^1, x^2, etc., to zero, we get

$c_2 = c_0$,
$c_3 = 0$,
$c_4 = -\frac{1}{3}c_2 = -\frac{1}{3}c_0$,
$c_5 = -\frac{1}{2}c_3 = 0$,
$c_6 = -\frac{3}{5}c_4 = \frac{1}{5}c_0$, etc.

Therefore, the series solution (3) becomes

$$y = c_0 + c_1 x + c_0 x^2 - \frac{c_0}{3}x^4 + \frac{c_0}{5}x^6 - \frac{c_0 x^8}{7} + \cdots$$
$$= c_1 x + c_0(1 + x^2 - \frac{x^4}{3} + \frac{x^6}{5} - \frac{x^8}{7} + \cdots)$$

$$= c_1 x + c_0 + c_0 x \left[x - \frac{x^3}{3} + \frac{x^5}{5} - \frac{x^7}{7} + \cdots \right]. \tag{5}$$

The solution (5) it involves two arbitrary constants c_1 and c_0 and therefore represents the general solution of (2). The infinite series in brackets is the expansion for $\tan^{-1} x$. Such recognition, in general, is not possible. Of course, if a particular series solves an important equation which occurs often in applications, it itself may define a new function as the solution of the given equation. If we prescribe the initial conditions for a particular solution $y_1(x)$ of (2) as $y_1(0) = 0$, $y_1'(0) = 1$, then (5) implies that $c_0 = 0$ and $c_1 = 1$, so that the solution is $y = x$. On the other hand, if the solution $y_2(x)$ has the initial conditions $y_2(0) = 1$ and $y_2'(0) = 0$, then $c_0 = 1$ and $c_1 = 0$ and the solution is

$$y_2(x) = \left[1 + x^2 - \frac{x^4}{3} + \frac{x^6}{5} + \cdots + (-1)^{k+1} \frac{x^{2k}}{2k-1} + \cdots \right]$$

$$= 1 + x \tan^{-1} x.$$

Once a formal solution has been obtained about an ordinary point, there remains the task of finding the interval (or radius) of convergence of the series. This point was settled by Fuchs (see Birkhoff and Rota 1978), who proved that all the n independent solutions of (1) are analytic in the neighborhood of an ordinary point and that if any solution is expanded in a Taylor series about the ordinary point x_0 in the form $y(x) = \sum_{n=0}^{\infty} a_n (x - x_0)^n$, then the radius of convergence of this series is at least as large as the distance to the nearest singularity of the coefficient functions of the DE in the complex plane. Moreover, the location of a singularity of a solution must coincide with the location of a singularity of a coefficient function. Thus the singularities of a linear DE are fixed by the coefficients and cannot occur at any other point.

If equation (2) is written as

$$y'' + \frac{2x}{1+x^2} y' - \frac{2}{1+x^2} y = 0, \tag{6}$$

the coefficients of y' and y show that the distance of $x = 0$ to the singularities at $x = \pm i$ is 1; so the radius of convergence of this series is at least 1. The singularities become apparent only when we consider the independent variable x as complex. Of course, the solution in the present case has an infinite radius of convergence on the real line, as we know from the actual solutions (5).

Series Solutions

Regular Singular Points

Ordinary points are not the only points about which the solution of a linear DE can be obtained in a power series. This can be done even about some singular points, that is, those points about which not all coefficient functions of the differential equation can be expanded in Taylor series or at which the coefficient of the highest derivative vanishes. A specially important subclass consists of the regular singular points. The point x_0 is a *regular singular point* of the linear homogeneous nth-order DE if it can be written in the form

$$(x - x_0)^n y^{(n)} + (x - x_0)^{n-1} b_1(x) y^{(n-1)} + (x - x_0)^{n-2} b_2(x) y^{(n-2)}$$
$$+ \cdots + (x - x_0) b_{n-1}(x) y^{(1)} + b_n(x) y = 0, \qquad (7)$$

where $b_1(x), b_2(x), \ldots, b_n(x)$ are analytic at $x = x_0$. Indeed, if the coefficients $b_1, b_2, \ldots, b_n$ are constant, then (7) reduces to the Euler–Cauchy type and, hence, $x = x_0$ is a regular singular point, and the method of solution of (7) is suggested very directly by the solution of the Euler–Cauchy equation as we presently indicate.

EXAMPLE 3

i. $(x - 1) y''' = xy$ can be written in the form (7) as

$$(x - 1)^3 y''' - [(x - 1)^2 + (x - 1)^3] y = 0$$

Hence, it has a regular singular point at $x = 1$.

ii. $x^2 y' - (x + 1) y = 0$ cannot be written in the form (7), since in the equivalent form $xy' - (1 + 1/x) y = 0$, the coefficient of y is not analytic at $x = 0$.

Now Fuchs has proved that at a regular singular point $x = x_0$ a linear differential equation has at least one solution of the form

$$y_1(x) = (x - x_0)^\alpha \sum_{k=0}^{\infty} c_k (x - x_0)^k \equiv (x - x_0)^\alpha A(x), \qquad (8)$$

say, where the expansion is valid in some neighborhood of x_0. Specifically, the series expansion (8) is valid for $0 < |x - x_0| < R$ if the Taylor series for the coefficient functions $b_1(x), \ldots, b_n(x)$ in (7) are valid in $|x - x_0| < R$. The number α in (8) is called the *indicial exponent*.

If in (7), $n \geq 2$, then there is a second linearly independent solution with one of two possible forms:

$$y_2 = (x - x_0)^\beta B(x) \qquad (9a)$$

or

$$y_2 = (x - x_0)^\alpha A(x) \ln(x - x_0) + C(x)(x - x_0)^\beta$$
$$= y_1(x) \ln(x - x_0) + C(x)(x - x_0)^\beta, \qquad (9b)$$

where $A(x)$, $B(x)$, and $C(x)$ are functions analytic at $x = x_0$, having radii of convergence at least as large as the distance of x_0 from the nearest singularity of the coefficient functions. The function $A(x)$ in (9b) is the same as the infinite series in (8). The exponent β in (9) depends on the nature of roots of the indicial equation (see Birkhoff and Rota, 1978, p. 234, and Examples 7 and 12). For a second-order DE, for example, if the two roots are distinct and do not differ by an integer, then the solution is given by (8) and (9a). If they are repeated and are (α, α), say, then the two solutions are given by (8) and (9b), wherein $\beta = \alpha$. If the two roots α and β are such that $\alpha - \beta = n$, a positive integer, then the solutions are given by (8) and (9b) with $\beta = \alpha - n$. For each new linearly independent solution, there is a new analytic function of x and either a new indicial exponent or another power of $\ln(x - x_0)$. Indeed, the nth solution has at worst the form

$$y(x) = (x - x_0)^\gamma \sum_{i=0}^{n-1} [\ln(x - x_0)]^i A_i(x), \qquad (10)$$

where the functions $A_i(x)$, $i = 0, 1, \ldots, (n - 1)$ are analytic at x_0.

This should be compared with the Euler–Cauchy DE

$$x^n \frac{d^n y}{dx^n} + a_1 x^{n-1} \frac{d^{n-1} y}{dx^{n-1}} + a_2 x^{n-2} \frac{d^{n-2} y}{dx^{n-2}} + \cdots + a_n y = 0, \qquad (11)$$

which has n solutions of the form x^{k_i}, where k_i are distinct roots of the equation

$$k(k-1) \cdots (k-n+1) + a_1 k(k-1) \cdots (k-n+2) + \cdots + a_n = 0. \qquad (12)$$

This may be seen by direct substitution of $y = x^k$ in (11) or by transforming (11) into a DE with constant coefficients by writing $x = e^t$. If a root k of (12) is repeated, say α times, then (11) has α linearly independent solutions,

$$x^k, x^k \ln x, x^k (\ln x)^2, \ldots, x^k (\ln x)^{\alpha - 1}.$$

Thus, the forms (8)–(10) of the Frobenius series are straightforward generalizations of the types of solutions that the Euler–Cauchy form of equation (7), for which b_1, b_2, ..., b_n are constant, possesses. The generalization consists in the variable analytic functions $A(x)$, $B(x)$ $C(x)$, and $A_i(x)$, etc., to account for the variable coefficients in (7) (see Section 2.2). The nature of solutions about the regular singular points depends on the exponents α, β, and γ, etc., as well as the logarithmic terms that may appear.

Series Solutions 75

For example if α in (8) is a positive integer or zero, one of the solutions is analytic. If α is a fraction, then it has an algebraic branch point. If α is a negative integer, the singularity $x = x_0$ is a pole. The forms (9b) and (10) of the solution indicate that it has a logarithmic branch point at $x = x_0$.

If the singular point is not regular, it is called *irregular*. It can then manifest a structure quite different from that for a regular singular point.

EXAMPLE 4 $xy'' + 2y' + xy = 0$.

This DE can be written as

$$x^2 y'' + 2xy' + x^2 y = 0,$$

which accords with (7), so $x = 0$ is a regular singular point. The substitution $z = xy$ changes this equation into

$$z'' + z = 0.$$

The solutions are $z_1 = \sin x$, $z_2 = \cos x$. The two solutions of the original DE are $y_1 = (\sin x)/x$, $y_2 = (\cos x)/x$. While y_1 is analytic at $x = 0$, y_2 has a simple pole there.

EXAMPLE 5 $xy'' - y' + 4x^3 y = 0$.

This equation, written as

$$x^2 y'' - xy' + 4x^4 y = 0,$$

shows that $x = 0$ is a regular singular point. The indicial equation is $m(m-1) - m = 0$ and has roots $m = 0, 2$. The equation remains invariant when $x \to -x$. This suggests the change of variable $x^2 = X$, so the equation assumes the form

$$\frac{d^2 y}{dX^2} + y = 0.$$

Therefore, the two linearly independent solutions are $y_1 = \sin x^2$ and $y_2 = \cos x^2$, which are analytic at $x = 0$. The series for y_1 starts with x^2 while that for y_2 starts with unity.

EXAMPLE 6 $xy'' + y' - xy = 0$.

This equation, in the form

$$x^2 y'' + xy' - x^2 y = 0, \tag{13}$$

shows that $x = 0$ is a regular singular point. The indicial equation is $m(m - 1) + m = 0$, with roots $m = 0, 0$. If we substitute

$$y = \sum_{n=0}^{\infty} a_n x^n$$

into (13) and equate coefficients of different powers of x to zero, we find that

$$y_1 = 1 + \frac{x^2}{2^2} + \frac{x^4}{(2^2)(4^2)} + \cdots.$$

The second solution is found by writing

$$y_2 = y_1 \ln x + \sum_{n=0}^{\infty} a_n x^n.$$

Substituting this into (13) and equating the coefficients of various powers of x to zero, we find that

$$y_2 = y_1 \ln x - \frac{x^2}{4} - \frac{3}{128} x^4 - \frac{11}{13824} x^6 + \cdots.$$

While y_1 is analytic at $x = 0$, y_2 has a logarithmic branch point there.

EXAMPLE 7 $x^2 y'' + x(1 - x) y' - (1 + 3x) y = 0.$

Here, $x = 0$ is a regular singular point. The indicial equation is $m(m - 1) + m - 1 = 0$. Its roots are $m = 1, -1$. The solution corresponding to $m = 1$ is easily found to be

$$y_1 = -3x - 4x^2 - \cdots,$$

while the second solution, sought in the form

$$y_2 = y_1 \ln x + \sum_{i=-1}^{\infty} a_i x^i,$$

is

$$y_2 = y_1 \ln x + \frac{1}{x} - 2 - x + 3x^2 \cdots.$$

The solution y_1 is analytic at $x = 0$ while y_2 has a logarithmic branch point as well as a simple pole at $x = 0$.

EXAMPLE 8 $x^2 y' - y = 0.$

This equation can be written as

$$xy' - \frac{1}{x} y = 0,$$

Series Solutions

and has an irregular singular point at $x = 0$. If we attempt a Frobenius series

$$y = \sum_{n=0}^{\infty} a_n x^{\rho+n}, \quad a_0 \neq 0,$$

we find that

$$\sum_{n=0}^{\infty} a_n(\rho + n) x^{\rho+n+1} - \sum_{n=0}^{\infty} a_n x^{\rho+n} = 0.$$

The lowest-degree term x^ρ has a_0 as its coefficient. Therefore, $a_0 = 0$. This contradicts the basic assumption that $a_0 \neq 0$; therefore, no Frobenius series solution exists. It is a simple matter to solve the equation directly so that $y = ce^{-1/x}$, where c is a constant, showing that $x = 0$ is an essential singularity; that is, it has an infinite number of terms with negative powers in the Laurent series.

EXAMPLE 9 $x^4 y'' + 2x^2 y' - (2x - 1)y = 0.$

This equation, written as

$$x^2 y'' + \frac{2}{x^2} y' - \left(\frac{2}{x^3} - \frac{1}{x^4}\right) y = 0,$$

shows that $x = 0$ is an irregular singular point. If we substitute $x = 1/X$, we get

$$\frac{d^2 y}{dX^2} + 2\left(\frac{1}{X} - 1\right) \frac{dy}{dX} + \left(1 - \frac{2}{X}\right) y = 0.$$

Since the sum of the coefficients is zero, $y = e^X = e^{1/x}$ is a solution. The second solution is $y = xe^{1/x}$. Both solutions have an essential singularity at the origin.

EXAMPLE 10 $x^3 y'' + xy' - y = 0.$

Here, again, $x = 0$ is an irregular singular point. The Frobenius series $y = \sum_{n=0}^{\infty} a_n x^{n+\rho}$ leads to

$$\sum_{n=0}^{\infty} (\rho + n)(\rho + n - 1) a_n x^{\rho+n+1} + \sum_{n=0}^{\infty} (\rho + n - 1) a_n x^{\rho+n} = 0.$$

The lowest-degree term x^ρ gives $(\rho - 1)a_0 = 0$; that is, since $a_0 \neq 0$, $\rho = 1$. The recurrence relation is

$$a_n = -(n - 1) a_{n-1},$$

implying that

$$a_1 = a_2 = a_3 = \cdots = 0.$$

Therefore, one solution, with $a_0 = 1$, may be chosen as $y_1 = x$. The second solution is

$$y_2 = x \int \frac{1}{x^2} e^{-\int^x dt/t^2} \, dx = -xe^{1/x}.$$

While y_1 is analytic at $x = 0$, y_2 has an essentially singularity there.

EXAMPLE 11 $x^2 y'' + (3x - 1)y' + y = 0.$

This equation, written in the form

$$x^2 y'' + \left(3 - \frac{1}{x}\right) xy' + y = 0,$$

shows that $x = 0$ is an irregular singular point. The Frobenius series form leads to

$$\sum_{n=0}^{\infty} a_n [(\rho + n)(\rho + n + 2) + 1] x^{\rho+n} - \sum_{n=0}^{\infty} a_n (\rho + n) x^{\rho+n-1} = 0.$$

The coefficient of $x^{\rho-1}$ on the left is $a_0 \rho$. Therefore, since $a_0 \neq 0$, $\rho = 0$. The recurrence relation in this case

$$a_{n+1} = \frac{(\rho + n)(\rho + n + 2) + 1}{\rho + n + 1} a_n$$

becomes $(n + 1)a_n$ for $\rho = 0$. Therefore, taking $a_0 = 1$, we get

$$y = \sum_{n=0}^{\infty} n! x^n.$$

This series has a zero radius of convergence:

$$R = \lim_{n \to \infty} \left| \frac{a_n}{a_{n+1}} \right| = \lim_{n \to \infty} \frac{n!}{(n+1)!} = 0.$$

EXAMPLE 12 (Jones, 1977).

$$\frac{d^2 y}{dx^2} - \frac{2}{x} \frac{dy}{dx} - \left(1 - \frac{x^2}{\tau^2}\right) y = 0. \tag{14}$$

For a fixed τ, this equation has a regular singular point at $x = 0$. The indicial equation is $m(m - 1) - 2m = 0$, and the roots are $m = 0, 3$. It is not difficult

Series Solutions

to find that the two linearly independent solutions of (14) are

$$y_1 = x^3 + \frac{1}{10}x^5 + \frac{1}{28}\left(\frac{1}{10} - \frac{1}{\tau^2}\right)x^7 + \cdots,$$

$$y_2 = 1 - \frac{1}{2}x^2 - \frac{1}{4}\left(\frac{1}{2} + \frac{1}{\tau^2}\right)x^4 + \cdots.$$

These expansions converge for all finite complex τ, apart from $\tau = 0$, which is an essential singularity. Actually, the solutions of (14) can be written in terms of confluent hypergeometric functions (see Murphy, 1960).

EXAMPLE 13 $(x-1)(2x-1)y'' + 2xy' - 2y = 0.$ (15)

This DE has regular singular points at $x = 1/2$, 1, and ∞ (for the latter see Section 3.3). One solution of (15) is easily found to be $y = 1/(x-1)$. A Taylor series form of this solution about the ordinary point $x = 0$ converges beyond the nearest singular point $x = 1/2$, but ceases to converge at $x = 1$. A linearly independent solution is $y = x$, whose Taylor series about $x = 0$ converges for all finite x.

3.3 THE POINT AT INFINITY

When the range of the independent variable is infinite, it is desirable to know the behavior of the solution near the point at infinity. This can be accomplished in a simple manner by changing the independent variable. We write

$$t = 1/x \tag{1}$$

in the given DE, say,

$$y'' + p(x)y' + q(x)y = 0. \tag{2}$$

Then large values of x correspond to small values of t. The change of variable (1) implies that

$$y' = \frac{dy}{dt}\frac{dt}{dx} = -t^2\frac{dy}{dt},$$

$$y'' = \frac{d}{dt}\left[-t^2\frac{dy}{dt}\right]\frac{dt}{dx} = \left(-t^2\frac{d^2y}{dt^2} - 2t\frac{dy}{dt}\right)(-t^2),$$

so (2) becomes

$$y'' + \left(\frac{2}{t} - \frac{p(1/t)}{t^2}\right)y' + \frac{q(1/t)}{t^4}y = 0. \tag{3}$$

The prime in (3) denotes differentiation with respect to t. The point $x = \infty$ is an ordinary point, a regular singular point, or an irregular singular point for (2), if the point $t = 0$ has the corresponding character for the transformed equation (3).

EXAMPLE 1 $y'' - y = 0$.

Here $p = 0$, $q = -1$, so this equation transforms to

$$\frac{d^2y}{dt^2} + \frac{2}{t}\frac{dy}{dt} - \frac{1}{t^4}y = 0.$$

Evidently, $t = 0$ is an irregular singular point. This follows more easily from the fact that the solutions of the original DE are e^x and e^{-x}, which have essential singularities at $x = \infty$ and $x = -\infty$, respectively.

EXAMPLE 2 The Euler–Cauchy equation

$$y'' + \frac{4}{x}y' + \frac{2}{x^2}y = 0$$

transforms via (1) to

$$y'' - \frac{2}{t}y' + \frac{2}{t^2}y = 0, \qquad ' = \frac{d}{dt}.$$

Here $t = 0$ is obviously a regular singular point with exponents given by $m(m-1) - 2m + 2 = 0$; thus, $m = 1, 2$. Hence the point $x = \infty$ is a regular singular point with exponents 1 and 2.

EXAMPLE 3 $y'' + \dfrac{3y'}{2x} - \dfrac{y}{4x^3} = 0$. $\qquad(4)$

This equation has an irregular singular point at $x = 0$. The two solutions in this neighborhood are $\sin(x^{-1/2})$ and $\cos(x^{-1/2})$, which have an essential singularity at $x = 0$. The first solution also has branch points at $x = 0$ and $x = \infty$. The second has no branch point and is analytic at $x = \infty$. The point at $x = \infty$ can be considered by changing (4) to

$$t^2\frac{d^2y}{dt^2} + \frac{1}{2}t\frac{dy}{dt} - \frac{1}{4}ty = 0,$$

where $t = 1/x$. The point $t = 0$ is a regular singular point of this equation and so is $x = \infty$ for (4).

Series Solutions

3.4 SINGULARITIES OF NONLINEAR DIFFERENTIAL EQUATIONS

The singularities of the linear DE that we studied in Sections 3.2 and 3.3 are fixed; that is, they are completely determined by the coefficients, and their nature can be identified without solving the equations; they do not depend on initial or boundary conditions. By contrast, singularities of nonlinear DE almost always depend on the initial or boundary conditions, and, therefore, they move as these conditions are changed, as we shall see. Besides, as we observed, the solutions of nonlinear DE are best studied in the complex plane because the isolated singular points there are surrounded by connected domains in the complex plane. This enables one to continue solutions beyond and around isolated singular points; they seem to terminate abruptly at singular points on the real line.

EXAMPLE 1 Consider $du/dx = u^2$ for real u and x. The formula $u = -1/x$ defines two real solutions of this equation; one, for $x > 0$, tends to $-\infty$ as $x \to +0$, while the other, for $x < 0$, tends to $+\infty$ as $x \to -0$. The two seem to be unrelated.

On the other hand, the DE $dw/dz = w^2$ in the complex plane has a single solution $w = -1/z$, where $w = u + iv$ and $z = x + iy$. This function defines one solution everywhere, including the points on the positive and negative real axes, except at the isolated singularity at $z = 0$. The general solution of this first-order equation is $w = 1/(c - z)$, which is defined in the entire z-plane except at $z = c$. Since the domain is connected, the solution can be continued analytically from one region to another. This process of analytic continuation can be uniquely carried out for any given path of continuation. The solution $w = -1/z$ (corresponding to $u = -1/x$) is equivalent to

$$w = -\frac{1}{r}(\cos\theta - i\sin\theta), \qquad u = -\frac{x}{x^2 + y^2}, \qquad v = \frac{y}{x^2 + y^2},$$

so the solution on the real positive axis, $\theta = 0$, is $w = -1/x$ and continuously changes to become $w = 1/x$ on $\theta = \pi$. The singularity of the general solution, a simple pole at $z = c$, depends on the constant of integration c and moves as c does.

Example 1 illustrates the continuation principle, namely, the function obtained by analytic continuation of any solution of an analytic DE, along any path in the complex plane, is a solution of analytic continuation of the DE along the same path.

EXAMPLE 2 The first-order Euler–Cauchy DE

$$z\frac{dw}{dz} - \gamma w = 0, \qquad \gamma = \alpha + i\beta, \quad \text{with } \alpha \text{ and } \beta \text{ real},$$

has solution

$$w = z^\gamma = e^{\gamma \ln z} = e^{(\alpha + i\beta)(\ln r + i\theta)}$$
$$= e^{(\alpha \ln r - \beta\theta)}[\cos(\beta \ln r + \alpha\theta) + i\sin(\beta \ln r + \alpha\theta)],$$

if the constant of integration is chosen to be unity This coincides with the real solution $u = x^\alpha$ on the positive x-axis, $\theta = 0$, when $\beta = 0$, and $\gamma = \alpha$ is real. The analytic continuation of this real solution (on the positive x-axis) through the upper half-plane to the negative real axis is obtained by putting $\theta = \pi$, so $u = (\cos \pi\alpha + i\sin \pi\alpha)|x|^\alpha$, and this is not equal to the real solution $|x|^\alpha$ on $x < 0$ except when α is an even integer.

The solution $w = z^\gamma$ displays several types of singularities at $z = 0$, whose nature depends on γ. When γ is a positive integer, it is analytic; when γ is a negative integer, it has a pole at $z = 0$. When $\gamma = m/n$, the solution $w = z^{m/n}$, where $n = 2, 3, 4, \ldots$ and m is an integer without common factors with n, has branch points $z = 0$ and $z = \infty$, each of order n: consequently, the greater the denominator n, the greater the number of different values of the function (i.e., the higher the order of the branch point).

The function $w = z^k = e^{k \ln z}$, where k is an irrational number, is infinite-valued, and when the point z passes around $z = 0$ or $z = \infty$, the values of the function go into one another but never come back to original value. Such points are called *logarithmic branch points* or *branch points of infinite order*. Actually, we can write $z^\gamma = e^{\gamma \ln z}$ and then $z = re^{i\theta}$. We note that $|z^i| = e^{-\theta}$ and $\arg(z^i) = \ln r$. By writing $\gamma = \gamma' + i\gamma''$, we find that as z traverses the circle $|z| = r$ once in the counterclockwise direction, z^γ changes by the factor $e^{2\pi i(\gamma' + i\gamma'')} = e^{-2\pi\gamma''}(\cos 2\pi\gamma' + i\sin 2\pi\gamma')$. We conclude that z is single-valued only if γ is an integer.

Example 2 shows that DE possessing single-valued functions as coefficients can have multivalued solutions in the complex plane. Thus, even a linear equation with single-valued coefficient functions can exhibit a variety of solutions in the complex plane.

For analyzing and classifying the singularities, we first consider the first-order DE

$$\frac{dw}{dz} = f(w, z), \tag{1}$$

where, for the present, we require $f(w, z)$ to be single-valued so that while continuing analytically the solution $w = w(z)$ of this equation we continue to

Series Solutions

deal with the same DE. Let $f(w,z)$ be analytic in a domain $A(w,z)$. Then, in the neighborhood of a point $(w_0, z_0) \in A(w,z)$, we have, by Cauchy's theorem, a unique holomorphic solution in the form of a convergent Taylor series

$$w = w_0 + \sum_{k=1}^{\infty} a_k (z - z_0)^k.$$

What happens to this solution as it is analytically continued?

1. Does the set of solutions remain single-valued?
2. If the solutions are single-valued, is there a solution $w = w(z)$ having no singular points whatever, say, when $f(w,z)$ is an entire function with no singular points in the finite plane, a polynomial, for example?
3. Is there a solution $w = w(z)$ having only poles as singularities?
4. Do any of the solutions have essential singularities?
5. Can there occur solutions whose singularities are not isolated?

Once we classify the singular points, we may proceed to find solutions in the neighborhood of each of these points. As we have seen in the case of linear equations, the method of solution of an equation may be suggested by the type of singularity. We have distinguished two main types: fixed and movable.

In response to queries (1)–(5), we classify equations, following Hille (1969), as we continue $w(z, z_0; w_0)$ along a path Γ in the z-plane, proceeding from $z = z_0$ to $z = z_1$. If at a point $\zeta \in \Gamma$, we set $w = w(\zeta; z_0, w_0)$ and suppose that $f(z, w)$ is holomorphic at each point (ζ, w) for $\zeta \in \Gamma$, $\zeta \neq z_1$, then the continuation of $w(z; z_0; w_0)$ along Γ is a solution of (1). As we approach z_1, various possibilities arise. We just state them here (see Hille, 1969, for proofs).

CASE 1 If $w(z; z_0, w_0) \to w_1$ as $z \to z_1$ along Γ and $f(z, w)$ is holomorphic at (z_1, w_1), then the continuation of $w(z; z_0, w_0)$ along Γ is holomorphic at $z = z_1$, and in some neighborhood of this point it coincides with $w(z; z_1, w_1)$. This is known as the theorem of Painlevé.

CASE 2 If $w(z; z_0, w_0) \to w_1$ as $z \to z_1$ along Γ, and $f(z, w)$ is not holomorphic at (z_1, w_1) but $[f(z, w)]^{-1}$ is holomorphic and $[f(z_1, w)]^{-1} \neq 0$, then the point $z = z_1$ is an algebraic branch point of $w(z; z_0, w_0)$, where a finite number of branches are permuted when z makes a circuit about $z = z_1$. Besides, such a singularity is movable; that is, the position of the branch point varies with the solution. For example, in the DE

$$w' = \frac{w - 2z}{w - z},$$

the right-hand side is not holomorphic at $w = z$, but its reciprocal is. The solution is easily found to be

$$w = z + (c^2 - z^2)^{1/2}.$$

This function has simple (movable) branch points at $z = \pm c$. If $z \to c$, $w \to c$, $w - z \to 0$, but $w - c \neq 0$, so the point (c,c) is covered by case 2.

CASE 3 If $w(z;z_0,w_0) \to w_1$ as $z \to z_1$ along Γ, $f(z,w)$ is not holomorphic at (z_1,w_1), and $[f(z_1,w)]^{-1} \equiv 0$, then no general statement can be made, since the only holomorphic solution of (1), written as

$$\frac{dz}{dw} = [f(z,w)]^{-1},$$

is $z(w) \equiv z_1$. This solution gives no information about the solution $w(z;z_0,w_0)$, starting from (z_0,w_0).

We again refer to the example

$$\frac{dw}{dz} = \lambda \frac{w}{z}$$

where λ is a constant. Here $\lambda w/z = f(z,w)$ is not holomorphic at $(0,0)$, and the inverse function $z/\lambda w \equiv 0$ for $z = 0$ and any $w \neq 0$. The solution of this equation is $w = Cz^\lambda$. As we remarked earlier in this section, this solution has no singularity, a pole, an algebraic branch point, or a transcendental critical point, depending on whether λ is a positive integer, a negative integer, a nonintegral rational number, or an irrational or complex number, respectively.

CASE 4 If $w(z;z_0,w_0) \to w_1$ as $z \to z_1$ along Γ and neither $f(z,w)$ nor $[f(z,w)]^{-1}$ is holomorphic at (z_1,w_1), then $z = z_1$ may or may not be a singularity of the solution.

The Riccati equation

$$\frac{dw}{dz} = 1 + z^{3/2}w - z^{1/2}w^2$$

has a fixed singularity at $z = 0$. Neither f nor f^{-1} is holomorphic about it (the functions $z^{1/2}$, $z^{3/2}$ making them nonholomorphic). The solution $w = z$, however, is holomorphic about $z = 0$.

In the Riccati equation

$$\frac{dw}{dz} = \frac{3}{2}z^{1/2} + z^{3/2}w - w^2,$$

f and f^{-1} are again nonholomorphic at $z = 0$. In this case, one solution, $w = z^{3/2}$, is singular, having a branch point at $z = 0$.

Series Solutions

CASE 5 If $w(z;z_0,w_0) \to \infty$ as $z \to z_1$, then $z = z_1$ is a singular point of the solution and its nature can be studied by putting $w = v^{-1}$. We then have

$$v' = -v^2 f(z, v^{-1}). \tag{1a}$$

Now, $v \to 0$ as $z \to z_1$, and the nature of the solution depends on the RHS of (1a). If the RHS is holomorphic at $(z_1, 0)$, we infer that v is holomorphic at $z = z_1$ and has a zero there. In turn, w has a pole at $z = z_1$. Similarly, if (1a) conforms to case 2 at $(z_1, 0)$, then w has an algebraic branch point there.

CASE 6 As $z \to z_1$, w does not tend to any limit. Such a situation can arise at a singular point of $f(w,z)$. As an example, we consider the DE

$$w' = w/z^2.$$

Its solution is

$$w = C \exp(-1/z),$$

where C is the constant of integration. This solution along the imaginary axis $z = iy$ has the form $w = C[\cos(1/y) + i \sin(1/y)]$ and does not tend to any limit as y tends to zero. This case may also arise for particular integrals at an arbitrarily preassigned point. This is illustrated by the equation

$$w' = -w(\log w)^2,$$

whose solution $w = \exp(z - a)^{-1}$ again does not tend to any limit as z approaches a along a line parallel to the imaginary axis.

Indeed, we have a very useful theorem regarding case 6, due to Painlevé, which simplifies the task for first-order DE of the type

$$\frac{dw}{dz} = \frac{P(z,w)}{Q(z,w)} \tag{2}$$

(Hille, 1969). This theorem states that if $f(z,w)$ in (1) is of type (2)—that is, it is rational in w with coefficients polynomials in z—then the movable singularities of the solutions are poles and/or algebraic points. This means that the solution in the neighborhood of these singular points can be written in power series of $(z - z_0)^{1/n}$, where $n \geq 1$ is a positive integer; the number of terms with negative powers in this series is finite. To be sure, finding the number of movable points and their position is a difficult task. On the other hand, if we consider a system of two nonlinear DE or a second-order nonlinear DE, we have no theorem corresponding to Painlevé's for first-order DE, and so the question of singular points for such equations is considerably more complicated.

The solution $w(z; z_0, w_0)$ of (2) always tends to a definite limit, finite or infinite. Moreover, the points of essential singularity, if any, have to be fixed. As an example, we consider

$$\frac{dw}{dz} = -\frac{w}{z^2}.$$

This equation is of type (2) and has a singularity at $w = 0$ and $z = 0$, where the numerator and the denominator vanish. The general solution of this DE is

$$w = ce^{1/z}, \quad c \text{ a constant}$$

which shows that $z = 0$ is a fixed essential singularity. The movable poles and branch points in solutions of DE of type (2) are met with quite frequently, even when it assumes a simple form with $Q(z, w) = 1$. For example, for

$$w' = 1 + w^2, \tag{3}$$

the solution $w(z) = \tan(z - a)$, with a as an arbitrary constant, has an (countably) infinite number of poles at $z = a + (k + 1/2)\pi$, where $k = 0, \pm 1, \pm 2, \ldots$. As an example of movable branch points, we may consider the DE

$$w' = -\frac{1}{n} w^{n+1}, \quad n > 1,$$

with solution $w(z) = (z - a)^{-1/n}$.

If we wish to restrict ourselves to DE (2) whose solutions have poles as the only movable singularities (not even movable branch points being admitted), we are led to the condition that (2) must be a Riccati equation. In view of the importance of the latter equation and the implication of this theorem for nonlinear second-order DE of Painlevé type, we give its proof.

If there is a point (w_0, z_0) such that $Q(z_0, w_0) = 0$ but $P(z_0, w_0) \neq 0$, then this is a singular point of (2) but a regular point of

$$\frac{dz}{dw} = \frac{Q(z, w)}{P(z, w)}. \tag{4}$$

If $Q(z, w)$ is not independent of w, then for a given w_0, $Q(z, w_0)$ may have q roots. Let z_0 be one of these roots. In the neighborhood of the point (z_0, w_0), we have the solution of (4) in the form

$$z - z_0 = a_1(w - w_0) + a_2(w - w_0)^2 + \cdots. \tag{5}$$

Since $Q(z_0, w_0) = 0$, $dz/dw = 0$ at (z_0, w_0), therefore, $a_1 = 0$, and the expansion (5) can be written as

$$z - z_0 = \sum_{r=2}^{\infty} a_r (w - w_0)^r, \tag{6}$$

Series Solutions

where some of the coefficients a_r, $r = n \geq 2$, will be different from zero. We can invert the series (6) to obtain the solution of (2), namely, $w - w_0$, in power series in $(z - z_0)^{1/n}$. This shows that z_0 is an n-fold branch point. Moreover, the point (w_0, z_0) is such that $Q(w_0, z_0) = 0$ so that it moves along a certain curve.

To get a solution for which there are no movable branch points, we therefore require that $Q(w, z)$ be independent of w. Thus, (2) can be written as

$$\frac{dw}{dz} = P_0(z) + P_1(z)w + P_2(z)w^2 + \cdots + P_m(z)w^m. \tag{7}$$

If $m > 2$, we set $w = 1/v$ to obtain

$$v'(z) = -[P_0(z)v^m + P_1(z)v^{m-1} + \cdots + P_m(z)]v^{2-m} \equiv \frac{P_1(z,v)}{Q_1(z,v)} \tag{8}$$

Here $Q_1 = v^{m-2}$, which, for $m > 2$, vanishes for $v = 0$ for any values of z. Taking a nonsingular value z_0 of z, where $P_m(z_0) \neq 0$, we have $P_1(z_0, v) \neq 0$ and $Q_1(z_0, v) = 0$. Equation (8) has a solution which vanishes at $z = z_0$ and has a branch point there. Since $w(z) = v(z)^{-1}$, the solution $w(z)$ has a branch point at a nonsingular point z_0. Hence we conclude that $m \leq 2$ and (7) is a Riccati equation.

We have already considered an example (see equation (3), which is a special case of the Riccati equation). Its solution manifests an infinite number of movable poles.

3.5 FIRST-ORDER BINOMIAL EQUATIONS

First-order binomial equations are described by

$$[w'(z)]^m = f(z, w), \tag{1}$$

where m is a positive integer greater than 1 and $f(z, w)$ is analytic in z and w. The case $m = 2$ is of special importance. We content ourselves here with a discussion of this case only. We thus consider

$$[w'(z, w)]^2 = R(z, w) = \frac{P(z, w)}{Q(z, w)}, \tag{2}$$

where

$$P(z, w) = P_0(z) \prod_{\mu=1}^{p} [w - \eta_\mu(z)],$$

$$Q(z,w) = Q_0(z) \prod_{\nu=1}^{q} [w - \zeta_\nu(z)]. \tag{3}$$

Here $P_0(z)$ and $Q_0(z)$ are polynomials in z, $\eta_\mu(z)$ are normally the p branches of an algebraic function $\eta(z)$ while $\zeta_\nu(z)$ are the q branches of an algebraic function $\zeta(z)$.

We quote a theorem (Hille, 1969) regarding the absence of movable branch points for (2): A necessary condition for the absence of movable branch points for (2) is that $R(z,w)$ be a polynomial in w of degree ≤ 4. It is further proved that the necessary and sufficient condition for $(w')^2 = P(z,w)$ to be without movable branch points is that $P(z,w) = P_0(z)W(w)$, where $W(w)$ has one of the following forms ($P_0(z)$ is a polynomial):

i. $w - a$,
ii. $(w-a)(w-b)$,
iii. $(w-a)^3$,
iv. $(w-a)^2(w-b)$,
v. $(w-a)(w-b)(w-c)$,
vi. $(w-a)^2(w-b)(w-c)$,
vii. $(w-a)(w-b)(w-c)(w-d)$,
viii. $[w - \eta(z)]^2(w-b)$,
ix. $[w - \eta(z)]^2(w-b)(w-c)$.

Here $\eta(z)$ is a polynomial, and a, b, c, and d denote distinct constants. Now, we give some examples of binomial equations.

EXAMPLE 1 $(w')^2 = w^3$. The solution is $w = 4(z-c)^{-2}$ with a pole of order 2.

EXAMPLE 2 $(w')^2 = (w-a)(w-b)^2$. Substituting $(w-b)^{-1/2} = W$, we have

$$\left(\frac{dW}{dz}\right)^2 = \frac{1}{4}[1 + (b-a)W^2].$$

This equation is easily integrated. There are two branches of the solution. The form depends on the sign of $b - a$. If $b - a > 0$, the solution is

$$\pm \frac{(b-a)^{1/2}}{2} z = \ln\left[W + (W^2 + \frac{1}{b-a})^{1/2}\right] + C,$$

Series Solutions

where $W = (w - b)^{-1/2}$, and C is a constant.

EXAMPLE 3 $(w')^2 = 4(w - e_1)(w - e_2)(w - e_3)$, $e_1 + e_2 + e_3 = 0$. The solution of this equation is expressible in terms of elliptic functions.

Since first-order nonlinear DE with solutions expressible in terms of elliptic functions have been discussed in great detail by Davis (1962) and Ames (1968), we omit their discussion here (see also Hille, 1976).

3.6 SINGULARITIES OF THE BRIOT–BOUQUET EQUATION

Now we consider in some detail solutions of case 6 of Section 3.4 for which the first-order equation has the form

$$\frac{dw}{dz} = \frac{P(z,w)}{Q(z,w)}, \tag{1}$$

where $P(z,w)$, and $Q(z,w)$ are holomorphic functions of the complex variables (z,w) in the neighborhood of the points $z = z_0$, $w = w_0$ at which $P(z_0, w_0) = 0$, $Q(z_0, w_0) = 0$, the singular points of (1). We shall, in particular, consider the analytic behavior of the special form of (1) which occurs most frequently in applications and to which many equations reduce after (several) transformations. This particular system was studied by Briot and Bouquet (1875). We consider special cases of the Briot and Bouquet equation which can be explicitly integrated in order to show the rich variety of solutions this equation gives rise to. We shall also indicate how several complicated systems reduce either to

$$z\frac{dw}{dz} = pz + qw + F(z,w), \tag{2}$$

where

$$F(z,w) = \sum\sum_{j+k \geq 2} c_{jk} z^j w^k \tag{3}$$

is holomorphic in the domain $|z| < R_1$, $|w| < R_2$, or to

$$z^{m+1}\frac{dw}{dz} = pz + qw + F(z,w), \qquad m \geq 1, \tag{4}$$

where $F(z,w)$ is as in (3).

EXAMPLE 1 The special case of (2) with $F \equiv 0$, namely

$$z \frac{dw}{dz} = pz + qw, \tag{5}$$

being linear has, for $q \neq 1$, the explicit solution

$$w = \frac{pz}{1-q} + cz^q, \tag{6}$$

where c is an arbitrary constant. For $c = 0$, (6) represents a holomorphic solution. If q is not a positive integer, then the only holomorphic solution which vanishes at $z = 0$ corresponds to $c = 0$. The same is true if $p = 0$. If q is a positive integer not equal to 1, all solutions are holomorphic and vanish at $z = 0$. If $q = 1$, the solution of (5) is

$$w = pz \log z + cz,$$

where c is a constant. In this case, if $p \neq 0$, there are no holomorphic solutions which vanish at $z = 0$. In the case $p = 0$, all solutions are holomorphic and vanish at $z = 0$.

EXAMPLE 2 $z^2 \frac{dw}{dz} = \lambda z - w$. This equation has the formal solution

$$w = \lambda \sum_{n=1}^{\infty} (-1)^{n-1} (n-1)! z^n,$$

which converges only for $z = 0$. This series may have use as an asymptotic solution.

EXAMPLE 3 A slightly more general form than that in Example 2 is

$$z^2 \frac{dw}{dz} = qw + pz, \qquad p \neq 0. \tag{7}$$

Written as

$$\frac{dw}{dz} - \frac{q}{z^2} w = \frac{p}{z},$$

the equation has the elementary solution

$$w(z) = \exp\left(-\frac{q}{z}\right) \left[p \int \exp\left(\frac{q}{t}\right) \frac{dt}{t} + C \right], \tag{8}$$

where C is the constant of integration. If $p = 0$, the solution has an essential singularity at the origin. If $p \neq 0$, a logarithmic term is also present, and (8) can be written (by expanding $\exp(q/t)$ and integrating) as

$$w(z) = \exp\left(-\frac{q}{z}\right) \left[p \left(\log z - \frac{q}{1! z} - \frac{q^2}{2! 2 z^2} - \cdots \right) + C \right]. \tag{9}$$

Series Solutions

This has a mixed singularity at $z = 0$: essential and (logarithmic) branch point. There are no solutions which are holomorphic in a domain containing the origin.

EXAMPLE 4 Taking $F \equiv 0$ in (4), we again have a linear equation

$$z^{m+1}\frac{dw}{dz} = qz + pw,$$

which can be integrated to give

$$w = e^{-(p/m)(z^{-m})}\left(q \int e^{(p/m)z^{-m}} z^{-m} dz + C\right),$$

where C is a constant.

Now, we consider some special cases of (1), which can be transformed into the Briot–Bouquet form.

i. The equation

$$\frac{dw}{dz} = \frac{b_1 z + b_2 w + \Sigma_2}{a_1 z + a_2 w + \Sigma_2'}, \tag{10}$$

where a_1, a_2, b_1, b_2 are real or complex constants, $a_1 b_2 - a_2 b_1 \neq 0$, and Σ_2 and Σ_2' represent terms of degree equal to or greater than 2. If we write $w = zv$, (10) becomes

$$z\frac{dv}{dz} = \frac{b_1 + (b_2 - a_1)v - a_2 v^2 + z\phi(z,v)}{a_1 + a_2 v + z\psi(z,v)} \tag{11}$$

with $\phi(z,v)$ and $\psi(z,v)$ holomorphic at $z = v = 0$. If $v(z)$ is any holomorphic solution of (11) and v_0 denotes $v(0)$, then letting $u = v - v_0$ in (11) we have

$$z\frac{du}{dz} = \frac{b_1 + (b_2 - a_1)v_0 - a_2 v_0^2 + \phi_1(z,u)}{a_1 + a_2 v_0 + \psi_1(z,u)}, \tag{12}$$

where $u(0) = 0$, and ϕ_1 and ψ_1 are holomorphic about $z = u = 0$; moreover, $\phi_1(0,0) = \psi_1(0,0) = 0$. Now, if we choose $v(0) = v_0$ such that

$$b_1 + (b_2 - a_1)v_0 - a_2 v_0^2 = 0$$

and

$$a_1 + a_2 v_0 \neq 0,$$

then (12) is obviously of Briot–Bouquet type.

ii. We now consider the system

$$\frac{dz}{dt} = P(z,w), \quad \frac{dw}{dt} = Q(z,w), \tag{13}$$

where P and Q are polynomials in z and w, which may be written as the sum of the homogeneous polynomials:

$$P(z,w) = \sum_{\nu=0}^{n} p_\nu(z,w), \qquad Q(z,w) = \sum_{\nu=0}^{n} q_\nu(z,w). \tag{14}$$

Here p_ν and q_ν are homogeneous of degree ν in z and w. Now, if we let

$$\xi = \frac{1}{z}, \qquad Z = \frac{w}{z} \tag{15}$$

in (13), we get

$$\frac{d\xi}{dt} = -\xi^2 P\left(\frac{1}{\xi}, \frac{Z}{\xi}\right), \qquad \frac{dZ}{dt} = \xi P\left(\frac{1}{\xi}, \frac{Z}{\xi}\right) - \xi Z Q\left(\frac{1}{\xi}, \frac{Z}{\xi}\right). \tag{16}$$

Now, if ξ is taken as the independent variable, we have

$$\frac{dt}{d\xi} = \frac{-1}{\xi^2 P(1/\xi, Z/\xi)}, \qquad \frac{dZ}{d\xi} = \frac{ZP(1/\xi, Z/\xi) - Q(1/\xi, Z/\xi)}{P(1/\xi, Z/\xi)}. \tag{17}$$

The system (17) has the advantage that its second equation involves only two variables, Z and ξ. If we can solve it, we can substitute the solution into the first equation and integrate to obtain a solution $t(\xi), Z(\xi)$ of (17). By inverting, $t = t(\xi)$, we get a solution $\xi(t), Z(t)$ of (16), from which we can obtain a solution $z(t), w(t)$ of (13) via the relations $z = 1/\xi$, $w = Z/\xi$.

Substituting (14) into (17) and using the homogeneity of p_ν and q_ν, we find that

$$\frac{dt}{d\xi} = \frac{-\xi^{n-2}}{p_n + \xi p_{n-1} + \xi^2 p_{n-2} + \cdots + \xi^n p_0}, \tag{18}$$

$$\xi \frac{dZ}{d\xi} = \frac{(Zp_n - q_n) + \xi(Zp_{n-1} - q_{n-1}) + \cdots + \xi^n(Zp_0 - q_0)}{p_n + \xi p_{n-1} + \cdots + \xi^n p_0}, \tag{19}$$

where p_j and q_j stand for $p_j(1, Z)$ and $q_j(1, Z)$, respectively. Let a root of the polynomial $\pi \equiv Zp_n(1, Z) - q_n(1, Z)$ be denoted by β. We expand (19) in the neighborhood of β by writing $Z = \beta + \zeta$, and get

$$\xi \frac{d\zeta}{d\xi} = \lambda \zeta + \mu \xi + B(\xi, \zeta), \tag{20}$$

where

$$\lambda = \frac{\pi'(\beta)}{p_n(1,\beta)}, \qquad \mu = \frac{\beta p_{n-1}(1,\beta) - q_{n-1}(1,\beta)}{p_n(1,\beta)},$$

Series Solutions

and the power series $B(\xi,\zeta)$ involves all terms of degree greater than 1. The same procedure applied to (18) yields

$$\frac{dt}{d\xi} = \frac{\xi^{n-2}}{p_n(1,\beta)}(1+P_1(\xi,\zeta)), \tag{21}$$

where P_1 is a convergent power series in ξ and ζ without a constant term. Now, equation (20) is in Briot–Bouquet form.

Actually the above reduction forms an important step in the proof of the following theorem for the solution of system (13). If $\lambda = \pi'(\beta)/p_n(1,\beta)$ is not a positive integer, then (13) has one and only one solution of the form

$$z = u^{-1}\left(1+\sum_{j=1}^{\infty}a_j u^j\right), \quad w = u^{-1}(1+\sum_{j=1}^{\infty}b_j u^j), \tag{22}$$

where $u = [(n-1)p_n(1,\beta)(t_0-t)]^{1/(n-1)}$, and the power series have positive radii of convergence. If $\lambda > 0$, then (13) has also a family of solutions

$$z = u^{-1}\{1+P_1[u,u^\lambda(c+h\ln u)]\},$$
$$w = u^{-1}\{1+P_2[u,u^\lambda(c+h\ln u)]\}, \tag{23}$$

where c is an arbitrary constant and $P_1(u,v)$ and $P_2(u,v)$ are convergent power series without constant terms, whose coefficients are independent of c. Here, $h = 0$ unless λ is a positive integer. In the latter case, h is a rational function of λ and the coefficients of P and Q (see Hille 1976).

iii. Indeed, Bendixson's reduction of the most general equation (13), where $P = A'w^{\alpha'}z^{\beta'} + \cdots$, $Q = Aw^\alpha z^\beta + \cdots$, and the indicated expressions for P and Q denote the term or one of the terms of the lowest degree in z, leads to one of the reduced forms

$$z\frac{dw}{dz} = pz + qw + \Sigma_2,$$
$$z\frac{dw}{dz} = pz + \Sigma_2,$$

or

$$z^{m+1}\frac{dw}{dz} = pz + qw + \Sigma_2$$

(see Sansone and Conti, 1952). The notation for Σ_2 is the same as in (10).

3.7 SOLUTIONS OF THE BRIOT–BOUQUET EQUATION

Now we treat analytically different cases that arise in the solution of the equation

$$zw' = pz + qw + F(z,w),$$
$$F(z,w) = \sum\sum_{j+k\geq 2} c_{jk}z^j w^k, \tag{1}$$

where $w(0) = 0$. First we note that if there is a solution $w = w(z)$ which is holomorphic at $z = 0$, then $zw'(z)$ is also holomorphic and evidently zero there. The right-hand side of the first of (1) with the substitution $z = 0$ represents a power series in w with qw as the first term. So, if there is a holomorphic solution w at $z = 0$, equation (1) requires that it vanish there. Suppose such a solution exists and has the representation

$$w(z) = \sum_{j=1}^{\infty} a_j z^j. \tag{2}$$

Substituting (2) into (1) gives

$$\sum_{j=1}^{\infty} j a_j z^j = pz + q \sum_{j=1}^{\infty} a_j z^j + \sum\sum_{j+k\geq 2} c_{jk} z^j \left(\sum_{n=1}^{\infty} a_n z^n\right)^k. \tag{3}$$

Expanding (3), rearranging it, and equating various powers of z on both sides, we have

$$(1-q)a_1 = p,$$
$$(2-q)a_2 = c_{20} + c_{11}a_1 + c_{02}(a_1)^2,$$
$$\vdots$$
$$(n-q)a_n = M_n(c_{jk}, a_1, \ldots, a_{n-1}), \tag{4}$$

where M_n is a multinomial in the indicated quantities, and $j + k \leq n$.

Several cases arise.

i. q is not a positive integer. Then the left-hand sides of (4) do not vanish, a_j can be successively and uniquely determined, and the series (2) represents a holomorphic solution at $z = 0$ provided it has a positive radius of convergence. This is shown by a Cauchy majorant argument involving an implicit function rather than a DE. To take care of the factors on the left-hand sides in (4), let B be a constant such that

$$\min_n |n - q| = B \quad n = 1, 2, \ldots. \tag{5}$$

Series Solutions

Choose a function $G(z,w)$ such that
$$pz + qw + F(z,w) \ll G(z,w). \tag{6}$$
One such function is
$$G(z,w) = M\left(1-\frac{z}{a}\right)^{-1}\left(1-\frac{w}{b}\right)^{-1} - M$$
$$= M\sum\sum_{j+k\geq 1}\left(\frac{z}{a}\right)^j\left(\frac{w}{b}\right)^k$$
$$= M\left[\frac{z}{a} + \frac{w}{b} + \sum\sum_{j+k\geq 2}\left(\frac{z}{a}\right)^j\left(\frac{w}{b}\right)^k\right], \tag{7}$$
where $0 < a < R_1$, $0 < b < R_2$, $M > \max(bB, M_0)$, where
$$M_0 = \max |c_{jk}|a^j b^k, \quad c_{10} = p, \quad c_{01} = q. \tag{8}$$
The constants R_1 and R_2 define the domain $|z| < R_1$, $|w| < R_2$ where $F(z,w)$ is holomorphic. Equations (8) ensure that G dominates F. The constant B in the inequality $M > \mathrm{Max}(Bb, M_0)$ will appear presently (see (10)).

Now, consider the implicit equation
$$BY = G(z,Y) = M\left(1-\frac{z}{a}\right)^{-1}\left(1-\frac{Y}{b}\right)^{-1} - M. \tag{9}$$
This equation is a quadratic in Y, namely,
$$\frac{B}{b}Y^2 + \left(\frac{M}{b} - B\right)Y + M\left(1-\frac{z}{a}\right)^{-1} - M = 0,$$
and the root which vanishes at $z = 0$ is given by
$$Y(z) = \frac{M-bB}{2b}\left\{-1 + (a-z)^{-1/2}\left[a - \left(\frac{M+bB}{M-bB}\right)^2 z\right]^{1/2}\right\}. \tag{10}$$
The square roots are taken to be real and positive for $z = 0$. The right-hand side of (10) is a holomorphic function and can be expanded about $z = 0$ provided
$$|z| < a_1 = a\left(\frac{M-bB}{M+bB}\right)^2 < a. \tag{11}$$
This avoids the branch point at $|z| = a_1$ in (10).

We write the solution of (9) in the form
$$Y = \sum_{n=1}^{\infty} A_n z^n,$$

which, after substitution in (9), gives the following relations among the coefficients:

$$BA_1 = C_{10},$$
$$BA_2 = C_{20} + C_{11}A_1 + C_{02}A_1^2,$$
$$\vdots$$
$$BA_n = M_n(C_{jk}; A_1, \ldots, A_{n-1}), \tag{12}$$

where M_n is the same multinomial in C_{jk} and A_m's as it is in c_{jk} and a_m's (see (4)). Moreover, all numerical coefficients are positive here.

Comparing (4) and (12) and taking (5) and (8) into account, we find that $|a_1| \leq A_1$, $|a_2| \leq A_2$, and, by complete induction,

$$|a_n| \leq A_n \tag{13}$$

for all n. We conclude that $Y(z)$ dominates the solution of (1), and the latter therefore has the (holomorphic) series solution (2), which converges absolutely for $|z| < a$. Thus, under the hypothesis concerning (1) and for q not a positive integer, (1) has a solution holomorphic in the neighborhood of $z = 0$ and satisfies the initial condition $w(0) = 0$. Besides, since the coefficients in the series (2) are determined in a unique way, this solution is unique.

ii. When q is a positive integer, we consider the argument in two parts.

(a) $q = 1$, so that (1) becomes

$$z\frac{dw}{dz} = pz + w + F(z,w), \tag{14}$$

with $F(z,w)$ satisfying the same conditions as in (1). Now, the first equation of (4) implies that $p = 0$. Therefore, we conclude that if $p \neq 0$, (1) does not have a holomorphic solution $w(z)$ vanishing at $z = 0$.

On the other hand, if $q = 1$, $p = 0$, equation (1) becomes

$$z\frac{dw}{dz} = w + F(z,w). \tag{15}$$

Suppose it has the series solution (2) with coefficients determined by (4). Then, the first equation of (4) shows that a_1 is arbitrary. For a given value of a_1, the series (2) can be shown to have a nonzero radius of convergence. Therefore, for $q = 1, p = 0$, (1) admits infinitely many holomorphic solutions of the form (2).

(b) If we assume that q in (1) is a positive integer greater than 1, we change the dependent variable w according to

$$w = \frac{pz}{1-q} + zv \tag{16}$$

Series Solutions

so that (1) becomes

$$z^2 v' + (1-q)zv = F\left[z, \left(\frac{p}{1-q} + v\right)z\right]. \tag{17}$$

Here the F term (starting with quadratic terms in its arguments) is divisible by z^2. We divide (17) by z and get

$$zv' = p_1 z + (q-1)v + F_1(z,v). \tag{18}$$

(The term $p_1 z$ has been taken out of F, so that $z[p_1 z + F_1(z,v)] = F[z,(p/(1-q)+v)z]$.) This has the same form as (1) with q replaced by $q-1$. If $q > 2$, we repeat this process of reduction until after $q-1$ steps we have an equation with $q = 1$. Finally, one of two possibilities arises: first, p_{q-1}, after $q-1$ steps, is zero. Case (a) applies, and (18), and, therefore, (1) has infinitely many holomorphic solutions. The likelihood of this case arising is very small. Second, $p_{q-1} \neq 0$ in the (finally) reduced equation. Then there are no solutions holomorphic in the neighborhood of the origin.

Example 1 of Section 3.6 indicates that the general (nonholomorphic) solution may admit representations in terms of psi series (Hille, 1976), a double series in z and z^q, if q is not a positive integer, or a logarithmic psi series if q is a positive integer. We quote the following theorem of Malmquist for (1): In the neighborhood of $z = 0$, the general solution of (1) is given by a convergent psi series

$$w(z) = \sum\sum a_{mn} c^n z^{m+nq}$$

if q is not a positive integer, and by

$$w(z) = \sum\sum a_{mn} z^{m+nq}(c + h \ln z)^n$$

in the integral case. Here, c is an arbitrary constant and h is a fixed constant.

We shall encounter in the following sections and some subsequent chapters this kind of solution in our discussion of series solutions of nonlinear DE.

3.8 POWER SERIES SOLUTIONS OF NONLINEAR DE

As pointed out in Section 3.1, when closed-form solutions of nonlinear DE are not obtained (which is usually the case), one avenue frequently open is the method of indeterminate coefficients or a power series solution. This was one of the earliest approaches to solutions of DE when mathematicians did not particularly worry about the "nature" of the formal series that satisfied a given DE. This was fraught with dangerous consequences since such formal series "solutions" could lead to absurd answers when initial or boundary conditions required by a physical problem were put in. The situation was subsequently retrieved by enquiring whether the series was convergent and then finding its radius of convergence. As we noted earlier, the study of linear equations via series solutions is now well established, and the coefficients appearing in the DE fully determine the analytic nature of the series and their radii of convergence. This is not true of nonlinear equations. For example, the radius of convergence for the series solution of the first-order nonlinear equation

$$\frac{dy}{dx} = F(x,y) = \sum_{i=0}^{\infty}\sum_{j=0}^{\infty} G_{ij} x^i y^j \qquad (1)$$

is not determined by that of F, in contrast to linear equations with analytic coefficients. For example, the Riccati equation

$$y' = 1 + y^2 \qquad (2)$$

has the solution

$$y = \tan(x + c), \qquad y(0) = \tan c = y_0, \qquad (3)$$

say, whose radius of convergence is $x_0 = \pi/2 - c$. This radius can be made as small as we please by taking c close to $\pi/2$, that is, by choosing y_0 sufficiently large; in contrast, the radius of convergence of the right-hand side of (2), $F(x,y) = 1 + y^2$, is infinite. Indeed, $F(x,y)$ is an entire function.

There are other cautions with regard to nonlinear DE. A simple power series solution may not exist. The solution may involve fractional or logarithmic terms, as we indicated in Section 3.7. Thus, considerable ingenuity is required in selecting the form of the series. There is also a practical aspect, namely, the computation of the series. If the series is slowly convergent, it may be of little use from a practical point of view. We may have to use an exponential series to take this aspect into account. On the other hand, we may sometimes obtain series which solve a boundary value problem over an infinite domain, the computational effort lying mainly in the numerical solution of one or several (coupled) transcendental equations.

We hope to clarify most of these points with the help of some nontrivial examples drawn from actual physical applications. However, once a "proper" series solution has been constructed, which satisfies given initial/boundary conditions, we have a constructive way of proving the existence and uniqueness of the solution. Thus, an important element in the analysis of DE via series is the proof of convergence of the series either directly or by the majorant methods of Cauchy and Lindelöf. We discuss these in Sections 3.9 and 3.10. Here, we discuss a few examples, where the

Series Solutions

radius of convergence of the series can be found more directly, say, by the ratio test or by the Weierstrass comparison test.

EXAMPLE 1

$$y' = \frac{y^2}{1-xy}, \quad y(0) = 1. \tag{4}$$

We seek a solution in the form

$$y(x) = \sum_{n=0}^{\infty} a_n x^n, \quad a_0 = 1. \tag{5}$$

Substituting (5) into (4), we have

$$(a_1 + 2a_2 x + 3a_3 x^2 + \cdots + na_n x^{n-1} + \cdots)$$
$$\times (1 - x - a_1 x^2 - a_2 x^3 - \cdots - a_n x^{n+1} - \cdots)$$
$$= 1 + 2a_1 x + (2a_2 + a_1^2)x^2 + (2a_1 a_2 + 2a_3)x^3 + \cdots. \tag{6}$$

Comparing the coefficients on both sides of (6), we have

$$a_1 = 1,$$
$$-a_1 + 2a_2 = 2a_1,$$
$$-a_1^2 - 2a_2 + 3a_3 = 2a_2 + a_1^2, \quad \text{etc.} \tag{7}$$

Therefore, $a_1 = 1$, $a_2 = 3/2$, $a_3 = 8/3$. By induction, we can show that

$$a_n = \frac{(n+1)^{(n-1)}}{n!}. \tag{8}$$

The radius of convergence is obtained by the ratio test:

$$R = \lim_{n \to \infty} \left| \frac{a_n}{a_{n+1}} \right| = \lim_{n \to \infty} \left(\frac{n+1}{n+2} \right)^n$$
$$= \lim_{n \to \infty} \left(1 - \frac{1}{n+2} \right)^n = \frac{1}{e}. \tag{9}$$

This radius of convergence is not suggested by the coefficients in (4). While it is satisfying that the coefficients of the series even for a nonlinear DE can be found by a step-by-step method in terms of the previous coefficients, the relations (7) in the coefficients are nonlinear (unlike those for the linear equations), and, in general, the radius of convergence by the Cauchy (or any other) test would be difficult to find directly. Of course, it could still be found computationally by considering the ratio $|a_n/a_{n+1}|$ for sufficiently large n. Since the coefficients a_n are all positive, the singularity of the series (5) would occur at a positive value of x. This value turns out to be $1/e$.

EXAMPLE 2 (Treve, 1967).

$$\dddot{x} + x\ddot{x} = 0,$$
$$x(\infty) = 1, \quad \dot{x}(\infty) = 0, \quad \ddot{x}(\infty) = 0. \tag{10}$$

To find the solution of this "initial" value problem for the Blasius equation, we write the "Picard" or exponential series to satisfy the conditions at infinity:

$$x = \sum_{n=0}^{\infty} a_n (ce^{-t})^n = \sum_{n=0}^{\infty} a_n \mu^n, \tag{11}$$

where $\mu = ce^{-t}$ and $a_0 = 1$. The constant c will be chosen presently. This series automatically satisfies the other conditions $\dot{x}(\infty) = 0$ and $\ddot{x}(\infty) = 0$. Substituting (11) into (10), we have

$$\sum_{n=1}^{\infty} n^3 a_n \mu^n = \left(\sum_{n=0}^{\infty} a_n \mu^n \right) \left(\sum_{n=1}^{\infty} n^2 a_n \mu^n \right). \tag{12}$$

Equating the coefficients, we have

$$a_1 = a_1,$$
$$2^3 a_2 = 2^2 a_0 a_2 + a_1^2,$$
$$\vdots$$
$$a_n = \frac{1}{n^2(n-1)} \sum_{k=1}^{n-1} k^2 a_k a_{n-k}. \tag{13}$$

Here, a_1 is arbitrary; we may choose it to be unity. Then,

$$a_2 = 1/4, \quad a_3 = 5/72, \ldots. \tag{14}$$

We shall show that the series (11) converges for $|\mu| < 3$. To this end, we write

$$x = 1 + \mu K(\mu), \tag{15}$$

where

$$K(\mu) = \sum_{n=0}^{\infty} b_n \mu^n \tag{16}$$

and

$$b_n = a_{n+1}. \tag{17}$$

Series Solutions

Then

$$b_n < \left(\frac{1}{3}\right)^n. \tag{18}$$

This is true for $m = 1, 2$ as is easily seen from (14). To show that it holds for $m > 2$, let $b_m < (1/3)^m$ for $m = 3, 4, \ldots, n-1$. Using the last of (13), we have

$$b_n = \frac{1}{n(n+1)^2} \sum_{k=1}^{n} b_{k-1} b_{n-k} k^2. \tag{19}$$

According to our assumptions, $b_0 = a_1 = 1$, and $b_{k-1} b_{n-k} < (1/3)^{k-1+n-k} = (1/3)^{n-1}$, so

$$b_n < \frac{1}{n(n+1)^2} \left(\frac{1}{3}\right)^{n-1} \sum_{k=1}^{n} k^2 = \frac{2n+1}{2n+2} \left(\frac{1}{3}\right)^n < \left(\frac{1}{3}\right)^n, \tag{20}$$

which proves the assertion. It therefore follows that the series $K(\mu)$ is "dominated" by the series

$$K_1(\mu) = \sum_{n=0}^{\infty} \left(\frac{\mu}{3}\right)^n = \frac{1}{1-\mu/3} \tag{21}$$

and therefore converges for $|\mu| < 3$.

This is an example of the method of dominant series. In the following section, we shall describe a more general approach in which the differential equation, rather than the solution series, is majorized.

EXAMPLE 3 The nonlinear nonharmonic motion of an oscillator is described by

$$\ddot{x} + w^2 x = -\alpha x^2, \tag{22}$$

α and w being constants, with initial conditions

$$x(0) = A, \quad \dot{x}(0) = 0. \tag{23}$$

We shall give two series solutions of this problem due to Shidfar and Sadeghi (1986). Noted that, if $A = -w^2/\alpha$, then

$$x(t) = -\frac{w^2}{\alpha} \tag{24}$$

is a solutions of (22)–(23). We seek a series solution of (22)–(23) which includes (24) as a special case. By writing

$$x = u - \frac{w^2}{2\alpha},$$

the problem (22)–(23) becomes

$$\ddot{u} + \alpha u^2 = \frac{w^4}{4\alpha}, \tag{25}$$

$$u(0) = A + \frac{w^2}{2\alpha}, \qquad \dot{u}(0) = 0. \tag{26}$$

First we solve (25)–(26) in the form

$$u(t) = c_0 + c_1 \sin wt + c_2 \sin^2 wt + \cdots, \tag{27}$$

where c_i, $i = 0, 1, \ldots$, are coefficients to be determined by the substitution of (27) into (25). We thus find that

$$2w^2 c_2 + \alpha c_0^2 = \frac{w^4}{4\alpha},$$

$$c_{n+2} = \frac{n^2}{(n+2)(n+1)} c_n$$

$$\quad - \frac{\alpha}{w^2(n+2)(n+1)} (c_0 c_n + c_1 c_{n-1} + \cdots + c_n c_0),$$

$$n \geq 1. \tag{28}$$

Equations (26b) and (27) imply that $c_1 = 0$. Equations (28) then yield $c_3 = 0$, $c_5 = 0$, etc. The even-order coefficients now simply are

$$c_0 = A + \frac{w^2}{2\alpha},$$

$$c_2 = -\frac{A}{2w^2}(w^2 + A\alpha),$$

$$c_4 = -\frac{1}{6w^2} A(w^2 + A\alpha) \left(\frac{3}{4} - \frac{A\alpha}{2w^2} \right),$$

$$c_6 = -\frac{1}{180w^2} A(w^2 + A\alpha) \left(\frac{3}{4} - \frac{A\alpha}{2w^2} \right) \left(15 - \frac{24 A\alpha}{w^2} \right)$$

$$\quad - \frac{\alpha A^2}{120 w^6}(w + A\alpha)^2, \quad \text{etc.} \tag{29}$$

The coefficient c_0 follows from the condition (26a). The solution for (22)–(23) can now be written as

$$x(t) = A - \frac{A}{2w^2}(w^2 + A\alpha)\sin^2 wt + \cdots. \tag{30}$$

Equations (29) and further induction show that c_{2i}, $i = 1, 2, \ldots$, all vanish for $A = -w^2/\alpha$, and hence the solution (24) is included in (30) as a special case.

Series Solutions

We now show that the series (27) is absolutely convergent for all t. First, we note that if $c_0 > 0$, $c_2 > 0$, and $\alpha = -\beta^2 < 0$, then all coefficients c_n in (28) are positive. Indeed, we may write

$$\sum_{n=2}^{\infty}(n+2)(n+1)c_{n+2} = \sum_{n=2}^{\infty} n^2 c_n + \frac{\beta^2}{w^2}\left(\sum_{n=0}^{\infty} c_n\right)^2 - \frac{\beta^2}{w^2} c_0^2 \qquad (31)$$

or

$$\beta^2 \left(\sum_{n=0}^{\infty} c_n\right)^2 + w^2 \sum_{n=0}^{\infty} n c_n = \beta^2 c_0^2 - 2w^2 c_2. \qquad (32)$$

Since the right-hand side of (32) is finite and c_i are positive, the series $\sum_{n=0}^{\infty} c_n$ converges. Now, if we put

$$c_0' = |c_0|, \quad c_1' = 0, \quad c_2' = |c_2|, \qquad (33)$$

$$c_{n+2}' = \frac{n^2}{(n+1)(n+2)} c_n' + \frac{|\alpha|}{w^2(n+1)(n+2)} (c_0' c_n' + \cdots + c_n' c_0'),$$

$$n \geq 2, \qquad (34)$$

then the series $\sum_{n=0}^{\infty} c_n'$ can be shown to converge. Since $|c_n| \leq c_n'$, it follows that the solution series (27) is absolutely convergent, and hence the series solution of (22)–(23) converges for all t.

It is interesting to write the power series solution for (25)–(26), namely,

$$u(t) = \sum_{n=0}^{\infty} a_n t^n. \qquad (35)$$

We find again that

$$a_{2i+1} = 0, \quad i = 0, 1, 2, \ldots, \qquad (36)$$

while

$$a_0 = A + \frac{w^2}{2\alpha},$$

$$2a_2 + \alpha a_0^2 = \frac{w^4}{4\alpha},$$

$$a_{n+2} = \frac{-\alpha}{(n+2)(n+1)} (a_0 a_n + a_2 a_{n-2} + \cdots + a_n a_0),$$

$$n = 2k, \quad k \geq 1. \qquad (37)$$

The coefficients a_{2i}, $i = 0, 1, 2, \ldots$, again vanish for $A = -w^2/\alpha$. The solutions (35) and (27) can be verified to be identical.

EXAMPLE 4 We discuss an initial value problem for the equation describing the vibration of the simple pendulum (Jindia and Sachdeva, 1984):

$$\frac{d^2\theta}{dt^2} + k^2 \sin\theta = 0, \tag{38}$$

$$\theta = \alpha, \quad \frac{d\theta}{dt} = 0, \quad \text{at } t = 0, \tag{39}$$

where θ is the angular displacement of the pendulum of length L, g measured from the vertical, at time t, and $k^2 = g/L$, g being the acceleration due to gravity.

Before giving the series solution of (38) and (39), we express it in terms of elliptic functions. Multiplying (38) by $(d\theta/dt)dt = d\theta$ and integrating, we obtain

$$\frac{1}{2}(d\theta/dt)^2 - k^2 \cos\theta = C. \tag{40}$$

Using (39), we find that $C = -k^2 \cos\alpha$. Equation (40) can be integrated once again. To express the solution in terms of an elliptic function, we make the substitution $K \sin\phi = \sin(\theta/2)$, where $K = \sin(\alpha/2)$. Furthermore, noting that $\cos\theta = 1 - 2\sin^2(\theta/2)$, we can write (40) as

$$kt = \frac{-1}{\sqrt{2}} \int_\alpha^\theta \frac{d\theta}{(\cos\theta - \cos\alpha)^{1/2}}$$

or

$$kt = \int_\phi^{\pi/2} \frac{d\phi}{(1 - K^2 \sin^2\phi)^{1/2}}$$

$$= \int_0^{\pi/2} \frac{d\phi}{(1 - K^2 \sin^2\phi)^{1/2}} - \int_0^\phi \frac{d\phi}{(1 - K^2 \sin^2\phi)^{1/2}}. \tag{41}$$

The complete and incomplete elliptic integrals of the first kind appearing in (41) are well tabulated as functions of $\alpha = 2\sin^{-1}K$, and $\phi = \sin^{-1}(K^{-1}\sin(\theta/2))$ (see Standard Mathematical Tables, 1969). The natural *period of oscillation* T of the pendulum is defined as the time required to make a complete oscillation between positions of maximum displacement, that is, when $d\theta/dt = 0$. For now, we note that $\cos\theta - \cos\alpha = 2K^2 \cos^2\phi$, so $d\theta/dt = 0$ when $\theta = \alpha$ and $\phi = \pi/2$ (see (40)). Therefore, the period of oscillation is

$$P(K) = 4\left(\frac{L}{g}\right)^{1/2} \int_0^{\pi/2} \frac{d\phi}{(1 - K^2 \sin^2\phi)^{1/2}} = \left(\frac{4}{k}\right)\mathcal{K}(K),$$

Series Solutions

where $K(K)$ is the complete elliptic integral of the first kind. Here we have used the result that

$$dt = \left(\frac{1}{2k^2}\right)^{1/2} \frac{d\theta}{(\cos\theta - \cos\alpha)^{1/2}} = \frac{1}{k} \frac{d\phi}{(1 - K^2 \sin^2\phi)^{1/2}}.$$

Now we attempt a series solution of the initial value problem (38)–(39) in the form

$$\theta = \sum_{n=0}^{\infty} a_n t^n, \tag{42}$$

$$\sin\theta = \sum_{n=0}^{\infty} S_n t^n, \tag{43}$$

$$\cos\theta = \sum_{n=0}^{\infty} C_n t^n. \tag{44}$$

The coefficients S_n and C_n can be expressed in terms of a_n as follows. From (42)–(44), we have

$$\sin\left(\sum_{n=0}^{\infty} a_n t^n\right) = \sum_{n=0}^{\infty} S_n t^n, \tag{45}$$

$$\cos\left(\sum_{n=0}^{\infty} a_n t^n\right) = \sum_{n=0}^{\infty} C_n t^n. \tag{46}$$

Differentiating (45) and (46), we have

$$\left(\sum_{n=0}^{\infty} C_n t^n\right)\left(\sum_{n=1}^{\infty} n a_n t^{n-1}\right) = \sum_{n=1}^{\infty} n S_n t^{n-1}$$

or

$$\left(\sum_{n=0}^{\infty} C_n t^n\right)\left(\sum_{n=0}^{\infty} (n+1) a_{n+1} t^n\right) = \sum_{n=0}^{\infty} (n+1) S_{n+1} t^n, \tag{47}$$

and

$$\left(\sum_{n=0}^{\infty} S_n t^n\right)\left(\sum_{n=1}^{\infty} n a_n t^{n-1}\right) = -\sum_{n=1}^{\infty} n C_n t^{n-1}$$

or

$$\left(\sum_{n=0}^{\infty} S_n t^n\right)\left(\sum_{n=0}^{\infty} (n+1) a_{n+1} t^n\right) = -\sum_{n=0}^{\infty} (n+1) C_{n+1} t^n. \tag{48}$$

Equating coefficients of equal powers of t^n on both sides of (47) and (48), we obtain

$$S_{n+1} = \frac{1}{n+1} \sum_{r=0}^{n} (r+1)a_{r+1}C_{n-r}, \tag{49}$$

$$C_{n+1} = \frac{-1}{n+1} \sum_{r=0}^{n} (r+1)a_{r+1}S_{n-r}. \tag{50}$$

From (45) and (46), $S_0 = \sin a_0$ and $C_0 = \cos a_0$; therefore, equations (49) and (50) show that all the coefficients S_n and C_n are known in terms of a_i, $i = 0, 1, 2, \ldots, n - 1$. Now substituting (42) and (43) into (38), we have

$$\sum_{n=0}^{\infty} [(n+1)(n+2)a_{n+2} + k^2 S_n] t^n = 0. \tag{51}$$

The recurrence relation for the coefficients a_n, therefore, is

$$(n+1)(n+2)a_{n+2} + k^2 S_n = 0, \qquad n \geq 0. \tag{52}$$

The initial conditions (39) give $a_0 = \alpha$ and $a_1 = 0$. The higher coefficients are given by (52) and (49)–(50). The solution (42) is

$$\alpha - \theta = \frac{\sin \alpha (kt)^2}{2!} - \frac{\sin \alpha \cos \alpha (kt)^4}{4!}$$

$$+ \frac{\sin \alpha (1 - 4 \sin^2 \alpha)(kt)^6}{6!}$$

$$- \frac{\sin \alpha \cos \alpha (1 - 34 \sin^2 \alpha)(kt)^8}{8!}$$

$$+ \frac{\sin \alpha (1 - 308 \sin^2 \alpha + 496 \sin^4 \alpha)(kt)^{10}}{10!}$$

$$- \frac{\sin \alpha \cos \alpha (1 - 2768 \sin^2 \alpha + 11056 \sin^4 \alpha)(kt)^{12}}{12!} + \cdots. \tag{53}$$

Now, to compare the series solution (53) with (41), we invert the former to express kt as a function of $\alpha - \theta$. We write

$$y = \alpha - \theta, \qquad (kt)^2 = x, \tag{54}$$

so (53) takes the form

$$y = c_1 x + c_2 x^2 + c_3 x^3 + c_4 x^4 + c_5 x^5 + c_6 x^6 + \cdots, \tag{55}$$

which on inversion becomes (Abramowitz and Stegun, 1964, p. 15)

$$x = b_1 y + b_2 y^2 + b_3 y^3 + b_4 y^4 + b_5 y^5 + b_6 y^6 + \cdots \tag{56}$$

with

$$b_1 = \frac{1}{c_1}, \quad b_2 = \frac{-c_2}{c_1^3}, \quad b_3 = \frac{2c_2^2 - c_1 c_3}{c_1^5},$$

$$b_4 = \frac{5c_1 c_2 c_3 - c_1^2 c_4 - 5c_2^3}{c_1^7},$$

$$b_5 = \frac{6c_1^2 c_2 c_4 + 3c_1^2 c_3^2 + 14c_2^4 - c_1 c_5 - 21 c_1 c_2^2 c_3}{c_1^9},$$

$$b_6 = \frac{7c_1^3 c_2 c_5 + 7c_1^3 c_3 c_4 + 84 c_1 c_2^3 c_3 + c_1^4 c_6 - 28 c_1^2 c_2^2 c_4 - 28 c_1^2 c_2 c_3^2 - 42 c_2^5}{c_1^{11}},$$

etc. $\hspace{6cm}$ (57)

Therefore, the solution (53) has the explicit form

$$(kt)^2 = \frac{2}{\sin \alpha}(\alpha - \theta) + \frac{\cos \alpha}{3 \sin^2 \alpha}(\alpha - \theta)^2$$

$$+ \frac{4 - \sin^2 \alpha}{45 \sin^3 \alpha}(\alpha - \theta)^3$$

$$+ \frac{\cos \alpha (36 + \sin^2 \alpha)}{1260 \sin^4 \alpha}(\alpha - \theta)^4$$

$$+ \frac{64 - 42 \sin^2 \alpha - \sin^4 \alpha}{6300 \sin^5 \alpha}(\alpha - \theta)^5$$

$$+ \frac{\cos \alpha (320 - 106 \sin^2 \alpha + \sin^4 \alpha)}{83{,}160 \sin^6 \alpha}(\alpha - \theta)^6 + \cdots. \hspace{1cm} (58)$$

Table 3.8.1 shows the values of kt as computed from (41). Those obtained from (58) are the same with the exception of the entries marked +. In these cases, the values obtained by the series (58) are less than those in the table by one unit in the fourth significant place.

This example shows that even with a small number of terms in the series, the solution is quite accurate. The only price in this problem is the inversion of the series (53).

A series solution requiring inversion was found for a more complex boundary value problem, arising in laminar convection over a linear heat source, by Sevruk (1958):

$$f''' + 3ff'' = f'^2 - \phi, \quad ' = \frac{d}{d\eta} \hspace{2cm} (59)$$

$$3\sigma (f\phi)' = -\phi'', \quad \sigma \text{ a constant}, \hspace{2cm} (60)$$

Table 3.8.1 kt as a Function of α and ϕ.

ϕ^0 \ α^0	10	20	30	40	50	60	70	80	90
0	1.574+	1.583+	1.598+	1.620+	1.649+	1.686+	1.731+	1.787+	1.854+
10	1.400+	1.408	1.423	1.445	1.474	1.511	1.556	1.612	1.679
20	1.225	1.234	1.249	1.270	1.299	1.335	1.380	1.435	1.501
30	1.050	1.059	1.073	1.094	1.121	1.156	1.200	1.253	1.318
40	0.875	0.883	0.896	0.916	0.941	0.974	1.015	1.066	1.227
50	0.700	0.707	0.719	0.736	0.758	0.788	0.824	0.870	0.926
60	0.525	0.531	0.540	0.554	0.572	0.596	0.626	0.664	0.712
70	0.350	0.354	0.361	0.370	0.384	0.400	0.422	0.450	0.484
80	0.175	0.177	0.181	0.186	0.192	0.201	0.212	0.227	0.246
90	0.000	0.000	0.000	0.000	0.000	0.000	0.000	0.000	0.000

These values are computed using equation (41) involving elliptic integrals. The values computed using series solution (58) are identical, except that entries marked + have values decreased by one unit in the fourth significant digit.
Source: From Jindia and Sachdeva (1984).

Series Solutions

$$f = f'' = 0, \quad \phi' = 0 \quad \text{at } \eta = 0;$$
$$f' = 0, \quad \phi = 0 \quad \text{at } \eta = \infty.$$

The procedure, though straightforward, involves considerable detail. It may be applied with advantage to this class of problems in boundary layer theory.

3.9 METHOD OF MAJORANTS

In the method of majorants we identify a DE which has a series solution with a known radius of convergence and which helps by comparison to give the radius of convergence of the series solution of a given DE. Let a function $f(z)$ have the power series representation

$$f(z) = \sum_{n=0}^{\infty} c_n z^n, \tag{1}$$

where the series has a positive radius of convergence, and let

$$g(z) = \sum_{n=0}^{\infty} C_n z^n \tag{2}$$

be a power series with nonnegative coefficients, having a radius of convergence R. The function $g(z)$ *majorizes* $f(z)$ ($g(z) \gg f(z)$) if $|c_n| \leq C_n$ for all n. Therefore, the radius of convergence of (1) is at least R. The relation $\ll$ is transitive, and other operations, such as raising the series by a power, differentiation, and integration, preserve it.

The majorant concept easily extends to functions of several variables. In particular, if

$$F(z,w) = \sum_{j=0}^{\infty} \sum_{k=0}^{\infty} c_{jk} z^j w^k \tag{3}$$

and

$$G(z,w) = \sum_{j=0}^{\infty} \sum_{k=0}^{\infty} C_{jk} z^j w^k \tag{4}$$

are holomorphic functions of (z,w) in the region $D: |z| \leq a, |w| \leq b$, we say that G is a *majorant* of F; that is, $F(z,w) \ll G(z,w)$, provided

$$|c_{jk}| \leq C_{jk} \quad \text{for all } j \text{ and } k. \tag{5}$$

The important fact with respect to the DE

$$w'(z) = F(z,w) = \sum_{j=0}^{\infty}\sum_{k=0}^{\infty} c_{jk} z^j w^k \tag{6}$$

is that a majorant relation for the right-hand side implies the corresponding relation for the solution (Hille, 1976). We shall use this fact to prove the existence of solutions for initial value problems for (6).

THEOREM 3.9.1 Let $F(z,w)$ be defined by the series (3), and let $G(z,w)$ be a majorant defined by (4) and (5). Suppose that the problem

$$W'(z) = G[z, W(z)], \qquad W(0) = 0, \tag{7}$$

has a solution

$$W(z) = \sum_{j=1}^{\infty} C_j z^j \tag{8}$$

convergent for $|z| < r$. Further, suppose that

$$w(z) = \sum_{j=1}^{\infty} c_j z^j \tag{9}$$

is a formal solution of

$$w'(z) = F[z, w(z)], \qquad w(0) = 0. \tag{10}$$

Then

$$w(z) \ll W(z), \tag{11}$$

the series (9) is absolutely convergent for $|z| < r$, and is the unique solution of (10).

By "formal" we mean that the operations to get the form of the series from the DE have been performed without due justification. However, if the series is absolutely convergent, the steps can be a posteriori justified. Substituting (9) into (10), we have

$$\sum_{j=1}^{\infty} j c_j z^{j-1} = \sum_{j=0}^{\infty}\sum_{k=0}^{\infty} c_{jk} z^j [\sum_{p=1}^{\infty} c_p z^p]^k$$

$$= c_{00} + (c_{10} + c_{01}c_{00})z$$

$$+ \left[c_{20} + c_{11}c_{00} + c_{02}(c_{00})^2\right.$$

Series Solutions

$$+ \frac{1}{2} c_{01} c_{10} + \frac{1}{2} (c_{01}^2) c_{00} \Big] z^2 + \cdots \tag{12}$$

(where the substitution for c_j in terms of c_{jk} has been done recursively on the right-hand side of (12)). Equating equal powers of z on both sides, we have

$$c_1 = c_{00},$$

$$c_2 = \frac{1}{2}(c_{10} + c_{01} c_{00}),$$

$$c_3 = \frac{1}{3}[c_{20} + c_{11} c_{00} + c_{02}(c_{00})^2 + \frac{1}{2} c_{01} c_{10} + \frac{1}{2}(c_{01})^2 c_{00}],$$

$$\vdots$$

$$(n+1) c_{n+1} = M_n(c_{jk}; c_p); \tag{13}$$

where M_n are polynomials in c_{jk} and $c_1, c_2, \ldots, c_{n-1}$.

From these recursion formulas, c_n can be found. It is clear that each c_n is polynomial in c_{jk} and the previous c_n, which in turn are expressed in terms of c_{jk} only. The numerical constants that enter, when the kth power of $\sum_{p=1}^{\infty} c_p z^p$ is found, are positive integers. Taking the kth power of the series in (12), taking its product with $c_{jk} z^j$ and summation with respect to j and k, and rearrangement of the series are all permissible since the series for $F(z, w)$ is absolutely convergent.

If the same procedure is applied to the majorant equation (7), we get exactly the same formulas for determining C_n's provided that we replace the lowercase letters in M_n by the corresponding capitals. It is then clear that

$$|c_n| \leq C_n \quad \text{for all } n, \tag{14}$$

so that $W(z)$ is a majorant for $w(z)$ provided that (8) has a finite disk of convergence. If this is so, the series (9) for $w(z)$ has a radius of convergence at least as large as that for $W(z)$. Then the operations performed on the series are justified, and the formal series is an actual solution. Besides, the coefficients c_n are uniquely determined, so that (9) is the only solution which is holomorphic in some neighborhood of $z = 0$.

The task that remains is to construct a suitable majorant equation having an absolutely convergent series solution. We shall give two methods for this purpose in the next section, following Hille (1976). The proof of Theorem 3.9.1 is easily extended to systems of nonlinear DE (see Hochstadt, 1963).

3.10 CONSTRUCTION OF MAJORANT EQUATIONS

For the problem

$$w'(z) = F(z, w(z)), \qquad w(0) = 0, \tag{1}$$

with

$$F(z, w) = \sum_{j=0}^{\infty} \sum_{k=0}^{\infty} c_{jk} z^j w^k, \tag{2}$$

where the double series is absolutely convergent in the region

$$D: |z| \leq a, \qquad |w| \leq b, \tag{3}$$

we wish to prove the convergence of the formal solution

$$\sum_{n=1}^{\infty} c_n z^n. \tag{4}$$

This is accomplished by constructing a suitable majorant $G(z, w)$ of the right-hand side of (1), namely, $F(z, w)$, such that the problem

$$W'(z) = G[z, W(z)], \qquad W(0) = 0, \tag{5}$$

has an absolutely convergent series solution

$$W(z) = \sum_{n=1}^{\infty} C_n z^n \tag{6}$$

with a certain radius of convergence. We give two methods for the construction of the majorant.

Cauchy's Majorant

We note that, because of the absolute convergence of the series representation (2) for $F(z, w)$ in D, the series

$$M(a, b) = \sum_{r=0}^{\infty} \sum_{j=0}^{\infty} |c_{jk}| a^j b^k \tag{7}$$

is convergent. Therefore, its terms are bounded. The bound may be taken to be $M(a, b) \equiv M$ so that

$$|c_{jk}| \leq M a^{-j} b^{-k} \equiv C_{jk}, \qquad \text{say, for all } j \text{ and } k. \tag{8}$$

Series Solutions

Thus, we choose

$$G(z,w) = M \sum_{j=0}^{\infty} \sum_{k=0}^{\infty} \left(\frac{z}{a}\right)^j \left(\frac{w}{k}\right)^k = M\left(1 - \frac{z}{a}\right)^{-1}\left(1 - \frac{z}{b}\right)^{-1}. \qquad (9)$$

Evidently,

$$F(z,w) = \sum_{j=0}^{\infty} \sum_{k=0}^{\infty} c_{jk} z^j w^k \ll M \sum_{j=0}^{\infty} \sum_{k=0}^{\infty} \left(\frac{z}{z}\right)^j \left(\frac{w}{b}\right)^k$$
$$= G(z,w). \qquad (10)$$

We therefore write the majorant problem as

$$W'(z) = M\left(1 - \frac{z}{a}\right)^{-1}\left(1 - \frac{W}{b}\right)^{-1}, \qquad W(0) = 0,$$

or

$$W' - \frac{1}{b} WW' = \frac{Ma}{a-z}, \qquad W(0) = 0. \qquad (11)$$

This can be explicitly integrated. We have

$$W(z) - \frac{1}{2b}[W(z)]^2 = -MA \ln\left(1 - \frac{z}{a}\right). \qquad (12)$$

Solving for $W(z)$ and choosing the negative sign for the square root so that $W(0) = 0$, we have

$$W(z) = b - \left[b^2 + 2abM \ln\left(1 - \frac{z}{a}\right)\right]^{1/2}. \qquad (13)$$

Here the principal value of the logarithm (which corresponds to real $\ln u$ for $u > 0$) is taken. Also the square root in (13) is positive. The term in brackets can be expanded in a Maclaurin's series and has positive coefficients. The radius of convergence of this series is less than the distance from the origin of its nearest singularity. It has a logarithmic singularity at $z = a$ and a branch point singularity at the point where

$$b^2 + 2abM \ln\left(1 - \frac{z}{a}\right) = 0$$

or

$$z = a\left(1 - \exp\left(-\frac{b}{2aM}\right)\right) \equiv R. \qquad (14)$$

The singularity nearer $z = 0$ is obviously $z = R$, since $R < a$.

This leads to the result that the system (1) has a series solution (4) which is absolutely convergent in the disk $|z| < R$ with R given by (14). The coefficients are uniquely determined according to (12)–(13) of Section 3.9.

The Lindelöf Majorant

The majorant of $F(z,w)$ is obtained simply by replacing the coefficients c_{jk} by their absolute values:

$$C_{jk} = |c_{jk}| \quad \text{for all } j,k. \tag{15}$$

With this choice, the whole discussion can be carried out in the real domain and the questions of convergence can be replaced by those of boundedness, which are simpler. Thus, we choose

$$G(x,y) = \sum_{j=0}^{\infty} \sum_{k=0}^{\infty} C_{jk} x^j y^k \tag{16}$$

for the majorant of $F(x,y)$. Here the variables are real and positive.

Consider the problem

$$y'(x) = G(x, y(x)), \quad y(0) = 0. \tag{17}$$

Let the nth partial sum of the formal series solution of (17) be denoted by

$$P_n(x) = \sum_{m=1}^{n} c_m x^m.$$

Let the series (16) be convergent for $x = a$ and $y = b$ so that

$$\sum_{j=0}^{\infty} \sum_{k=0}^{\infty} C_{jk} a^j b^k \equiv M(a,b) < \infty. \tag{18}$$

It is easily seen that

$$P'_{n+1}(x) \le \sum_{j=0}^{n} \sum_{k=0}^{n} C_{jk} x^j [P_n(x)]^k \quad \text{for all } n. \tag{19}$$

This follows from the form of the recursion formula for C_p, which depends on $C_1, C_2, \ldots, C_{p-1}$ and C_{jk} with $0 \le j + k \le p$ (see Section 3.9): while all the terms on the left of the equality in (19) cancel due to the recursion relation, the right-hand side still has some additional positive terms.

Let

$$r = \min\left[a, \frac{b}{M(a,b)}\right]. \tag{20}$$

Suppose that, for some n,

$$P_n(x) \le b \quad \text{for } 0 \le x \le r. \tag{21}$$

Series Solutions

Then

$$P'_{n+1}(x) \leq \sum_{j=0}^{n} \sum_{k=0}^{n} C_{jk} r^j [P_n(r)]^k$$

$$\leq \sum_{j=0}^{n} \sum_{k=0}^{n} C_{jk} a^j b^k \leq M(a,b), \tag{22}$$

using (21) and (18). For $0 \leq x \leq r$, (22) gives, by integration and use of (20) and (21), the inequality

$$P_{n+1}(x) \leq xM(a,b) \leq rM(a,b) \leq b. \tag{23}$$

Since

$$P_1(x) = C_{00} x \leq rM(a,b) \leq b, \tag{24}$$

the estimate (21) holds for all n. The sequence $P_n(x)$ is nondecreasing for a fixed x and is bounded for $0 \leq x \leq r$. The partial sums of the formal series solution converge to a limit; that is, the formal solution is the actual series solution of (17) with a radius of convergence at least equal to r. The latter is given explicitly by (20).

How should a and b be chosen to obtain the largest possible radius of convergence r? In the absence of any further information, one may choose the two terms on the right-hand side of (20) to be equal,

$$aM(a,b) = b, \tag{25}$$

and maximize r with this side condition.

EXAMPLE 1

$$\frac{dw}{dz} = (1 - z - w)^{-1}. \tag{26}$$

Here

$$F(z,w) = (1 - z - w)^{-1} = \sum_{j=0}^{\infty} \sum_{k=0}^{\infty} \frac{(j+k)!}{j!k!} z^j w^k.$$

For the Cauchy majorant, we may take $a = p$, $b = 1-p$, $0 < p < 1$. The series

$$M(a,b) = \sum_{j=0}^{\infty} \sum_{k=0}^{\infty} \frac{(j+k)!}{j!k!} p^j (1-p)^k$$

diverges for this choice of *a* and *b*, but its terms are bounded. Actually, we may take M to be unity. This follows from the expansion

$$1 = [p + (1-p)]^n = \sum_{j=0}^{n} \frac{n!}{j!(n-j)!} p^j (1-p)^{n-j}.$$

Equation (14) for the radius of convergence becomes

$$r(p) = p\left[1 - \exp\left(-\frac{1-p}{2p}\right)\right],$$

$$\frac{dr}{dp} = \left[1 - \exp\left(-\frac{1-p}{2p}\right)\right] - \frac{1}{2p}\exp\left(-\frac{1-p}{2p}\right).$$

This is zero for p_0, $1/2 < p_0 < 1$. At this point, the maximum of r is approximately 0.212. The actual location of the singularity of the solution of (26) with $w(0) = 0$ can be easily found.

Writing (26) (with a change of notation) as

$$\frac{dx}{dy} + x = 1 - y,$$

the solution, satisfying $x(0) = 0$, is

$$x = 2 - y - 2e^{-y}.$$

At the singularity $x = x_0$, $dy/dx = \infty$ or $dx/dy = 0$. This leads to $1 = 2e^{-y_0} = \ln 2$ and $x_0 = 1 - \ln 2 = 0.30685$. This value of the point of singularity is quite different from 0.212 found by using Cauchy's method.

EXAMPLE 2

$$\frac{dw}{dz} = (1 - z - w)^{-1}, \qquad w(0) = 0.$$

Now we treat this example by Lindelöf's method. Here, $M(a,b) = (1-a-b)^{-1}$. According to (20), we have

$$r = \min\left[a, \frac{b}{M(a,b)}\right].$$

Here, in the absence of further information, we put

$$a = \frac{b}{M(a,b)} = b(1 - a - b)$$

or

$$a = \frac{b - b^2}{1 + b}.$$

Series Solutions

The maximum of a as a function of b is attained at $b = 2^{1/2} - 1$ and is equal to $3 - (2)(2^{1/2}) = 0.1716$. This is the maximum r (see (20) and (25)) as obtained by the Lindelöf method and is considerably underestimated in comparison to the actual value 0.30685.

EXAMPLE 3 Thomas–Fermi equation.

$$y''(x) = x^{-1/2} y(x)^{3/2},$$
$$y(0) = 0, \qquad y'(0) = 1. \tag{27}$$

Substituting

$$y(x) = a_1 x + a_2 x^2 + a_3 x^3 + a_4 x^4 + a_5 x^5 + \cdots,$$
$$y''(x) = 2a_2 + 6a_3 x + 12a_4 x^2 + 20a_5 x^3 + \cdots,$$

into the equation we have

$$x[2a_2 + 6a_3 x + 12a_4 x^2 + \cdots]^2 = [a_1 x + a_2 x^2 + a_3 x^3 + \cdots]^3.$$

Equating the coefficients of different powers of x, and using the initial conditions, we have

$a_1 = 1, \qquad a_4 = 0,$
$a_2 = 0, \qquad a_5 = 1/80,$
$a_3 = 1/6, \qquad$ etc.

so

$$y = x + \frac{1}{6} x^3 + \frac{1}{80} x^5 + \cdots.$$

In fact, all even-power terms are absent.

To construct the majorant, we first set

$$y = x(1 + v)$$

in (27) so that the DE for v is

$$v'' + \frac{2}{x} v' = (1 + v)^{3/2}. \tag{28}$$

We use a slight variation of the Lindelöf method. The majorant equation for (28) is taken to be

$$V'' + \frac{2}{x} V' = (1 - V)^{-3/2}, \qquad \text{with } V(0) = V'(0) = 0. \tag{29}$$

The initial conditions here are the same as for $v = y/x - 1$. The coefficients for the v series are possibly all nonnegative; those for the V series certainly have this property. Let V_n denote the nth partial sum of the series

$$V = \frac{1}{6}x^2 + \frac{1}{80}x^4 \cdots, \tag{30}$$

so $V_2 = (1/6)x^2$.

Suppose there is a function $r(\alpha) > 0$ for a given α, $0 < \alpha < 1$, and for an integer m such that

$$V_m(x) < \alpha, \qquad 0 < x < r(\alpha). \tag{31}$$

Then by (29),

$$V''_{m+1}(x) + \frac{2}{x} V'_{m+1}(x) < [1 - V_m(x)]^{-3/2} < (1 - \alpha)^{-3/2}. \tag{32}$$

This follows (cf. Section 3.10) from substituting the expansion for V_m, etc., in the equation obtained by replacing the inequality sign in (32): all coefficients are positive, and all terms on the left are canceled by some terms on the right (there are still some positive terms on the right). Integrating (32) and using the initial conditions in (29), we have

$$V'_{m+1}(x) < \frac{1}{3}(1-\alpha)^{-3/2} x$$

and

$$V_{m+1}(x) < \frac{1}{6}(1-\alpha)^{-3/2} x^2.$$

Then $V_{m+1}(x)$ is at most α if

$$x < [6\alpha(1-\alpha)^{3/2}]^{1/2} \equiv r(\alpha). \tag{33}$$

For such values of x, we have $V_2(x) = (1/6)x^2 < \alpha$ since

$$\frac{1}{6}[6\alpha(1-\alpha)^{3/2}] = \alpha(1-\alpha)^{3/2} < \alpha \qquad \text{for } 0 < \alpha \leq 1.$$

By induction (31) holds for all m. To get the optimal value of r in (33), we note that $r(\alpha)$ is a maximum at $\alpha = 2/5$, so the radius of convergence of (29) is at least

$$r = \left[6 \left(\frac{2}{5} \right) \left(\frac{3}{5} \right)^{3/2} \right]^{1/2} = \left(\frac{12}{5} \right)^{1/2} \left(\frac{3}{5} \right)^{3/4}.$$

3.11 SERIES SOLUTION OF BOUNDARY VALUE PROBLEMS

We discuss boundary value problems over an infinite domain. Typically, the steps involved are selecting a suitable formal power series satisfying the given DE, finding its radius of convergence if it is convergent, or demonstrating that it is asymptotic. The given boundary conditions are then imposed; the resulting equations, which may be transcendental in nature, are solved numerically. This approach assumes that there are no singularities in the domain of the solution. If singularities exist, then one has to find the location of the singularities as a part of the solution.

EXAMPLE 1 Consider the class of DE

$$F''' + cFF'' + mF'^2 = 0, \qquad ' \equiv \frac{d}{dx}, \tag{1}$$

where $c \neq 0$, and m are constants, with boundary conditions

$$F(0) = 0, \quad F''(0) = -1, \quad F'(\infty) = 0. \tag{2}$$

System (1)–(2) describes many problems in boundary layer theory. We shall subsequently consider a different set of boundary conditions in lieu of (2). Following Kravchenko and Yablonskii (1965), we give a representation of the solution of the boundary value problem (1) and (2) in the form of a uniformly and absolutely convergent series on the half-axis $x > -\epsilon$, where ϵ is some positive number.

A formal series solution is sought in the form

$$\begin{aligned} F(x) &= \frac{\gamma}{c} + \gamma \sum_{i=1}^{\infty} b_i a^i e^{-i\gamma x} \\ &= \frac{\gamma}{c} + \gamma \sum_{i=1}^{\infty} b_i (ae^{-\gamma x})^i. \end{aligned} \tag{3}$$

This series is sometimes called a *Dirichlet* or *Picard* series. The form (3) satisfies $F'(\infty) = 0$ and gives the condition $F(\infty) = \gamma/c$. There are two arbitrary constants γ and a in (3) which will be chosen to satisfy the two conditions at $x = 0$. Because of the exponential form of (3), its terms rapidly decrease in magnitude as i increases. We easily check that

$$F'(x) = -\gamma^2 \sum_{i=1}^{\infty} i b_i a^i e^{-i\gamma x},$$

$$F''(x) = \gamma^3 \sum_{i=1}^{\infty} i^2 b_i a^i e^{-i\gamma x},$$

$$F'''(x) = -\gamma^4 \sum_{i=1}^{\infty} i^3 b_i a^i e^{-i\gamma x}. \tag{4}$$

Substituting (4) into (1), we have

$$-\gamma^4 \sum_{i=1}^{\infty} i^3 b_i a^i e^{-i\gamma x} + \gamma^4 \sum_{i=1}^{\infty} i^2 b_i a^i e^{-i\gamma x} + c\gamma^4 \sum_{i=2}^{\infty} \sum_{l=1}^{i-1} l^2 b_{i-l} b_l a^i e^{-i\gamma x}$$

$$+ m\gamma^4 \sum_{i=2}^{\infty} \left(\sum_{l=1}^{i-1} l(i-l) b_{i-l} b_l a^i e^{-i\gamma x} \right) = 0.$$

The factor $\gamma^4 \neq 0$ can be canceled. We therefore have

$$-\sum_{i=1}^{\infty} i^3 b_i a^i e^{-i\gamma x} + \sum_{i=1}^{\infty} i^2 b_i a^i e^{-i\gamma x} + c \sum_{i=2}^{\infty} \left(\sum_{l=1}^{i=1} l^2 b_{i-l} b_l a^i e^{-\gamma x} \right)$$

$$+ m \sum_{i=2}^{\infty} \left(\sum_{l=1}^{i-1} l(i-l) b_l b_{i-l} a^i e^{-i\gamma x} \right) = 0. \tag{5}$$

For $i = 1$, we have

$$-b_1 a + b_1 a = 0,$$

which is identically satisfied. Since a is to be determined by the boundary conditions, we may choose $|b_1| = 1$ for convenience. For $i \geq 2$, equation (5) yields

$$\sum_{i=2}^{\infty} \left\{ -i^2(i-1)b_i + \sum_{l=1}^{i-1} [cl^2 + ml(i-l)] b_l b_{i-l} \right\} = 0,$$

$$i = 2, 3, \ldots, \tag{6}$$

or the recursion formula

$$b_i = \frac{1}{i^2(i-1)} \sum_{l=1}^{i-1} [cl^2 + m(i-l)l] b_l b_{i-l}, \quad i = 2, 3, \ldots. \tag{7}$$

If we choose $|a| < 1$ and show that $|b_i| \leq 1$, $i \geq 2$, then the series (3) converges uniformly and absolutely for any $\gamma > 0$ and $x > -\epsilon$, where $\epsilon = -(\ln|a|/\gamma + \delta) > 0$, and $\delta > 0$ is a sufficiently small number depending on

Series Solutions

a and γ. The latter relation follows from requiring that

$$|ae^{+\gamma\epsilon}| = e^{-\gamma\delta} \quad \text{or} \quad \epsilon = -\left(\frac{\ln|a|}{\gamma} + \delta\right).$$

Thus terms in (3) decrease for all i and $x \geq -\epsilon$.

Now, the requirement that $|b_i| \leq 1$, $i = 2, 3, \ldots$, imposes certain conditions on the coefficients c and m in (1). We have already chosen $|b_1| = 1$ so, from (7),

$$|b_2| = \frac{1}{4}|c + m|. \tag{8}$$

We therefore require that

$$|b_2| = \left|\frac{1}{4}(c + m)\right| \leq 1. \tag{9}$$

Again from (7), we have

$$b_3 = \frac{1}{18}(5c + 4m),$$

so the second condition is

$$|b_3| = \left|\frac{1}{18}(5c + 4m)\right| \leq 1$$

or

$$\frac{1}{18}|4(c + m) + c| \leq 1. \tag{10}$$

In view of (9), condition (10) becomes

$$|c| \leq 2. \tag{11}$$

From (9) and (11) it follows that the coefficients c and m must satisfy the conditions $|c| \leq 2$ and $|c + m| \leq 4$. If these conditions are violated, the series (3) may not be convergent and therefore the solution of (1)–(2) may not exist on the entire positive real line, signaling the existence of singularities between 0 and ∞. We do not consider this case here.

We now prove by induction that $|b_i| \leq 1$ for all i. Let this be true for $i = n$. We therefore have

$$|b_{n+1}| \leq \frac{1}{n(n+1)^2}\left|\sum_{l=1}^{n}[cl^2 + ml(n+1-l)]\right|$$

$$= \frac{1}{n(n+1)^2}\left|(c-m)\sum_{l=1}^{n}l^2 + m(n+1)\sum_{l=1}^{n}l\right|$$

$$= \frac{1}{n(n+1)^2} \left| \frac{1}{6} n(n+1)(2n+1)(c-m) + (n+1)m\frac{n(n+1)}{2} \right|$$

$$= \frac{1}{6(n+1)} |(c+m)(n+2) + c(n-1)|$$

$$\leq \frac{1}{6(n+1)} [4(n+2) + 2(n-1)] = 1, \qquad (12)$$

using (9) and (11).

Hence, for $|c + m| \leq 4$ and $|c| \leq 2$, $|b_i| \leq 1$, $i = 1, 2, \ldots$. Therefore, we obtain a two-parameter (γ and a) family of solutions of (1) defined on the half-axis $x > -\epsilon$ by the series (3) and satisfying the condition $\lim_{x \to \infty} F'(x) = 0$.

The other conditions in (2) require

$$F(0) = \frac{\gamma}{c} + \gamma \sum_{i=1}^{\infty} b_3 \dot{a} = 0, \qquad (13)$$

$$F''(0) = \gamma^3 \sum_{i=1}^{\infty} i^2 b_i a^i = -1. \qquad (14)$$

These are two uncoupled transcendental equations for γ and a. If equation (13) (after canceling γ) is solved for a, with the assumption $|a| < 1$, then (14) at once provides γ. We have yet to determine the signs of the b_i's. We may rewrite (7) as

$$b_i = \frac{1}{2i^2(i-1)} \sum_{l=1}^{i-1} [ci^2 - 2(c-m)il + 2(c-m)l^2] b_l b_{i-l}$$

$$= \frac{1}{2i^2(i-1)} \sum_{l=1}^{i-1} l^2 [c\frac{i^2}{l^2} - 2(c-m)\frac{i}{l} + 2(c-m)] b_l b_{i-l}. \qquad (15)$$

The coefficient of $b_l b_{i-l}$ in the summation, namely, $c\lambda^2 - 2(c-m)\lambda + 2(c-m)$ with $\lambda = i/l$, retains the same sign as c provided that

$$4(c-m)^2 - 8c(c-m) < 0$$

or

$$4(m^2 - c^2) < 0. \qquad (16)$$

Two cases arise:

i. $c > 0$ and $b_1 = 1$, $b_i > 0$ ($i = 2, 3, \ldots$),
ii. $c < 0$, $b_1 = -1$, $b_i < 0$ ($i = 2, 3, \ldots$).

Series Solutions

Now, it is clear from equation (13) that, since c and b_i have the same sign, it can have only a negative root $a = a_0$, say. Thus, the whole problem reduces to finding a negative root $a = a_0$ of the transcendental equation (13) and, hence, evaluating γ from (14). Then the series (3) provides a unique solution of the boundary value problem (1)–(2) over the entire half-line $x \geq -\epsilon$, $\epsilon > 0$, as discussed earlier. The conditions that the parameters occurring in the problem must satisfy are (9), (11), and (16).

3.12 ASYMPTOTIC SERIES SOLUTION OF BOUNDARY VALUE PROBLEMS

We shall again consider boundary value problems for the boundary layer equations dealt with in Section 3.11. However, now we shall transform the basic equation, which is autonomous third order, to a second-order one whose series (solution) turns out to be asymptotic rather than convergent. For a thorough discussion of asymptotic series, see Copson (1967) and Bender and Orszag (1978). Here we content ourselves with a few general remarks regarding asymptotic series. For a convergent series, the convergence is an intrinsic property of the expansion coefficients, and we may prove the convergence of a series without knowing the function to which it converges. By contrast, asymptoticity is a relative property of the expansion coefficients and the function $f(x)$ to which the series is asymptotic. Thus, to prove that a power series is asymptotic to $f(x)$, one must consider both $f(x)$ and the expansion coefficients. Actually, all power series are asymptotic since they are asymptotic to some continuous function $f(x)$ as $x \to x_0$, say; the Taylor series is a special case of the asymptotic series.

Now we mention a few properties of asymptotic series. There is only one asymptotic power series for each function; specifically, if a function $f(x)$ can be expanded as $f(x) \sim \sum_{n=0}^{\infty} a_n (x - x_0)^n$, $x \to x_0$, then the expansion coefficients are unique. Arithmetic operations on asymptotic series can be performed term by term. The series $\sum_{n=0}^{\infty} a_n (x - x_0)^n \sim \sum_{n=0}^{\infty} b_n (x - x_0)^n$ only if both of them are asymptotic to the same class of functions as $x \to x_0$. In such a case, $a_n = b_n$ for all n. An asymptotic series can be integrated term by term if the function $f(x) \sim \sum_{n=0}^{\infty} a_n (x - x_0)^n$ is integrable.

Asymptotic series cannot be differentiated termwise without additional restrictions. One statement regarding the derivative of an asymptotic series is the following: Suppose $f'(x)$ exists, is integrable, and $f(x) \sim \sum_{n=0}^{\infty} a_n (x - x_0)^n$, $x \to x_0$. Then it follows that

$$f'(x) \sim \sum_{n=1}^{\infty} n a_n (x - x_0)^{n-1}.$$

Thus, formal procedures for solutions to differential equations in terms of asymptotic expansions require justification. However, once it is known that a function $f(x)$ is expandable in an asymptotic power series, the differential equation itself ensures that derivatives of $f(x)$ can also be expanded. It remains only to show that the solutions have asymptotic series expansions. We again consider the DE

$$F''' + FF'' + mF'^2 = 0, \tag{1}$$

where m is a parameter, with two different sets of boundary conditions arising from different physical contexts (Merkin, 1984):

$$F = 0, F' = 1, \quad \text{at } x = 0; \quad F'(\infty) = 0. \tag{2}$$
$$F = 0, F'' = -1 \quad \text{at } x = 0; \quad F'(\infty) = 0. \tag{3}$$

First we consider the problem (1)–(2). From (2) it follows that $F \to C$ as $x \to \infty$, where C is some constant to be determined. The method of solution is to transform DE (1) into one which has $\phi = C - F$ as the independent variable and $p = -dF/dx = d\phi/dx$ as the dependent variable. The order of (1) thus decreases by 1. The function p is then expanded as a power series in ϕ. The application of the boundary conditions at $x = 0$, namely $p = 1$ and $F = 0$ or $\phi = C$, determines C. Then $(d^2F/dx^2)_{x=0}$ is easily found by integrating (1) and using the boundary conditions (2):

$$\left.\frac{d^2F}{dx^2}\right|_0^\infty = (m-1)\int_0^\infty \left(\frac{dF}{dx}\right)^2 dx = -(m-1)\int_C^0 p\, d\phi$$

$$= (m-1)\int_0^C p\, dF. \tag{4}$$

To find C, we proceed as follows. Writing (1) in terms of $\phi = C - F$ and $p = -dF/dx = d\phi/dx$, we have

$$\frac{d}{dx}\left(p\frac{dp}{d\phi}\right) + (\phi - C)p\frac{dp}{d\phi} + mp^2 = 0.$$

Now using the definition $p = d\phi/dx$, we get

$$\frac{d}{d\phi}\left(p\frac{dp}{d\phi}\right) + (\phi - C)\frac{dp}{d\phi} + mp = 0. \tag{5}$$

We expand p in the form

$$p = A_1\phi + A_2\phi^2 + A_3\phi^3 + A_4\phi^4 + A_5\phi^5 + \cdots. \tag{6}$$

Substituting (6) into (5), we have

$$\frac{d}{d\phi}[(A_1\phi + A_2\phi^2 + A_3\phi^3 + A_4\phi^4 + \cdots)$$

Series Solutions

$$\times (A_1 + 2A_2\phi + 3A_3\phi^2 + 4A_4\phi^3 + \cdots)]$$
$$+ (\phi - C)(A_1 + 2A_2\phi + 3A_3\phi^2 + 4A_4\phi^3 + \cdots)$$
$$+ m(A_1\phi + A_2\phi^2 + A_3\phi^3 + \cdots) = 0. \tag{7}$$

Equating the coefficients of different powers of ϕ to zero, we have

$$A_1 = C, \quad A_2 = -\frac{1+m}{4}, \quad A_3 = \frac{1-m^2}{72C},$$

$$A_4 = \frac{(1-m^2)(1+2m)}{576C^2},$$

$$A_5 = \frac{(1-m^2)(11 + 81m + 88m^2)}{86,400C^3}, \quad \text{etc.} \tag{8}$$

Substituting the coefficients A_i from (8) into (6) and imposing the condition that $p = 1$ when $\phi = C$, we obtain

$$C^2 \left[1 - \frac{1+m}{4} + \frac{1-m^2}{72} \right.$$
$$\left. + \frac{(1-m^2)(1+2m)}{576} + \frac{(1-m^2)(11 + 81m + 88m^2)}{86,400} + \cdots \right] = 1. \tag{9}$$

It is a fortunate circumstance of equation (1) that C is explicitly obtainable from (9). For more general equations, one may be required to solve a transcendental equation for C for given m.

Now substituting the series (6) for p in the integral on the right-hand side of (4) (with coefficients A_i given by (8)), we obtain

$$\left(\frac{d^2F}{dx^2} \right)_0 = C^3(m-1) \left[\frac{1}{2} - \frac{1+m}{12} + \frac{1-m^2}{288} + \frac{(1-m^2)(1+2m)}{2880} \right.$$
$$\left. + \frac{(1-m^2)(11 + 18m + 88m^2)}{518,400} + \cdots \right], \tag{10}$$

where C is given by (9). We see that the terms in the series (9) and (10) rapidly decrease, suggesting that good estimates for C and $(d^2F/dx^2)_0$ (the so-called skin friction) can be obtained by taking only a few terms. Table 3.12.1 gives these values for different numbers of terms. The values approach those by exact numerical solution, $C = 1.14277$ and $(d^2F/dx^2)_0 = -0.62755$, with four or five terms. Indeed the error in the series solution changes sign between the fourth and fifth approximations, suggesting the asymptotic nature of the series. This point was examined by considering the coefficient A_6 of the term of $O(\phi^6)$ in (6) for the choice $m = 0$. This was found to be $-(115,200C^4)^{-1}$, which is smaller than A_5 and differs from it

Table 3.12.1 Values of C and $(d^2F/dx^2)_0$ for $\lambda = 0$ Obtained by Increasing the Number of Terms in the Series (9) and (10)

Number of terms	C	$\left(\dfrac{d^2F}{dx^2}\right)_0$
1	1.00000	−0.50000
2	1.15470	−0.64150
3	1.14416	−0.62929
4	1.14286	−0.62766
5	1.14276	−0.62754

Source: From Merkin (1984).

in sign. When this additional term is also included in (9) and (10), we find that $C = 1.14277$ and $F''(0) = -0.62755$.

Values of C and $(d^2F/dx^2)_0$, as computed from (9) and (10) for various values of m, are given in Table 3.12.2. This table also contains the numerical solution of the two-point boundary value problems (1)–(2). We find that the series method gives correct values for $m = 1$ and $m = -1$, and the agreement with the computed values is good for most values of m in the range $-2 \leq m \leq 1.8$. For example, for $m = -2$, the errors in C and $(d^2F/dx^2)_0$ are 0.1% and 0.2%, respectively. This error decreases as m increases from $m = -2$. The agreement, however, deteriorates as m approaches 2. For $m = 1.8$, the errors in C and $(d^2F/dx^2)_0$ are 7% and 19%, respectively; actually, problem (1)–(2) has no solution for $m = 2$, as we now show.

Taking $m = 2$, multiplying (1) by F, and integrating, we get

$$F \frac{d^2F}{dx^2} - \frac{1}{2}\left(\frac{dF}{dx}\right)^2 + F^2 \frac{dF}{dx} = F_0, \tag{11}$$

say. The conditions at $x = \infty$ fix the constant F_0 to be zero, while those at $x = 0$ give $F_0 = -1/2$. Hence the problem has no solution for this choice of m.

For problem (1) and (3), we merely summarize the results, since the procedure for the solution is exactly the same as for (1)–(2). First we note that if $m = -1$, there is an exact solution $F = 1 - e^{-x}$. For $m = 1$, we can again verify that (1) and (3) have no solution. Assuming the series (6) for

Series Solutions

Table 3.12.2 Values of C and $(d^2F/dx^2)_0$ for Various λ Obtained from Series (9) and (10) and Those Obtained Numerically

	Series		Numerically	
	C	$\left(\dfrac{d^2F}{dx^2}\right)_0$	C	$\left(\dfrac{d^2F}{dx^2}\right)_0$
−2.0	0.90648	−1.28461	0.90563	−1.28181
−1.8	0.92249	−1.23150	0.92204	−1.23009
−1.6	0.93974	−1.17693	0.93957	−1.17632
−1.4	0.95832	−1.12046	0.95824	−1.12026
−1.2	0.97835	−1.06164	0.97833	−1.06160
−1.0	1.00000	−1.00000	1.00000	−1.00000
−0.8	1.02348	−0.93502	1.02348	−0.93501
−0.6	1.04908	−0.86610	1.04908	−0.86609
−0.4	1.07715	−0.79253	1.07715	−0.79254
−0.2	1.10816	−0.71341	1.10817	−0.71343
0	1.14276	−0.62754	1.14277	−0.62755
0.2	1.18179	−0.53326	1.18175	−0.53323
0.4	1.22642	−0.42818	1.22629	−0.42808
0.6	1.27827	−0.30869	1.27800	−0.30853
0.8	1.33969	−0.16911	1.33933	−0.16900
1.0	1.41421	0.00000	1.41421	0.00000
1.2	1.50739	0.21539	1.50945	0.21616
1.4	1.62858	0.50944	1.63839	0.51774
1.6	1.79503	0.95274	1.83302	1.00913
1.8	2.04319	1.72912	2.20605	2.14634

Source: From Merkin (1984).

p for equation (1) and proceeding as before, we have

$$(1-m)\int_0^C p\,d\phi = 1, \tag{12}$$

corresponding to (4), and

$$(1-m)C^3\left[\frac{1}{2} - \frac{1+m}{12} + \frac{1-m^2}{288} + \frac{(1-m^2)(1+2m)}{2880}\right.$$

$$+ \frac{(1-m^2)(11+18m+88m^2)}{518{,}400} + \cdots \bigg] = 1, \tag{13}$$

corresponding to (9). Two results follow immediately from (13): first that the case $m = 1$ has no solution, and second that the coefficients in the denominators of the terms in (13) are much larger than those in (9), promising better accuracy. The first derivative at $x = 0$ is

$$\left(\frac{dF}{dx}\right)_0 = C^2 \bigg[1 - \frac{1+m}{4} + \frac{1-m^2}{72} + \frac{(1-m^2)(1+2m)}{576}$$
$$+ \frac{(1-m^2)(11+18m+88m^2)}{86{,}400} + \cdots \bigg]. \tag{14}$$

Table 3.12.3 Values of C and $(dF/dx)_0$ for Various λ as Obtained from (13) and (14) and Those Obtained Numerically

	Series		Numerically	
λ	C	$\left(\dfrac{dF}{dx}\right)_0$	C	$\left(\dfrac{dF}{dx}\right)_0$
−2.0	0.83388	0.84622	0.83369	0.84746
−1.8	0.86063	0.87038	0.86053	0.87105
−1.6	0.89007	0.89708	0.89002	0.89739
−1.4	0.92266	0.92698	0.92264	0.92709
−1.2	0.95903	0.96091	0.95902	0.96093
−1.0	1.00000	1.00000	1.00000	1.00000
−0.8	1.04667	1.04581	1.04666	1.04581
−0.6	1.10057	1.10058	1.10057	1.10058
−0.4	1.16400	1.16768	1.16396	1.16767
−0.2	1.24020	1.25249	1.24020	1.25247
0	1.33478	1.36429	1.33478	1.36427
0.2	1.45734	1.52069	1.45733	1.52074
0.4	1.62715	1.76028	1.62713	1.76056
0.6	1.89138	2.18935	1.89132	2.19012
0.8	2.42259	3.27003	2.42250	3.27151
0.9	3.07906	5.01432	3.07897	5.01608

Source: From Merkin (1984).

Series Solutions

The numerical solution of (1) and (3) as well as that by the series (6) with C and $(dF/dx)_0$ given by (13) and (14) are detailed in Table 3.12.3. In the present case, the agreement is even better than that for problem (1)–(2). The largest percentage error belongs to the case $m = -2$, for which the values of C and $(dF/dx)_0$ are in error by 0.02% and 0.15%, respectively. The agreement is good even when m approaches the critical value 1. This accuracy, as pointed out earlier, is due to the rapid decrease of the coefficients in the series solution for p for the present case.

3.13 COMPUTATION OF RADIUS OF CONVERGENCE OF A SERIES

While it is straightforward to write a Taylor series solution for an initial value problem for a nonlinear DE, the determination of its radius of convergence and the location of singularities pose hardships. The recurrence relations are quite involved, and an analytic evaluation of the radius of convergence is usually not possible. Here, we briefly discuss some numerical approaches to this aspect of the problem; we shall, however, just quote some results and illustrate them with a few examples.

The simplest approach is the evaluation of the radius of convergence by the ratio test (see Example 1 of Section 3.8), namely,

$$R_c = \lim_{n \to \infty} \left| \frac{a_n}{a_{n+1}} \right|. \tag{1}$$

While this is a useful test, it often overestimates the radius of convergence of the series, as we shall show. We treat the series in the complex plane in the following discussion, since sometimes the singularity on the real line is farther from the initial point than it is when measured in the complex plane, and hence the radius of convergence is not correctly found if we restrict ourselves to the real line.

We quote a theorem of Golomb (1943), which gives the order of the singularities if the given function is meromorphic with poles as its only singularities.

THEOREM 3.13.1 If $f(z) = \sum_{n=0}^{\infty} a_n z^n$ is meromorphic and has no singularities on the circle of convergence $|z| = R_c$ except poles, and if $z_1 = \lim_{n \to \infty} (a_n/a_{n+1})$ exists, then f has order

$$m = 1 + \lim_{n \to \infty} \frac{\ln |a_n| + n \ln R_c}{\ln n} \tag{2}$$

on $|z| = R_c$ and z_1 is the only pole of order m on $|z| = R_c$.

Thus, this theorem locates the only pole at $z = z_1$ and determines the order of this pole. The proof of this theorem does not hold for functions which have logarithmic or algebraic branch points (e.g., $\ln z$ or $z^{2/3}$). For the latter, Chang and Corliss (1980) give what they call three-term and five-term tests. These are useful for determining poles, logarithmic branch points, or algebraic branch points, but not essential singularities. Here we content ourselves with the statement of the simpler three-term test and refer the reader to Chang and Corliss (1980) for higher-term tests.

The three-term ratiolike test is contained in the following theorem.

THEOREM 3.13.2 Let $\sum_{i=1}^{\infty} a_i$ be a nonzero series of real numbers such that

$$\lim_{i \to \infty} \left[i \left(\frac{a_{i+1}}{a_i} \right) - (i-1) \left(\frac{a_i}{a_{i-1}} \right) \right] \tag{3}$$

exists and equals L. Then

i. If $|L| < 1$, $\sum a_i$ is absolutely convergent.
ii. If $|L| = 1$, the test fails.
iii. If $|L| > 1$, $\sum a_i$ is divergent.

Under the hypothesis of Theorem 3.13.2, it can be proved that $\lim_{i \to \infty}(a_{i+1}/a_i)$ exists and equals L. Thus, the three-term test is weaker than the usual ratio test (see the limit in (3)).

Table 3.13.1 gives estimates for the 30-term series for $(1-z)^{-\alpha}$, as obtained from ratio and three-term tests and computed with single-precision

Table 3.13.1 Estimates from 30-Term Series for $(1-z)^{-\alpha}$

Order α	Ratio test R_c	Three-term test R_c	Three-term order
10.0	0.7632	1.0000	9.9995
5.5	0.8657	1.0000	5.5002
2.0	0.9667	1.0000	2.0008
1.0	1.0000	1.0000	1.0001
0.333	1.0235	1.0000	0.3333
0.0	1.0357	1.0000	0.0000
−1.0	1.0741	1.0000	−1.0001
−5.5	1.2889	1.0000	−5.5000
−10.0	1.6111	1.0000	−9.9997

Source: From Chang and Corliss (1980).

Series Solutions

arithmetic. The order is obtained by using another expression, which requires considerable explanation; see Chang and Corliss (1980) for further details.

Table 3.13.1 shows that the ratio test underestimates R_c if $\alpha > 1$; it overestimates it if $\alpha < 1$. In the latter case, the algorithm for summation may proceed to sum a divergent series, with disastrous results.

Chang and Corliss (1980) solved an initial value problem for the first Painlevé transcendent,

$$y'' = 6y^2 + x, \quad y(0) = 1, \quad y'(0) = 0$$

(see Section 8.5 for the general Taylor series solution for arbitrary initial conditions). The results are given in Table 3.13.2. Poles at $x = 1.2068$ and $x = -1.256$ are part of a sequence of poles of order 2 on the real axis. The table contains a comparison of the estimates for the radius of convergence R_c and order of the poles given by the three-term test with those by the ratio test. The three-term test for $x_0 = 0$ failed due to the proximity of the second pole at $x = 1.2068$. The value 1.2230 given in Table 3.13.2 is from a five-term test. This is the usual difficulty encountered in estimating the radius of convergence when the test either fails or deteriorates if the two singularities are rather contiguous.

A few additional remarks may be in order. In the Taylor series solution of an IVP for a DE, it is now "standard" (see Chang and Corliss (1980) and references there) to write compiler-like programs which accept the differential equations as input; the Leibnitz rule for differentiating a product is used to enumerate long Taylor series, using recurrence relations for higher derivatives. For many test problems, these compiler-generated

Table 3.13.2 Pole Location and Order for $y'' = 6y + x$, $y(0) = 1$, $y'(0) = 0$

x_0	Three-term test			Ratio test		
	R_c	Pole	Order	R_c	Pole	Order
0.0	1.2230	1.2230	—	1.3636	1.3636	3.1835
0.1	1.0962	1.1962	2.4030	1.0514	1.1514	1.0785
0.2	0.9969	1.1969	2.0033	0.9636	1.1636	1.0837
0.3	0.9037	1.2037	2.0001	0.8736	1.1736	1.0904
0.4	0.8083	1.2083	2.0000	0.7814	1.1813	1.0981

Source: From Chang and Corliss (1980).

programs achieve a high degree of accuracy and have proved competitive with standard methods of solutions of DE.

There are other difficulties associated with series solutions, such as the slow convergence of the series. These can be handled by standard techniques, such as the Shanks transformation or Padé approximation (see Bender and Orszag, 1978, chapter 8), and by continuous analytic continuation when it is deemed necessary (see Davis, 1962, chapter 9).

4
Local and Global Analysis of Nonlinear Differential Equations

4.1 INTRODUCTION

We have seen in Chapter 2 that in spite of a variety of transformations that have been discovered, the exact closed-form solution of a nonlinear DE is a matter of chance. Given a nonlinear DE (arising out of a physical problem, for example), there is no guarantee that it will admit an exact explicit solution. Even when it does, the solution may be so implicit and intricate that it gives little direct information (see Sections 2.11–2.12). It is therefore imperative that we have access to approximate methods, both local and global. We have discussed in Chapter 3 some cases for which the solution is found about an ordinary point; the Taylor series proves very useful and can sometimes be made to provide even global solutions. It is, however, the solution in the neighborhood of a singular point which is more interesting and difficult and provides important clues to the physical problem. This can often be found by local analysis.

There is another aspect to local analysis. The solution is often needed over (practically) an infinite interval. Even in this high-speed computer era, this can be quite expensive. Besides, maintaining the accuracy over an infinite interval is a formidable task. This is because the error in a numerical solution started at $t = t_0$ tends to mount as $t - t_0$ increases, so as t becomes large the accuracy deteriorates: thus, the larger the interval $t - t_0$, the greater the cost of accuracy. Here, a local analysis for large time can prove very useful. Indeed, it can provide a whole family of solutions with the required number of arbitrary constants in the coefficients of the expansion. Thus, we have valuable information for a family of solutions.

This chapter deals with the local behavior of solutions about singular points and points at infinity and, following the work of Levinson (1970), will provide in the latter case even the order of error term in the asymptotic expansion. This will be accomplished by formulating an integral equation equivalent of the given DE and proving the existence and uniqueness of its solution. For analysis about the singularities, we shall follow the work of Ablowitz, Ramani, and Segur (1980), which has proved particularly useful in unraveling singularities of equations of Painlevé type and those which describe chaotic behavior. It has proved to be a simple and useful alternative to the α-method of Painlevé and his co-workers (see Ince, 1956). In fact the Painlevé property of nonlinear DE has proved invaluable in the categorization of nonlinear dynamical systems with respect to their singular structure and integrability. We may note that special exact (usually singular) solutions of nonlinear DE are not isolated mathematical curiosities. They are often embedded in larger families of solutions and constitute their asymptotics in some limit.

Besides the local solutions, we shall deal with global solutions of nonlinear DE, usually over an infinite interval. Here, again, the integral equation formulation of the nonlinear DE will be found useful, and iterative methods with the appropriate linear solution as the zeroth iterate give quick and quite accurate results. We shall first discuss a few simple examples to motivate the subsequent discussion.

EXAMPLE 1 (Hille, 1969).

$$\frac{dy}{dx} = x^3 - y^3. \tag{1}$$

We study a single-valued solution of this equation, which tends to infinity as $x \to +0$ and as $x \to +\infty$, and show that over this semiline the solution is real positive, has a single positive minimum, and becomes infinite with x.

Assuming $y = Ax^m$, we have

$$Amx^{m-1} = x^3 - A^3 x^{3m}.$$

Considering the balancing of terms, we have

i. $m - 1 = 3$ or $m = 4$. In this case the term x^{3m} is ignored for a solution which tends to zero as x tends to zero.
ii. $m - 1 = 3m$ or $m = -1/2$. This gives a branch point.
iii. $3m = 3$ or $m = 1$ and $A = 1$. This gives the first term of the series in x, which goes to infinity as $x \to \infty$. We shall show that the descending

Local and Global Analysis

series starting with this term satisfies the appropriate condition at the origin, namely, $y(0) = \infty$. Assuming that

$$y = x + a_0 + \frac{a_1}{x} + \frac{a_2}{x^2} + \frac{a_3}{x^3} + \frac{a_4}{x^4} + \frac{a_5}{x^5} + \cdots$$

and substituting it in (1), we have

$$1 - \frac{a_1}{x^2} - \frac{2a_2}{x^3} - \frac{3a_3}{x^4} - \frac{4a_4}{x^5} - \frac{5a_5}{x^6} - \cdots$$

$$= x^3 - x^3 \left\{ 1 + \frac{a_0}{x} + \frac{a_1}{x^2} + \frac{a_2}{x^3} + \frac{a_3}{x^4} + \frac{a_4}{x^5} + \cdots \right\}^3$$

$$= x^3 - x^3 \left\{ 1 + \frac{3a_0}{x} + \frac{3a_1}{x^2} + \frac{3a_2}{x^3} + \frac{3a_3}{x^4} + \frac{3a_4}{x^5} \right.$$

$$+ \cdots + \frac{3 \times 2}{2 \times 1} \left(\frac{a_0}{x} + \frac{a_1}{x^2} + \frac{a_2}{x^3} + \cdots \right)^2$$

$$\left. + \frac{3 \times 2 \times 1}{3 \times 2 \times 1} \left(\frac{a_0}{x} + \frac{a_1}{x^2} + \frac{a_2}{x^3} + \cdots \right)^3 + \cdots \right\}.$$

Equating the coefficients of various powers of x on both sides, we find that $a_0 = a_1 = 0$, $a_2 = -1/3$, $a_3 = 0$, $a_4 = 0$, $a_5 = -2/9$, etc., so the solution is

$$y = x - \frac{1}{3}x^{-2} + O(x^{-5}) \quad \text{as } x \longrightarrow \infty.$$

At the minimum of this solution, $y'(x_m) = 0$, so $x_{\min}^3 = y_{\min}^3$. At this point, $y''(x) = 3x^2 - 3y^2 y' = 3x^2 > 0$. Therefore, this point is a minimum. Moreover, this minimum is positive, since otherwise at this point,

$$y'_{\min} = x_{\min}^3 + |y_{\min}|^3 > 0,$$

so $y'_{\min} \neq 0$. Thus, we have shown that the solution of (1) goes to infinity as $x \to +0$ and as $x \to +\infty$, and has a positive minimum at some finite point.

EXAMPLE 2

$$\frac{dy}{dx} = x^2 + y^2. \tag{2}$$

We study the behavior of this Riccati equation as $x \to +\infty$. First we inquire what kind of singularities this equation has. Substituting $y(x) \sim A(x-a)^m$, where a is the location of the singularity, we find that the terms dy/dx and y^2 balance and give $A = -1$ and $m = -1$. Thus, the solution has simple poles, and we may find a Laurent series about any such pole, which is convergent in a disk not containing any other singularity. Here, however, it is easier to find an approximate solution of this equation for large x and discover that it has an infinite number of first-order poles with an accumulation point at $x = \infty$.

The homogeneity of the right side of (2) suggests the substitution

$$y(x) = xu(x)$$

so (2) becomes

$$u' = (1 + u^2)x - \frac{u}{x}.$$

The term u/x is much smaller than $(1 + u^2)x$, since, for large x,

$$u \leq 1 + u^2$$

for all u and

$$x^{-1} \ll x^2,$$

so we have an approximate equation for u:

$$u' \sim (1 + u^2)x.$$

The variables are separable for this equation, and the solution is

$$y = xu(x) \sim x \tan \frac{x^2}{2}. \tag{3}$$

This suggests that, for large x, the solution of (2) behaves like a tangent function and has an infinite number of first-order poles with a point of accumulation at $x = \infty$. It is a simple matter to estimate the distance between two consecutive poles, say at x and $x + \Delta x$. The solution (3) in the $x \tan(x^2/2)$ form suggests that the distance between two consecutive poles is π, so that

$$\frac{1}{2}(x + \Delta x)^2 - \frac{1}{2}x^2 \sim \pi.$$

Expanding by the binomial theorem and assuming Δx to be small yield

$$\Delta x \sim \frac{\pi}{x}.$$

(see Bender and Orszag, 1978 for a similar discussion for a related equation, $y' = x + y^2$, and comparison with the exact numerical solution).

EXAMPLE 3

$$y'' + cy' + y(1 - y) = 0. \tag{4}$$

Local and Global Analysis

This equation describes the steady (traveling wave) solution of the Fisher's equation

$$u_t = u_{xx} + u(1-u) \tag{4a}$$

(see Ablowitz and Zeppetella, 1979); equation (4) follows immediately from (4a) if we substitute $u = y(x-ct)$, where c is the speed of the traveling wave.

We study the movable singularities of (4) by assuming $y \sim A(x-a)^{-m}$, $m > 0$. The two most dominant terms, y'' and y^2, must balance. We have

$$A(-m)(-m-1)(x-a)^{-m-2} = A^2(x-a)^{-2m}.$$

This requires that $m = 2$ and $A = 6$ for a nontrivial dominant behavior. We now check whether a Laurent series solution of (4) with this leading behavior exists. Let

$$y = \frac{6}{x^2} + \frac{a_{-1}}{x} + a_0 + a_1 x + a_2 x^2 + a_3 x^3 + a_4 x^4 + \cdots. \tag{5}$$

(We henceforth set $a = 0$ for convenience.) Substituting (5) into (4) and equating coefficients of $x^{-4}, x^{-3}, x^{-2}, \ldots, x^1, x^2$, etc., to zero, we get

$$0 = 0,$$

$$a_{-1} = -\frac{6}{5}c,$$

$$a_0 = \frac{1}{2} - \frac{c^2}{50},$$

$$a_1 = -\frac{c^3}{250},$$

$$a_2 = -\frac{7}{5000}c^4 + \frac{1}{40},$$

$$a_3 = -\frac{79}{75,000}c^5 + \frac{11c}{600},$$

$$(0)(a_4) + \frac{100}{40}\left(\frac{c}{5}\right)^2 - \frac{720}{8}\left(\frac{c}{5}\right)^6 = 0. \tag{6}$$

The last of equations (6) shows that Laurent's expansion is valid only if either $c = 0$ or $c = 5/\sqrt{6}$. In either case, the coefficient a_4 remains undetermined. For other values of c, it would be necessary to introduce logarithmic terms in the expansion, thus obtaining a branch point rather than a pole. The case $c = 0$ is completely integrable, and we can express the solution of (4) in terms of elliptic functions. We also note that we need consider only positive values of c, since changing c to $-c$ and x to $-x$ leaves (4) invariant.

The solutions of biological interest require the boundary conditions

$$y(-\infty) = 1, \qquad y(+\infty) = 0. \tag{7}$$

We seek the solution of (4) and (7) for $c = 5/\sqrt{6}$. We first note that the linearized form of (4) is

$$y'' + cy' + y = 0, \tag{8}$$

with the general (real) solution

$$y = a\exp(\lambda_1 x) + b\exp(\lambda_2 x), \tag{9}$$

where

$$\lambda_1 = \frac{-c + \sqrt{c^2 - 4}}{2}, \qquad \lambda_2 = \frac{-c - \sqrt{c^2 - 4}}{2}, \quad \text{and} \quad c > 2. \tag{10}$$

Here a and b are arbitrary constants. The roots λ_1 and λ_2 are negative, and therefore, $(0, 0)$ is a stable node in the (y, y') phase plane (see Chapter 6). This suggests a solution of (4) in the form

$$y = \sum_{\substack{m,n \geq 0 \\ m+n \geq 1}} a_{mn} \exp(m\lambda_1 + n\lambda_2)x. \tag{11}$$

For $c = 5/\sqrt{6}$, $\lambda_1 = -2/\sqrt{6}$ and $\lambda_2 = -3/\sqrt{6}$, and for this special choice of λ_1 and λ_2 the expansion (11) simplifies and starts with a minimum index $n = 2$:

$$y = \sum_{n=2}^{\infty} a_n \exp\left(-\frac{nx}{\sqrt{6}}\right). \tag{12}$$

Substituting (12) into (4) and equating the coefficients of $\exp(-nx/\sqrt{6})$ to zero, we obtain the recurrence relation

$$\frac{(n-2)(n-3)}{6} a_n = \sum_{j=2}^{n-2} a_j a_{n-j} \qquad (n \geq 4). \tag{13}$$

We note that n assumes values greater than or equal to 4 in (13); the coefficients a_2 and a_3 are arbitrary. The physical problem requires that $u > 0$ for all x, so we take $a_2 = a^2$. Then, on choosing $a_3 = a^3 b_3$, (13) shows that $a_n = a^n b_n$, where $b_2 = 1$, so (13) assumes the form

$$b_n = \frac{6\sum_{j=2}^{n-2} b_j b_{n-j}}{(n-2)(n-3)} \qquad (n \geq 4). \tag{14}$$

The solution (12) now becomes

$$y = \sum_{n=2}^{\infty} a^n b_n \exp\left(-\frac{nx}{\sqrt{6}}\right) = \sum_{n=2}^{\infty} b_n \exp\left\{\frac{n(-x + \sqrt{6}\ln a)}{\sqrt{6}}\right\}. \tag{15}$$

We see, therefore, that a is only a translation parameter. We have the solution (15) of (4) involving two parameters, b_3 and a. We consider the special case $a = 1$. Different choices for b_3 give different trajectories moving into the node (0, 0). The particular choice $b_3 = -2$ gives $b_4 = 3$, and, by induction, (14) gives $b_n = (-1)^n(n-1)$. This special solution therefore has the form

$$y = \sum_{n=2}^{\infty}(-1)^n(n-1)\exp\left(\frac{-nx}{\sqrt{6}}\right) = \frac{1}{(1+\exp(x/\sqrt{6}))^2}. \tag{16}$$

This traveling wave solution of Fisher's equation describes a profile which assumes the value 1 at $x = -\infty$ and decreases monotonically to 0 at $x = +\infty$.

If we instead take $b_3 = 2$, we find that $b_4 = -1$, and the solution corresponding to (16) may again be found. If we do not choose $a = 1$ in (15), the latter would describe a one-parameter family of solutions

$$y(x) = \frac{1}{[1 - r\exp(x/\sqrt{6})]^2}, \tag{17}$$

satisfying the conditions $y(-\infty) = 1$ and $y(+\infty) = 0$. The solutions for $r > 0$ all blow up at some finite real x. The exponential series (11) for equations with equilibrium points works quite generally for autonomous systems of ordinary differential equations (Lefschetz, 1963; see also Section 3.11).

We have shown that for $c = 5/\sqrt{6}$, equation (4) has a second-order pole. Indeed, we can in this special case transform (4) such that it is solvable in terms of elliptical functions. Writing

$$y(x) = e^{\lambda_1 x} w(z), \qquad z = \alpha \exp[-(c + 2\lambda_1)x], \tag{18}$$

we can change (4) to

$$w'' = \frac{6}{\alpha^2} w^2 e^{(5\lambda_1 + 2c)z}, \tag{19}$$

which, for $c = 5/\sqrt{6}$, $\lambda_1 = -2/\sqrt{6}$, and $\alpha = 1$, simplifies to

$$w'' = 6w^2. \tag{20}$$

The solution of (20) is $w = \wp(z-k; 0, g_3)$, where $\wp(z; g_2, g_3)$ is the Weierstrass $\wp$ function with invariants g_2 and g_3 (Abramowitz and Stegun, 1964). Here, k and g_3 are arbitrary constants. In terms of original variables, we have

$$y(x) = \exp\left(\frac{-2x}{\sqrt{6}}\right)\wp\left(\exp\left(\frac{-x}{\sqrt{6}}\right) - k; 0, g_3\right).$$

For the special case $g_3 = 0$, $\wp(s; 0, 0) = s^{-2}$, and

$$z = \alpha(\exp[-(c + 2\lambda_1)x]) = \exp\left(-\frac{x}{\sqrt{6}}\right),$$

$$y(x) = \frac{\exp(-2x/\sqrt{6})}{[\exp(-x/\sqrt{6}) - k]^2}$$
$$= [1 - k \exp(x/\sqrt{6})]^{-2},$$

which is the same as (17). For $g_3 \neq 0$, the $\wp$ function is doubly periodic with an infinite number of poles on the real axis. Therefore, all solutions other than (16) blow up at finite real values of x.

4.2 LOCAL ANALYSIS ABOUT SINGULAR POINTS—THE PAINLEVÉ PROPERTY

We discussed in Chapter 3 the types of singularities that the nonlinear DE can display. Here we give a simple algorithm due to Ablowitz, Ramani, and Segur (1980) (ARS for short) designed to determine whether a given nonlinear DE has a Painlevé property, that is, whether its only movable singularities (see Section 3.4 for definitions) in the complex plane are poles. This was the criterion adopted by Painlevé to categorize all nonlinear DE of second order—hence the name. By a curious turn of events, this property reappears in the study of nonlinear evolutionary PDE. The latter class of PDE admits an exact linearization in the form of an integral equation of Gelfand–Levitan–Marchenko type through the use of an inverse scattering transform (IST) (Whitham, 1974). It is also true that precisely this class of PDE reduces through similarity transformations to nonlinear ODE enjoying the Painlevé property, either directly or by the use of some further transformations. Indeed, Ablowitz, Ramani, and Segur (1978) conjectured that every nonlinear ODE obtained by an exact reduction of a nonlinear PDE of IST class is of P type. A PDE, as we pointed out earlier, belongs to IST if nontrivial solutions of it can be found by way of solving a linear integral equation of Gelfand–Levitan–Marchenko form. No rigorous proof of this conjecture is yet available, but it seems to cover a large number of evolutionary nonlinear PDE. Conversely, Painlevé equations have now been shown to reduce to linear integral equations.

Thus, the Painlevé property can be used to test whether a given PDE may be exactly solved by IST. Before we give the algorithm for testing an ODE for the Painlevé property and hence finding its solution in the neighborhood of a pole, requiring the study of "resonances" in the solution, we give some examples of reduction of PDE to ODE of Painlevé (P) type.

The modified Korteweg–deVries equation

$$u_t - 6u^2 u_x + u_{xxx} = 0$$

Local and Global Analysis

reduces through the similarity transformation

$$u(x,t) = (3t)^{-1/3} w(z), \qquad z = \frac{x}{(3t)^{1/3}}$$

to

$$w''' - 6w^2 w' - (zw)' = 0.$$

This equation can be integrated at once to give

$$w'' = 2w^3 + zw + \alpha, \qquad (1)$$

where α is constant of integration. This is the second of the six Painlevé equations.

The Sine–Gordon equation

$$u_{xt} = \sin u$$

admits the self-similar form

$$u(x,t) = f(z), \qquad z = xt.$$

We therefore have

$$zf'' + f' = \sin f = \frac{e^{if} - e^{-if}}{2i}.$$

If we set $w(z) = \exp(if)$, then $w(z)$ satisfies the equation

$$w'' = \frac{1}{w}(w')^2 - \frac{1}{z} w' + \frac{1}{2z}(w^2 - 1). \qquad (2)$$

This is again of P type. We shall discuss the Painlevé equations in great detail in Chapter 8.

Singular Point Analysis

A necessary condition for an ODE to be of P type is that it have no movable branch points—algebraic or logarithmic. The algorithm to test this property is due to Ablowitz, et al. (1980), which is a little simpler than the α method of Painlevé (see Ince, 1956), and makes two assumptions:

i. The nth order system of ODE has the form

$$\frac{d}{dz} w_j = F_j(z;, w_1, w_2, \ldots, w_n), \qquad j = 1, 2, \ldots, n, \qquad (3)$$

where each F_j is analytic in z and rational in other arguments. The equivalent case of (3) is the nth-order ODE

$$\frac{d^n w}{dz^n} = F\left(z; w, w' \ldots, \frac{d^{n-1} w}{dz^{n-1}}\right), \qquad (4)$$

where F is analytic in z and rational in its other arguments.
ii. The dominant behavior of the solution in a sufficiently small neighborhood of the (movable) singularity is algebraic, namely

$$w_j \sim \alpha_j(z-z_0)^{p_j} \qquad \text{as } z \longrightarrow z_0.$$

This assumption of the leading behavior does not exclude logarithmic branch points appearing later in the expansion, but it does exclude branch points for which the dominant behavior is logarithmic.

We note that an ODE without movable branch points might still admit movable essential singularities. Moreover, the algorithm that follows does not identify essential singularities and, therefore, it provides only necessary conditions for an ODE to be of P type. We shall illustrate these points with the help of several examples.

There are three main steps in the algorithm.

1. Assume the solution in the form

$$w \sim \alpha(z-z_0)^p, \tag{5}$$

where $\operatorname{Re} p < 0$ and z_0 is arbitrary, and substitute it into the given DE.

Different sets of terms may balance to provide the values of p. When the terms balance, they in general determine the value of the coefficient α as well. In exceptional cases, α may remain undetermined (arbitrary) at this stage of balancing. For each choice of p, the terms that balance in the DE are called the *leading terms*. Thus, we may have several expansions possible corresponding to different values of p and α. If any p is a fraction, we have a branch point and the equation does not possess the P property. If all p are negative integers, we may proceed to check whether the DE, for each p, has a Laurent series expansion about z_0. If it requires logarithmic terms at some stage to make terms balance appropriately, we have a (logarithmic) branch point and the DE does not possess the P property. If, however, for each p, (5) represents the first term in a Laurent series, valid in a deleted neighborhood of a movable pole, then a solution of (4) is

$$w(z) = (z-z_0)^p \sum_{j=0}^{\infty} a_j(z-z_0)^j, \qquad 0 < |z-z_0| < R, \tag{6}$$

where z_0 is an arbitrary constant. If, besides z_0, $n-1$ of the coefficients a_j in (6) are arbitrary, then there are n constants of integration and (6) is the general solution in the deleted neighborhood. The powers at which the arbitrary constants a_j appear are called *resonances*.

2. Resonances. For each (p, α) in step 1, the original DE is truncated to retain only the leading terms, and a two-term expansion

$$w = \alpha(z-z_0)^p + \beta(z-z_0)^{p+r} \tag{7}$$

Local and Global Analysis

is sought. Substituting (7) into the leading terms of (4) and retaining the terms of order β, we have

$$Q(r)\beta(z-z_0)^q = 0, \qquad q \geq p + r - n. \tag{8}$$

(The terms of order α vanish because of step 1.) Since the term $\beta(z-z_0)^{p+r}$ has been differentiated n times, the degree of (8) in $z - z_0$ is $q = p + r - n$ and the degree of $Q(r)$ is n. This is the case if the highest-order derivative in (4) is a leading term. In the opposite circumstance, the degree of the polynomial $Q(r)$ equals the order of the highest-order derivative among the leading terms and is less than n.

The roots of the polynomial $Q(r)$ determine the resonances. This is reminiscent of the indicial equation in the method of Frobenius for finding solutions of a linear ODE about a regular singular point. The following may be noted.

i. One root of $Q(r)$ is always -1. This represents the arbitrariness of the position of singularity z_0:

$$\begin{aligned} \alpha(z-z_0)^p &= \alpha(z-z_1+z_1-z_0)^p \\ &= \alpha(z-z_1)^p + \alpha(z-z_1)^{p-1} p(z_1-z_0) + \cdots \\ &= \alpha(z-z_1)^p + \beta(z-z_1)^{p-1} + \cdots, \text{ say, etc.} \end{aligned}$$

ii. If α is arbitrary in step 1, $r = 0$ is another root.

iii. If there is a root with $\text{Re}\, r < 0$, it may be ignored since it violates the hypothesis that $(z-z_0)^p$ is the most dominant term (with negative power) in the expansion near $z = z_0$.

iv. Any root with $\text{Re}(r) > 0$ but r not a real negative integer indicates a (movable) branch point at $z = z_0$. The algorithm may be terminated. Further analysis may be needed to confirm that the equation actually has a branch point. Sometimes a transformation may change the DE into one of P type.

v. If, for every possible (p, α) from step 1, all the roots of the polynomial $Q(r)$ (except -1 and possibly 0) are positive integer, then there are no algebraic branch points. One may proceed for each (p, α) to step 3 to find further terms in the series and rule out the presence of logarithmic branch points at a later stage in the expansion.

vi. If, for a given (p, α), the Laurent series solution for the nth order DE (4) is to represent a general solution, the polynomial equation $Q(r) = 0$ must have $n - 1$ nonnegative real distinct roots, each introducing an arbitrary constant at its resonance. If, for every (p, α) from step 1, $Q(r)$ has fewer than $n - 1$ roots, then none of the local solutions is general. In this circumstance, (5) misses an essential part of the solution.

3. Find the constants of integration. For a given (p,α) from step 1, arrange the positive integer roots of $Q(r) = 0$ in the order $r_1 \leq r_2 \leq \cdots \leq r_s$, where $s \leq n - 1$. Substitute the form

$$w = \alpha(z - z_0)^p + \sum_{j=1}^{r_s} a_j(z - z_0)^{p+j} \qquad (9)$$

into the full equation (4). Here again the procedure is reminiscent of the Frobenius method for linear DE except that (p,α) in (9) are determined from the balance of dominant terms. The second term on the right-hand side of (9) has r_s terms, r_s being the largest root of the equation $Q(r) = 0$, since for these nonlinear problems it is usually difficult to go beyond a few terms; the recurrence relation is generally very complicated. The coefficient of $(z - z_0)^{p+j-n}$ must vanish identically so that

$$Q(j)a_j - R_j(z_0, \alpha, a_1, \ldots, a_{j-1}) = 0, \qquad (10)$$

R_j arising from terms other than $d^n w/dz^n$ in (4).

The following points may be noted:

i. For $j < r_1$, $Q(j) \neq 0$ and (10) determines a_j.
ii. For $j = r_1$, (10) becomes

$$(0)(a_{r_1}) - R_{r_1}(z_0, \alpha, a_1, \ldots, a_{r_1-1}) = 0. \qquad (11)$$

If $R_{r_1}(z_0, \alpha, a_1, \ldots, a_{r_1-1}) \neq 0$, (11) cannot be satisfied. There is no solution of the form (9), and we must introduce logarithmic terms into the expansion:

$$w = \alpha(z - z_0)^p + \sum_{j=1}^{r_1-1} a_j(z - z_0)^{p+j}$$
$$+ [a_{r_1} + b_{r_1} \ln(z - z_0)](z - z_0)^{p+r_1} + \cdots. \qquad (12)$$

Substitution of (12) into (4) shows that the coefficient of $(z - z_0)^{p+r_1-n} \ln(z - z_0)$ is

$$Q(r_1)b_{r_1} = 0, \qquad (13)$$

where b_{r_1} is determined by requiring that the coefficient of $(z - z_0)^{p+r_1-n}$ vanish. It will be found that a_{r_1} is arbitrary. The expansion (12) will have more and more logarithmic terms in its higher-order terms. Thus, $R_{r_1}(z_0, \alpha, a_1, \ldots, a_{r_1-1}) \neq 0$ is symptomatic of a movable logarithmic branch point.

iii. If $R_{r_1} = 0$, then a_{r_1} is arbitrary and one may proceed to the next coefficient.

Local and Global Analysis

iv. Any resonance which corresponds to the multiple root of $Q(r)$ represents a movable logarithmic branch point (again there is some resemblance with Frobenius series solution). The equation is again not of P type.

v. At each nonresonant power ($j \neq r_i$), (10) determines a_j. At each resonance, either $R_{r_j} \neq 0$, logarithmic terms must be introduced into (9), and the given equation is not of P type, or $R_{r_j} = 0$, and a_{r_j} is an arbitrary constant of integration.

vi. If no logarithmic terms are introduced at any of the resonances, one could proceed to compute all the terms in the series. However, the calculations soon become cumbersome and it becomes difficult to determine the region of convergence of the series.

vii. If no logarithms are introduced at any of the resonances for all possible (p, α) from step 1, then DE (4) satisfies all the necessary conditions for the Painlevé property.

This is the end of the algorithm. The above discussion relates to the case $p < 0$ so that the function becomes infinite at the singularity. Other cases for which $p > 0$, so that the function has a zero or a branch point and its higher derivatives possibly become infinite, may be treated in the same manner.

EXAMPLE 1

$$w'' = \frac{2w(w')^2}{w^2 - 1}. \tag{14}$$

Let $w \sim \alpha(z - z_0)^p$. The terms $w^2 w''$ and $2w(w')^2$ balance if

$$\alpha^3 p(p-1)(z - z_0)^{3p-2} = 2\alpha^3 p^2 (z - z_0)^{3p-2}$$

or $p^2 - p = 2p^2$; i.e., $p = -1$. The coefficient α remains unrestricted. This is the only balance possible. The given equation has movable simple poles. Actually this equation can be easily integrated by writing it as

$$\frac{w''}{w'} = \frac{2ww'}{w^2 - 1}$$

so that

$$\frac{w'}{w^2 - 1} = A,$$

where A is the constant of integration. A second integration gives

$$w = \tanh(Az + B),$$

where B is another constant of integration. This is the general solution of (14) with two arbitrary constants. Equation (14) therefore has no movable critical points. It has simple movable poles as shown earlier and is of P type.

EXAMPLE 2

$$w'' - 10w^4 = 0. \tag{15}$$

Let $w = \alpha(z - z_0)^p$. Then

$$\alpha p(p-1)(z-z_0)^{p-2} - 10\alpha^4(z-z_0)^{4p} = 0.$$

This gives $p = -2/3$ and $\alpha = (1/9)^{1/3}$. We write $w = v^{-2/3}$ so that (15) becomes

$$\frac{3}{5}vv'' = (v')^2 - (3)^2. \tag{16}$$

Equation (16) has a solution which is regular at $z = z_0$ if $v(z_0) = 0$, $v'(z) = \pm 3$, and $v''(z_0)$ is finite. In such a case, $v(z)$ is analytic at z_0. Equation (15) has a movable branch point of order $-2/3$. It is not of P type.

EXAMPLE 3

$$ww'' = \frac{5}{2}w'^2. \tag{17}$$

Let $w = \alpha(z - z_0)^p$. Then the two terms in (17) balance if

$$\alpha^2 p(p-1)(z-z_0)^{2p-2} = \frac{5}{2}\alpha^2 p^2 (z-z_0)^{2p-2}.$$

This gives $p = -2/3$. α remains unrestricted. For $w = v^{-2/3}$, (17) reduces to $v'' = 0$. This is a linear equation and therefore of P type. But equation (17) has a movable branch point of order $-2/3$ and is not of P type.

EXAMPLE 4

$$w'' + 4ww' + 2w^3 = 0. \tag{18}$$

Again with $w \sim \alpha(z - z_0)^p$, we have

$$\alpha p(p-1)(z-z_0)^{p-2} + 4\alpha^2 p(z-z_0)^{2p-1} + 2\alpha^3(z-z_0)^{3p} = 0.$$

All the terms balance if $p = -1$ and α satisfies the equation

$$2 - 4\alpha + 2\alpha^2 = 0;$$

that is, $\alpha = 1, 1$. Substituting

$$w = (z - z_0)^{-1} + \beta(z - z_0)^{-1+r}$$

into (18) and considering only terms of order β, we have

$$(r-1)(r-2) + 4(r-1-1) + (2)(3) = 0;$$

Local and Global Analysis

that is,

$$r^2 + r = 0.$$

The root $r = -1$ corresponds to the arbitrariness of z_0. The root 0 corresponds here not to the arbitrariness of α but to the fact that $\alpha = 1$ is a double root. Therefore, $(z - z_0)^{-1}$ cannot be the first term in a Laurent series form of a general solution of (18). In fact, peeling off the factor $(z - z_0)^{-1}$, we write

$$w = \frac{f(z - z_0)}{z - z_0}$$

so that f satisfies the equation

$$(z - z_0)^2 f'' + (4f - 2)(z - z_0)f' + 2f(f - 1)^2 = 0.$$

This equation is of Cauchy–Euler type with respect to the independent variable $z - z_0$ so that the substitution $\ln(z - z_0) = t$ transforms it to

$$\frac{d^2 f}{dt^2} + (4f - 3)\frac{df}{dt} + 2f(f - 1)^2 = 0. \tag{19}$$

Equation (19) clearly has a variety of regular solutions, including those satisfying $f(0) = 1$. The latter corresponds to the solution of (18) with $(z - z_0)^{-1}$ as the first term. Every nonexponential solution of (19) corresponds to one of (18) with movable logarithmic branch point. Thus, (18) is not of P type.

EXAMPLE 5

$$w'' - z^m w = 2w^3. \tag{20}$$

Here it is easy to check that w'' and $2w^3$ balance and the leading behavior is given by $w \sim \pm 1/(z - z_0)$. Choosing the positive sign and writing $\xi = z - z_0$, we may transform (20) by putting

$$w = \frac{1}{\xi} + \rho(\xi).$$

The function $\rho(\xi)$ satisfies the equation

$$\rho'' - \frac{6}{\xi^2}\rho = \left(z_0^m + m z_0^{m-1}\xi + \frac{m(m-1)}{2} z_0^{m-2}\xi^2 + \cdots\right)$$
$$\times \left(\frac{1}{\xi} + \rho\right) + \frac{6}{\xi}\rho^2 + 2\rho^3. \tag{21}$$

Assuming the form

$$\rho(\xi) = a_0 + a_1\xi + a_2\xi^2 + \cdots,$$

and substituting in (21), we find by equating the coefficients of ξ^m, etc. that

$$a_0 = 0, \quad a_1 = \frac{-z_0^m}{6}, \quad a_2 = \frac{-mz_0^{m-1}}{4}.$$

At order ξ^3, we have

$$0 = 0, \quad 0 \times a_3 = \frac{1}{2}m(m-1)z_0^{m-2}. \tag{22}$$

If $m \neq 0, 1$, then logarithmic terms have to be introduced. The solution then has the form

$$\rho(\xi) = a_0 + a_1\xi + a_2\xi^2 + (b_3\xi^3 \ln \xi + a_3\xi^3) + \cdots.$$

The solution in this case has logarithmic branch points and is therefore not of P type. Indeed, the higher terms would contain higher powers of both ξ and $\ln \xi$. The cases $m = 0, 1$ would have a_3 arbitrary, and we may proceed to obtain higher terms in the series for ρ. In these special cases, the necessary conditions for the Painlevé property are satisfied, and equation (20) has no movable critical points; we have two arbitrary constants z_0 and a_3, and the Laurent series starting with $(z - z_0)^{-1}$ represents the general solution in the neighborhood of the movable pole. This may formally be proved by using the α-method of Painlevé (see Ince, 1956, § 14.41). Since no other algebraic singularity is possible, there are no movable algebraic branch points.

Now we turn to the ARS algorithm for systems of ODE and illustrate it with an example (see Ramani, Grammaticos, and Bountis, 1988). We consider the system of ODE

$$\frac{dw_i}{dz} = F_i(w_1, w_2, \ldots, w_n, z), \quad i = 1, 2 \ldots, n. \tag{23}$$

The algorithm involves the following steps.

1. Dominant behavior is sought in the form

$$w_i = \alpha_i(z - z_0)^{p_i} \tag{24}$$

where $\text{Re}(p_i) < 0$ and z_0 is arbitrary. We substitute (24) into (23) and find all the possible cases such that two or more terms balance in each of the equations of the system (23) and the rest of the terms are ignored as less dominant. For each such choice of p_i's, the balance of the terms also determines the values of α_i's. Here it is important to emphasize that *all* possible dominant behaviors are to be considered; an omission my lead to erroneous results.

If, in this step, any of the p_i are not integers, one would have an algebraic branch point behavior and may conclude the Painlevé property does not

Local and Global Analysis

hold. However, a simple change of variables may sometimes transform the system into one which has the Painlevé property, having no branch point dominant behavior.

If, in rare circumstances, all p_i's are integers, then for each dominant behavior one may seek the Laurent series

$$w_i = (z-z_0)^{p_i} \sum_{m=0}^{\infty} a_i^{(m)} (z-z_0)^m, \qquad i = 1, 2, \ldots, n, \qquad (25)$$

where $a_i^{(0)} = \alpha_i$ and z_0 is the first free constant of integration for the system (23). For an nth-order system, if one can find $n-1$ other free constants in the expansion (25), the latter would be referred to as generic. The powers m at which these free constants arise are called *resonances* of the series, to be obtained in the next step.

2. Resonances are obtained by first retaining only the leading terms in (23) for each of the dominant balances and writing the solution as

$$w_i = \alpha_i \tau^{p_i}(1 + \gamma_i \tau^r), \qquad r > 0, \quad i = 1, 2, \ldots, n,$$
$$\tau = z - z_0. \qquad (26)$$

Substituting (26) in the truncated system of (23) with the leading terms only and retaining terms linear in γ_i, we arrive at the homogeneous algebraic system

$$Q(r)\gamma = 0, \qquad \gamma \equiv (\gamma_1, \gamma_2, \ldots, \gamma_n), \qquad (27)$$

where $Q(r)$ is an $n \times n$ matrix, with r entering only in its diagonal elements at most linearly. For the system (27) to have nontrivial solution and hence permit some free constants, we have

$$\det Q(r) = (r+1)(r^{n-1} + A_2 r^{n-2} + \cdots + A_n) = 0, \qquad (28)$$

where $r = -1$ (as for the scalar equations) refers to one free constant already present in $z - z_0$, the location of the movable singularity.

If $r = 0$ is a root, it corresponds to the coefficients of the leading term being arbitrary. The root $r = -1$ corresponds to the free constant z_0, as mentioned earlier. If $\operatorname{Re} r < 0$ for some resonances (other than $r = -1$), we must ignore these, since in writing the balances we assume that the leading terms are the most singular. Such an exigency points to the fact that the expansion (25) is not generic. If for any r, $\operatorname{Re}(r) > 0$ is not an integer, the algorithm terminates; however, we must check whether this algebraic point, for rational r, can be transformed into a pole by some change of variables. If p_i in the dominant behavior itself is rational, and later r appears as rational with the same denominator as in p_i, we have a finite branching with multiplicity determined by the leading singularity. This special branch point behavior is defined as the *weak Painlevé property*.

3. To discover the presence of nondominant logarithmic branch points and free constants in the expansions involving logarithmic terms, we write for every leading-order behavior (24) a finite expansion

$$w_i = \alpha_i \tau^{p_i} + \sum_{m=0}^{r_s} a_i^{(m)} \tau^{p_i + m}, \qquad (29)$$

where $\tau = z - z_0$ and r_s is the largest possible root of (28), and obtain, by equating different powers of τ, etc., the system of inhomogeneous equations

$$Q(m) a^{(m)} = R^m(z_0; a^{(j)}), \qquad j = 1, \ldots, (m-1), \qquad (30)$$

with $m = 1, \ldots, r_s$, $R = (R_1, R_2, \ldots, R_n)^T$.

The following cases arise.

i. If $m < r_1$, the smallest positive resonance, (30) determines $a^{(m)}$.
ii. If $m = r_1$, $\det Q(m) = 0$, the system of inhomogeneous equations (30) is compatible if $\operatorname{rank} Q(m) = \operatorname{rank}(Q(m), R^m)$ so that

$$\det Q^{(k)}(r_1) = 0, \qquad k = 1, 2, \ldots, n, \qquad (31)$$

where $Q^{(k)}(r_1)$ is the matrix $Q(r_1)$ with its kth column replaced by $R^{(r_1)}$.
iii. If (31) is satisfied, then (30) determines $a^{(m)}$ for $r_1 < m < r_2$, where r_2 is the next smallest positive resonance.
iv. This procedure is applied successively at each resonance, continuing thus to the largest resonance.
v. If the system of equations (30) for some resonance r becomes inconsistent so that the conditions (31) are not satisfied, we must introduce the logarithmic terms in one or more of expansions (25) so that

$$w_i = \sum_{m=0}^{r-1} a_i^{(m)} \tau^{p_i + m} + (a_i^{(r)} + b_i^{(r)} \ln \tau) \tau^{p_i + r} + \cdots, \qquad (32)$$

with higher powers of $\ln \tau$ possibly occurring at higher orders. The complete series form for such an eventuality does not seem to have been investigated (see, however, Hille, 1976). We determine the coefficients $b_i^{(r)}$ of the logarithmic term by demanding that the coefficient of the appropriate powers of τ vanish, while $a_i^{(r)}$ are free.

Local and Global Analysis

EXAMPLE 6 Consider the second-order system

$$\frac{dx}{dz} = x(a - x - y),$$
$$\frac{dy}{dz} = y(x - 1). \tag{33}$$

Let the leading-order behavior be

$$x = \alpha \tau^p, \quad y = \beta \tau^q \quad \text{with } \tau = z - z_0. \tag{34}$$

Substitution of (34) into (33) gives two distinct behaviors:

i. $p = -1, q = 1, \alpha = 1$, and β free. In this case the leading terms are

$$\frac{dx}{dz} = -x^2, \quad \frac{dy}{dz} = xy. \tag{35}$$

The dominant behavior therefore is $x = \tau^{-1}, y = \beta\tau$.

ii. $p = -1, q = -1, \alpha = -1$, and $\beta = 2$. The leading-order terms here are

$$\frac{dx}{dz} = -x^2 - xy, \quad \frac{dy}{dz} = xy. \tag{36}$$

The dominant behavior of the solution is

$$x = -\tau^{-1}, \quad y = 2\tau^{-1} \tag{37}$$

To get the resonances, we write for case (i)

$$x = \tau^{-1}(1 + \gamma\tau^r), \quad y = \beta\tau(1 + \delta\tau^r). \tag{38}$$

Substituting (38) in the leading order terms (35) and retaining only linear terms in γ and δ, we have

$$\begin{bmatrix} r+1 & 0 \\ 1 & -r \end{bmatrix} \begin{bmatrix} \gamma \\ \delta \end{bmatrix} = \begin{bmatrix} 0 \\ 0 \end{bmatrix}. \tag{39}$$

The determinant of the matrix in (39) on vanishing gives the resonances $r = 0, -1$.

Similar calculations for case (ii) lead to the matrix system

$$\begin{bmatrix} 1-r & -2 \\ 2 & 2r \end{bmatrix} \begin{bmatrix} \gamma \\ \delta \end{bmatrix} = \begin{bmatrix} 0 \\ 0 \end{bmatrix}, \tag{40}$$

giving the resonance equation $(1-r)(2r) + 4 = 0$ so that $r = -1, 2$.

To take account of the free constants, we note that for case (i), the resonances are $r = 0, -1$ and the two (free) integration constants are z_0 and β. The Laurent series for this case contains only integer powers of τ throughout, so no compatibility conditions need be checked. For case (ii), we recall that one resonance is at $r = 2$. Therefore, we must carry out the expansion solution to second order to determine whether logarithmic terms are needed. Thus, we write

$$x = -\frac{1}{\tau} + a_1 + a_2\tau + \cdots,$$
$$y = \frac{2}{\tau} + +b_1 + b_2\tau + \cdots.$$
(41)

Substituting (41) into (33), we find that

$$b_1 = a, \qquad a_1 = \frac{a}{2} + 1.$$
(42)

Equating the terms τ^0 gives

$$a_2 - b_2 = aa_1 - a_1 b_1 - a_1^2,$$
$$2a_2 - 2b_2 = b_1 - a_1 b_1.$$
(43)

This system of equations is compatible only if

$$2(aa_1 - a_1 b_1 - a_1^2) = b_1 - a_1 b_1,$$
(44)

i.e., if $a = -1$, where we have used (42). We thus find that unless $a = -1$, the system (43) is not compatible and logarithmic terms must be introduced in (41) to bring in the second arbitrary constant. Following the scheme (32), we easily find that the solution in this case is

$$x = -\frac{1}{\tau} + \left(\frac{a}{2} + 1\right) + a_2\tau - \frac{2(a+1)}{3}\tau\ln\tau + \cdots,$$
$$y = \frac{2}{\tau} + a + \left(a_2 + \frac{a+1}{3} + \frac{a^2}{4}\right)\tau - \frac{2(a+1)}{3}\tau\ln\tau + \cdots.$$
(45)

Here a_2 is the second arbitrary constant, apart from z_0. We conclude that the system (33) is Painlevé only if $a = -1$.

An excellent review of the Painlevé property, singularity analysis of integrable and nonintegrable systems, and the possibility of chaos for the latter systems, has been given by Ramani, Grammaticos, and Bountis (1988). It also has a fairly exhaustive bibliography on this topic.

4.3 LARGE-TIME (ASYMPTOTIC) BEHAVIOR

In Section 4.2, we considered asymptotic solutions of nonlinear DE in the neighborhood of singular points. Here, we present solutions for large time and demonstrate that formal approximate solutions of nonlinear DE are the leading terms in an asymptotic representation of actual solutions (Bellman,

Local and Global Analysis

1953; Levinson, 1970). We shall formally derive an integral equation equivalent to a given DE and prove that the method of successive approximations gives the asymptotic solution, with the formal (approximate) solution as the first iterate. This approach is similar to one that is used in proving the existence and uniqueness of solutions of initial value problems of DE, namely Picard's iterative method. We quote one form of Picard's theorem (Birkhoff and Rota, 1978).

THEOREM 4.3.1 Let the vector function $X(\mathbf{x};t)$ be continuous and satisfy the Lipschitz condition

$$|X(\mathbf{x},t) - X(\mathbf{y},t)| \leq L|\mathbf{x}-\mathbf{y}| \quad \text{if } (\mathbf{x},t) \in \mathcal{R}, (\mathbf{y},t) \in \mathcal{R},$$

on the interval $|t-a| \leq T$ for all $\mathbf{x},\mathbf{y}$. Then, for any constant vector $\mathbf{c}$, the vector DE $\mathbf{x}'(t) = X(\mathbf{x};t)$ has a solution defined on the interval $|t-a| \leq T$, which satisfies the initial condition $\mathbf{x}(a) = \mathbf{c}$.

EXAMPLE 1 $\quad \dfrac{dy}{dx} = xy(y-2), \quad y_0 = 1, \quad x_0 = 0.$ \hfill (1)

This equation is equivalent to the integral equation

$$y = 1 + \int_0^x xy(y-2)\,dx, \tag{2}$$

which also satisfies the given initial conditions. We choose the first iterate as $y_0 = 1$ so that the integrand in (2) becomes $-x$. Subsequent iterates are given by

$$y_1 = y_0 + \int_0^x -x\,dx = 1 - \frac{1}{2}x^2,$$

$$y_2 = y_0 + \int_0^x x\left(1 - \frac{x^2}{2}\right)\left(-1 - \frac{1}{2}x^2\right) dx$$

$$= y_0 + \int_0^x \left(-x + \frac{1}{4}x^5\right) dx$$

$$= 1 - \frac{x^2}{2} + \frac{x^6}{24},$$

$$y_3 = y_0 + \int_0^x -x\left(1 - \frac{x^4}{4} + \frac{x^8}{24} - \frac{x^{12}}{576}\right) dx$$

$$= 1 - \frac{x^2}{2} + \frac{x^6}{24} - \frac{x^{10}}{240} + \frac{x^{14}}{8064},$$

$$y_4 = y_0 + \int_0^x -x\left(1 - \frac{x^4}{4} + \frac{x^8}{24} - \frac{17x^{12}}{2880} + \cdots\right) dx$$

$$= 1 - \frac{x^2}{2} + \frac{x^6}{24} - \frac{x^{10}}{240} + \frac{17x^{14}}{40,3020} + \cdots. \tag{3}$$

Since (1) is separable, it can be readily integrated to give

$$y = \frac{2}{1 + e^{x^2}} \tag{4}$$

as the solution of the IVP (1). It is easily checked by expanding (4) that (3) is exact to the last term. The solution (4) has poles in the complex x plane where $x^2 = \ln(-1) = \pi i$, giving $x_1 = \sqrt{\pi/2}(1+i)$, $x_2 = \sqrt{\pi/2}(1-i)$. The series solution (3) therefore has a radius of convergence equal to $\sqrt{\pi}$.

EXAMPLE 2

$$\frac{dy}{dx} = 1 + y^2, \qquad y(0) = 0. \tag{5}$$

Here, we choose $y_0 = x$ instead of 0 as the first iterate to see what change comes about. This choice satisfies the initial condition as well. Thus, we have

$$y_1 = \int_0^x (1 + x^2)\,dx = x + \frac{x^3}{3},$$

$$y_2 = \int_0^x \left(1 + x^2 + \frac{x^6}{9} + \frac{2x^4}{3}\right) dx$$

$$= x + \frac{x^3}{3} + \frac{x^7}{63} + \frac{2x^5}{15},$$

$$y_3 = \int_0^x \left(1 + x^2 + \frac{2x^4}{3} + \frac{17x^6}{45} + \frac{38}{315}x^8 \right.$$

$$\left. + \frac{134}{(21)(225)}x^{10} + \frac{4}{945}x^{12}\right) dx$$

$$= x + \frac{x^3}{3} + \frac{2x^5}{15} + \frac{17x^7}{315} + \frac{38x^9}{(9)(315)}$$

$$+ \frac{134 x^{11}}{(11)(21)(225)} + \frac{4x^{13}}{(13)(945)}.$$

The exact solution of this problem is

$$y = \tan x.$$

The choice $y_0 = x$ in place of 0 generates more terms in the successive iterates and approaches the exact solution faster.

The motive for an integral equation formulation for the study of asymptotic behavior of nonlinear DE is rather different, as we now explain. We

Local and Global Analysis

look at the plausible solution of a DE, say,

$$xx'' + x' + tx = t^2, \tag{6}$$

where $x' = dx/dt$, in the form $x = ct^a$. Then the left side of (6) is

$$c^2 a(a-1)t^{2a-2} + cat^{a-1} + ct^{a+1}. \tag{7}$$

Two balances are possible: (i) $2a - 2 = a + 1$ or $a = 3$. Then $6c^2 + c = 0$ or $c = -1/6$. The leading-order behavior, therefore, is $x = -t^3/6$. It balances the terms xx'' and tx, and the neglected terms of $O(t^2)$ are smaller than those retained $O(t^4)$ for large t. (ii) $a + 1 = 2$ or $a = 1$, and $c = 1$ Here terms of order t^2 (t^2 and tx in (6)) are dominant in (7), and those $O(1)$ are small in comparison for large t. This gives $x = t$. The questions then arise: What is the significance of each of these approximate solutions? Do they form leading terms in asymptotic expansions for large time? How can we construct these expansions and show that they are asymptotic? We shall, following Levinson (1970), answer some of these questions. We note in passing that any trial solution of (6) in the exponential form $x \sim ce^{at}$ will not be appropriate, since the term t^2 on the right will not be balanced by those on the left.

Consider the second-order equation

$$x'' = f(t, x, x'), \tag{8}$$

where $f(t, x, y)$ is assumed to possess continuous first-order partial derivatives with respect to x and y for those values of t, x, and y which enter the discussion here.

Let $x = \varphi(t)$ be a plausible approximate solution of (8) for large t (see the discussion of equation (6)). If we define

$$E(t) = \varphi''(t) - f(t, \varphi(t), \varphi'(t)) \tag{9}$$

(the error in the solution), then φ is a plausible approximate solution of (8) if $E(t)$ is "small" (compared to the approximate solution) as $t \to \infty$. If we substitute $x = \varphi(t) + u$ in (8), we get the corresponding equation in the variable u:

$$\begin{aligned} u'' &= f(t, \varphi + u, \varphi' + u') - \varphi''(t) \\ &= f(t, \varphi + u, \varphi' + u') - f(t, \varphi(t), \varphi'(t)) \\ &\quad + f(t, \varphi(t), \varphi'(t)) = \phi''(t) \\ &= f(t, \varphi + u, \varphi' + u') - f(t, \varphi(t), \varphi'(t)) - E(t). \end{aligned} \tag{10}$$

We linearize $f(t, \varphi + u, \varphi' + u')$ about (t, φ, φ') by a Taylor series expansion in u and u', assuming the latter to be small. Then (10) can be written as a "linear inhomogeneous" DE in u by transferring the rest of the terms to the right-hand side:

$$u'' - f_y(t, \varphi, \varphi')u' - f_x(t, \varphi, \varphi')u$$
$$= f(t, \varphi + u, \varphi' + u') - f_y(t, \varphi, \varphi')u'$$
$$- f_x(t, \varphi, \varphi')u - f(t, \varphi, \varphi') - E(t)$$
$$\equiv Q(t, u, u') - E(t), \quad \text{say.} \tag{11}$$

Here, Q is evidently of second degree or higher in u and u'. We treat (11) as an inhomogeneous second-order DE with a linear operator on the left, namely,

$$u'' - f_y(t, \varphi, \varphi')u' - f_x(t, \varphi, \varphi')u \equiv u'' + 2\tilde{\alpha}(t)u' + \tilde{\beta}(t)u$$
$$\equiv \tilde{L}u, \quad \text{say.} \tag{12}$$

If the linear DE $\tilde{L}u = 0$ admits an exact solution, we may proceed to construct an integral equation for (11) in the usual way by varying the parameters. Otherwise, we try to find a linear differential operator L "close" to $\tilde{L}$ such that $Lu = 0$ can be solved explicitly. This can be accomplished by the WKB method (see Bender and Orszag, 1978). Let this operator be defined as

$$L = \frac{d^2}{dt^2} + 2\alpha(t)\frac{d}{dt} + \beta(t), \tag{13}$$

where L is close to $\tilde{L}$ in the sense that $|\alpha - \tilde{\alpha}|$ and $|\beta - \tilde{\beta}|$ are small compared to $\tilde{\alpha}$ and $\tilde{\beta}$, respectively. Both φ and L may be refined subsequently. With this choice of L, (11) can be rewritten as

$$Lu = R(t, u, u'), \tag{14}$$

where

$$R = Q(t, u, u') - E(t) + a(t)u + b(t)u'$$

and

$$a = 2(\alpha - \tilde{\alpha}), \quad b = 2(\beta - \tilde{\beta}).$$

Since two explicit linearly independent solutions $\psi_1(t)$ and $\psi_2(t)$ of $lu = 0$ are known, the "solution" of the inhomogeneous equation (14) can be written in the form

$$u(t) = C_1\psi_1(t) + C_2\psi_2(t) + \psi_1(t)\int_a^t \theta_1(s)R(s, u(s), u'(s))\,ds$$
$$+ \psi_2(t)\int_a^t \theta_2(s)R(s, u(s), u'(s))\,ds, \tag{15}$$

Local and Global Analysis

where C_1 and C_2 are constants and

$$\theta_1(s) = \frac{\psi_2(s)}{\psi_1'(s)\psi_2(s) - \psi_1(s)\psi_2'(s)},$$

$$\theta_2(s) = \frac{\psi_1(s)}{\psi_2'(s)\psi_1(s) - \psi_2(s)\psi_1'(s)} \qquad (16)$$

(see Birkhoff and Rota, 1978).

It may sometimes be necessary to replace an integral over (a,t) by the negative integral over (t,∞). Sometimes (depending on the requirements of the physical problem) one or both of the constants C_j may have to be taken to be zero.

We shall prove the existence of the solution of the integral equation (15) in u. Since the right-hand side of (15) contains u', we differentiate (15) to get a second equation for u'. We can then consider the (coupled) vector integral equation in u and u'. The norm of a vector y with components y_j will be defined as

$$|y| = \sum_{j=1}^{2} |y_j|. \qquad (17)$$

The following theorem of Levinson (1970) gives an existence proof for the solution of a coupled system of integral equations of the type under consideration. The proof employs successive iterates.

THEOREM 4.3.2 Let $h(t)$, $K_1(t,s)$, and $K_2(t,s)$ be continuous vector functions and $g(y,t)$ a continuous scalar function. Consider the integral equation

$$y(t) = h(t) + \int_a^t K_1(t,s)g(y(s),s)\,ds$$
$$+ \int_t^\infty K_2(t,s)g(y(s),s)\,ds. \qquad (18)$$

Let $g(0,t) = 0$. Let there be a continuous scalar function $H(t)$ such that

$$|h(t)| \le H(t). \qquad (19)$$

Moreover, for y and $\tilde{y}$ satisfying

$$|y| \le 2H(t), \qquad |\tilde{y}| \le 2H(t), \qquad (20)$$

let there be a scalar function $r(t)$ for which

$$|g(y,t) - g(\tilde{y},t)| \le r(t)|y - \tilde{y}|, \qquad (21)$$

and

$$\int_a^t |K_1(t,s)| r(s) H(s) \, ds \leq \frac{1}{4} H(t), \tag{22}$$

$$\int_t^\infty |K_2(t,s)| r(s) H(s) \, ds \leq \frac{1}{4} H(t). \tag{23}$$

Then (18) has a solution which can be obtained by successive approximations, starting with $y = 0$ as the zeroth iterate on the right side. The process will converge uniformly in a norm involving $H(t)$. The solution $y(t)$ will satisfy $|y(t)| \leq 2H(t)$ and will be unique. If $h(t)$ depends continuously on some parameters, then y will also do so.

Proof. The zeroth iterate is $y^{(0)} = 0$. Let

$$y^{(k+1)}(t) = h(t) + \int_a^t K_1(t,s) g(y^{(k)}(s),s) \, ds$$

$$+ \int_t^\infty K_2(t,s) g(y^{(k)}(s),s) \, ds. \tag{24}$$

Then

$$y^{(1)}(t) = h(t) \qquad \text{since } g(0,t) = 0.$$

By (19),

$$|y^{(1)}(t) - y^{(0)}(t)| = |y^{(1)}(t)| = |h(t)| \leq H(t).$$

Now we use induction. If $y^{(k)}(t)$ is continuous, then continuity of h, K_1, K_2, and g implies that $y^{(k+1)}$ in (24) is also continuous. Suppose that for $k \leq n$,

$$|y^{(k)}(t) - y^{(k-1)}(t)| \leq \frac{1}{2^{k-1}} H(t), \qquad 1 \leq k \leq n; \tag{25}$$

then we show that (25) is also true for $k = n+1$, and hence $y^{(k)}(t)/H(t)$ converges uniformly for $t \geq a$ as $k \to \infty$. Using (25) for $k \leq n$, we have

$$|y^{(k)}(t)| = |y^{(k)}(t) - y^{(k-1)}(t) + y^{(k-1)}(t) - y^{(k-2)}(t) + \cdots$$

$$+ y^{(1)}(t) - y^{(0)}(t)|$$

$$\leq |y^{(k)}(t) - y^{(k-1)}(t)| + |y^{(k-1)}(t) - y^{(k-2)}(t)| + \cdots$$

$$+ |y^{(1)}(t) - y^{(0)}(t)|$$

$$\leq \left(1 + \frac{1}{2} + \cdots + \frac{1}{2^{k-1}}\right) H(t) < 2H(t).$$

Local and Global Analysis

Writing $k = n$ and $n - 1$, respectively, in (24) and subtracting, we have

$$y^{(n+1)}(t) - y^{(n)}(t) = \int_a^t K_1(t,s)[g(y^{(n)}(s),s) - g(y^{(n-1)}(s),s)]\,ds$$
$$+ \int_t^\infty K_2(t,s)[g(y^{(n)}(s),s) - g(y^{(n-1)}(s),s)]\,ds.$$

Using the condition (21) for g, and (25) for $k = n$, we have

$$|y^{(n+1)}(t) - y^{(n)}(t)| \le 2^{-n+1} \int_a^t |K_1(t,s)|r(s)H(s)\,ds$$
$$+ 2^{-n+1} \int_t^\infty |K_2(t,s)|r(s)H(s)\,ds$$
$$\le 2^{-n+1}\left(\frac{1}{4}H(t) + \frac{1}{4}H(t)\right) = 2^{-n}H(t). \tag{26}$$

The last inequality in (26) follows from (22) and (23). Thus, (25) is true for $k = n + 1$. Since it is true for $k = 1$ ($|y^{(1)}(t)| \le H(t)$), the induction is complete. The uniformity of the convergence of the sequence $y^{(k)}(t)/H(t)$ proves that the limit function $y(t)/H(t)$ and hence $y(t)$ are continuous. Following the usual arguments for continuous dependence of the solution on the parameters, it can be proved that $y(t)$ will depend continuously on any parameters on which h depends continuously, so long as they range over a region for which $|h(t)| \le H(t)$.

EXAMPLE 3

$$xx'' + x' + tx = t^2. \tag{27}$$

This is equation (6) which we discussed earlier. We here pursue the asymptotic approximate solution $-t^3/6$. Substituting $x = -t^3/6 + u$ into (27), we have

$$u'' - \frac{6}{t^3}\left(1 - \frac{6u}{t^3}\right)^{-1} u' + \frac{9}{t}\left(1 - \frac{6u}{t^3}\right)^{-1} = 0$$

or

$$u'' - \frac{6}{t^3}u'\left(1 - \frac{6u}{t^3}\right)^{-1} + \frac{54u}{t^4}\left(1 - \frac{6u}{t^3}\right)^{-1} = \frac{-9}{t}. \tag{28}$$

We define the operator $\tilde{L}$ by

$$\tilde{L}u = u'' - \frac{6u'}{t^3} + \frac{54u}{t^4}.$$

Now, if $u \sim t^{3+m}$, $m < 0$, then u/t^4 and u'/t^3 are smaller than u'' by a factor t^{-2} for large t. Therefore, a good first approximation $\tilde{L}$ to $\hat{L}$ may simply be taken as d^2/dt^2. The two solutions of

$$\tilde{L}u = u'' = 0$$

are $\psi_1 = 1$ $\psi_2 = t$; therefore

$$\frac{\psi_1(t)\psi_2(s) - \psi_1(s)\psi_2(t)}{\psi_1'(s)\psi_2(s) - \psi_1(s)\psi_2'(s)} = t - s. \tag{29}$$

Equation (28) can be reformulated as an integral equation:

$$u(t) = \tilde{C}_1 t + C_2$$

$$+ \int_t^\infty (s-t)\left\{-\frac{9}{s} + [6su'(s) - 54u(s)]\left(1 - \frac{6u(s)}{s^3}\right)^{-1}\right\}\frac{ds}{s^4}$$

$$= C_1 t + C_2 - 9t \ln t$$

$$+ 6\int_t^\infty (s-t)\left(1 - \frac{6u(s)}{s^3}\right)^{-1} [su'(s) - 9u(s)]\frac{ds}{s^4}, \tag{30}$$

where the term $-9t \ln t$ has arisen from the integration of $(t-s)(-9/s)$ with respect to s. C_1 and C_2 are arbitrary constants. A second equation is obtained by differentiating (30) with respect to t and multiplying it by t. This makes $tu'(t)$ of the same order as $u(t)$, and the right-hand sides of these two equations have essentially the same balance. Thus, the second equation is

$$tu'(t) = C_1 t - 9t \ln t - 9t$$

$$- 6t \int_t^\infty \left(1 - \frac{6u(s)}{s^3}\right)^{-1} [su'(s) - 9u(s)]\frac{ds}{s^4}. \tag{31}$$

In the notation of Theorem 4.3.2,

$$y(t) = \begin{bmatrix} u(t) \\ tu'(t) \end{bmatrix}, \quad h(t) = \begin{bmatrix} C_1 t + C_2 - 9t \ln t \\ (C_1 - 9)t - 9t \ln t \end{bmatrix}, \tag{32}$$

so that

$$h(t) \leq |C_1 t + C_2 - 9t \ln t| + |(C_1 - 9)t - 9t \ln t|$$

$$\leq (2|C_1| + |C_2| + 27)t \ln t \leq At \ln t.$$

Therefore,

$$|h(t)| \leq H(t) = At \ln t, \quad \text{where } |A| \geq 2|C_1| + |C_2| + 27. \tag{33}$$

The scalar function

$$g(u(s), su'(s), s) = \left(1 - \frac{6u(s)}{s^3}\right)^{-1} (su'(s) - 9u(s))\frac{1}{s^4}. \tag{34}$$

Local and Global Analysis

The function $r(t)$ is easily found by finding the partial derivatives of (34):

$$\frac{\partial g}{\partial u} = \frac{9}{s^4}, \qquad \frac{\partial g}{\partial (su')} = \frac{1}{s^4}.$$

Therefore, $r(t)$, according to (21), may be found to be $2(9+1)/t^4 = 20/t^4$. We may therefore estimate (23) as

$$\int_t^\infty |K_2(t,s)| r(s) H(s)\, ds = (6+6) \int_t^\infty s \frac{20}{s^4} As \ln s\, ds$$

$$\leq H(t) \frac{240(1 + \ln t)}{t^2 \ln t}$$

$$\leq \frac{1}{4} H(t),$$

for large enough t. Therefore, we have verified that Theorem 4.3.2 is valid and the system (30)–(31) has a solution $u(t)$ and $tu'(t)$ depending continuously on C_1 and C_2. This solution satisfies the bound

$$|u(t)| + |tu'(t)| \leq 2H(t) = 2At \ln t. \tag{35}$$

Using (35) in the integrals of (30) and (31) and integrating we easily verify that the "remainder" terms in these expansions are $O((A \ln t)/t)$ and $O((A \ln t)/t^2)$, respectively. Thus,

$$u(t) = C_1 t + C_2 - 9t \ln t + O\left(\frac{A \ln t}{t}\right),$$

$$u'(t) = (C_1 - 9) - 9\ln t + O\left(\frac{A \ln t}{t^2}\right), \tag{36}$$

where

$$A \geq (2|C_1| + |C_2| + 27).$$

We therefore have a two-parameter family (36) of solutions, which may be checked to satisfy (28) by differentiating the former, etc.

The solution (36) may be refined by expanding $(1 - 6u/s^3)^{-1}$ and substituting it into the integrals in (30) and (31). The parameters C_1 and C_2 may be used to meet the initial conditions $u(t_0))$ and $u'(t_0)$ for a given t_0 with rather mild restrictions on the magnitudes of $|u(t_0)|$, $|u'(t_0)|$ as compared to $9t_0 \ln t_0$ (see (35)). One may have to ascertain roughly how large t is in relation to A.

In the next example, we study the second asymptotic solution of (27), which behaves like t for large t. We shall get another two-parameter family of solutions. This variety of possibilities commonly exists for nonlinear DE.

EXAMPLE 4

$$xx'' + x' + tx = t^2. \tag{27}$$

Now we consider the second solution of this equation, which is asymptotic to t for large t. In this case $\phi(t) = t$. Substituting $x = t + u(t)$ in (27), we have

$$(t+u)u'' + 1 + u' + t(t+u) = t^2$$

or

$$u'' + \frac{u'}{t+u} + \frac{tu+1}{t+u} = 0. \tag{37}$$

This can be expanded for small u and u'; retaining the linear terms in u, u' on the left, shifting $-1/t$ to the right, and readjusting, we have

$$u'' + \frac{u'}{t} + u\left(1 - \frac{1}{t^2}\right) = -\frac{1}{t} + \frac{uu'}{t(t+u)} + \frac{u^2}{t+u}\left(1 - \frac{1}{t^2}\right). \tag{38}$$

This suggests the linear operator

$$\tilde{L}u = u'' + \frac{1}{t}u' + u\left(1 - \frac{1}{t^2}\right). \tag{39}$$

One could by inspection (see Murphy, 1960) find an operator L close to $\tilde{L}$, which can be explicitly solved. Or, more systematically (see Bender and Orszag, 1978, for a detailed discussion and examples), one may use the WKB method to find approximate solutions of $\tilde{L}u = 0$ and hence the linear operator L close to $\tilde{L}$. This is done by the ansatz $u = \exp[\rho(t)]$ in $\tilde{L}u = 0$; we get

$$\rho'' + \rho'^2 + \frac{1}{t}\rho' + \left(1 - \frac{1}{t^2}\right) = 0. \tag{40}$$

This is a nonlinear DE in ρ. If ρ is a positive power of t, for example, then ρ'' is much smaller than ρ'^2. One may ignore the former and seek solutions of

$$t(\rho')^2 + \rho' + t - \frac{1}{t} = 0;$$

we thus have

$$\rho' = \frac{-1 \pm (5 - 4t^2)^{1/2}}{2t}$$

$$= -\frac{1}{2t} \pm i\left(1 - \frac{5}{8t^2} + \cdots\right)$$

for large t. Integrating gives

$$\rho(t) = -\frac{\ln t}{t} \pm it + O\left(\frac{1}{t}\right). \tag{41}$$

Local and Global Analysis

We verify that $\rho''(t) = O(1/t^2)$ and hence is smaller than the terms ρ'^2, ρ'/t, and 1 in (40). Therefore, our assumption regarding $\rho''(t) \ll \rho'^2$ is correct. One has to validate the assumption each time by obtaining the approximate solution. The approximations to the solutions of $\tilde{L}u = 0$ are

$$\phi_1 = t^{-1/2} e^{it}, \qquad \phi_2 = t^{-1/2} e^{-it}.$$

We now form the DE having these functions as its solution. If this DE is

$$Lu = p(t)u'' + q(u)u' + r(t)u' = 0, \tag{42}$$

then ϕ_1 and ϕ_2 satisfy this equation. For nontrivial $p(t)$, $q(t)$, and $r(t)$, the determinant of the homogeneous system (42), $L\phi_1 = 0$ and $L\phi_2 = 0$, should be zero. This gives the required DE:

$$\begin{vmatrix} u & u' & u'' \\ \phi_1 & \phi_1' & \phi_1'' \\ \phi_2 & \phi_2' & \phi_2'' \end{vmatrix} = 0.$$

Substituting for ϕ_1 and ϕ_2, we have

$$Lu = u'' + \frac{1}{t} u' + u \left(1 - \frac{1}{4t^2}\right) = 0. \tag{43}$$

This is a Bessel equation with solutions

$$J_{1/2} = \frac{2}{(\pi t)^{1/2}} \sin t \equiv \psi_1,$$

$$J_{-1/2} = \frac{2}{(\pi t)^{1/2}} \cos t \equiv \psi_2.$$

Thus we have

$$\frac{\psi_1(t)\psi_2(s) - \psi_2(t)\psi_1(s)}{\psi_1'(s)\psi_2(s) - \psi_1(s)\psi_2'(s)} = \left(\frac{s}{t}\right)^{1/2} \sin(t-s). \tag{44}$$

Equation (38) can be rewritten as

$$Lu = \frac{-1}{t} + \frac{uu'}{t(t+u)} + \frac{u}{t(t+u)} - \frac{u}{4t^2} + \frac{u^2}{t+u} \equiv r(t), \tag{45}$$

whose solution is given by

$$u = t^{-1/2} \int_a^t s^{1/2} \sin(t-s) r(s) \, ds. \tag{46}$$

Here, it is convenient to take $a = \infty$. The term $1/t$ in $r(t)$ contributes to the solution (46) the component

$$t^{-1/2} \int_t^\infty \frac{\sin(t-s)}{s^{1/2}} \, ds = \frac{-1}{t} + \frac{1}{2} t^{-1/2} \int_t^\infty \frac{\cos(t-s)}{s^{3/2}} \, ds,$$

where an integration by parts has been carried out. We therefore have the solution of (45) in the form

$$u(t) = C_1 t^{-1/2} \sin t + C_2 t^{-1/2} \cos t - \frac{1}{t}$$
$$+ \frac{1}{2} t^{-1/2} \int_t^\infty \frac{\cos(t-s)}{s^{3/2}} ds$$
$$- t^{-1/2} \int_t^\infty s^{1/2} \sin(t-s) g(u, u', s) ds, \qquad (47)$$

where

$$g(u, u', s) = \frac{u(s)}{s(s+u(s))} - \frac{u(s)}{4s^2} + \frac{uu'}{s(s+u)} + \frac{u^2}{s+u}. \qquad (48)$$

Since $u'(t)$ is $O(t^{-1/2})$, we obtain the second equation by simply differentiating (47) with respect to t:

$$u'(t) = C_1 t^{-1/2} \cos t - C_2 t^{-1/2} \sin t - \frac{1}{2} C_1 t^{-3/2} \sin t$$
$$- \frac{1}{2} C_2 t^{-3/2} \cos t + \frac{1}{t^2}$$
$$- \frac{1}{4} t^{-3/2} \int_t^\infty \frac{\cos(t-s)}{s^{3/2}} ds - \frac{1}{2} t^{-1/2} \int_t^\infty \frac{\sin(t-s)}{s^{3/2}} ds$$
$$+ \frac{1}{2} t^{-3/2} \int_t^\infty s^{1/2} \sin(t-s) g(u, u', s) ds$$
$$- t^{-1/2} \int_t^\infty s^{1/2} \cos(t-s) g(u, u', s) ds,$$

$$h(t) = \begin{bmatrix} C_1 t^{-1/2} \sin t + C_2 t^{-1/2} \cos t \\ C_1 t^{-1/2} \cos t - C_2 t^{-1/2} \sin t \end{bmatrix},$$

retaining only the most dominant terms. Therefore,

$$|h(t)| \leq A t^{-1/2} \equiv H(t),$$

where A is large; it may be easily verified that

$$|K_2(t,s)| \leq 3 \left(\frac{s}{t} \right)^{1/2},$$
$$r(t) \leq 10 A t^{-3/2}.$$

Thus, $|u(t)|$ and $|u'(t)|$ are each less than or equal to $2At^{-1/2}$. Using these estimates again in (47), we find that

$$u(t) = (C_1 \sin t + C_2 \cos t) t^{-1/2} + O\left(\frac{1}{t} \right). \qquad (49)$$

Local and Global Analysis

Further refinements may be made as indicated in Example 3. The solutions (49) are also a two-parameter family.

Levinson (1970) has also treated the following equations for their large-time behavior

$$x''^2 = x' + x \tag{50}$$

and

$$x^2 x'' = t + 1. \tag{51}$$

Bellman (1953) has studied in great detail the Emden–Fowler equation

$$\frac{d}{dt}\left(t^\rho \frac{du}{dt}\right) \pm t^\sigma u^n = 0, \tag{52}$$

after suitable transformations, in the simplified form

$$\frac{d^2 u}{dt^2} \pm t^\sigma u^n = 0. \tag{53}$$

He gave a thorough discussion of the large-time asymptotic nature of the real and continuous solutions of (53) (referred to by him as proper solutions), which have the special exact form

$$u = ct^w, \tag{54}$$

where

$$w = -\frac{\sigma + 2}{n - 1}$$

and

$$c = \left[\mp \frac{(\sigma + 2)(\sigma + n + 1)}{(n - 1)^2}\right]^{1/(n-1)}$$

are real constants. Although an integral equation formulation of (53) is used often for the purpose, other devices employed are rather different from Levinson's and are based on the original work of Fowler (see references in Bellman, 1953). These techniques may profitably be used for other nonlinear DE as well.

4.4 EMBEDDING OF ASYMPTOTIC SOLUTIONS IN A LARGER FAMILY—THE THOMAS–FERMI EQUATION

We now discuss the Thomas–Fermi equation

$$y''(x) = x^{-1/2}[y(x)]^{3/2} \tag{1}$$

to show that its special exact solution

$$y_s(x) = \frac{144}{x^3} \tag{2}$$

though a poor approximation by itself as a leading term for the large-distance solution of (1), plays a crucial role in its qualitative analytic study. This solution is easily found by writing $y = Ax^m$ and balancing terms in (1) so that $A = 144$ and $m = -3$. It serves as the important first term in the families of solutions which either go to infinity at $x = 0$ or to zero at $x = \infty$. Moreover, it demarcates the quarter-plane $x > 0, y > 0$ into two regions in which the analytic character of the solution changes drastically. We follow Hille (1970) and show how this special solution may be used to build up the asymptotic solution directly in the series form rather than through an integral equation formulation.

First, we note that (1), subject to the single boundary condition $y(0) = 1$, has the formal series solution

$$y(x) = 1 + b_2 x + b_3 x^{3/2} + \cdots + b_k x^{k/2} + \cdots, \tag{3}$$

where $b_i (i \geq 3)$ are polynomials in the arbitrary parameter b_2. A given value of the coefficient b_2 specifies a particular curve. Hille quotes a critical value $b_2 = -1.588$ (the numerical value -1.5880710 given by Bender and Orszag, 1978, is correct to seven decimal places) that leads to $y_\infty(x)$, a solution which satisfies the other boundary condition $y_\infty(\infty) = 0$. As computations of Bender and Orszag (1978) show, a slightly lower value of the slope $y'(0) = b_2$ makes the solution tend to zero at a finite positive value of x, while a slightly higher value of $y'(0)$ makes it blow up at a finite positive value of x. The leading asymptotic solution $y(x) = 144 x^{-3}$ is shown there to be a poor approximation at large distances.

We note that the substitution

$$v(x) = x^3 y(x) \tag{4}$$

changes (1) to

$$x^2 v'' - 6xv' + 12v = v^{3/2}, \tag{5}$$

an equation which plays a crucial role in the following analysis. We write

$$v = 144 + u \tag{6}$$

to peel off the asymptotic solution $144 x^{-3}$ from $y(x)$. Equation (5) becomes

$$x^2 u'' - 6xu' + 12u + 12^3 = [12^2 + u]^{3/2}. \tag{7}$$

Local and Global Analysis

This may be written as

$$x^2 u'' - 6xu' - 6u = 12^3 \left[(1 + 12^{-2}u)^{3/2} - 1 - \frac{3}{2} 12^{-2} u \right]. \tag{8}$$

This step is similar to one in Levinson's method (1970) (see Section 4.3); the right-hand side in (8) is now small $O(u^2)$ while the left side is a familiar Cauchy–Euler (linear) form. The homogeneous equation

$$x^2 u_0'' - 6x u_0' - 6u_0 = 0 \tag{9}$$

has the two linearly independent solutions

$$x^{-\sigma} \quad \text{and} \quad x^\tau \quad \text{with } \sigma = \frac{1}{2}[\sqrt{73} - 7], \quad \tau = \frac{1}{2}[\sqrt{73} + 7]. \tag{10}$$

It is these quadratic irrationalities which occur in the approximate solution of (1) attributed to Sommerfeld (1932), namely,

$$y_\infty^{(x)} \sim 144 x^{-3} [1 - (12^{2/3} x^{-1})^\sigma]^{-\tau/2}. \tag{11}$$

The solution (11) turns out to be a good approximation to the asymptotic solution, though its derivation is rather obscure.

Since (1) is nonlinear, it may have movable singularities, besides the two fixed ones at $x = 0$ and $x = \infty$. To show this, we pose the initial value problem

$$y(a) = 0, \quad y'(a) = -b, \quad b > 0. \tag{12}$$

Changing the dependent variable in (1) to v through

$$y = b(a - x) + v, \tag{13}$$

we have

$$v'' = [a - (a - x)]^{-1/2} [b(a - x) + v]^{3/2}, \quad v(a) = v'(a) = 0. \tag{14}$$

Writing $v \sim A(a - x)^m$ and balancing terms show that $A = a^{-1/2} b^{3/2}$ and $m = 7/2$. Therefore, we may write the series starting out with the term $a^{-1/2} b^{3/2} (a-x)^{7/2}$. This series converges for sufficiently small values of $|a-x|$, and its radius of convergence depends on a and b; it has a movable branch point singularity. This solution becomes complex-valued for $x > a$. If $y'(a) = b > 0$, the solution is real positive to the right of $x = a$ and complex-valued to the left. The preceding series holds to the right of $x = a$ if $a - x$ is replaced by $x - a$.

However, our major task is to investigate what role the asymptotic solution plays in the families of solutions which either go to infinity at a finite x or tend to zero at infinity. First, we prove that under certain conditions the solution to an initial value problem for (1) tends to infinity from the left. We also find its analytic behavior near the singularity.

THEOREM 4.4.1 Let $y(x)$ be a solution of (1) defined by an initial condition of the form

$$y(a) = b_0, \quad y'(a) = b_1, \qquad a \geq 0, \, b_0 \geq 0, \, b_1 \geq 0, \, b_0 + b_1 > 0. \tag{15}$$

Then there exists a number c such that $a < c < \infty$ and $y(x) \to +\infty$ as $x \to c$. Further, for $c - x$ small

$$y(x) < 400c(c-x)^{-4}. \tag{16}$$

If, in particular,

$$5b_1^2 \geq 4a^{-1/2} b_0^{5/2}, \tag{17}$$

then

$$c < \left[a^{3/4} + \frac{3}{2} \sqrt{5} b_0^{-1/4} \right]^{4/3} \tag{18}$$

and (16) holds for $a < x < c$.

Proof. We multiply (1) by $2y'$ and integrate to get

$$[y'(x)]^2 = \int_a^x x^{-1/2} y^{3/2} y' \, dx + b_1^2. \tag{19}$$

An integration by parts gives

$$[y'(x)]^2 = \frac{4}{5} x^{-1/2} [y(x)]^{5/2} + \frac{2}{5} \int_a^x s^{-3/2} [y(s)]^{5/2} \, ds$$
$$+ b_1^2 - \frac{4}{5} a^{-1/2} b_0^{5/2}. \tag{20}$$

Here, the initial conditions (15) have been used. We assume that (17) holds. Then (20) immediately leads to

$$[y'(x)]^2 > \frac{4}{5} x^{-1/2} [y(x)]^{5/2}, \qquad x > a, \tag{21}$$

or

$$[y(x)]^{-5/4} y'(x) > \frac{2}{5} \sqrt{5} x^{-1/4}, \tag{22}$$

which is valid in the largest interval to the right of $x = a$ where $y(x)$ remains finite. We note that, for $x > 0$, if $y(x) > 0$, then $y'(x) > 0$, and (1) implies that $y''(x) > 0$, so that $y(x)$ is positive and strictly increasing, and its graph is

Local and Global Analysis

concave upward. Therefore, $y(x)$ ultimately becomes infinite. The solution may possibly exist for all $x > 0$. We now show that this cannot happen.

Integrating (22) and using the initial conditions (15), we have

$$b_0^{-1/4} - [y(x)]^{-1/4} > 2\frac{\sqrt{5}}{15}(x^{3/4} - a^{3/4}). \tag{23}$$

This inequality cannot hold since the right-hand side becomes infinite as x does, while $y(x)$ is assumed to go to infinity there. It follows, therefore, that the interval of existence $[a,c)$ of finite values of $y(x)$ is finite. An estimate for c is obtained by putting $x = c$ and $y(x) = \infty$ in (23). This leads to the inequality (18) for c. To get an upper bound for the growth rate of $y(x)$ as x increases to c, we integrate (22) from x to c. We have

$$[y(x)]^{-1/4} < \frac{2}{15}\sqrt{5}[c^{3/4} - x^{3/4}]$$

$$= \left(\frac{3}{4}\right)\frac{2}{15}\sqrt{5}x_0^{-1/4}(c-x), \quad x < x_0 < c, \tag{24}$$

by the mean value theorem. Hence it follows that

$$y(x) < 400x_0(c-x)^{-4} < 400c(c-x)^{-4}.$$

If now (17) does not hold, we find that the integral in (20) is an unbounded function of x on the interval where $y(x)$ exists. Therefore, (21) holds for $x > A$ for some $A > a$. The interval of (finite) existence of $y(x)$ is finite, and (16) holds with the inequality (18) replaced by

$$c < \left\{A^{3/4} + \frac{3}{5}\sqrt{5}[y(A)]^{-1/4}\right\}^{4/3}. \tag{25}$$

Inequality (16) suggest that $y(x)$ has a pole of order 4 with leading term given by the right-hand side of (16). The singularity of $y(x)$ at $x = c$ raises the question of analytic continuation beyond c. If $c > 0$, three possibilities arise: (1) $y(x)$ exists for all $x > c$; it strictly decreases to the limit zero (as we have noted earlier), and the graph is concave upward. (2) $y(x)$ decreases to zero for some finite value a Then $x = a$ is a branch point, and $y(x)$ is complex-valued in some interval (a,b) where possibly $b = +\infty$. (3) $y(x)$ decreases to some positive minimum and must have another infinitude at $x = c_1$ as ensured by the theorem (see Figure 4.4.1). Hille (1970) remarks that existence of infinitely may infinitudes is a possibility which cannot be dismissed.

We now consider the role of the singular (asymptotic) solution (2) as a dividing curve between two families of solutions: one which remains bounded

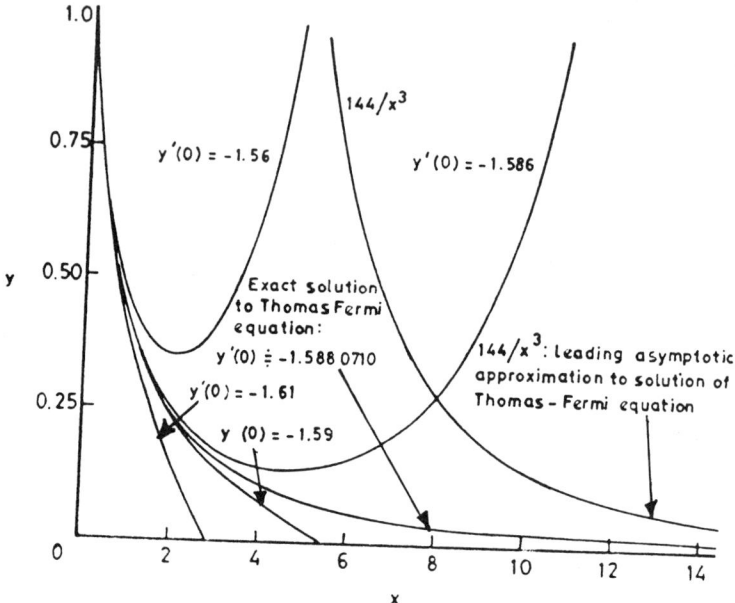

Figure 4.4.1 A computer plot of the solution to the Thomas–Fermi equation (1) with $y(0) = 1$, $y(\infty) = 0$ along with its leading asymptotic behavior $144x^{-3}$: the leading behavior is a poor estimate of $y(x)$ even when $x = 15$. Also shown are four functions satisfying equation (1) and $y(0) = 1$, but having initial slopes which are slightly different from the initial slope needed to give $y(+\infty) = 0$. Two of these functions cross the x-axis and become complex: the other two become infinite before $x = 15$ (From Bender and Orszag 1978.)

and tends to zero at infinity, and the second which also tends to zero at infinity but has a vertical tangent at the left end.

THEOREM 4.4.2 The boundary value problem for (1) with

$$y(a) = b, \quad \lim_{x \to \infty} y(x) = 0, \quad a \geq 0, \quad b > 0, \tag{26}$$

has a unique solution bounded at infinity. The singular solution $y_s(x) = 144x^{-3}$ satisfies (26) for any choice of (a,b) on its graph C. This curve separates the points of the first quadrant into two domains: D_- below C and D_+ above. For $(a,b) \in D_-$, there is a unique value of $\beta > 0$ such that

$$y = \beta^{-3} y_\infty(\beta^{-1} x) \tag{27}$$

Local and Global Analysis

defines the solution of (26). All these integral curves stay in D_- and are defined for $x \geq 0$. They can be extended to the left of (a,b) to the y-axis. If $(a,b) \in D_+$, then the corresponding solution of (26) stays in D_+. Its graph has a vertical asymptote.

Proof. First we recall that $y_\infty(x)$ is the solution of (1) with boundary conditions $y(0) = 1$, $\lim_{x \to \infty} y(x) = 0$. Moreover, if $y_\infty(x)$ is a solution, then a simple calculation shows that $\beta^{-3} y_\infty(\beta^{-1} x)$ is also a solution for $\beta > 0$. The solution $y_\infty(x)$ itself is given by (3) with $b_2 = -1.5880710$.

The existence and uniqueness of (1) and (26) are guaranteed by the theorem of Membrioni (1929). The uniqueness of the solution implies that the integral curves cannot intersect; in particular, they cannot intersect the singular solution curve C. Consider a particular curve of (27). It satisfies the condition at infinity. If it is to pass through (a,b), then

$$b\beta^3 = y_\infty(\beta^{-1} a). \tag{28}$$

If this equation has a root β, it is obtained by the intersection of the curves

$$v = bu^3 \quad \text{and} \quad v = y_\infty\left(\frac{a}{u}\right) \tag{29}$$

in the (u,v) plane. From Sommerfeld's approximate solution (11), we get

$$y_\infty\left(\frac{a}{u}\right) = 144 \left(\frac{u}{a}\right)^3 [1 + O(u)] \tag{30}$$

for $u \gtrsim 0$, where the $O(u)$ term is negative. Therefore, the curves $v = bu^3$ and $v = y_\infty(a/u)$ touch each other at the origin in the (u,v) plane. Now, by definition, $(a,b) \in D_-$ and lies below the curve C; therefore, there is a maximal interval $(0,\beta)$ in which

$$y_\infty\left(\frac{a}{u}\right) = \frac{144 u^3}{a^3} [1 + O(u)] > bu^3. \tag{31}$$

For large values of u, the inequality is reversed. Therefore, there is a value $u = \beta$ for which the equality sign holds, and for this value (27) represents the solution of (1) and (26).

If, on the other hand, $(a,b) \in D_+$, (28) has no solution. There is a unique solution of (1) satisfying (26). The graph of this solution lies above C. It can be extended to the left, but has a vertical asymptote $x = c$ where $c \geq 0$. The proof that $c > 0$ seems difficult, but this fact can be verified numerically.

Now we prove the following theorem, which characterizes all the solutions of (1), strictly decreasing to zero.

THEOREM 4.4.3 If $y(x)$ is a positive solution of (1) defined and finite for $x \geq a \geq 0$, then $y(x)$ is strictly decreasing to zero. Moreover, the function

$$v(x) = x^3 y(x) \tag{32}$$

is bounded and monotone with the limit 144 as $x \to +\infty$. The solution $y(x)$ coincides with the solution of the boundary value problem (26) where $b = y(a)$. If $(a,b) \in D_-$, then $v(x)$ is strictly increasing, and if $(a,b) \in D_+$, it is strictly decreasing. If (a,b) lies on C, the graph of the singular solution, then $v(x) \equiv 144$.

Proof. Since $y''(x) > 0$, the graph of $y(x)$ is concave upward. Hence, $y'(x)$ is increasing. If there were a point x_0 where $y'(x_0) \geq 0$, then by Theorem 4.4.1 there is a point $x = c$, $x_0 < c < \infty$, such that $y(x) \uparrow +\infty$ as $x \uparrow c$. This violates the assumption that $y(x)$ is defined and finite for $x \geq a \geq 0$. It follows that $y'(x) < 0$ for $x > a$. Hence, $\lim_{x \to \infty} y(x) = l \geq 0$. If $l > 0$, it follows from (1) that

$$y''(x) > l^{3/2} x^{-1/2}. \tag{33}$$

Integrating this inequality would show that $y'(x)$ becomes infinite with x. This again violates the assumption. Hence $l = 0$ and $y'' \downarrow 0$, $y'(x) \uparrow 0$, $y(x) \downarrow 0$. Therefore, if $y(a) = b$, we find that our solution $y(x)$ coincides with the solution of the boundary value problem (1) and (26). Since these integral curves are unique, they cannot intersect. We conclude that, in terms of the function $v(x)$ in (32), this implies that

$$v(x) > 144, \ = 144, \text{ or } < 144, \quad x \geq a, \tag{34}$$

according as the point (a,b) lies in D_+, on C, or in D_-. We have already seen that the function v satisfies equation (5). This DE can be rewritten as

$$[x^{-6} v'(x)]' = x^{-8} v(x) \{[v(x)]^{1/2} - 12\}. \tag{35}$$

On integration, this gives

$$x^{-6} v'(x) = -\int_x^\infty v(s) \{[v(s)]^{1/2} - 12\} s^{-8} ds. \tag{36}$$

In view of (32) and (26), the integral exists and the left member goes to zero as $x \to +\infty$. If $v(x) > 144$, for $x > a$, the right-hand side of (36) is negative and, therefore, $v(x)$ is decreasing. It increases if $v(x) < 144$. It follows then that $v(x)$ tends to a limit, say B. This limit must be 144. If $B \neq 144$, then (36) gives

$$\lim_{x \to \infty} x v'(x) = -\frac{1}{7} B(B^{1/2} - 12) \neq 0. \tag{37}$$

Local and Global Analysis

This would imply, on integration, that $v(x)$ becomes (logarithmically) infinite with x. This is absurd. Hence $B = 144$, and the theorem is proved.

Now we prove an embedding theorem for the singular solution $144x^3$.

THEOREM 4.4.4 Equation (1) has a one-parameter family of solutions of the form

$$y_\infty(x,a) = x^{-3}[144 + \sum_{n=1}^{\infty} a_n x^{-n\sigma}], \tag{38}$$

where $a_1 = a$ is arbitrary and the other coefficients are uniquely determined in terms of a. The series converges for large values of x. In particular, $a = 0$ gives the singular solution (2).

Proof. We start with DE (8) for u where $x^3 y = v = 144 + u$. Of the two solutions (10) of the homogeneous equation (9), the first, $x^{-\sigma}$, is small for large x. We, therefore, seeks a power series solution of (8) in the form

$$u(x) = \sum_{n=1}^{\infty} a_n x^{-n\sigma}. \tag{39}$$

Substituting (39) into (8) and equating coefficients of $x^{-n\sigma}$ to zero, we find that

$$(\sigma^2 - 7\sigma - 6)a_1 = 0. \tag{40}$$

Since σ is a root of $\sigma^2 - 7\sigma - 6$ (see equation (10)), a_1 is arbitrary. Other coefficients are uniquely determined by recurrence relations of the form

$$(n^2\sigma^2 - 7n\sigma - 6)a_n = P_n(a_1, a_2, \ldots, a_{n-1}) \quad n > 1. \tag{41}$$

The multiplier of a_n on the left is not equal to zero for all $n > 1$. The right-hand side P_n is a multinomial in the indicated arguments. On scrutinizing (8), we find that P_n is linear in the binomial coefficients of order 3/2. All other coefficients in P_n are powers of 12, positive and negative, and of positive integers arising from the expansions of the powers of u.

To prove the convergence of the series (39), we use the Lindelöff method (see Section 3.10). The dominant function is chosen to be

$$U(x) = \sum_{n=1}^{\infty} A_n x^{-n\sigma}, \quad A_1 = |a_1|, \tag{42}$$

such that it satisfies the equation

$$x^2 U'' - 6xU' - 6U = 12^3\left[(1 - 12^{-2}U)^{-3/2} - 1 - \frac{3}{2} 12^{-2} U\right]. \tag{43}$$

Comparing (43) with (8), we note that all terms in the expansion of the right-hand side will be positive because of the negative sign in $1 - 12^{-2}U$. Using the same argument as in Section 3.10, we show that all A_n's are positive and

$$|a_n| \le A_n \quad \text{for all } n.$$

Writing the partial sum of (42) as

$$U_n(x) = \sum_{k=1}^{n} A_k x^{-k\sigma}, \tag{44}$$

we find that U_n can be made as small as we please by choosing x large enough. We choose α and r such that, for a given n,

$$U_n(x) < \alpha, \quad x > r. \tag{45}$$

Again referring to Section 3.10, we observe that the partial sums U_{n+1} and U_n satisfy the inequality

$$x^2 U''_{n+1} - 6x U'_{n+1} - 6 U_{n+1} < 12^3[(1 - 12^{-2}U_n)^{-3/2} - 1 - \frac{3}{2} 12^{-2} U_n]$$

$$< 12^3[(1 - 12^{-2}\alpha)^{-3/2} - 1 - \frac{3}{2} 12^{-2}\alpha] \equiv B(\alpha), \tag{46}$$

where, for the second inequality, we have used (45). Substituting (44) in the left of (46), we have

$$\sum_{k=2}^{n+1}(k^2\sigma^2 + 7k\sigma - 6)A_k x^{-k\sigma} < B(\alpha). \tag{47}$$

(Note the range of k from 2 to $n + 1$.) Multiplying both sides of (47) by $x^{-\sigma-2}$ and integrating twice from r to infinity, we get

$$\sum_{k=2}^{n+1} A_k R_k r^{-(k+1)\sigma} < B(\alpha)[\sigma(\sigma + 1)]^{-1} r^{-\sigma}, \tag{48}$$

where

$$R_k = \frac{(k\sigma)^2 + 7k\sigma - 6}{(k + 1)\sigma[(k + 1)\sigma + 1]} \tag{49}$$

Evidently, R_k tends to 1 as $k \to \infty$ and, with $\sigma \simeq 0.772$, has a positive lower bound R for $k > 1$. Therefore, the left member of (48) satisfies the inequality

$$\sum_{k=2}^{n+1} A_k R_k r^{-(k+1)\sigma} > R[U_{n+1}(r) - A_1 r^{-\sigma}]. \tag{50}$$

Local and Global Analysis

Therefore, if r and α can be chosen such that

$$B(\alpha)[\sigma(\sigma+1)R]^{-1} = 1, \quad (A_1+1)r^{-\sigma} = \alpha, \tag{51}$$

then we have $U_n(r) \le \alpha$ for all n. Since σ and R are fixed, the first of (51) determines a unique value of $\alpha < 12^2$. The second of (51) then gives r. Thus, the series for $U(x)$ converges for $x > r$. By the dominant series test, the series for $u(x)$ also converges for $x > r$. Hence the theorem.

It is evident from (38) and the relevant recurrence relations that the case $a_1 = a = 0$ corresponds to the singular solution. Also, for $a > 0$, the integral curves $y_\infty(x;a)$ lie in D_+ above C while for $a < 0$, they lie in D_- below C. Numerical computations carried out by Bender and Orszag (1978) show (see Fig. 4.4.1) that retaining the first two terms in (38) gives a rather poor approximation in the asymptotic limit. The constant a_1 was found by numerical curve fitting to be -13.2709738, which is rather large in magnitude. However, the expansion (38) is convergent, even though not very useful for computational purposes.

Finally, we quote theorems for a second embedding of the singular solution in a one-parameter family of solutions unbounded at the origin. The embedding uses the second solution x^τ in (10), which is small for small values of x.

THEOREM 4.4.5 The singular boundary value problem

$$\lim_{x \downarrow 0} y(x) = +\infty, \quad y(a) = b, \quad a > 0, b > 0, \tag{52}$$

has a unique solution which is strictly decreasing. The function

$$v(x) = x^3 y(x) \tag{53}$$

is bounded and monotone with the limit 144 as $x \downarrow 0$. If $(a,b) \in C$, the graph of the singular solution, then $v(x) = 144$. If $(a,b) \in D_-$, then $y(x)$ stays in D_- and $v(x)$ increase to 144 as $x \downarrow 0$. If $(a,b) \in D_+$ then $y(x)$ stays in D_+, and $v(x)$ decreases to 144 as $x \downarrow 0$.

THEOREM 4.4.6 Equation (1) has a one-parameter family of solutions

$$y_0(x;c) = x^{-3}\left\{144 + \sum_{n=1}^{\infty} c_n x^{n\tau}\right\}, \tag{54}$$

where $c_1 = c$ is arbitrary; the other coefficients are uniquely determined, and the series converges for small values of x.

The proofs are similar to those for the first embedding theorem and are omitted here. However, reference may be made to Hille's (1970) work for details of this case. Hille (1970a) has also carried out a similar detailed analysis for the more general Emden–Fowler equation

$$y'' = x^{1-m}[y(x)]^m. \tag{55}$$

4.5 GLOBAL SOLUTIONS OF NONLINEAR BOUNDARY VALUE PROBLEMS BY ITERATION

Since most nonlinear boundary value problems do not admit exact solutions, one has to put together all analytic elements—local and global—be aided and guided by numerical solutions, and discover all the features of the solutions. For example, a local series solution may provide the solution up to the nearest singularity, delimiting its radius of convergence. The boundary value problem for a DE may be changed to an integral equation, which may provide solutions in all or parts of the domain. We shall illustrate this approach by two nonlinear boundary value problems over an infinite domain (Miles, 1982, 1978).

1. A Nonlinear Bessel Equation

The equation

$$F'' + r^{-1}F' - F + F^3 = 0, \qquad 0 < r < \infty, \tag{1}$$

governs a similarity solution of the axis-symmetric, cubic Schrödinger equation (Zakharov and Synakh, 1976), and also appears in the context of water waves. Equation (1) is to be solved subject to the conditions

$$F(r) \longrightarrow a, \qquad r \downarrow 0, \tag{2}$$

$$F(r) \sim AK_0(r) \sim \left(\frac{\pi}{2}\right)^{1/2} Ar^{-1/2}e^{-r} \equiv \left(\frac{8R}{r}\right)^{1/2} e^{R-r}, \qquad r \uparrow \infty, \tag{3}$$

where a and A are finite constants, K_0 is a modified Bessel function of the second kind, and R is an asymptotic scale. We fix the sign of A as positive and observe that equation (1) is invariant under $F \to -F$.

We note that the solution of the one-dimensional analog of (1), where the term $r^{-1}F'$ does not appear, has an exact solution $F = \sqrt{2}\,\text{sech}\,r$ which decays like e^{-r} as $r \uparrow \infty$. This describes a soliton centered about zero.

We follow Miles (1982). Numerical solution of (1)–(3) starting with conditions F and F' from (3) at a large distance to the right and integrating to the left shows (see Table 4.5.1 and Figures 4.5.1 and 4.5.2) that the solution is logarithmically singular as $r \downarrow 0$ for all but a discrete set of eigenvalues:

Local and Global Analysis

$R = R_n$, $0 < R_0 < R_1 \cdots$ or $a = a_n$, $0 < a_0 < -a_1 < a_2 < \cdots$. The eigenfunction F_0 corresponding to R_0 (or a_0) decays monotonically as $r \to \infty$ and is similar to the one-dimensional solution $\sqrt{2}\,\text{sech}\,r$ mentioned earlier. Other eigenfunctions F_n, $n \geq 1$, have n zeros in $0 < r < \infty$ and are qualitatively similar to the cnoidal wave solution of the one-dimensional equation. Since only F_0 with its monotonic behavior is of physical interest, we consider this case.

Before we discuss the integral equation formulation, we note that (1) has the following power series solutions near $r = 0$, where $F(0) = a$:

$$F(r) = a\left[1 + (1-a^2)\left(\frac{r}{2}\right)^2 + \frac{(1-a^2)(1-3a^2)}{(2!)^2}\left(\frac{r}{2}\right)^4 \right.$$
$$\left. + \frac{(1-a^2)(1-18a^2+21a^4)}{(3!)^2}\left(\frac{r}{2}\right)^6 + \cdots\right]. \tag{4}$$

This is a generalization of the $I_0(r)$ solution of the special linear form of (1), where F^3 is absent.

The series (4) describes $F_0(r)$ in $r \leq 1$, and the terms shown explicitly in (4) give an accuracy with 1% error for $r \leq 0.5$.

We consider (1) in the form

$$F'' + r^{-1}F' - F = -F^3 \tag{5}$$

Table 4.5.1 Results of Numerical Integration of Equations (1)–(3): Eigenvalues R_n and Amplitudes A_n and a_n

n	A_n	a_n	R_n	$\int_0^\infty F^2 r\,dr$
0	2.806	2.206	0.533	1.862
1	66.37	−3.332	2.856	12.28
2	1540	4.150	5.659	31.17
3	3.570×10^4	−4.829	8.594	58.47
4	8.270×10^5	5.424	11.587	94.16
5	1.923×10^7	−5.959	14.617	138.3
6	4.434×10^8	6.449	17.660	190.7
7	1.026×10^{10}	−6.905	20.722	251.6
8	2.376×10^{11}	7.332	23.795	320.8
9	5.499×10^{12}	−7.736	26.876	398.4
10	1.273×10^{14}	8.119	29.964	484.4

Source: From Miles (1982).

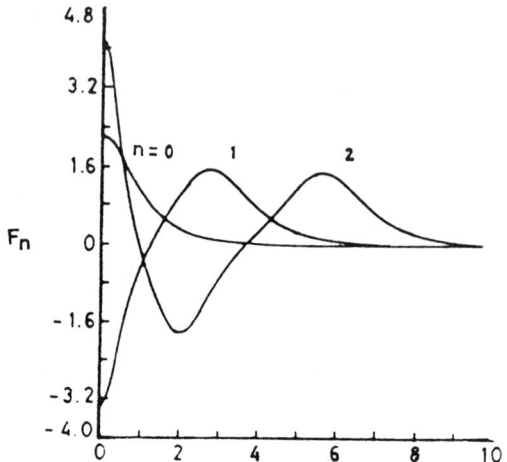

Figure 4.5.1 $F_n(r)$ as determined by numerical integration of (1)–(3). (From Miles 1982.)

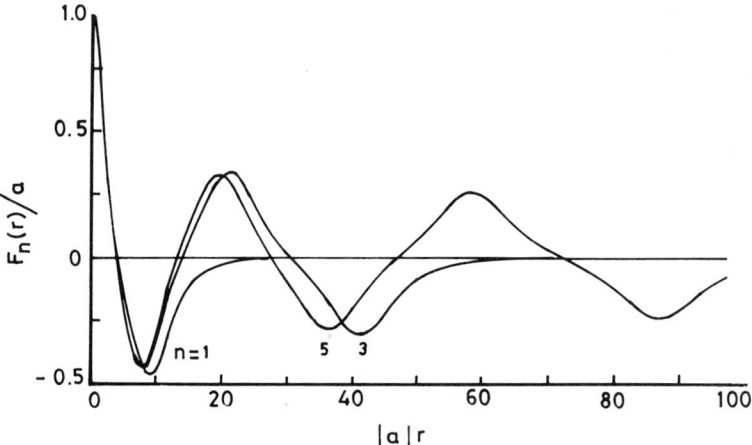

Figure 4.5.2 $F_n(r)/a$ versus $|a|r$, as determined by numerical integration. (From Miles 1982.)

Local and Global Analysis 179

and regard it as an inhomogeneous equation. On the left we have a Bessel operator with solutions $I_0(r)$ and $K_0(r)$. The linear solution which satisfies (3) is $AK_0(r)$. Therefore, the solution of the system (1)–(3) can be written by variation of parameters as

$$F(r) = AK_0(r) + \int_r^\infty [I_0(r)K_0(\eta) - I_0(\eta)K_0(r)]F^3(\eta)\eta\, d\eta, \tag{6}$$

where we have used the Wronskian value $W(I_0(r), K_0(r)) = -1/r$ in the integral term. The integral equation (6) in F is not solvable exactly. It may be solved iteratively starting with $F_0 = AK_0(r)$ as the zeroth iterate. One would get an expansion for F/A in powers of A^2. In view of the limits in the integral in (6), this expansion would be useful only for $r \gg 1$. Writing the first two terms in the asymptotic expansion for $I_0(r)$ and $K_0(r)$, namely,

$$I_0(r) = e^r (2\pi r)^{-1/2} \left(1 + \frac{1}{8r} + \cdots\right)$$

$$K_0(r) = e^{-r} \left(\frac{\pi}{2r}\right)^{1/2} \left(1 - \frac{1}{8r} + \cdots\right)$$

and substituting $4(R/\pi)^{1/2} e^R$ for A in (6) (see equation (3)), we get, after an integration,

$$F = \left(\frac{8R}{r}\right)^{1/2} e^{R-r} \left[1 - \frac{1}{8r} + O\left(\frac{1}{r^2}\right)\right]$$
$$\times \left\{1 - \frac{Re^{2R-2r}}{r}\left[1 - \frac{1}{r} + O\left(\frac{1}{r^2}\right)\right] + O\left(\frac{R^2}{r^2}\right)e^{4R-4r}\right\}. \tag{7}$$

Carrying out further iteration is difficult. However, (7) suggests an outer expansion that is uniformly valid with respect to R as $r \to \infty$. The variable that is natural to this expansion is

$$\zeta = \ln\left[\left(\frac{R}{r}\right)^{-1/2} e^{-R+r}\right] = r - R + \frac{1}{2}\ln\left(\frac{r}{R}\right). \tag{8}$$

We therefore seek a solution of (1) and (3) in the form

$$F \sim \sum_{n=0}^\infty r^{-n} f_n(\zeta), \tag{9}$$

$$F \sim AK_0(r) \sim A\left(\frac{\pi}{2r}\right)^{1/2} e^{-r} \sum_{n=0}^\infty C_n r^{-n}$$

$$= 2^{3/2}\left(\frac{R}{r}\right)^{1/2} e^{R-r} \sum_{n=0}^\infty C_n r^{-n} \quad \text{as } r \uparrow \infty, \tag{10}$$

where
$$C_n = (-1)^n 2^{-5n}(n!)^{-3}[(2n)!]^2.$$
Substituting (9) into (1) and equating coefficients of r^{-n} to zero, we have
$$f_0'' - f_0 + f_0^3 = 0, \tag{11}$$
$$f_n'' - f_n + 3f_0^2 f_n = g_n, \quad n \geq 1, \tag{12}$$
where
$$g_1 = -f_0' - f_0'', \quad g_2 = -\frac{1}{4}f_0'' + f_1' - f_1'' - 3f_0 f_1^2, \text{ etc..} \tag{13}$$
From (10),
$$f_n(\zeta) \sim 2^{3/2} C_n e^{-\zeta}, \quad \zeta \uparrow \infty. \tag{14}$$
The solution of (11), with (14) as the boundary condition as $\zeta \uparrow \infty$, is simply
$$f_0 = 2^{1/2} \operatorname{sech} \zeta, \tag{15}$$
the centered soliton for the one-dimensional case we noted earlier. Substituting (15) into (9) shows that
$$F = 2^{1/2} \operatorname{sech}\left(r - R + \frac{1}{2}\ln\frac{r}{R}\right) + O(r^{-1}), \tag{16}$$
which, Miles (1982) averred, is within 1% of the numerical solution for F_0 in $r > 1$ (see Figure 4.5.3).

To get the next-order terms, we first note that differentiation of (11) immediately shows that f_0' is a solution of
$$(f_0')'' - f_0' + 3f_0^2 f_0' = 0, \tag{17}$$

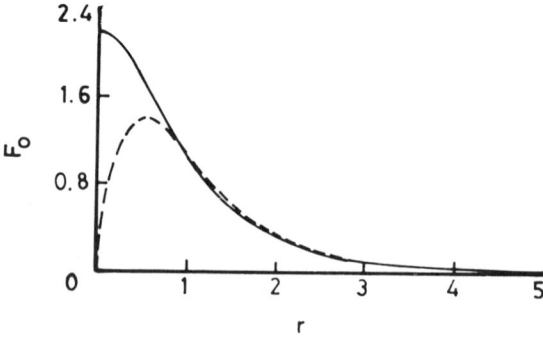

Figure 4.5.3 $F_0(r)$, as determined by numerical integration (—) and the asymptotic approximation (16) (- - -). (From Miles 1982.)

Local and Global Analysis

the homogeneous part of (12) with $n = 1$. Multiplying (17) by f_n and (12) by f_0' and subtracting, we obtain

$$f_n(f_0')'' - f_0' f_n'' = f_0' g_n. \tag{18}$$

This can be integrated to give

$$-\left(\frac{f_n}{f_0'}\right)' = \frac{1}{f_0'^2} \int_\zeta^\infty f_0' g_n \, d\eta, \tag{19}$$

where we have used the condition $f_n \to 0$ as $\zeta \uparrow \infty$. A second integration of (19) gives

$$f_n = f_0' \left[\int_\zeta^\infty \frac{d\xi}{f_0'^2(\xi)} \int_\xi^\infty f_0' g_n(\eta) \, d\eta - C_n \right], \quad n \geq 1, \tag{20}$$

where again the boundary condition (14) as $\zeta \uparrow \infty$ has been employed. To get f_1, we note that

$$f_0 = 2^{1/2} \operatorname{sech} \zeta, \quad f_0' = -2^{1/2} \operatorname{sech} \zeta \tanh \zeta,$$
$$g_1 = 2^{1/2} \operatorname{sech} \zeta \tanh \zeta + 2^{1/2} \operatorname{sech} \zeta \tanh^2 \zeta - 2^{1/2} \operatorname{sech}^3 \zeta,$$
$$C_1 = -\frac{1}{8}.$$

Substituting these quantities in (20), we get for $n = 1$

$$f_1 = 2^{1/2} \left[\frac{1}{2} e^{-\zeta} \left(\operatorname{sech}^2 \zeta - \frac{1}{3} \right) - \frac{1}{24} \operatorname{sech} \zeta \tanh \zeta \right],$$

so that the solution for F is

$$F = 2^{1/2} \left\{ \operatorname{sech} \zeta + r^{-1} \left[\frac{1}{2} e^{-\zeta} \left(\operatorname{sech}^2 \zeta - \frac{1}{3} \right) \right. \right.$$
$$\left. \left. - \frac{1}{24} \operatorname{sech} \zeta \tanh \zeta \right] + O(r^{-2}) \right\}, \tag{21}$$

which matches (7) in $\zeta \gg 1$ and also serves to give the explicit error in the first approximation (16).

Finally, we remark that the integral equation (6) in the limit $r \to 0$ yields, on comparison of coefficients, etc., the identities (see equations (2)–(3))

$$a = \int_0^\infty K_0(r) F^3(r) r \, dr, \quad A = \int_0^\infty I_0(r) F^3(r) r \, dr, \tag{22}$$

providing useful checks on the numerical integration.

2. Second Painlevé Transcendent

Now we consider the following problem for the second Painlevé transcendent (see also Section 8.6 for other properties of this equation):

$$F'' - 2F \pm 2F^3 = 0, \tag{23}_\pm$$

$$F \sim aAi(z) \sim \frac{1}{2}\pi^{-1/2}az^{-1/4}\exp\left(-\frac{2}{3}z^{3/2}\right), \quad z \uparrow \infty, \tag{24}$$

and

$$|F| < \infty, \quad -\infty < z < \infty. \tag{25}$$

Both forms $(23)_\pm$ can be handled simultaneously and should be considered vertically ordered. Here, only the behavior at the "front" $z = \infty$ is known. We want to find the solution in the entire domain $-\infty < z < \infty$, and, in particular, at the "tail" $z = -\infty$. We shall treat $(23)_\pm$ by an integral equation formulation and examine what information can be garnered. Equations $(23)_\pm$ occur in many nonlinear wave motions (see Section 4.2 for relation of (23) to the modified KdV equation). We shall follow Miles's (1978) analysis in the following. We choose a to be positive, since for negative a we have the relation

$$F_\pm(z; -a) = -F_\pm(z; a).$$

Again, the restriction to real a is analytically convenient since

$$F_\pm(z; ia) = iF_\mp(z; a).$$

By the Painlevé property, the second Painlevé transcendents (that is, solutions of $(23)_\pm$; see Section 8.5) have movable poles in the complex z-plane, but no branch points.

To get some understanding of the solution, Miles (1978) numerically integrated $(23)_\pm$ for $0 < a \leq 1$ by taking the values of F and F' from (24) at a sufficiently large value of z and found a one-parameter family of solutions with a as the parameter. This numerical procedure was found to be satisfactory from the stability point of view. Some of the results for $(23)_+$ are shown in Figure 4.5.4. As expected, the graph of $F_+(z)$ moves closer to $aA_i(z)$ as z decreases, but departs appreciably as a increases or z decreases. The largest zeros of $F_+(z)$, $F'_+(z)$, and $F''_+(z)$, denoted by z_0, z_1 and z_2, respectively, and the maximum $F_1 = F_+(z_1)$ are given in Table 4.5.2.

As for the nonlinear Bessel equation, we transform $(23)_\pm$ and (24) into an integral equation. The appropriate linear solution is $aAi(z)$. The term $\mp 2F^3$ is regarded as the inhomogeneous term. The Wronskian for the pair of solutions of the homogeneous (Airy) equation

$$F'' - zF = 0 \tag{26}$$

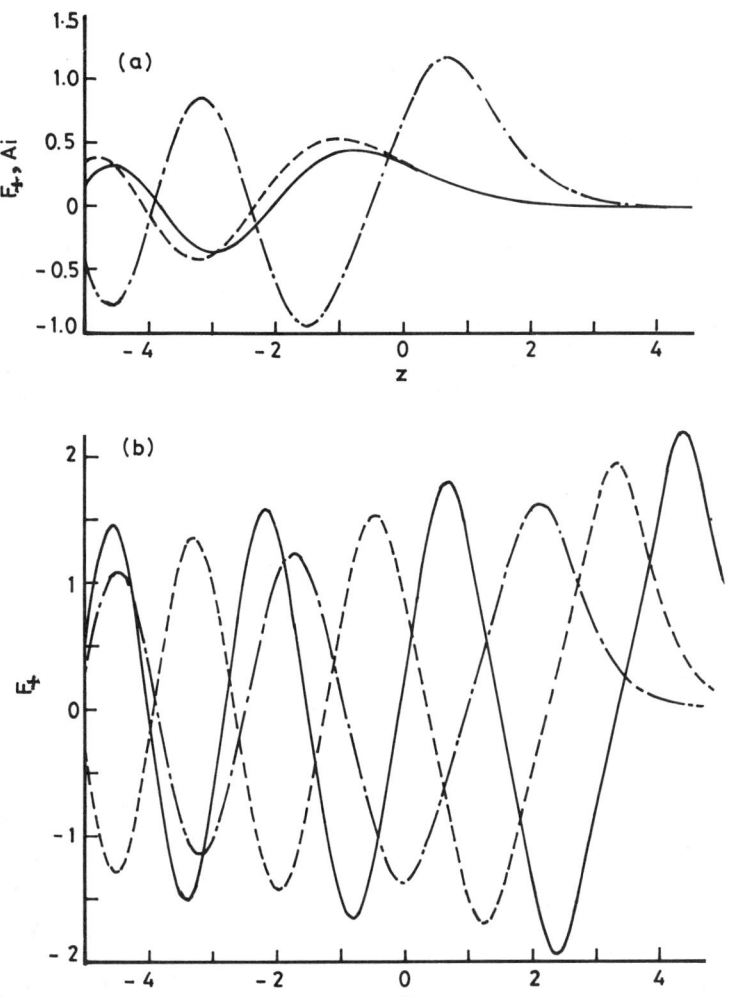

Figure 4.5.4 The solution of $(23)_+$ and (24) for (a) $a = 1$ (—) and (10) (—.—) compared with $Ai(z)$ (- - -); (b) $A = 10^2$ (—.—), $a = 10^3$ (- - -), $a = 10^4$ (—). (From Miles 1978.)

Table 4.5.2 Some Results of Numerical Integration of $(23)_+$ and (24)

a	z_0	z_1	z_2	F_1
0	−2.338	−1.019	0	—
0.1	−2.334	−1.015	0.004	0.053
0.3	−2.303	−0.985	0.022	0.158
0.5	−2.246	−0.932	0.057	0.254
0.7	−2.174	−0.863	0.100	0.340
0.9	−2.095	−0.788	0.155	0.416
1	−2.054	−0.749	0.182	0.450
10	−0.524	0.691	1.333	1.178
30	0.231	1.402	1.964	1.421
10^2	0.983	2.113	2.616	1.634
10^3	2.250	3.322	3.754	1.951
10^4	3.364	4.392	4.781	2.197
10^5	4.374	5.368	5.726	2.402
10^6	5.309	6.275	6.609	2.579

z_0 = the largest zero of $F_+(z)$; z_1 = the largest zero of $F'_+(z)$; z_2 = the largest zero of $F''_+(z)$; $F_1 = F_+(z_1)$.
Source: From Miles (1978).

is $1/\pi$, so using variation of parameters for the particular solution we have

$$F(z) = aAi(z) \mp 2\pi \int_z^\infty G(z,t) F^3(t)\, dt. \tag{27}$$

Now, the zeroth approximation to the solution of the integral equation (27) in the limit $a \to 0$ is just $aAi(z)$. The first iterate is obtained by writing $F(t) = aAi(t)$ in the integral in (27). Therefore, we have

$$F_\pm(z) = aAi(z) \pm a^3 [A(z)Bi(z) - B(z)Ai(z)] + O(a^5), \tag{28}$$

where

$$A(z) = 2\pi \int_z^\infty Ai^4(t)\, dt, \qquad B(z) = 2\pi \int_z^\infty Ai^3(t) Bi(t)\, dt. \tag{29}$$

Miles compared the solution (28) with the numerical solution and found that the error is much less than 1% for $z \geq 0$ and less than 2% for $z_2 \leq z < 0$, if $a < 1$, where we recall that z_2 is the largest zero of $F''_+(z)$.

Local and Global Analysis

The approximation (28) is not uniformly valid as $z \downarrow -\infty$ because of the logarithmic singularity of $\mathcal{A}(z)$ in that limit, as we now show. From Abramowitz and Stegun (1964), the asymptotic form of $Ai(-z)$ and $Bi(-z)$ for $|\arg z| < \frac{2\pi}{3}$ is

$$Ai(-z) \sim \pi^{-1/2} z^{-1/4} \left[\sin\left(\zeta + \frac{\pi}{4}\right) \sum_0^\infty (-1)^k c_{2k} \zeta^{-2k} \right.$$

$$\left. - \cos\left(\zeta + \frac{\pi}{4}\right) \sum_0^\infty (-1)^k c_{2k+1} \zeta^{-2k-1} \right],$$

$$|\arg z| \leq \frac{2\pi}{3}; \quad (30)$$

$$Bi(-z) \sim \pi^{-1/2} z^{-1/4} \left[\cos\left(\zeta + \frac{\pi}{4}\right) \sum_0^\infty (-1)^k c_{2k} \zeta^{-2k} \right.$$

$$\left. + \sin\left(\zeta + \frac{\pi}{4}\right) \sum_0^\infty (-1)^k c_{2k+1} \zeta^{-2k-1} \right],$$

$$|\arg z| \leq \frac{2\pi}{3};$$

$$\zeta = \frac{2}{3} z^{3/2}.$$

Using (30) in (29), one arrives at the result

$$\pi \mathcal{A}(z) = \frac{3}{4} \ln(-z) + \text{const} + O(z^{-3/2}), \quad z \downarrow -\infty. \quad (31)$$

$$\mathcal{B}(z) = \text{const} + O(z^{-3/2}).$$

Knowing the asymptotic form (31) of the functions $\mathcal{A}(z)$ and $\mathcal{B}(z)$, we evaluate the constants herein by an accurate computation of the integrals $\mathcal{A}(z)$ and $\mathcal{B}(z)$ defined by (29). The final form is determined as

$$\pi \mathcal{A}(z) = \frac{4}{3} + \frac{3}{4} \ln(-z) + O(z^{-3/2}) \equiv \pi \tilde{\mathcal{A}}(z) \quad (32)$$

and

$$\mathcal{B}(z) = \frac{1}{4} + O(z^{-3/2}), \quad z \downarrow -\infty. \quad (33)$$

Now using (30), (32), and (33) in (28), we get the asymptotic form

$$F_\pm(z) \sim \pi^{-1/2} a(-z)^{-1/4} \left[\left(1 \mp \frac{1}{4} a^2\right) \sin \mathcal{X} \pm a^2 \tilde{\mathcal{A}}(z) \cos \mathcal{X} \right]$$

$$(z \downarrow -\infty), \quad (34)$$

where

$$\mathcal{X} = \frac{2}{3}(-z)^{3/2} + \frac{\pi}{4}.$$

The approximation (34) can be recast, for $a \ll 1$, into the form

$$F_\pm(z) \sim \pi^{-1/2} a(1 \mp \frac{1}{4}a^2)(-z)^{-1/4} \sin(\mathcal{X} \pm a^2 \tilde{A}). \qquad (35)_\pm$$

The approximate solutions (35) are compared with the results of numerical integrations in Figures 4.5.5 and 4.5.6 for $z \leq -1$ and $a = 1$, 0.9, and 0.5. The agreement is remarkable. Miles used perturbation analysis to find other segments of the solutions—those for $a \uparrow \infty$ near z_1, the largest zero of $F'_\pm(z)$, and the oscillatory tail of F_+ as $a \uparrow \infty$ (see Miles, 1978a).

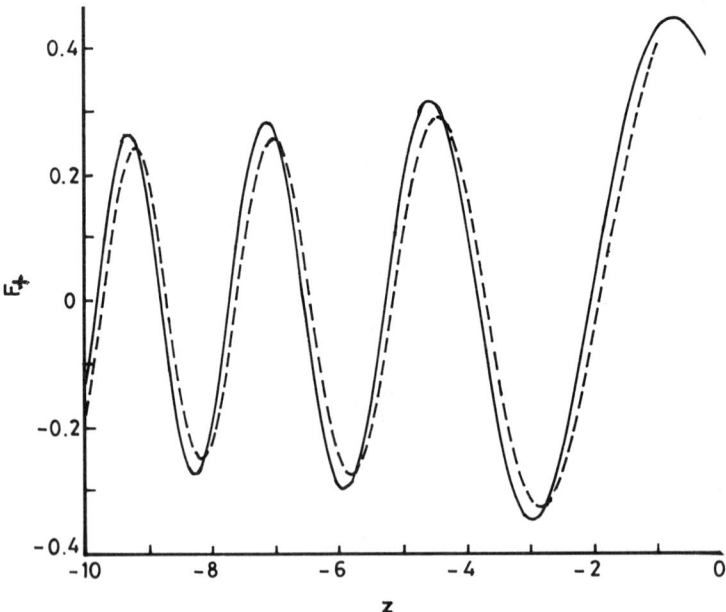

Figure 4.5.5 Comparison of the asymptotic approximation $(35)_+$ for $z \leq -1$ (- - -) with the numerical integration of $(23)_+$ and (24) (—) for $a = 1$. (From Miles 1978.)

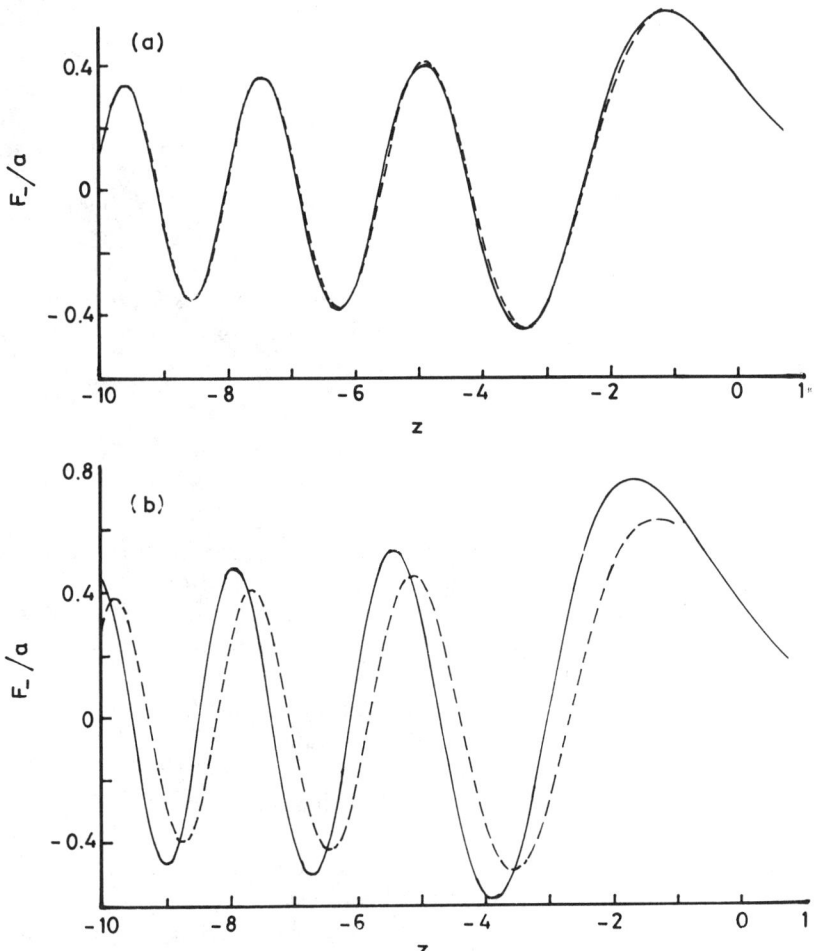

Figure 4.5.6 Comparison of the asymptotic approximation $(35)_-$ for $z \leq -1$ (- - -) with the numerical integration of $(23)_-$ and (24) (—) for (a) $a = 0.5$ and (b) $a = 0.9$. (From Miles 1978.)

4.6 THE CONNECTION PROBLEM FOR EULER–PAINLEVÉ EQUATIONS

Here we treat a class of nonlinear ODE which arise from group-theoretic consideration (or similarity transformation) of the generalized Burgers equations (GBE), and seem to characterize these PDE in the same manner as Painlevé equations represent the Korteweg–deVries type of equations. This class of equations may be written as

$$yy'' + ay'^2 + f(x)yy' + g(x)y^2 + by' + c = 0, \tag{1}$$

where $f(x)$ and $g(x)$ are (sufficiently) smooth arbitrary functions and a, b, and c are real constants. We shall refer to the solutions of (1) as *Euler–Painlevé transcendents*, since the special case $b = 0$, $c = 0$ of (1), namely,

$$yy'' + ay'^2 + f(x)yy' + g(x)y^2 = 0, \tag{2}$$

which contains nonlinear terms of degree 2, was exactly solved by Euler and Painlevé (see Kamke, 1943); indeed (2) can be exactly linearized by the simple transformation

$$y = v^{1/(a+1)}$$

to

$$v'' + f(x)v' + (a+1)g(x)v = 0. \tag{3}$$

In contrast, the addition of linear terms to (2) renders the (resulting) more general form (1) generally nonintegrable in closed form. There are special cases of (1) which, however, can be solved in closed form, and the solutions satisfy physically meaningful boundary conditions at $x = \pm\infty$ or $x = \infty$ and $x = x_0$, a finite point.

Actually, we shall treat equations of the form (1), and another equivalent form which arises directly from the similarity transformation of GBE. It is, in fact, for the latter equation that we pose a connection problem: to find the solution of certain nonlinear ODE which satisfy definite asymptotic conditions at $x = +\infty$ and at $x = -\infty$. The solution of the connection problem will exist only for certain ranges of parameters occurring in the problem and the boundary conditions. It is best to treat these problems directly with reference to PDE and show when ODE form the intermediate asymptotics (see Barenblatt, 1979, and Sachdev, 1987), to which a large class of solutions of a certain class of initial value problems asymptote in the limit of large time. The intermediate asymptotics will solve the connection problem and provide the link between the solutions of IVP of PDE and the connection problems of the ODE. We consider two distinct generalized Burgers equations separately.

1. A Generalized Burgers Equation with Damping

We consider the GBE

$$u_t + u^\beta u_x + \lambda u^\alpha = \frac{\delta}{2} u_{xx} \tag{4}$$

where α, β, and λ are real numbers and δ is a small parameter, called the *diffusivity of sound*. Equation (4) reduces to the standard Burgers equation when $\beta = 1$ and $\lambda = 0$. Equation (4) has a self-similar form of the solution

$$u = t^{1/(1-\alpha)} f(\eta), \qquad \eta = x(2\delta t)^{-1/2}, \tag{5}$$

provided

$$\beta = \frac{\alpha - 1}{2}. \tag{6}$$

Under this condition, (4) reduces to the ODE

$$f'' + 2\eta f' - \frac{4}{1-\alpha} f - 4(2\delta)^{-1/2} f^{(\alpha-1)/2} f' - 4\lambda f^\alpha = 0, \tag{7}$$

where the prime denotes differentiation with respect to η. The "reciprocal" transformation

$$H = \delta^{1/2} f^{(1-\alpha)/2} \tag{8}$$

removes the fractional powers in (7) and changes it to a simpler form

$$HH'' - 2(1 + \alpha_1)H'^2 + 2\eta HH' - 2H^2 - 2^{3/2} H' - 2\lambda_1 = 0, \tag{9}$$

where

$$\alpha_1 = \frac{1}{2} \frac{3-\alpha}{\alpha - 1}, \qquad \lambda_1 = \lambda \delta(1 - \alpha). \tag{10}$$

Equation (9) is a special case of (1) with $a = -2(1+\alpha_1) = (1+\alpha)/(1-\alpha)$, $f(x) = 2x$, $g(x) = -2$, $b = -2^{3/2}$ and $c = -2\lambda_1$. For the standard Burgers equation, $\alpha_1 = 0$, and $\lambda_1 = 0$, corresponding to $\alpha = 3$, $\lambda = 0$. In this case, (9) reduces to

$$HH'' - 2H'^2 + 2\eta HH' - 2H^2 - 2^{3/2} H' = 0. \tag{11}$$

This equation has a solution corresponding to the "single hump" solution of the Burgers equation, namely,

$$H_B = \frac{(2\pi)^{1/2}}{e^R - 1} \exp(\eta^2) + \left(\frac{\pi}{2}\right)^{1/2} \exp(\eta^2) \operatorname{erfc} \eta, \tag{12}$$

where

$$\operatorname{erfc} z = 1 - \operatorname{erf} z$$

$$= 1 - \frac{2z}{(\pi)^{1/2}} e^{-z^2} \sum_{k=0}^{\infty} \frac{(2z^2)^k}{1 \cdot 3 \cdots (2k+1)}, \qquad |z| < \infty. \tag{13}$$

In (12) R is the Reynolds number equal to $(1/\delta)\int_{-\infty}^{\infty} u\,dx$, which is constant for standard Burgers equation. The solution of (4) in this case is

$$u = \left(\frac{\delta}{t}\right)^{1/2} \frac{1}{H_B(\eta)}.$$

We now attempt to find the decaying solutions (5) of (4) so that $\alpha > 1$. In this case, the solutions of (9) are sought in the series form

$$H = \sum_{n=0}^{\infty} a_n \eta^n. \tag{14}$$

Substituting (14) into (9) leads to the following values of the coefficients:

$$a_2 = \frac{1}{a_0}[(a_0^2 + 2^{1/2}a_1 + a_1^2) + \lambda_1 + \alpha_1 a_1^2],$$

$$a_3 = \frac{1}{3a_0}[(a_0 a_1 + 2^{3/2}a_2 + 3a_1 a_2) + 4\alpha_1 a_1 a_2],$$

$$\vdots$$

$$a_{k+2} = \frac{2a_k}{(k+1)(k+2)} + \frac{2a_{k+1}}{(k+2)a_0}(2^{1/2} + \alpha_1 a_1 + a_1)$$

$$+ \frac{2}{(k+1)(k+2)a_0}\sum_{i=1}^{k}\left\{-\frac{(k+1-i)(k+2-i)}{2}a_i a_{k-i+2}\right.$$

$$+ (1+\alpha_1)(i+1)(k+1-i)a_{i+1}a_{k-i+1} + a_i a_{k-i}$$

$$\left. - (k+1-i)a_{i-1}a_{k-i+1}\right\}, \quad k = 1, 2, \ldots. \tag{15}$$

Thus we have a two-parameter (a_0, a_1) family of series solutions. The convergence of this series by direct computation seems difficult to establish.

For the Burgers equation, the case with $\alpha = 3$, $\lambda = 0$, the function H_B, given by (12), follows from (14) if $a_1 = -2^{1/2}$. The free parameter a_0 in this case gives a single-parameter family of solutions and corresponds to the (constant) value of the Reynolds number, which fixes a definite single hump profile. For the Euler–Painlevé equation (9), it does not seem possible to fix a priori the range of parameters a_0 and a_1 such that the series (15) converges over $-\infty < \eta < \infty$.

We follow a rather different route for this purpose. We first treat equation (7) (connected to (9) by (8)). We find the asymptotic solution of the f equation (7) when η tends either to $+\infty$ or to $-\infty$; that is, we require

Local and Global Analysis

that $f \to 0$ as $\eta \to \pm\infty$. Thus, linearizing (7), we have

$$f'' + 2\eta f' - \frac{4}{1-\alpha}f = 0. \tag{16}$$

The solution of (16) is

$$f = A\exp(-\eta^2)H_\nu(\eta) \sim A\exp(-\eta^2)(2\eta)^{2\alpha_1} \quad \text{as } \eta \uparrow +\infty \tag{17}$$

and

$$f \sim O(\eta^{-2\alpha_1-1}) \quad \text{as } \eta \downarrow -\infty, \tag{18}$$

where

$$\nu \equiv 2\alpha_1 = \frac{3-\alpha}{\alpha-1}.$$

Here, H_ν is the Hermite function of order ν, and A is the amplitude parameter. Thus, the linear solution decays exponentially as $\eta \to +\infty$ and algebraically as $\eta \to -\infty$, provided $2\alpha_1 + 1 > 0$; that is, $\alpha > 1$.

We now pose the boundary value or connection problem for (7), namely,

$$f'' + 2\eta f' - \frac{4}{1-\alpha}f - 4(2\delta)^{-1/2}f^{(\alpha-1)/2}f' - 4\lambda f^\alpha = 0, \tag{7}$$

$$f \sim A\exp(-\eta^2)H_\nu(\eta) \sim A\exp(-\eta^2)(2\eta)^{2\alpha_1}, \quad \eta \uparrow \infty \tag{19a}$$

$$f \longrightarrow 0, \quad \eta \downarrow -\infty, \tag{19b}$$

and

$$|f| < \infty, \quad -\infty < \eta < \infty. \tag{20}$$

Before we solve (7), (19), and (20), we note two special exact solutions of (7). The first is the constant solution

$$f = [\lambda(\alpha-1)]^{1/(1-\alpha)} = f_m, \quad \text{say}; \tag{21}$$

f_m is, incidentally, also the maximum value of f that the maxima of the single hump solutions can attain. This follows easily from (7) if we note that, at the maximum, $f' = 0$, $f'' < 0$, etc.

The second exact solution is

$$f = \begin{cases} (A_+\eta)^{2/(1-\alpha)}, & \eta > 0, \\ (-A_-\eta)^{2/(1-\alpha)}, & \eta < 0, \end{cases} \tag{22}$$

where

$$A_+ = \frac{(2/\delta)^{1/2}(\alpha-1)}{(\alpha+1)[(1+\lambda\delta(1+\alpha))^{1/2}+1]},$$

$$A_- = \frac{(2/\delta)^{1/2}(\alpha-1)}{(\alpha+1)[(1+\lambda\delta(1+\alpha))^{1/2}-1]}. \tag{23}$$

The solution (22) is singular and tends, for $\alpha > 1$, to infinity as $|\eta| \to 0$.

We now give the results of the numerical study of the connection problem (7), (19), and (20). The integration was started from $\eta \sim 4$, with a certain value of A and initial conditions (19a) and continued to decreasing values of η until the solution became essentially zero. Figures (4.6.1)–(4.6.4) show the solution for $\alpha = 1.5, 2, 2.5$, and 3 in the similarity range of this parameter, with the corresponding values $\beta = (\alpha - 1)/2 = 0.25, 0.5, 0.75$, and 1.0, respectively. For each pair (α, β), there is a value $A = A_{\max}$ for which the solution does not tend to zero as $\eta \to -\infty$, but instead asymptotes to the (exact) constant solution (21). Table 4.6.1 gives the values of $A_{\max}$ for various (α, β) pairs. From the numerical solutions for these values of A between 0 and $A_{\max}$, we calculated the values of $f(0)$ and $f'(0)$ and, hence, the values of $a_0 = H(0)$, $a_1 = H'(0)$ via (8). Table 4.6.2 contains the relevant values of a_0 and a_1 for different sets of A, α, and β. With a_0 and a_1 thus determined, the series (14) was summed and compared with the (exact) numerical solution. Analytic continuation was used when the convergence of the series slowed down. Agreement of this analytic solution with the numerical one was excellent, the discrepancy being $O(10^{-7})$ (see Table 4.6.3).

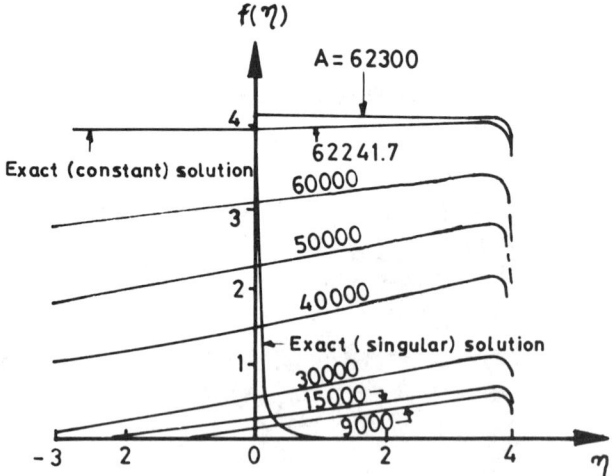

Figure 4.6.1 Solution of equations (7) and (17) for various values of A and for $\alpha = 1.5$, $\lambda = 1$. The constant solution (21) and the singular solution (22) are also shown. (From Sachdev, Nair, and Tikekar 1986.)

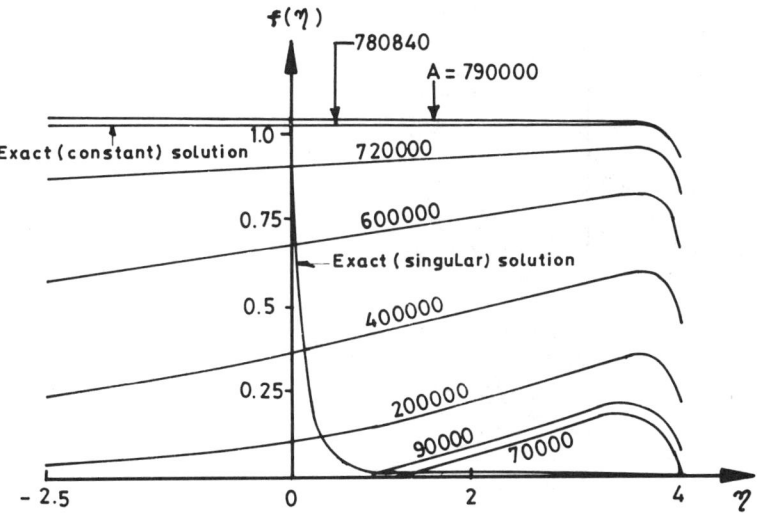

Figure 4.6.2 Same as in Figure 4.6.1 for $\alpha = 2$. (From Sachdev, Nair, and Tikekar 1986.)

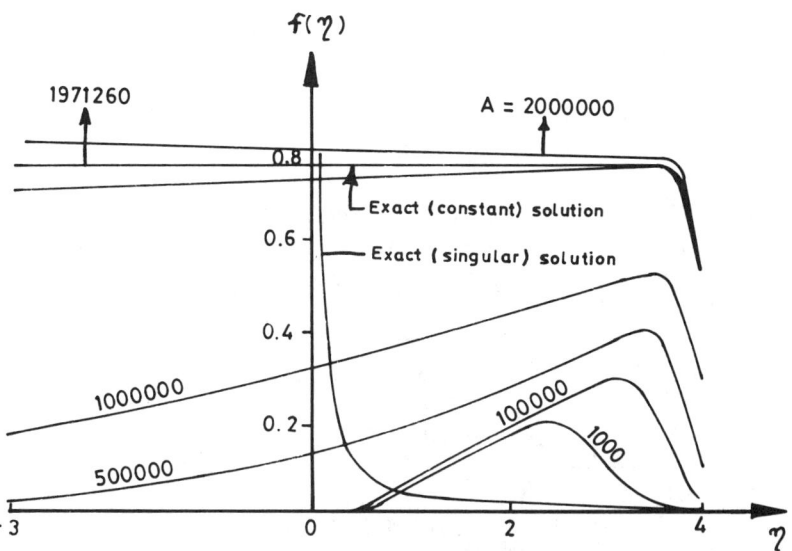

Figure 4.6.3 Same as in Figure 4.6.1 for $\alpha = 2.5$. (From Sachdev, Nair, and Tikekar 1986.)

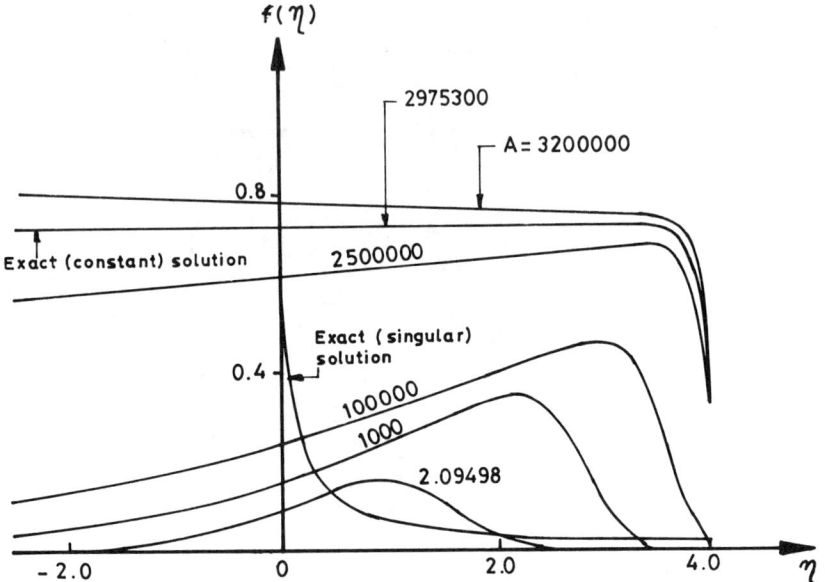

Figure 4.6.4 Same as in Figure 4.6.1 for $\alpha = 3$. (From Sachdev, Nair, and Tikekar 1986.)

Table 4.6.1 Critical Values of the Amplitude Parameter A and $f_{\max}$ for Different Choices of α and β in the Similarity Range for $\lambda = 1$ (see equations (19) and (21))

α	β	A_m	$f_{\max}$ Numerical	$f_{\max}$ Exact
1.5	0.25	62,241.75	4.0	4.0
2.0	0.5	780,840.6	1.0	1.0
2.5	0.75	1,971,256.0	0.763143	0.763143
3.0	1.0	2,975,300.0	0.707128	0.707107

Source: From Sachdev, Nair, and Tikekar (1986).

Table 4.6.2 Coefficients a_0 and a_1 in Series (14) for the Permissible (Similarity) Range of the Amplitude Parameter A Corresponding to Different Values of α and β for $\lambda = 1$, $\delta = 0.01$

No.	A	a_0	a_1
(i) $\alpha = 3$, $\beta = 1$			
1	0.0	0.0	0.0
2	2.095	1.2182	−1.1953
3	3.25	1.4233	−1.3305
4	1,000.0	0.70023	−0.36571
5	100,000.0	0.44529	−0.13077
6	2,500,000.0	0.16709	−0.00555
7	2,953,000.0	0.141414	0.0
(ii) $\alpha = 2.5$, $\beta = 0.75$			
1	0.0	0.0	0.0
2	500,000.0	0.49037	−0.17083
3	1,000,000.0	0.24186	−0.03108
4	1,800,000.0	0.13462	−0.0022
5	1,900,000.0	0.12728	−0.00085
(iii) $\alpha = 2$, $\beta = 0.5$			
1	200,000.0	0.32406	−0.07034
2	400,000.0	0.16960	−0.01336
3	600,000.0	0.12276	−0.00359
4	720,000.0	0.10651	−0.00095
5	780,841.0	0.1	0.0

Source: From Sachdev, Nair, and Tikekar (1986).

It is clear from the asymptotic form (17) (and we have confirmed it numerically) that the solution of the connection problem for (7) exists for all $\alpha > 1$. However, the similarity solution (5) of (4) is significant only in the range $1 < \alpha \leq 3$, where the solutions to the initial value problems for (4) with "suitable" vanishing initial conditions at $\pm\infty$ asymptote to the self-similar form. Actually the initial conditions for (4) should match with the asymptotic behavior (19) of the ODE (7). The solutions of the lat-

Table 4.6.3 Comparison of Series Solution (14) and Numerical Solution of (7) and (19) for $\alpha = 3$, $\beta = 1$, $\lambda = 1$, and $\delta = 0.01$

	Series solution		Numerical solution
η	$H(\eta)$	$f(\eta)$	$f(\eta)$
−3.0	36.77590	0.0027192	0.0027192
−2.5	29.15179	0.0034303	0.0034303
−2.0	19.82509	0.0050441	0.0050441
−1.5	10.26901	0.0097380	0.0097380
−1.0	4.541242	0.0220204	0.0220204
−0.5	2.165305	0.0461829	0.0461829
0.0	1.218223	0.0820868	0.0820867
0.5	0.8097551	0.1234941	0.1234941
1.0	0.6595272	0.1516237	0.1516237
1.5	0.8547518	0.1169930	0.1169928
2.0	2.926066	0.0341755	0.0341754
2.5	24.9906	0.0040015	0.0040015
3.0	387.044	0.0002583	0.0002584
3.5	9,993.162	0.0000100	0.0000100
4.0	336,869.9	0.0000003	0.0000003

The a_i in (14) from $i = 0$ to $i = 13$ are 1.2182, −1.1957, 0.98736, −0.60357, 0.34600, −0.14134, 0.05471, −0.00713, −0.00259, 0.00537, −0.00297, 0.00155, −0.000444, 0.0007.
Source: From Sachdev, Nair, and Tikekar (1986).

ter then constitute intermediate asymptotics. For numerical elucidation of these statements, see Sachdev, Nair, and Tikekar (1986). This paper also contains results regarding the non-self-similar solutions of (4) as well as some growing solutions for $\lambda < 0$.

2. Nonplanar Burgers Equations with General Nonlinearity

Now we treat the GBE

$$u_t + u^\alpha u_x + \frac{ju}{2t} = \frac{\delta}{2} u_{xx}, \tag{24}$$

where α is a real number; $j = 0, 1, 2$ for plane, cylindrical, and spherical symmetry, respectively. Equation (24) admits the similarity form of the solution

$$u = t^{-1/2\alpha} f(\eta), \qquad \eta = x(2\delta t)^{-1/2}. \tag{25}$$

Substituting (25) into (24), we get the nonlinear ODE

$$f'' - 2^{3/2} \delta^{-1/2} f^\alpha f' + 2\eta f' + \frac{2(1-\alpha j)}{\alpha} f = 0, \tag{26}$$

where prime denotes differentiation with respect to η.
  Setting

$$H = \delta^{1/2} f^{-\alpha} \tag{27}$$

transforms (26) to

$$HH'' - \frac{\alpha+1}{\alpha} H'^2 + 2\eta HH' - 2(1-\alpha j)H^2 - 2^{3/2} H' = 0. \tag{28}$$

Equation (28) is a special case of the Euler–Painlevé form (1) with $a = -(\alpha+1)/\alpha$, $f(x) = 2x$, $g(x) = -2(1-\alpha j)$, $b = -2^{3/2}$, and $c = 0$. The solution of (28) can be written in the form of a Taylor series

$$H(\eta) = \sum_{n=0}^{\infty} a_n \eta^n, \tag{29}$$

where the coefficients a_k can be found by substitution into (28) as

$$a_2 = \frac{1}{a_0} \left[\frac{\alpha+1}{2\alpha} a_1^2 + (1-\alpha j) a_0^2 + \sqrt{2} a_1 \right], \tag{30}$$

$$a_{k+2} = \frac{2}{(k+1)(k+2)a_0}$$

$$\times \left\{ \frac{\alpha+1}{2\alpha} (k+1) a_1 a_{k+1} + (1-\alpha j - k) a_0 a_k + \sqrt{2} (k+1) a_{k+1} \right.$$

$$+ \sum_{i=1}^{k} \left[-\frac{1}{2}(k+2-i)(k+1-i) a_i a_{k+2-i} \right.$$

$$+ \frac{\alpha+1}{2\alpha}(i+1)(k+1-i) a_{i+1} a_{k+1-i} \tag{31}$$

$$\left. \left. + (1-\alpha j) a_i a_{k-i} - (k-1) a_i a_{k-i} \right] \right\}, \qquad k = 1, 2, \ldots.$$

Thus, we have a two-parameter (a_0, a_1) family of solutions of (28). However, for the special choice $\alpha = 1/(j+1), j = 0, 1, 2,$ and $a_1 = -2^{3/2}\alpha/(\alpha + 1)$, the coefficient a_2 in (30) depends only on a_0 and we have a single-parameter family of solutions. This series can be summed up and the solution written in terms of exponential functions and erfc. We shall show this more directly when we find some special exact solutions (see equation (40)). The free parameter a_1 in this special case may correspond to the amplitude or the Reynolds number,

$$R = \frac{1}{\delta} \int_{-\infty}^{\infty} u \, dx,$$

of the profile, as for the standard Burgers equation, which is a special case of (24) with $j = 0$, $\alpha = 1$.

For the damped Burgers equation (4), we again find the asymptotic solutions for large η under the condition that $f \to 0$ as $\eta \to \pm\infty$. The linearized form of (26) is

$$f'' + 2\eta f' + \frac{2(1-\alpha j)}{\alpha} f = 0, \tag{32}$$

with the solution form

$$f(\eta) = A e^{-\eta^2} H_\nu(\eta) \quad \text{for } \eta > 0, \tag{33a}$$

$$f(\eta) \sim \frac{B\pi^{1/2}}{\Gamma(-\nu)} |-\eta|^{j-1/\alpha} \quad \text{for large negative } \eta, \tag{33b}$$

provided $\alpha j < 1$. Here, $\nu = 1/\alpha - (j+1)$, $H_\nu(\eta)$ is the Hermite function of order ν, and A and B are amplitude parameters. Thus, here again the linear solution decays exponentially as $\eta \to +\infty$ and algebraically as $\eta \to -\infty$.

We now pose the boundary value or connection problem for (26):

$$f'' - 2^{3/2}\delta^{-1/2}f^\alpha f' + 2\eta f' + \frac{2(1-\alpha j)}{\alpha} f = 0, \tag{26}$$

$$f \sim A \exp(-\eta^2) H_\nu(\eta), \quad \eta \uparrow \infty, \tag{34a}$$

$$f \longrightarrow 0, \quad \eta \downarrow -\infty, \tag{34b}$$

and

$$|f| < \infty, \quad -\infty < \eta < \infty. \tag{35}$$

The study of this connection problem is aided considerably by the following analysis, which yields some special exact solutions and identifies the ranges of the parameter α for which the solutions vanish either at $\eta = -\infty$ or at a finite value of $\eta = \eta_0$, say, as we proceed backward from $\eta = \infty$.

Local and Global Analysis

For the special choice $\alpha = 1/(j + 1)$, (26) can be written as

$$f + \eta f' + \frac{1}{2}f'' = \left(\frac{2}{\delta}\right)^{1/2} f^\alpha f' \tag{36}$$

and hence can be immediately integrated. Using the conditions that both f and f' tend to zero as $\eta \to +\infty$, we get

$$\eta f + \frac{1}{2}f' = \frac{1}{\alpha + 1}\left(\frac{2}{\delta}\right)^{1/2} f^{\alpha+1}. \tag{37}$$

Putting $f^{-\alpha} = G$ in (37), we have

$$G' - 2\alpha\eta G = -\frac{2\alpha}{\alpha + 1}\left(\frac{2}{\delta}\right)^{1/2}. \tag{38}$$

This equation can be integrated to yield

$$G = \left(C - \frac{2}{\alpha + 1}\left(\frac{2\alpha}{\delta}\right)^{1/2} \int_0^{\alpha^{1/2}\eta} e^{-t^2}\, dt\right) e^{\alpha\eta^2}, \tag{39}$$

where C is the constant of integration. Using (39), the function f can be written as

$$f(\eta) = \exp(-\eta^2)\left[C - \frac{2}{\alpha + 1}\left(\frac{2\alpha}{\delta}\right)^{1/2} \int_0^{\alpha^{1/2}\eta} e^{-t^2}\, dt\right]^{-1/\alpha}, \tag{40}$$

where C may be identified as $f^{-\alpha}(0)$. The solution $u = t^{-1/2\alpha} f(\eta)$, $\alpha = 1/(j+1)$, generalizes the exact single hump solution of the standard Burgers equation for other geometries: $j = 0$, $\alpha = 1$; $j = 1$, $\alpha = 1/2$; $j = 2$, $\alpha = 1/3$. Just as for (7) (or 9)), we have here for $\alpha = 1/j$ another exact solution of (26), namely, $f = c$, an arbitrary nonzero constant.

Now we deduce some inequalities involving the parameter α, for which the single hump solutions starting from zero at $\eta = +\infty$ vanish again either at $\eta = -\infty$ or $\eta = \eta_0$, a finite number. We introduce the function $F = f^\alpha$ in (26) so that it becomes

$$\frac{1}{2}FF'' - \frac{\alpha - 1}{2\alpha}F'^2 + (1 - \alpha j)F^2 + \eta FF' - \left(\frac{2}{\delta}\right)^{1/2} F^2 F' = 0. \tag{41}$$

We integrate (41) from $\eta = -\infty$ to $\eta = +\infty$ and assume that both f and f' (and hence F and F') vanish at $\eta = \pm\infty$. We thus have

$$(2\alpha j - 1)\int_{-\infty}^{\infty} F^2\, d\eta = \frac{1 - 2\alpha}{\alpha}\int_{-\infty}^{\infty} F'^2\, d\eta$$

or
$$r \equiv \frac{\int_{-\infty}^{\infty} F^2 \, d\eta}{\int_{-\infty}^{\infty} F'^2 \, d\eta} = \frac{1 - 2\alpha}{\alpha(2\alpha j - 1)} > 0. \tag{42}$$

Inequality (42) yields the following results for different geometries.

i. $j = 0$. In this case, $r = -(1 - 2\alpha)/\alpha$ is greater than zero provided $\alpha > 1/2$. This, therefore, is the condition for the existence of single hump solution vanishing at $\eta = \pm\infty$.
ii. $j = 1$. The only possible choice of α is $1/2$, and this corresponds to the exact solution (40) with $\alpha = 1/2$. Thus we have an isolated single value of α, for which the single hump solutions vanishing at $\eta = \pm\infty$ exist.
iii. $j = 2$. In this case $r = (1 - 2\alpha)/\alpha(4\alpha - 1) > 0$ if $1/4 < \alpha < 1/2$. Thus, we have a finite range of α (unlike cases (i) and (ii)) for which the single hump solutions vanishing at $\eta \pm \infty$ exist.

Now we consider the case when the solution starting at $\eta = +\infty$ vanishes not at $-\infty$ but at a finite point η_0, $-\infty < \eta_0 < \infty$. It is clear that $f'(\eta_0) > 0$. Integrating (26) from $\eta = \eta_0$ to $\eta = \infty$, we find that

$$\frac{\alpha(j + 1) - 1}{\alpha} \int_{\eta_0}^{\infty} f \, d\eta = -\frac{1}{2} f'(\eta_0) < 0. \tag{43}$$

Since $f > 0$ in $\eta_0 < \eta < \infty$, equation (43) implies that $\alpha < 1/(j + 1)$. Thus, single hump solutions of (26) vanishing at $\eta = +\infty$ and $\eta = \eta_0$ exist if $\alpha < 1/(j + 1)$. We may combine this result with those in (i)–(iii) to conclude that single hump solutions vanishing at $\eta = +\infty$ and again at $\eta = -\infty$ or $\eta = \eta_0$, a finite point, exist if $1/(j + 2) < \alpha < 1/(j + 1), j = 0, 2$.

The condition that a maximum exists for the single hump solution is that $f' = 0, f'' < 0$ at $\eta = \eta_{\max}$, say. Then, equation (26) immediately yields that $\alpha j < 1$. However, this condition is rather lax as far as the existence of single hump solutions is concerned. The conditions derived from (42) and (43) are more precise, as our numerical studies confirm. We summarize these results in Table 4.6.4.

We integrated (36) numerically subject to (34a), starting from $\eta \sim 4$, where the values of f and f' were $O(10^{-5})$. The integration was carried out towards decreasing values of η. First, we treated the strictly self-similar single hump solutions which exist over $-\infty \leq \eta \leq \infty$. From the numerical solution, the value of $f(0)$ and $f'(0)$ were determined, and $H(0)$ and $H'(0)$ were found using (27). With the values of $a_0 = H(0)$, $a_1 = H'(0)$ thus obtained, the series solution (29) was computed. It agreed very closely with the numerical solution of the connection problem in the entire range $-\infty < \eta < \infty$, the discrepancy being $O(10^{-7})$ only. The series was analytically continued as its convergence slowed down. We reconfirmed the accuracy of the numerical and series solutions by comparing them with special exact

Local and Global Analysis

Table 4.6.4 Single Hump, Monotonic, and Diverging Solutions of (26) and (34)–(35)

Behavior at left boundary	$j = 0$	$j = 1$	$j = 2$
(a) Solutions vanishing at $\eta = -\infty$	$\alpha \geq 1$	$\alpha = \frac{1}{2}$	$\frac{1}{3} \leq \alpha < \frac{1}{2}$
(b) Solutions vanishing at $\eta = \eta_0$	$\frac{1}{2} < \alpha < 1$	$\frac{1}{3} < \alpha < \frac{1}{2}$	$\frac{1}{4} < \alpha < \frac{1}{3}$
(c) Solutions monotonically approaching a constant at $\eta = -\infty$	—	$\alpha = 1$	$\alpha = \frac{1}{2}$
(d) Solutions diverging to infinity at $\eta = -\infty$	—	$\alpha > 1$	$\alpha > \frac{1}{2}$

solutions (40) for $\alpha = 1/(j+1)$. For the nonplanar Burgers equation (24), the solution in the similarity range of the parameter α, for different geometries, exists for all values of the amplitude parameter. The admissible range of the parameter α itself determines the existence (or otherwise) of the single hump solutions. This is in contrast to the case of the damped GBE (4), for which the self-similar solutions, for a given α in the permissible range, are restricted by the magnitude of the amplitude parameter.

The nature of the solution as summarized in Table 4.6.4 was completely checked numerically. The intermediate asymptotic nature of the self-similar solutions of the single hump type (a) in Table 4.6.4 was confirmed by solving (24) numerically, subject to a class of initial conditions which vanish at $\eta = \pm\infty$ in conformity with (34a), (34b), and behave "reasonably" in the intervening interval (see Sachdev and Nair, 1987).

It must be emphasized that we have verified our conjecture regarding the characterizing property of Euler–Painlevé transcendents (solutions of (1)) for GBE just for two equations (4) and (24). It is hoped that other GBE will be verified for their Euler–Painlevé property when they are encountered in applications (see also Sachdev, Nair, and Tikekar, 1988).

4.7 OTHER NONLINEAR DE RELATED TO EULER–PAINLEVÉ EQUATIONS

Many second-order nonlinear DE listed in Kamke (1943) and Murphy (1960) are special cases of (1) of Section 4.6, either directly or after some simple transformations. We list here these and other DE which are special cases of (1) of Section 4.6 if we allow the coefficients a, b, and c to vary with x. Kamke has appended some historical notes with the equations and has also given the geometrical or physical origin of the equations. Here, we content ourselves with listing the equations and the solutions. We note that a few common features characterize this set of equations: the equations are either autonomous or linearizable by a logarithmic or power law transformation, or are reducible to first-order equations such as of Riccati or Bernoulli form. These equations may be solved in closed form in terms of a quadrature or treated in the phase plane. We say that DE which require the coefficients of equation (1) of Section 4.6 to vary with x *belong to the class of generalized Euler–Painlevé equations* (GEPE).

We identify Kamke's equations by K and Murphy's by M. The constants of integration are denoted by c_1, c_2, etc. The constants in the DE are denoted by a, b, c, etc. The equations carry the same number as in Kamke and Murphy.

K6.104 $yy'' = a$

$$x = \int (2a \ln y + c_1)^{-1/2} \, dy + c_2.$$

K6.105 $yy'' = ax, a \neq 0$

The initial value problem with $y(0) = c_0$ and $y'(0) = c_1$ has the solution

$$y = \pm \left(\frac{4a}{3} x^3 \right)^{1/2}$$

if $c_0 = c_1 = 0$, and

$$y = c_1 x + c_2 x^2 + c_3 x^3 + \cdots$$

if $c_0 = 0$ and $c_1 \neq 0$. The other coefficients are

$$c_2 = \frac{a}{2c_1}, \quad c_3 = -\frac{c_2^2}{3c_1}, \quad c_4 = \frac{2c_2^3}{9c_1^2}, \quad c_5 = -\frac{17}{90} \frac{c_2^4}{c_1^3}, \text{ etc.}$$

This series converges for $|x| < c_1^2/|a|$.

K6.106 $yy'' = ax^2, a \neq 0$

As for K6.105, this equation has a solution with initial conditions $y(0) = c_0$ and $y'(0) = c_1$ in the form

$$y = \pm \left(\frac{a}{2} \right)^{1/2} x^2$$

Local and Global Analysis

if $c_0 = c_1 = 0$, and in the series form
$$y = c_1 x(1 + b_1 x^2 - b_2 x^4 + b_3 x^6 + \cdots),$$
where
$$b_1 = \frac{a}{6} c_1^2, \quad b_2 = \frac{3}{10} b_1^2, \quad b_3 = \frac{13}{70} b_1^3, \quad b_4 = \frac{25}{168} b_1^4, \text{ etc.,}$$
for $a > 0, c_1 > 0, x > 0, y > 0$. The series converges for $|x| < (3c_1^2/|a|)^{1/2}$.

K6.107 $yy'' + y'^2 - a = 0$
$$y^2 = ax^2 + c_1 x + c_2$$
and
$$y = \pm a^{1/2} x + c.$$

K6.108 $yy'' + y'^2 = ax + b$
This equation may be solved in a series form.

K6.109 $yy'' + y'^2 - y' = 0$
$$y = C$$
and
$$x = y + C_1 \ln|y - C_1| + C_2.$$

K6.110 $yy'' - y'^2 + 1 = 0$
$$C_1 y = \sinh(C_1 x + C_2),$$
and
$$C_1 y = \sin(C_1 x + C_2).$$

K6.111 $yy'' - y'^2 - 1 = 0$
$$C_1 y = \cosh(C_1 x + C_2).$$

K6.117 $yy'' - y'^2 + ayy' + by^2 = 0$
The substitution $u = y'/y$ renders this equation linear, besides lowering the order:
$$u' + au + b = 0.$$

K6.122 $yy'' - y'^2 + f(x)yy' + g(x)y^2 = 0$

This generalizes K6.117. The transformation $u = y'/y$ changes this equation to

$$u' + fu + g = 0.$$

K6.124 $yy'' - 3y'^2 + 3yy' - y^2 = 0$

With $y' = yu(x)$, we have

$$u' = 2u^2 - 3u + 1.$$

The final solution is

$$(2e^x - C_1)y^2 = C_2 e^{2x}.$$

K6.125 $yy'' = ay'^2$

This exact equation has the solution

$$y = \begin{cases} |C_1 x + C_0|^{1/(1-a)} & \text{for all } a \neq 1, \\ C_1 e^{Cx} & \text{for all } a = 1. \end{cases}$$

K6.126 $yy'' + a(y'^2 + 1) = 0$

$$x = \int (C_1 y^{-2a} - 1)^{-1/2} \, dy + C_2.$$

K6.128 $yy'' + ay'^2 + byy' + cy^2 + dy^{1-\alpha} = 0$

This DE is of Euler–Painlevé type only for $\alpha = 1$. The substitution

$$y = \begin{cases} e^u & \text{for } \alpha = -1, \\ u^{1/(\alpha+1)} & \text{for } \alpha \neq -1, \end{cases}$$

transforms it to

$$u'' + bu' + c + d = 0$$

and

$$u'' + bu' + (a+1)cu = -(a+1)d,$$

respectively.

Local and Global Analysis

K6.129 $yy'' + ay'^2 + f(x)yy' + g(x)y^2 = 0$

Let $y = u^{1/(a+1)}$. Then this equation becomes linear:

$u'' + fu' + (a+1)gu = 0$.

K6.131 $yy'' - \dfrac{a-1}{a}y'^2 - fy^2 y' + \dfrac{a}{(a+2)^2}f^2 y^4 - \dfrac{a}{a+2}f'y^3 = 0, \quad f = f(x)$

First, we change this equation to

$auu'' - (a-1)u'^2 = 0$

via

$$u(x) = y \exp\left(-\dfrac{a}{a+2}\int yf\,dx\right).$$

After further transformations, the final solution is

$$y = -\dfrac{(a+2)|x+C_1|^a}{a\int |x+C_1|^2 f\,dx + C_2}.$$

K6.133 $(y+x)y'' + y'^2 - y' = 0$

This is of EP type if we write $Y = y + x$. Its first integral is

$(y+x)y' - 2y = C$.

K6.134 $(y-x)y'' - 2y'(y'+1) = 0$

Again, $y - x = Y$ changes it to EP type. Its solutions are

$y = C, \quad y = C - x, \quad \text{or} \quad y = C_1 + \dfrac{C_2}{x - C_1}$.

K6.136 $(y-x)y'' + f(y') = 0$

We may write $y - x = Y$; we obtain the form

$YY'' + f(Y'+1) = 0$.

This is more general than the EP equation. Its solution is

$(y-x)\phi(y') = C_2$,

where
$$\phi(u) = \exp\left(\int \frac{u-1}{f(u)} du\right).$$

K6.137 $2yy'' + y'^2 + 1 = 0$

$$C_1 \arctan\left(\frac{y}{C_1 - y}\right)^{1/2} - (y(C_1 - y))^{1/2} = x + C_2,$$

or the parameter form of the cycloid

$$x = C_1(t - \sin t) + C_2, \qquad y = C_1(1 - \cos t).$$

K6.138 $2yy'' - y'^2 + a = 0$

Its first integral is

$$y'^2 - a = Cy,$$

which is again integrable in an elementary manner.

K6.139 $2yy'' - y'^2 + f(x)y^2 + a = 0, a > 0$

If u and v are solutions of the linear DE

$$4y'' + f(x)y = 0$$

such that

$$(uv' - u'v)^2 = a,$$

then $y = uv$ is a solution of the given DE (see Section 2.6).

K6.150 $2yy'' - 3y'^2 = 0$

$$y = C_1(x + C_2)^{-2} \quad \text{and} \quad y = C.$$

K6.151 $2yy'' - 3y'^2 - 4y^2 = 0$

$$y \cos^2(x + C_1) = C_2.$$

K6.152 $2yy'' - 3y'^2 + f(x)y^2 = 0$

The substitution $u(x) = |y|^{-1/2}$ renders this equation linear:

$$u'' - \frac{f(x)}{4} u = 0.$$

Local and Global Analysis

K6.155 $2(y - a)y'' + y'^2 + 1 = 0$

Writing $y - a = Y$, etc., one may arrive at the solution

$$2x = C_1 \pm [(y - a + C_2)(a - y)]^{1/2} \mp C_2 \arctan \left[\frac{a - y}{y - a + C_2} \right]^{1/2}$$

K6.156 $3yy'' - 2y'^2 = ax^2 + bx + c$

This is of EP type if c in equation (1) of Section (4.6) is assumed to depend on x. The solution given by Kamke is highly implicit.

K6.157 $3yy'' - 5y'^2 = 0$

$y^2 = (C_1 x + C_2)^{-3}$.

K6.158 $4yy'' - 3y'^2 + 4y = 0$

The substitution $y = \pm u^2$ transforms it to

$2uu'' - u'^2 \pm 1 = 0$.

This is a special case of K6.138.

K6.164 $nyy'' - (n - 1)y'^2 = 0$

$y = (C_1 x + C_2)^n$.

K6.166 $ayy'' + by'^2 - \dfrac{yy'}{\sqrt{x^2 + c^2}} = 0$ (GEPE)

The substitution $y' = yu(x)$ transforms to the Bernoulli form

$$u' - \frac{u}{a(x^2 + c^2)^{1/2}} + \left(1 + \frac{b}{a}\right) u^2 = 0,$$

with the solution

$y^{1+b/a} = C_1 + C_2 [x + (x^2 + c^2)^{1/2}]^{1/a} [(x^2 + c^2)^{1/2} - ax]$.

K6.168 $(ay + b)y'' + cy'^2 = 0$

Writing $ay + b = Y$, etc., we find the solution to be

$$ay + b = \begin{cases} (C_1 x + C_0)^{a/(a+c)} & \text{for } a + c \neq 0, \\ C_0 e^{C_1 x} & \text{for } a + c = 0. \end{cases}$$

K6.169 $xyy'' + xy'^2 - yy' = 0$ (GEPE)
$y^2 = C_1 x^2 + C_2$.

K6.170 $xyy'' + xy'^2 + ayy' + f(x) = 0$ (GEPE)
The substitution $u(x) = y^2$ changes it to the linear DE
$xu'' + au' + 2f(x) = 0$
with solution
$$u = C_1 + C_2 x^{1-a} - 2 \int x^{-a} \left(\int x^{a-1} f(x) dx \right) dx.$$

K6.173 $xyy'' + 2xy'^2 + ayy' = 0$ (GEPE)
With $u = y^3$, we obtain the linear DE
$xu'' + au' = 0$.
The solution is
$y^3 = C_1 + C_2 x^{1-a}$.

K6.174 $xyy'' - 2xy'^2 + (y+1)y' = 0$ (GEPE)
With $y(x) = \eta(\xi)$, $\xi = \ln|x|$ etc., this DE has solutions
$$y = C, \qquad y = \frac{1}{2} \ln|x|,$$
$$2Cy = \tan(C \ln|x|), \qquad 2Cy = \coth(C \ln|x|).$$

K6.175 $xyy'' - 2xy'^2 + ayy' = 0$ (GEPE)
With $u = 1/y$, we have
$xu'' + au' = 0$.
The solution is
$$y = \begin{cases} (C_1 + C_2 x^{1-a})^{-1} & \text{for } a \neq 1, \\ (C_1 + C_2 \ln x)^{-1} & \text{for } a = 1. \end{cases}$$

K6.176 $xyy'' - 4xy'^2 + 4yy' = 0$ (GEPE)
$y = (C_1 + C_2 x^{-3})^{-1/3}$.

Local and Global Analysis

K6.177 $xyy'' + \left(\dfrac{ax}{(b^2-x^2)^{1/2}} - x\right)y'^2 - yy' = 0$ (GEPE)

With $u = y'/y$, this DE transforms to Bernoulli type. The solutions are
$$y = C$$
and
$$y = C_1 \exp\left\{\dfrac{1}{a}(b^2 - x^2)^{1/2} + \dfrac{C_2}{a^2}\ln\left[C_2 - a(b^2 - x^2)^{1/2}\right]\right\}.$$

K6.178 $x(y+x)y'' + xy'^2 - (y-x)y' - y = 0$

With $y + x = u$, this DE goes to
$$xuu'' + xu'^2 - uu' = 0 \text{ (GEPE)}$$
This is K6.169. The solution is
$$(y+x)^2 = C_1 x^2 + C_2.$$

K6.179 $2xyy'' - xy'^2 + yy' = 0$ (GEPE)
$$y = C_1(|x|^{1/2} + C_2)^2.$$

The following equations are taken from Section 4 of Murphy (1960).

M129 $yy'' + y'^2 + y^2 = 0$
$$y^2 = C_1 \sin(x\sqrt{2} + C_2).$$

M130 $yy'' + y'^2 + 2a^2 y^2 = 0$
$$y^2 = C_1 \cos 2ax + C_2 \sin 2ax.$$

M133 $yy'' = y'^2 + yy'$
$$y = C_1 \exp(C_2 e^x).$$

M138 $yy'' = y'^2 - y'$
$$y + C_2 = C_1 e^{x/C_2}.$$

M140 $yy'' = y'^2 - 2y'$
$C_1 y + C_2 e^{C_1 x} + 2 = 0$
and
$y = C_0.$

M142 $yy'' + y'^2 + axy' = 0$
The transformations
$y = x^2 u(z), \quad z = \ln x, \quad u' = p$
change the DE to the first-order form
$upp'(u) + p^2 + (7u + a)p + 2u(a + 3u) = 0.$

M150 $yy'' = 2y'^2 + y^2$
With $u = y'/y$, etc., the solution is
$y = C_1 \csc(C_2 + x).$

M190 $3yy'' = 2y'^2 + 36y^2$
With $u = y'/y$, etc., the solution is
$y = (C_1 e^{2x} + C_2 e^{-2x})^3.$

M195 $5yy'' = y'^2$
$C_1 y^4 = (C_2 + x)^5.$

M199 $xyy'' + xy'^2 + yy' = 0$ (GEPE)
$y^2 = C_1 \ln x + C_2.$

M201 $xyy'' - xy'^2 + yy' = 0$ (GEPE)
With $u = y'/y$, etc., the solution is
$y = C_2 x^{C_1}.$

M203 $xyy'' + xy'^2 + 2yy' = 0$ (GEPE)
$x(C_1 + y^2) = C_2.$

Local and Global Analysis

M204 $xyy'' + xy'^2 - 3yy' = 0$ (GEPE)
$y^2 = C_1 + C_2 x^4$.

M219 $x^2 yy'' + (xy' - y)^2 = 0$ (GEPE)
With $y = ux$, this DE changes to
$xuu'' + 2uu' + xu'^2 = 0$ (GEPE)
The solution is
$y = \pm [x(C_1 + C_2 x)]^{1/2}$.

M220 $x^2 yy'' + (xy' - y)^2 - 3y^2 = 0$ (GEPE)
With $y^2 = u(x)$, this DE transforms to the linear equation
$x^2 u'' - 2xu' - 4u = 0$.
The solution is
$$y = \pm \left[\frac{C_1 + C_2 x^5}{x} \right]^{1/2}.$$

M221 $x^2 yy'' - (2x^2 y'^2 + axyy' + ay^2) = 0$ (GEPE)
$y = (C_1 x + C_2 x^a)^{-1}$.

M222 $x^2 yy'' + ax^2 y'^2 + bxyy' + cy^2 = 0$ (GEPE)
Writing $u = y'/y$, we have a first-order DE
$x^2 u' + c + bxu + (1+a)x^2 u^2 = 0$.

M227 $2x^2 yy'' - x^2 y'^2 + y^2 = 0$ (GEPE)
$y = x(C_1 + C_2 \ln x)^2$.

M228 $2x^2 yy'' - x^2 y'^2 - 2xyy' + 4y^2 = 0$.
$y = x^2 (C_1 + C_2 \ln x)^2$.

M229 $x^3yy'' + x^3y'^2 + 6x^2yy' + 3xy^2 - a = 0$

$$y = \pm \left[\frac{ax^2 + C_1x + C_2}{x^3}\right]^{1/2}.$$

M230 $x(1+x)^2yy'' - x(1+x)^2y'^2 + 2(1+x^2)yy' - a(2+x)y^2 = 0$ (GEPE)

Writing $u = y'/y$, we get a linear DE

$x(1+x)^2 u' + 2(1+x)^2 u - a(2+x) = 0.$

The solution is

$y = C_1(1+x)^a e^{C_2/x}.$

M231 $8(1-x^3)yy'' + 4(1-x^3)y'^2 - 12x^2yy' + 3xy^2 = 0$ (GEPE)

With $y = u^2(z)$, $z = x^3$, we get the linear DE

$48z(1-z)u'' - 8(4-7z)u' + u = 0.$

M233 $yy'' - y'^2 - \dfrac{1}{x}yy' - \dfrac{by'^2}{(a^2-x^2)^{1/2}} = 0.$

With $u = y'/y$, this becomes a first-order nonlinear DE

$xu' - u - \dfrac{bxu^2}{(a^2-x^2)^{1/2}} = 0.$

M234 $f_0(x)yy'' + f_1(x)y'^2 + f_2(x)yy' + f_3(x)y^2 = 0$ (GEPE)

With $u = y'/y$, it becomes a first-order nonlinear DE of Riccati type:

$f_0 u' + (f_0 + f_1)u^2 + f_2 u + f_3 = 0.$

5
Existence Theory for Boundary Value Problems via Shooting Techniques

5.1 INTRODUCTION

We have dealt with boundary value problems over an infinite domain, using exponential series in Section 3.11. These expansions are valid only for certain special kinds of boundary conditions. Ordinary Taylor series, which satisfy the initial conditions and which result in transcendental equations when boundary conditions are inserted, have even more severe problems than the exponential series. In each case, two difficulties arise. The first is the convergence of the series and the determination of the radius of convergence. If the radius of convergence is less than the distance between the two boundary points, the answers will make no sense. The situation may be retrieved by the method of analytic continuation, which may have to be employed at several successive points. The second difficulty arises from the computation of the series and the number of terms needed to achieve a desired accuracy. And, of course, in general some simultaneous transcendental equations (resulting from the boundary conditions) have to be solved to obtain the unknown coefficients before the series can be evaluated. However, in some special cases, the series may be so well behaved (see Section 3.11) that all these difficulties can be easily overcome. And then the convergence of the series provides not only the existence of the solution but also directly computable answers.

Keller (1968) has discussed accurate numerical methods for initial value problems, and shooting methods for boundary value problems. Indeed, here also the transcendental equations result from the boundary conditions, and, in that sense, the power series procedure is but a special case of initial value methods. But the numerical methods for initial value problems, such as high-order Runge–Kutta and finite difference schemes, are free from the limitations imposed by radius of convergence difficulties in the series method. Keller has proved rigorous (constructive) theorems for existence and uniqueness of solutions to boundary value problems, using initial value methods; in addition, computational aspects were also carefully analyzed.

We consider the existence of solutions to two-point boundary problems (generally over a semi-infinite domain) for several differential equations arising in applications. The existence proof is based on initial value or shooting techniques and is constructive in the sense that often the upper and lower bounds are also obtained. The essential steps require the study of (1) local existence theorems for initial value problems with assumed initial conditions, (2) continuous dependence of this solution on the initial conditions, (3) various possible conditions to which the solution can tend at the other boundary, (4) continuous dependence of the solution on the parameters involved for each of these cases, and (5) the proof that at least one solution exists in the class which satisfies the requisite boundary conditions. Each problem will be analyzed by steps, often suggested by its own special nature.

This chapter shows that using simple analysis, quite nontrivial boundary value problems can be solved. This is in consonance with the style and manner of the material in this book. We shall deal with three boundary value problems and one connection problem arising in physical applications, each of which we shall explain at the appropriate point (see Section 5.6 for definition of a connection problem.)

We shall see that the shooting argument (see Section 5.2) naturally breaks into two parts: one accessible to the analyst, the other to the computationalist. The asymptotic behavior near the initial part or at the other (possibly infinite) boundary and continuous dependence of the solution on the initial condition belong easily to the analyst, while the behavior of at least a finite number of solutions in the beginning, middle, and end of the interval can be carried through only with the help of a computer. The latter enables us to prove that the solutions so followed satisfy the inequalities that place them in specific subsets.

5.2 ELEMENTARY DISCUSSION OF BOUNDARY VALUE PROBLEMS

Unlike an initial value problem, the conditions for a boundary value problem for a DE are prescribed not at one point but at two or more points.

Existence Theory for BVP via Shooting Techniques

This makes the solution of the problem considerably more difficult. The solution may not exist at all, or there may be an infinity of solutions. (This may happen for an initial value problem also.) To illustrate this point, we consider the harmonic equation

$$y'' + y = 0, \tag{1}$$

subject to the boundary conditions

$$y(0) = 1, \quad y(2) = 0. \tag{2}$$

It is easy to verify that (1) and (2) have the unique solution

$$y(x) = \cos x - (\cot 2) \sin x. \tag{3}$$

In contrast, if the boundary conditions (2) are replaced by

$$y(0) = 1, \quad y(\pi) = 0, \tag{4}$$

one may easily convince oneself that no linear combination of the solutions $\sin x$ and $\cos x$ of (1) satisfies (4). Hence the problem (1) and (4) has no solution. Now, if we impose the conditions

$$y(0) = 0, \quad y(\pi) = 0 \tag{5}$$

on (1) instead of (2) or (4), then there is an infinity of solutions

$$y = c \sin x, \tag{6}$$

where c is an arbitrary constant.

Thus we find that the boundary conditions make the solutions behave so variedly for the linear DE (1). A nonlinear DE or nonlinear boundary conditions may exhibit even more diverse behavior.

The existence and uniqueness theory for pure initial value problems (IVP) is well established. In contrast, as the simple example shows, boundary value problems (BVP) pose serious difficulties. It is natural to attempt to relate the solution of a BVP to that of an IVP and to use the analytic results for the latter to solve the former.

We consider a BVP for a vector DE

$$y' = f(x,y), \quad 0 \leq x \leq 1,$$

where y is a (vector) function, depending on the variable x, with the (vector) boundary conditions

$$F(y(0), y(1)) = 0.$$

We solve the BVP subject to the initial conditions $y(0) = s$ and denote the solution by $y(x;s)$. This solution is now required to satisfy the conditions

$$F(s, y(1,s)) = 0.$$

This (vector) functional equation in s may possess no solution, a unique solution, or an infinity of solutions. If it has a unique (vector) solution $s = s_0$, say, then the boundary value problem has a unique solution. If the functional equation $F(s,y(1,s)) = 0$ is linear, one may be able to solve it explicitly. If it is not, one may have to use some iterative scheme such as Newton–Raphson (see Blum, 1972).

Turning specifically to a two-point BVP for a general second-order nonlinear DE

$$y'' + f(t,y(t),y'(t)) = 0, \tag{7}$$
$$y(a) = A, \quad y(b) = B, \tag{8}$$

we first consider the initial value problem for (7) with

$$y(a) = A, \quad y'(a) = \mu. \tag{9}$$

One may assume a trial value of μ and integrate (7) to see what $y(b)$ is. Some trial values of μ may lead to the correct value, $\mu = \mu_0$, say, for which $y(b) = B$, provided the solution of the BVP exists. Denoting the solution of the IVP (7) and (9) by $y(t,\mu)$, we must solve the functional equation

$$y(b,\mu) = B. \tag{10}$$

One may, by trial, determine two numbers μ_1 and μ_2 such that

$$y(b,\mu_1) < B,$$
$$y(b,\mu_2) > B. \tag{11}$$

If the BVP (7) and (8) has a unique solution, then by using the theorem on continuous dependence of solutions on initial conditions, one concludes that there exists at least one root $\mu = \mu_0$ of the functional equation (10). In practice, the interval (μ_1,μ_2) may be reduced by trial and the actual value μ_0 arrived at by interpolation. This implies that, if we shoot from $t = a$ with this value μ_0 of the derivative, we shall hit the required value at the other boundary, $y(b,\mu_0) = B$.

It is sometimes possible to exploit some inequalities to solve a boundary value problem via a shooting argument. First, we recall the statement of the existence theorem for IVP for (7). We know that, if (7) is subject to the initial conditions

$$y(t_0) = y_0, \quad y'(t_0) = y'_0, \tag{12}$$

and if the function f is assumed to be continuous on a domain D containing the point (t_0,y_0,y'_0), then there exists at least one solution on a nonempty interval containing the point t_0. If the function f satisfies a Lipschitz condition in the domain D, then there is at most one solution $y(t)$ of (7) which satisfies the given initial conditions (12). Moreover, this solution depends continuously on the initial conditions.

Existence Theory for BVP via Shooting Techniques

Now we consider equation (7) with the boundary conditions (8). We wish to establish the existence of solution(s) of (7) and (8). For this purpose we assume that f in (7) is continuous and satisfies a Lipschitz condition on $\bar{D}$: $[a,b] \times R^1 \times R^1$, and

$$|f(t,y,y')| \leq N, \tag{13}$$

where N is a constant, for all t, y, and y' on $\bar{D}$. Consider the initial conditions

$$y(a) = A, \quad y'(a) = m, \tag{14}$$

in relation to (7). Let $y(t,m)$ represent a solution of (7) and (14). We can extend this solution, under the given assumptions, to the point $t = b$. Now we may write

$$y'(t,m) = y'(a,m) + \int_a^t y''(\tau,m)\,d\tau$$

$$= m - \int_a^t f(\tau, y(\tau), y'(\tau))\,d\tau$$

$$\geq m - N(t-a), \tag{15}$$

in view of (7) and (13). Integrating (15), we have

$$y(t,m) \geq A + m(t-a) - \frac{N}{2}(t-a)^2. \tag{16}$$

For $t = b$ this inequality yields

$$y(b,m) \geq A + m(b-a) - \frac{N}{2}(b-a)^2. \tag{17}$$

Therefore, there exists m_1 sufficiently large and positive such that $y(b,m_1) > B$. In the same manner, since

$$y(b,m) \leq A + m(b-a) + \frac{N(b-a)^2}{2}, \tag{18}$$

there exists m_2 sufficiently large and negative such that $y(b,m_2) < B$. Now, noting that the solution $y(b,m)$ is a continuous function of m, there exists at least one m_3, $m_1 > m_3 > m_2$, such that $y(b,m_3) = B$. Thus a solution of (7) and (14) with $m = m_3$ is a solution of BVP (7) and (8).

More general existence theory for BVPs, which uses initial value methods, may be found in Bailey, Shampine, and Waltman (1968).

We conclude this section by illustrating the conversion of a BVP into two IVPs. This conversion requires a certain scale invariance of the given DE with respect to a parameter. This parameter in turn determines the missing initial condition which renders the BVP equivalent to an IVP. The advantages of the conversion of a BVP to two IVPs are analytical and numerical. We have already discussed the former aspect. The conversion to two IVPs is very convenient from the numerical point of view, since now it merely requires marching of the solution from an initial point. The equivalent formulation also obviates the need to carry out expensive interpolation to satisfy the boundary condition at the other end of the interval. We caution, however, that this method is strongly dependent on the invariance properties of the given DE and does not always apply.

EXAMPLE 1 The scaling method seems to have been first suggested by Blasius (see Schlichting, 1960) for the third-order nonlinear ODE, which arises from the similarity reduction of the boundary layer equations, with relevant boundary conditions. The so-called Blasius problem thus is

$$y''' + yy'' = 0, \qquad ' = \frac{d}{d\eta}, \tag{19}$$

$$y(0) = y'(0) = 0 \quad \text{and} \quad y'(\infty) = 2. \tag{20}$$

First, we note that equation (19) admits the invariance

$$y(\eta) = \lambda^{1/3} Y(\eta \lambda^{1/3}), \tag{21}$$

where λ is a finite parameter.

Now we consider the IVP for (19) with the conditions

$$y(0) = y'(0) = 0, \qquad y''(0) = \lambda \text{ (say)}. \tag{22}$$

We could march off or shoot from $\eta = 0$ if we knew the precise value of λ which would satisfy the condition $y'(\infty) = 2$. In the absence of that, we proceed as follows. First, we find the Taylor series solution of the IVP (19) and (22) by using $y(0)$, $y'(0)$, and $y''(0) = \lambda$ from the latter, and finding higher-order derivatives from the former. Thus, for example,

$$y'''(0) = -y(0)y''(0) = 0, \tag{23}$$
$$y^{iv}(0) = -(yy'')'(0) = -y'(0)y''(0) - y(0)y'''(0) = 0, \text{ etc.}$$

The solution of IVP (19)–(20) is thus

$$y(\eta) = \frac{\lambda \eta^2}{2!} - \frac{\lambda^2 \eta^5}{5!} + \frac{11\lambda^3 \eta^8}{8!} - \frac{375\lambda^4 \eta^{11}}{11!} + \cdots. \tag{24}$$

Making use of the invariance (21), the solution (24) becomes

$$Y(\bar{\eta}) = \frac{\bar{\eta}^2}{2!} - \frac{\bar{\eta}^5}{5!} + \frac{11\bar{\eta}^8}{8!} - \frac{375\bar{\eta}^{11}}{11!} + \cdots, \qquad \bar{\eta} = \eta \lambda^{1/3}. \tag{25}$$

Also, the boundary condition $y'(\infty) = 2$, by way of (21), becomes

$$2 = \lim_{\eta \to \infty} y'(\eta) = \lim_{\eta \to \infty} \left[\lambda^{1/3} Y(\eta \lambda^{1/3}) \right]'$$

Existence Theory for BVP via Shooting Techniques

$$= \lim_{\eta \to \infty} \lambda^{2/3} Y'(\eta \lambda^{1/3}) = \lambda^{2/3} Y'(\infty). \tag{26}$$

It follows that

$$\lambda = \left[\frac{2}{Y'(\infty)}\right]^{3/2}. \tag{27}$$

In view of (21), the initial value problem in the variable Y becomes

$$Y''' + YY'' = 0, \quad Y(0) = Y'(0) = 0, \quad Y''(0) = 1, \tag{28}$$

while the original IVP assumes the form

$$y''' + yy'' = 0, \quad y(0) = y'(0), \quad y''(0) = \left[\frac{2}{Y'(\infty)}\right]^{3/2}. \tag{29}$$

Thus, the BVP (19) and (20) has been "reduced" to the IVP (28) and (29), which may be solved in succession, so that $Y'(\infty)$ for the latter becomes known from the former. A straightforward Runge–Kutta scheme would give accurate results. Obviously, the scaling property is sufficient for the present method, and the solution in the form of a Taylor series is not necessary.

EXAMPLE 2 The Thomas–Fermi equation

$$\frac{d^2y}{dx^2} = y^{3/2} x^{-1/2}, \tag{30}$$

subject to the boundary conditions

$$y(0) = 0, \quad y(\infty) = 0, \tag{31}$$

arises in the problem of determining the effective nuclear charge in heavy atoms (see also Section 4.4). Writing $x = 1/t$, we transform the problem (30)–(31) into

$$\{t^4 D^2 + 2t^3 D\} y = y^{3/2} t^{1/2}, \quad D = \frac{d}{dt}, \tag{32}$$

$$y(0) = 0, \quad y(\infty) = 1. \tag{33}$$

We let $y'(0) = \lambda$ (which is to be determined), and verify that the transformation

$$y(t) = \lambda^{3/2} F(\lambda^{-1/2} t). \tag{34}$$

leaves (32) invariant.

Therefore, $F(\xi)$ with $\xi = \lambda^{-1/2} t$ satisfies the IVP

$$\{\xi^4 D^2 + 2\xi^3 D\} F = (F^3 \xi)^{1/2}, \quad D = \frac{d}{d\xi}, \tag{35}$$

$$F(0) = 0, \quad F'(0) = 1. \tag{36}$$

Now, we let $t \to \infty$ in (34) and obtain

$$\lambda = \{F(\infty)\}^{-2/3}. \tag{37}$$

Thus, the problem reduces to two IVPs, namely (30) with $y(0) = 0$, $y'(0) = \lambda = \{F(\infty)\}^{-2/3}$, and (35)–(36).

This example is due to Klamkin (1965). Ames (1968) has discussed in great detail the extension of the method of scaling to BVP for rather general second-order DE as well as for systems of DE, with applications to fluid mechanics, to which reference may be made.

5.3 LAGERSTROM MODEL FOR FLOW AT LOW REYNOLDS NUMBERS

We study the following mathematical model for N-dimensional viscous incompressible flow at low Reynolds numbers past an object (Tam 1975), described by the system

$$u'' + \frac{a}{r}u' + bu'^2 + uu' = 0, \tag{1}$$

$$u(\epsilon) = 0, \quad \epsilon > 0, \tag{2a}$$

$$u(\infty) = 1, \tag{2b}$$

where prime denotes differentiation with respect to r. Here $a > 0$ and $b \geq 0$ are constants, and ϵ is a small parameter. The problem is singular in the limit $\epsilon \to 0$ when a/r tends to infinity (the other singularity being the infinite extent of the domain).

We consider the initial value problem for (1):

$$u(\epsilon) = 0, \quad u'(\epsilon) = \alpha. \tag{3}$$

The solution of (1) subject to (3) clearly exists locally, since the right-hand side in

$$u'' = -\frac{a}{r}u' - bu'^2 - uu'$$

is regular about $r = \epsilon$. The solution depends continuously on α and can be continued for all $r > \epsilon$, provided u and u' remain uniformly bounded there. Three cases arise.

i. If $\alpha = 0$, then (1) and (3) show that $u \equiv 0$.
ii. If $u'(\epsilon) = \alpha > 0$, then $u'' = -au'/r - bu'^2 - uu' < 0$ for $r \geq \epsilon$ so that u cannot have a stationary point in the finite domain. This implies that $u > 0$ and $u' > 0$ for $r > \epsilon$.
iii. If $u'(\epsilon) = \alpha < 0$, then $u < 0$, and so $u' < 0$ for $r > \epsilon$.

Existence Theory for BVP via Shooting Techniques

Since we seek a solution for which $u(\infty) = 1$, we consider only case (ii) with $\alpha > 0$. In this case, (1) can be written as

$$u'' + uu' = -\frac{a}{r} - bu'^2 < 0, \tag{4}$$

and, therefore, by integration,

$$u' + \frac{u^2}{2} < \alpha. \tag{5}$$

Since u' is positive, (5) implies that

$$0 < u' < \alpha; \quad u < (2\alpha)^{1/2}; \quad r > \epsilon. \tag{6}$$

This provides an upper bound for u.

We now proceed to obtain a lower bound for u and the continuous dependence of $u(\infty, \alpha)$ on α. These two results will help us to choose an $\bar{\alpha}$ such that $u(\infty; \bar{\alpha}) = 1$ and thus prove Theorem 5.3.1. Following Tam, we prove these results via the following lemmas.

LEMMA 5.3.1 For $\alpha > 0$, the solution $u(r, \alpha)$ of the initial value problem (1) and (3) satisfies

$$u_b(r, \alpha) < u(r, \alpha) < (2\alpha)^{1/2}, \tag{7}$$

where

$$u_b = \frac{1}{b} \ln\left\{ 1 + b\alpha\epsilon^a \int_\epsilon^r t^{-a} e^{-(2\alpha)^{1/2}(t-\epsilon)} \, dt \right\}. \tag{8}$$

Proof. We have already obtained the upper bound in (6). For the lower bound, we write (1) as

$$u'' + u'\left(\frac{a}{r} + bu' + u\right) = 0. \tag{9}$$

This is easily put in the form

$$\frac{d}{dr}\left[u' \exp\left\{ \int_\epsilon^r \left(\frac{a}{t} + u + bu'\right) dt \right\} \right] = 0. \tag{10}$$

Integrating and using the boundary conditions (3), we have

$$e^{bu} u' = \alpha \left(\frac{\epsilon}{r}\right)^a \exp\left(\int_\epsilon^r u \, dt\right). \tag{11}$$

Integrating and using (3a), we write (11) as

$$e^{bu} = 1 + b\alpha\epsilon^a \int_\epsilon^r t^{-a} \exp\left(-\int_\epsilon^t u \, dz\right) dt. \tag{12}$$

Using the upper bound $u < (2\alpha)^{1/2}$ in (12), we get

$$e^{bu} > 1 + b\alpha\epsilon^a \int_\epsilon^r t^{-a} e^{-(2\alpha)^{1/2}(t-\epsilon)} dt. \tag{13}$$

Now taking the logarithm, we obtain (8). The case $b = 0$ can easily be treated by taking the limit $b \to 0$ in (8), where the logarithm term is expanded suitably.

LEMMA 5.3.2 For α in $\{\alpha_0, \alpha_1\}$ where $\alpha_0 \geq \delta > 0$, $u(\infty, \alpha)$ exists and is a continuous function of α.

Proof. It follows from the solution of the initial value problem (1) and (3) that $u(r, \alpha)$ is a continuous function of α for all finite r, and can be continued for all $r > \epsilon$. Moreover, $u(r, \alpha)$ is monotonically increasing ($u' > 0$) and bounded above by $(2\alpha)^{1/2}$. It follows that $u(\infty, \alpha)$ exists.

We now prove that $u(r, \alpha)$ is continuous in α. We first note from (8) that, for α in $\{\alpha_0, \alpha_1\}$,

$$u_b(r, \alpha) > \frac{1}{b} \ln \left\{ 1 - b\alpha_0 \epsilon^a \int_\epsilon^r t^{-a} \exp[-(2\alpha_1)^{1/2}(t - \epsilon)] dt \right\}$$

$$\equiv f(r), \text{ say.} \tag{14}$$

Equation (14) shows that $f(r)$ is a positive increasing function of r, and the integral there converges as $r \to \infty$, so $f(\infty)$ exists. Writing

$$S(\alpha) = \alpha \int_\epsilon^\infty t^{-a} \exp\left(-\int_\epsilon^t u(z, \alpha) dz\right) dt, \tag{15}$$

we have

$$S(\alpha) < \alpha_1 \int_\epsilon^\infty t^{-a} \exp\left(-\int_\epsilon^t f(z) dz\right) dt \equiv \int_\epsilon^\infty \mu(t) dt, \tag{16}$$

where we have used (14). Since the function $\mu(t)$ is integrable over (ϵ, ∞), and is independent of α, the improper integral $S(\alpha)$ is uniformly convergent and is therefore a continuous function of $\alpha \in \{\alpha_0, \alpha_1\}$. Hence, putting $r = \infty$ in (12), we find that $u(\infty, \alpha)$ is a continuous function of α.

The final step in the existence proof is to find α_0 and α_1 such that $u(\infty, \alpha_0) < 1$ and $u(\infty, \alpha_1) > 1$. The continuous dependence of $u(\infty, \alpha)$ on α ensures that there exists $\bar{\alpha}$, $\alpha_0 < \bar{\alpha} < \alpha_1$, for which $u(\infty, \bar{\alpha}) = 1$ and the boundary condition on u at $r = \infty$ is satisfied.

LEMMA 5.3.3 For $\alpha_0 = 1/2$ and

$$\alpha_1 = \left[\frac{e^b - 1}{\sqrt{2b}} + \left(\frac{1}{2b^2}(e^b - 1)^2 + \frac{a}{b\epsilon}(e^b - 1)\right)^{1/2}\right]^2, \tag{17}$$

Existence Theory for BVP via Shooting Techniques

there is at least one $\bar{\alpha}$, $\alpha_0 < \bar{\alpha} < \alpha_1$, for which $u(\infty, \bar{\alpha}) = 1$.

Proof. That $u(\infty, \alpha_0) < 1$ follows immediately from (7) if we choose $\alpha = \alpha_0 = 1/2$. From (7) and (8) we find that

$$u(\infty, \alpha) > \frac{1}{b} \ln \left\{ 1 + b\alpha \epsilon^a \int_\epsilon^\infty t^{-a} e^{-(2\alpha)^{1/2}(t-\epsilon)} \, dt \right\}. \tag{18}$$

Writing $\gamma = 1 - a$, $x = (2\alpha)^{1/2}\epsilon$, and $C = \alpha \epsilon^a (2\alpha)^{(a-1)/2} e^{(2\alpha)^{1/2}\epsilon}$, we can express the right-hand side of inequality (18) in terms of the Gamma function

$$\Gamma(\gamma, x) = \int_x^\infty t^{\gamma-1} e^{-t} \, dt, \qquad x > 0,$$

so the former can be written as

$$u(\infty, \alpha) > \frac{1}{b} \ln \left\{ 1 + bC\Gamma(1-a, (2\alpha)^{1/2}\epsilon) \right\} = u_b(\infty, \alpha). \tag{19}$$

The inequality

$$x^{1-\gamma} e^x \Gamma(\gamma, x) > \frac{x}{x + 1 - \gamma}, \qquad x > 0, \ \gamma < 1,$$

for the Gamma function (see Luke 1962) enables us to write (19) as

$$u(\infty, \alpha) > \frac{1}{b} \ln \left\{ 1 + \frac{b\alpha\epsilon}{(2\alpha)^{1/2}\epsilon + a} \right\}. \tag{20}$$

Now we can choose α_1 such that

$$\frac{1}{b} \ln \left\{ 1 + \frac{b\alpha_1 \epsilon}{(2\alpha_1)^{1/2}\epsilon + a} \right\} = 1. \tag{21}$$

Thus, $u(\infty, \alpha_1) > 1$. We solve (21) for α_1:

$$\alpha_1 = \left\{ \frac{e^b - 1}{\sqrt{2}b} + \left(\frac{1}{2b^2} (e^b - 1)^2 + \frac{a(e^b - 1)}{b\epsilon} \right)^{1/2} \right\}^2. \tag{22}$$

With the choice $\alpha_0 = 1/2$ and α_1 given by (22), the existence of an $\bar{\alpha}$, $\alpha_0 < \bar{\alpha} < \alpha_1$, follows immediately from Lemma 5.3.2.

The above lemmas together constitute the proof of the following existence theorem.

THEOREM 5.3.1 The boundary value problem (1)–(2) has at least one solution.

We remark that the monotonic behavior of the solution curve plays a crucial role in the existence proof of the above boundary value problem.

5.4 SIMILARITY SOLUTIONS OF THE POROUS MEDIA EQUATION

We consider the equation

$$u_t = (u^m)_{xx}, \quad u > 0, \tag{1}$$

which describes the one-dimensional flow of a polytropic gas through a homogeneous porous medium. Here, u is the density of the gas, and x and t are the space and time coordinates. The parameter m is assumed to be greater than 1. We study a class of similarity solutions of (1) in the domain $0 < x < \infty$, $0 < t \leq T$, where T is some positive constant. These solutions assume one of the following forms, wherein α and τ are real numbers:

i. $u_1(x,t) = (t + \tau)^\alpha f_1(\eta) \quad \eta = x(t + \tau)^{-[1+(m-1)\alpha]/2}$,
ii. $u_2(x,t) = (\tau - t)^\alpha f_2(\eta), \quad \eta = x(\tau - t)^{-[1+(m-1)\alpha]/2}$, \quad (2)
iii. $u_3(x,t) = e^{\alpha(t+\tau)} f_3(\eta), \quad \eta = x \exp\left[-\frac{1}{2}\alpha(m-1)(t+\tau)\right]$.

Substituting (2i)–(2iii) into (1) leads to the following ODE for f_1, f_2, and f_3, respectively.

i. $(f_1^m)'' + \dfrac{1}{2}\{1 + (m-1)\alpha\}\eta f_1' = \alpha f_1, \quad 0 < \eta < \infty,$

ii. $(f_2^m)'' - \dfrac{1}{2}\{1 + (m-1)\alpha\}\eta f_2' = -\alpha f_2, \quad 0 < \eta < \infty,$ \quad (2a)

iii. $(f_3^m)'' + \dfrac{1}{2}\alpha(m-1)\eta f_3' = \alpha f_3, \quad 0 < \eta < \infty.$

At the boundaries of the medium, we impose the conditions

$$f_i(0) = U \geq 0, \quad f_i(\infty) = 0, \quad i = 1,2,3, \tag{3}$$

so that $u_i(x,t)$ satisfy the lateral boundary conditions

$$u_1(0,t) = (t + \tau)^\alpha U, \quad u_2(0,t) = (\tau - t)^\alpha U, \quad u_3(0,t) = e^{\alpha(t+\tau)} U,$$

and

$$u_i(x,t) \longrightarrow 0 \quad \text{as } x \longrightarrow \infty, \quad i = 1,2,3,$$

for fixed $t \in [0, T]$.

Gilding and Peletier (1976) (see also the references to previous work in this paper) have considered in detail the existence of the solutions (2a) subject to (3) for $i = 1, 2, 3$, respectively, using a shooting method. Here we consider the specific problem

$$(f^m)'' + p\eta f' = qf, \quad 0 < \eta < \infty, \tag{4}$$
$$f(0) = U, \quad f(\infty) = 0, \tag{5}$$

Existence Theory for BVP via Shooting Techniques 225

where p and q are arbitrary constants. Evidently (4) covers the cases in (2a). We restrict ourselves to the case $U > 0$ and show that the problem (4)–(5) has a weak positive solution with compact support if and only if

$$p \geq 0 \quad \text{and} \quad p + 2q > 0. \tag{6}$$

Moreover, this solution is unique.

We now define a weak solution of (4). A function f is a *weak solution* of (4) if (a) f is bounded, continuous, and nonnegative on $[0, \infty)$, (b) $f^m(\eta)$ has a continuous derivative with respect to η on $(0, \infty)$, and (c) f satisfies the identity

$$\int_0^\infty \phi' \{(f^m)' + p\eta f\} \, d\eta + (p+q) \int_0^\infty \phi f \, d\eta = 0$$

for all $\phi \in C_0^1(0, \infty)$.

First, we find a necessary condition for the existence of a positive solution in the left neighborhood of a point $\eta = a$.

LEMMA 5.4.1 The existence of a nontrivial weak solution of equation (4) with compact support implies one of the following propositions. (i) $p > 0$ or (ii) $p = 0$ and $q > 0$.

Proof. Let f be a nontrivial weak solution of equation (4) with compact support. We can then find a positive number a such that $f > 0$ in $(a - \epsilon, a)$ for some $\epsilon > 0$ and $f = 0$ in $[a, \infty]$. The function f satisfies (4) in $(a - \epsilon, a)$ for some $\epsilon > 0$ and the conditions

$$f(a) = 0, \quad (f^m)'(a) = 0, \tag{7}$$

which follow from the requirement that f and $(f^m)'$ are continuous on $(0, \infty)$. Integrating (4) from $\eta \in (a - \epsilon, a)$ to a and using (7), we have

$$-(f^m)'(\eta) = p\eta f(\eta) + (p+q) \int_\eta^a f(\xi) \, d\xi. \tag{8}$$

Since f and $(f^m)'$ are continuous, we can find $\eta_0 \in (a - \epsilon, a)$ such that $f'(\eta_0) < 0$. The left-hand side of (8) is positive; therefore, not both p and $p + q$ on the right-hand side are negative. Thus if $p = 0$, then $q > 0$.

Now, suppose for contradiction that $p < 0$, then $(p+q) > 0$ and hence $q > 0$. In this case, (4) shows that $f'' > 0$ on $(a - \epsilon, a)$, and therefore, f cannot have a maximum on $(a - \epsilon, a)$; thus $f' < 0$ on $(a - \epsilon, a)$. Using the mean value theorem of integral calculus for the right-hand side of (8), noting that $f(\theta) \leq f(\eta)$ in $\eta < \theta < a$, and canceling the factor $f(\eta)$, we obtain

$$-m f^{m-2}(\eta) f'(\eta) - p\eta \leq (p+q)(a - \eta) \tag{9}$$

for all $\eta \in (a - \epsilon, a)$. Now allowing η to tend to a, we get

$$-mf^{m-2}(a)f'(a) \leq pa, \tag{10}$$

which is a contradiction, since the left-hand side is positive while the right-hand side is negative. Hence the lemma.

We first dispose of the case $p = 0$ and $q > 0$. In this case, the problem (4) with the boundary conditions $f(0) = U$ and (7) has an explicit exact (and unique) solution

$$f(\eta; a) = \left\{ \frac{q(m-1)^2}{2m(m+1)} (a - \eta)^2 \right\}^{1/(m-1)}, \quad 0 < \eta < a. \tag{11}$$

Uniqueness is proved as follows. The function $f(0; a)$ is a continuous, monotonically increasing function of a such that $f(0; 0) = 0$ and $f(0; \infty) = \infty$; therefore, the equation $f(0; a) = U$ has a unique solution $a = a(U)$ for every $U > 0$. Hence there exists a unique solution $f = f(\eta; a(U))$ of the boundary value problem for this special case.

Now we turn to the case $p > 0$. The existence and uniqueness of the boundary value problems (4)–(5) are proved through several lemmas. The shooting argument is applied from the point $\eta = a$ backward to $\eta = 0$ (referred to as reverse shooting). The elements in the proof are (i) a local existence result which is proved via an integral equation formulation (this requires a preparatory lemma to determine the sign of $f'(\eta)$ on the interval $(0, a)$); (ii) bounds for the solution and its backward continuation; (iii) dependence of f on the point a from which the solution is "shot" back to $\eta = 0$; (iv) the uniqueness proof using a scaling argument due to Barenblatt (1952).

LEMMA 5.4.2 Let $b \in (0, 1)$, and let f be a positive solution of (4) and (7) on $[b, a)$; if $p + q \geq 0$, then $f'(\eta) < 0$ on $[b, a)$. Moreover, if f is a positive solution of (4) and (7) on $[0, a)$, then $f'(0) < 0$ provided $p + q > 0$.

Proof. Integrating (4) from η to a, we get

$$-(f^m)'(\eta) = p\eta f(\eta) + (p + q) \int_\eta^a f(\eta) d\eta. \tag{12}$$

Since $p + q \geq 0$ (and $p > 0$), $(f^m)'(\eta) < 0$; hence $f'(\eta) < 0$ on $[b, a)$. If we put $\eta = 0$ in (12), we obtain

$$-(f^m)'(0) = (p + q) \int_0^a f(\xi) d\xi, \tag{13}$$

which shows that $f'(0) < 0$ for $p + q > 0$.

Existence Theory for BVP via Shooting Techniques

LEMMA 5.4.3 Let $p > 0$ and q be arbitrary. Then given $a > 0$, there exists an $\epsilon > 0$ such that in $(a - \epsilon, a)$ the problem (4) and (7) has a unique positive solution.

Proof. For an integral equation formulation of the problem, we assume that f is a positive solution in an interval $(a - \epsilon, a)$ for some $\epsilon > 0$. Lemma 5.4.2 then ensures that for some $\epsilon > 0$, $f' < 0$ in $(a - \epsilon, a)$. This monotonicity of f enables us to write the inverse function $\eta = \sigma(f)$, say. After integrating by parts the integral in (12), we have

$$(f^m)'(\eta) = q\eta f(\eta) + (p+q) \int_\eta^a \xi f'(\xi) d\xi. \tag{14}$$

Now we introduce $\eta = \sigma(f)$ in (14) and have

$$\frac{d\sigma}{df} = \frac{mf^{m-1}}{qf\sigma(f) - (p+q)\int_0^f \sigma(\phi) d\phi}, \tag{15}$$

where the condition $f = 0$ at $\eta = a$ has been used. Integrating (15) from 0 to f yields

$$\sigma(f) - a = m \int_0^f \frac{\phi^{m-1} d\phi}{q\phi\sigma(\phi) - (p+q)\int_0^\phi \sigma(\psi) d\psi}. \tag{16}$$

Writing

$$\tau(f) = 1 - a^{-1}\sigma(f) \tag{17}$$

in (16), we get an integral equation for τ:

$$\tau(f) = \frac{m}{a^2} \int_0^f \frac{\phi^{m-1} d\phi}{p\phi + q\phi\tau(\phi) - (p+q)\int_0^\phi \tau(\psi) d\psi}. \tag{18}$$

We now have to show that (18) has a unique solution in the right neighborhood of $f = 0$. This is achieved by the contraction mapping principle (see Hartman 1964). For this purpose, we define a set X of bounded functions $\tau(f)$ on $[0, \gamma]$, where $\gamma > 0$ such that

$$0 \leq \tau(f) \leq \rho = \frac{p}{2(|q| + |p+q|)}, \tag{19}$$

where the last inequality will become clear presently. The set X is a complete metric space with supremum norm (on X) denoted by $\|\cdot\|$. The right side of (18) defines an operator M on X:

$$M(\tau)(f) = \frac{m}{a^2} \int_0^f \frac{\phi^{m-1} d\phi}{p\phi + q\phi\tau(\phi) - (p+q)\int_0^\phi \tau(\psi) d\psi}. \tag{20}$$

Assuming that $\tau \in X$, we have

$$p\phi + q\phi\tau(\phi) - (p+q)\int_0^\phi \tau(\psi)d\psi \geq \{p - (|q| + |p+q|)\|\tau\|\}\phi$$

$$\geq \frac{1}{2}p\phi$$

(cf. (19)), so

$$M(\tau)(f) \leq \frac{2m}{pa^2}\int_0^f \phi^{m-2}d\phi \leq \frac{2m}{(m-1)pa^2}\gamma^{m-1}. \tag{21}$$

Thus, $M(\tau)$ is well defined on the whole of X. Therefore, $M(\tau): [0, \gamma] \to R$ gives a mapping which is nonnegative and continuous; moreover, there exists a $\gamma_0 > 0$ such that for $\gamma \leq \gamma_0$ and $\tau \in X$, $\|M(\tau)\| \leq \rho$. This implies that, if $\gamma \leq \gamma_0$, M maps X into X.

Now to prove the contraction, we choose $\tau_1, \tau_2 \in X$ and $\gamma \leq \gamma_0$ such that

$$\|M(\tau_1) - M(\tau_2)\|$$
$$\leq \frac{4m}{a^2p^2}\int_0^f \phi^{m-3}\left(|q|\phi\|\tau_1 - \tau_2\| + |p+q|\int_0^\phi \|\tau_1 - \tau_2\|d\psi\right)d\phi$$
$$\leq \frac{4m}{(m-1)p^2a^2}(|q| + |p+q|)\gamma^{m-1}\|\tau_1 - \tau_2\|. \tag{22}$$

Therefore, we can find a $\gamma_1 \in (0, \gamma_0]$ such that if $\gamma \leq \gamma_1$, M is a contraction on X (that is, the factor before $\|\tau_1 - \tau_2\|$ on the right of (22) is less than 1). Thus, by the Banach–Cacciopoli contraction principle, the operator M has a unique fixed point in X and so equation (18) has a unique solution.

The above proof implies the existence and uniqueness of a positive solution of (4) and (7) in a left neighborhood of $\eta = a$.

Having constructed a positive solution f in the left neighborhood of $\eta = a$, we continue it backward toward $\eta = 0$. This can be done uniquely so long as f remains positive and bounded. In the process of backward continuation, three possibilities arise:

A. $f(\eta) \to \infty$ as $\eta \downarrow \eta_1$ for some $\eta_1 \in [0, a)$.
B. $f(\eta)$ can be continued back to $\eta = 0$.
C. $f(\eta) \to 0$ as $\eta \downarrow \eta_2$ for some $\eta_2 \in (0, a)$.

We first rule out possibility A by finding an upper bound for f.

Existence Theory for BVP via Shooting Techniques

LEMMA 5.4.4 Let $b \in [0,a)$ and let f be a positive solution of (4) and (7) on (b,a). Then, if $p > 0$,

$$\sup_{(b,a)} f(\eta) \leq \left[\left(\frac{(m-1)}{2m}\right) a^2 \max\{p, 2p+q\}\right]^{1/(m-1)}. \qquad (23)$$

Proof. We assume that $p + q \geq 0$; since $p > 0$, this implies $2p + q > 0$. We have already proved in Lemma 5.4.2 that $f' < 0$ on (b,a). We therefore obtain from (8) the inequality

$$-(f^m)'(\eta) \leq p\eta f(\eta) + (p+q)f(\eta)(a-\eta)$$

or

$$-mf^{m-2}(\eta)f'(\eta) \leq (p+q)a - q\eta, \qquad b \leq \eta < a. \qquad (24)$$

Integrating (24) from η to a, we have

$$\left(\frac{m}{m-1}\right)f^{m-1}(\eta) \leq \left\{pa + \frac{1}{2}q(a-\eta)\right\}(a-\eta), \qquad b \leq \eta \leq a, \qquad (25)$$

and hence

$$\sup_{(b,a)} \left(\frac{m}{m-1}\right)f^{m-1}(\eta) \leq \frac{1}{2}(2p+q)a^2. \qquad (26)$$

The lemma therefore follows. A lower bound can be similarly obtained:

$$\left(\frac{m}{m-1}\right)f^{m-1}(\eta) \geq \frac{1}{2}p(a^2 - \eta^2), \qquad b \leq \eta \leq a.$$

We now show that $f(\eta) > 0$ in $[0,a)$ for $2p + q > 0$ so that only possibility B can occur in the present case.

LEMMA 5.4.5 Let f be the positive solution of (4) and (7) in a left neighborhood of $\eta = a$. Assume that $p > 0$; then $f(\eta) > 0$ on $[0,a)$ if $2p + q > 0$.

Proof. Integrating (8) from η to a, we have the integral equation

$$f^m(\eta) = p\eta \int_\eta^a f(\xi)d\xi + (2p+q) \int_\eta^a (\xi - \eta)f(\xi)d\xi. \qquad (27)$$

Since $p > 0$ and $2p + q > 0$, the positivity of f in $[0,a)$ follows immediately from (27).

We can, therefore, continue $f(\eta)$ back to $\eta = 0$ where $f(0) > 0$. Now we prove the continuous dependence of f on a, the right end of the interval where $f(a) = 0$. Subsequently we show that to each value of a corresponds a value of $f(0) = U$.

LEMMA 5.4.6 Let $p > 0$ and $2p + q \geq 0$. Suppose $f(\eta; a_1)$ and $f(\eta; a_2)$ are solutions of the problem (4) and (7) on $(0, a_1)$ and $(0, a_2)$, respectively. Then, if $a_1 > a_2$, $f(\eta; a_1) > f(\eta; a_2)$ everywhere on $(0, a_2)$.

Proof. For convenience, we write $f_i(\eta; a_i) = f_i$, $i = 1, 2$. For contradiction, let $\bar{\eta} \in (0, a_2)$ be a point where f_1 and f_2 intersect so that $f_1(\bar{\eta}) = f_2(\bar{\eta})$ and $f_1(\eta) > f_2(\eta)$ on $(\bar{\eta}, a_2)$. Then, writing $\eta = \bar{\eta}$, $f = f_i$, $a = a_i$, $i = 1, 2$, in (27), we have

$$f_i^m(\bar{\eta}) = p\bar{\eta} \int_{\bar{\eta}}^{a_i} f_i(\xi) d\xi + (2p + q) \int_{\bar{\eta}}^{a_i} (\xi - \bar{\eta}) f_i(\xi) d\xi. \tag{28}$$

Subtracting (28) for $i = 1$ from that for $i = 2$ and rearranging the limits in the integrals, we have

$$p\bar{\eta} \int_{\bar{\eta}}^{a_2} (f_1 - f_2) d\xi + (2p + q) \int_{\bar{\eta}}^{a_2} (\xi - \bar{\eta})(f_1 - f_2) d\xi$$

$$+ p\bar{\eta} \int_{a_2}^{a_1} f_1(\xi) d\xi + (2p + q) \int_{a_2}^{a_1} (\xi - \bar{\eta}) f_1 d\xi = 0, \tag{29}$$

where we have assumed that $f_1(\bar{\eta}) = f_2(\bar{\eta})$. Now from the other assumed conditions it follows that, for $2p + q > 0$, the second and fourth terms in (29) are nonnegative while the other two are positive. We arrive at a contradiction. Hence the lemma.

We now prove the main result, Theorem 5.4.1. Lemma 5.4.3 shows that for each $a > 0$ there exists a unique positive solution $f(\eta; a)$ of (4) and (7) in a left neighborhood of $\eta = a$. By Lemma 5.4.5, this solution can be continued back to $\eta = 0$ if $2p + q > 0$. The boundary condition at $\eta = 0$ is satisfied if we can find an $a > 0$ such that

$$f(0; a) = U. \tag{30}$$

this solution is unique if there is only one root a of equation (30). To prove that this is the case, we note (Barenblatt 1952) that if $f(\eta; a)$ is a solution of (4) and (7) on $(0, a)$, then, for any $\mu > 0$, the function $\mu^{-2/(m-1)} f(\mu\eta; \mu a)$ is a solution of (4) and (7) on $(0, \mu a)$. This scaling law can be easily checked by direct substitution. Choosing $\mu = a^{-1}$, we arrive at the boundary condition

$$f(0; a) = a^{2/(m-1)} f(0; 1) = U. \tag{31}$$

Since $f(0; 1) > 0$ for $2p + q > 0$, equation (31) has, for each $U > 0$, a unique solution $a(U)$. The function $f(\eta; a(U))$ now satisfies the DE (4) and the

Existence Theory for BVP via Shooting Techniques

boundary conditions (5) and (7). Since there is a unique value of a (U) for which these conditions are satisfied, the solution of the problem is unique. Combining this with the explicit results proved earlier for $p = 0$, we have the statement of the main theorem.

THEOREM 5.4.1 Let $U > 0$. Then there exists a unique $a > 0$ and a unique solution of problem (4), (5), and (7), which is positive on $(0, a)$ if and only if $p \geq 0$ and $2p + q > 0$.

Now we form the function

$$f(\eta) = \begin{cases} f(\eta; a), & 0 \leq \eta < a, \\ 0, & a \leq \eta < \infty \end{cases} \tag{32}$$

over the infinite line $0 \leq \eta < \infty$. This function is a weak solution of (4) and satisfies the boundary conditions (5). We have now merely to show that this is the only solution of the problem (4) and (5) with compact support. Suppose $f(\eta)$ is a weak solution of (4) and (5) with compact support. Then, Lemma 5.4.5 shows that if $U > 0$ this problem has such a solution only if $2p + q > 0$; this solution is of the form

$$f(\eta) > 0 \quad \text{on } [0, a),$$
$$f(\eta) = 0 \quad \text{on } [a, \infty),$$

for some $a > 0$. In other words, f must be of the type discussed above. By Theorem 5.4.1, this solution is unique.

We have thus proved the following theorem.

THEOREM 5.4.2 Let $U > 0$. Then there exists a unique weak solution with compact support of problem (4) and (5) if and only if $p \geq 0$, $2p + q > 0$.

The case $U = 0$ requires special treatment and is considered in Gilding and Peletier (1976).

5.5 A DIFFERENTIAL EQUATION ARISING IN ELECTROMAGNETIC THEORY

The boundary value problem associated with

$$y'' + \frac{2}{r}y' + \left[y - \left(1 + \frac{2}{r^2}\right)\right]y = 0, \tag{1}$$

requiring

$$y(0) = 0, \quad y(\infty) = 0, \tag{2}$$

arises in the electromagnetic theory of strong interaction, Bergström (1973), and has been treated by McLeod (1977), using a shooting argument. Positive solutions of (1)–(2) are sought, and shooting is carried out in the forward direction from $r = 0$. Equation (1) is singular at $r = 0$, and the second boundary is at $r = \infty$. Thus, again, numerical methods are not convenient. Equation (1) has $y = 0$ as one solution and $y = 1$ as the other in the limit $r \to \infty$. So the positive solution may tend to either of these as $r \to \infty$, or it may become zero at a finite distance r, depending on the initial slope $y'(0) = \alpha$. Each possibility will be discussed.

The initial value problem at $r = 0$ is discussed via an integral equation formulation, reminiscent of the method of Levinson (see Section 4.3). All possibilities except the one which points to the solution tending to 0 as $r \to \infty$ are ruled out. The existence proof of the solution is in the form of six lemmas and the main concluding theorem.

LEMMA 5.5.1 Given any $\alpha > 0$, there exists a unique solution of (1) which has the asymptotic behavior

$$y(r) \sim \alpha r \quad \text{as } r \longrightarrow 0; \tag{3}$$

plainly, this solution is nonnegative for sufficiently small r.

Proof. We can write (1) in the form

$$Ly \equiv y'' + \frac{2}{r}y' - \frac{2}{r^2}y = (1-y)y. \tag{4}$$

$Ly = 0$ has solutions r and $1/r^2$ with Wronskian $-3/r^2$. Treating (4) as an inhomogeneous equation, we can write the general solution by variation of parameters as

$$y = Ar + \frac{B}{r^2} + \frac{1}{3}\int_{r_0}^{r}\left(r - \frac{t^3}{r^2}\right)\{1-y(t)\}y(t)\,dt. \tag{5}$$

Since we are interested in solutions which are asymptotic to αr as $r \to 0$, we choose $B = 0$, $A = \alpha$, and $r_0 = 0$, so (5) assumes the form

$$y = \alpha r + \frac{1}{3}\int_{0}^{r}\left(r - \frac{t^3}{r^2}\right)\{1-y(t)\}y(t)\,dt. \tag{6}$$

The integral equation (6) in $y(r)$ can be solved by Picard iteration and has one and only one solution for the initial value problem for (4) with

$$y(0) = 0, \quad y'(0) = \alpha.$$

Hence the lemma is proved.

LEMMA 5.5.2 Any solution $y(r)$ of (1) which is initially nonnegative must possess one of three mutually exclusive properties:

Existence Theory for BVP via Shooting Techniques

i. It remains nonnegative for only a finite range of values of r.
ii. It remains nonnegative for all r, and $y(r) \to 1$ as $r \to \infty$.
iii. It remains nonnegative for all r, and $y(r) \to 0$ as $r \to \infty$.

Proof. Since $y(t) > 0$, it follows from (6) that

$$y \le \alpha r + \frac{1}{3} \int_0^r \left(r - \frac{t^3}{r^2} \right) y(t) \, dt. \tag{7}$$

We compare the solutions of this inequality with those of the equation

$$Y = \alpha r + \frac{1}{3} \int_0^r \left(r - \frac{t^3}{r^2} \right) Y(t) \, dt \tag{8}$$

with $Y \sim \alpha r$ as $r \to 0$. By differentiating twice, we can change the linear integral equation (8) into a linear second-order DE whose solution, $Y \sim \alpha r$ near $r = 0$, can be shown to exist for all r and is bounded in any compact set. Thus, any nonnegative solution of (7) is bounded below by zero and above by Y. It therefore follows that, since $y(r) < Y(r)$, y exists for all r and is bounded in any compact set. This rules out possibility (i), and the solution $y(r)$ exists for all r.

We have to prove that either $y \to 1$ or $y \to 0$ as $r \to \infty$. Now, two cases arise: either y is ultimately monotonic, or it oscillates. We consider each case separately.

A. If y is ultimately monotonic, we show that either $y \to 1$ or $y \to 0$. Suppose neither is true so that $y(\infty) = k \ne 0, 1$, where $y(\infty)$ exists since y is monotonic; then either $0 < k < 1$ or $k > 1$, possibly ∞. If $0 < k < 1$, then writing (4) as

$$(r^2 y')' = -r^2 y \left\{ y - \left(1 + \frac{2}{r^2} \right) \right\}$$

and carrying out the integrations, we find that, for $r \to \infty$,

$$(r^2 y')' \sim r^2 k (1 - k),$$

$$r^2 y' \sim \frac{1}{3} r^3 k (1 - k),$$

$$y' \sim \frac{1}{3} r k (1 - k).$$

This contradicts that $y \to k$ as $r \to \infty$. An exactly similar argument rules out the case $k > 1$. Thus, if y is ultimately monotonic, then $y \to 1$ or $y \to 0$.

B. Now we consider the case when y is not ultimately monotonic. Multiplying (4) by y' and integrating between r_0 and r, we get

$$\left[\frac{1}{2} y'^2 \right]_{r_0}^r + 2 \int_{r_0}^r \frac{y'^2}{t} \, dt - \int_{r_0}^r \frac{2}{t^2} y(t) y'(t) \, dt - \left[\frac{y^2}{2} \right]_{r_0}^r$$

$$+ \left[\frac{y^3}{3}\right]_{r_0}^{r} = 0. \tag{9}$$

We use the Cauchy inequality to write

$$y(r) = \int_0^r y'(t)\,dt \leq \left[\int_0^r dt\right]^{1/2} \left[\int_0^r y'^2\,dt\right]^{1/2}$$

$$= r^{1/2} \left[\int_0^r y'^2\,dt\right]^{1/2} \tag{10}$$

and estimate the integral

$$\int_{r_0}^{r} \frac{2yy'}{t^2}\,dt = \left[\frac{y^2}{t^2}\right]_{r_0}^{r} + \int_{r_0}^{r} \frac{2}{t^3} y^2\,dt$$

$$\leq \left[\frac{y^2}{t^2}\right]_{r_0}^{r} + 2 \int_{r_0}^{r} \left(\int_0^t y'^2(u)\,du\right) \frac{dt}{t^2}$$

$$= \left[\frac{y^2}{t^2}\right]_{r_0}^{r} - 2 \left[\frac{1}{t}\int_0^t y'^2(t)\,dt\right]_{r_0}^{r} + 2\int_{r_0}^{r} \frac{y'^2(t)}{t}\,dt. \tag{11}$$

Equation (9), in view of (11), becomes

$$\left[\frac{1}{2}y'^2 + \frac{1}{3}y^3 - \frac{1}{2}y^2 - \frac{y^2}{t^2} - \frac{2}{t}\int_0^t y'^2(u)\,du\right]_{r_0}^{r} \leq 0. \tag{12}$$

Inequality (12) shows that, since y^3 dominates y^2 for large y, the expression

$$\frac{1}{2}y'^2 + \frac{1}{3}y^3 + \frac{2}{t}\int_0^t y'^2\,du \tag{13}$$

is bounded as $r \to \infty$, and so are y and y' in this limit.

Now either y oscillates finitely and $y(\infty)$ does not exist, or, along with the oscillation, $y(\infty)$ exists. In the latter case we may use the same argument as for the monotonic case to show that either $y(\infty) = 0$ or $y(\infty) = 1$. We have already proved that y and y' are bounded; therefore, (9) shows that, since $\int_{r_0}^{r} 2yy'/t^2\,dt$ is evidently bounded, $\int_{r_0}^{r} y'^2(t)/t\,dt$ is also bounded as $r \to \infty$. Equation (9) then implies that

$$\frac{1}{2}y'^2 + \frac{1}{3}y^3 - \frac{1}{2}y^2 \longrightarrow L, \text{ say,} \quad \text{as } r \longrightarrow \infty. \tag{14}$$

Three cases arise, depending on whether $y \to 0$, $y \to 1$, or y tends to neither of these limits. We consider each of them separately.

Existence Theory for BVP via Shooting Techniques

a. If $y \to 1$, then $L \to -1/6$ as $r \to \infty$ so that extreme values of y are given by

$$\frac{1}{3}y^3 - \frac{1}{2}y^2 + \frac{1}{6} = 0. \tag{15}$$

The roots of (15) are $(1, 1, -1/2)$. Since y is nonnegative, the solution tends to 1 as $r \to \infty$.

b. If y does not tend to 0 or 1, $L \neq 0, -1/6$, the roots of

$$\frac{1}{3}y^3 - \frac{1}{2}y^2 - L = 0 \tag{15a}$$

are distinct and none of them is 0 or 1. In the limit $r \to \infty$, y oscillates between any two of the roots, which are the extreme values. When y is near the extreme values, y' is small, and vice versa. Equation (1), however, shows that, for large r, when the extreme value is neither 1 nor 0,

$$y'' \sim (y-1)y$$

and y'' is therefore not small. Therefore, y' is not small for most values of r of the oscillation, and hence the integral

$$\int_{r_0}^r \frac{y'^2(t)}{t} dt$$

does not converge as $r \to \infty$. This contradicts what we have proved earlier (see equation (14)).

c. There remains the case $L = 0$ for which, from (14), the extreme values are $0, 0, 3/2$. There is no immediate reason to rule out the oscillations in the limit $r \to \infty$ between 0 and $3/2$. The arguments of (a) and (b) do not apply. The analysis is more delicate, and we refer the reader to the original paper of McLeod (1977). However, the conclusion is that no oscillatory solution in the limit $r \to \infty$ is possible in this case either.

LEMMA 5.5.3 The set of α (> 0) for which the solution is given by property (i) of Lemma 5.5.2 is denoted by S_1 and is open in the topology of the positive semiaxis.

Proof. The solution $y(r, \alpha)$ of

$$y(r, \alpha) = \alpha r + \frac{1}{3} \int_0^r \left(r - \frac{t^3}{r^2}\right) \{1 - y(t)\} y(t) dt$$

is evidently continuous in α, so if $y(r, \alpha_0)$ is a solution with $y(r_0, \alpha_0) < 0$ we can find α close to α_0 such that $y(r_0, \alpha) < 0$.

LEMMA 5.5.4 The set of α (> 0) for which the solution given by Lemma 5.5.1 has property (ii) of Lemma 5.5.2 is denoted by S_2 and is open in the topology of the positive semiaxis.

Proof. Let the solution $y(r, \alpha_0)$ as given by Lemma 5.5.1 have the property that $y(r, \alpha_0) \to 1$ as $r \to \infty$. We have to show that, for α sufficiently close to α_0, we still have $y(r, \alpha) \to 1$ as $r \to \infty$.

We define two functions

$$F(r, \alpha) = \frac{1}{2} y'^2 + \frac{1}{3} y^3 - \frac{1}{2} y^2 \tag{16}$$

(see equation (14) for which the dependence on α has been made explicit) and

$$G(t, \alpha) = \frac{1}{2} y'^2 + \frac{1}{3} y^3 - \frac{1}{2} y^2 - \frac{y^2}{t^2} + \frac{2}{t} \int_0^t y'^2(u) \, du \tag{17}$$

(see inequality (12)). Since $y(r, \alpha_0) \to 1$, $y'(r, \alpha_0) \to 0$ as $r \to \infty$,

$$F(r, \alpha_0) \longrightarrow -\frac{1}{6} \quad \text{as } r \longrightarrow \infty,$$

as in the proof of Lemma 5.5.2. The function

$$G(t, \alpha_0) \longrightarrow -\frac{1}{6} \quad \text{as } t \longrightarrow \infty.$$

Therefore, given $\epsilon > 0$, we can find r_0 sufficiently large such that

$$G(r_0, \alpha_0) \leq -\frac{1}{6} + \epsilon. \tag{18}$$

If α is sufficiently close to α_0, then using the continuous dependence of y and y' on α, we have

$$G(r_0, \alpha) - G(r_0, \alpha_0) < \epsilon. \tag{19}$$

From (18) and (19),

$$G(r_0, \alpha) < -\frac{1}{6} + 2\epsilon. \tag{20}$$

The inequalities (12) and (20) imply that, for $r \geq r_0$,

$$G(r, \alpha) \leq G(r_0, \alpha) < -\frac{1}{6} + 2\epsilon. \tag{21}$$

We now claim that $y(r, \alpha)$ can become zero neither for finite nor infinite values of r, since in either eventuality, (17) would imply that $G(r, \alpha)$ becomes nonnegative at the corresponding finite or infinite value of r. This contradicts (21) if we choose $\epsilon < 1/12$. We thus conclude that $y(r, \alpha) \to 1$ as $r \to \infty$, and the lemma is proved.

Existence Theory for BVP via Shooting Techniques

LEMMA 5.5.5 For α positive and sufficiently large, the solution given by Lemma 5.5.1 has property (i) of Lemma 5.5.2 so that $\alpha \in S_1$.

Proof. To use the fact that α is large, we introduce the variables

$$t = \alpha^{1/3} r, \qquad Y = \alpha^{-2/3} y(r), \tag{22}$$

so (1) becomes

$$Y'' + \frac{2}{t} Y' + \left\{ Y - \left(\alpha^{-2/3} + \frac{2}{t^2} \right) \right\} Y = 0 \tag{23}$$

and has initial conditions

$$Y(0) = 0, \qquad Y'(0) = 1.$$

For large α we have a convenient comparison equation for (23), namely,

$$Y_0'' + \frac{2}{t} Y_0' + \left\{ Y_0 - \frac{2}{t^2} \right\} Y_0 = 0 \tag{24}$$

with

$$Y_0(0) = 0, \qquad Y_0'(0) = 1. \tag{25}$$

The lemma is proved if we can show that $Y_0(t)$ becomes negative for some finite t. For if $Y_0(t) < 0$, then for sufficiently large α, $Y(t)$ can be made as close to $Y_0(t)$ as we please in any compact t interval. To prove that Y_0 becomes negative, we note that Y_0 and Y_0' are initially positive and

$$\frac{d}{dt}(t^2 Y_0') = (2 - t^2 Y_0) Y_0 > 0 \qquad \text{if } Y_0 < \frac{2}{t^2}, \tag{26}$$

so $t^2 Y_0'$ increases until Y_0 meets the curve $z = 2/t^2$. What happens after Y_0 crosses $z = 2/t^2$? To see this, we note that

$$\frac{d}{dt}(t^2 Y_0') < 0 \qquad \text{for } Y_0 > \frac{2}{t^2}, \tag{27}$$

so $t^2 Y_0'$ decreases and must become negative; the latter follows from equation (24):

$$Y_0'' = -\frac{2}{t} Y_0' - \left\{ Y_0 - \frac{2}{t^2} Y_0 \right\} < 0, \tag{28}$$

showing that Y_0' decreases and becomes negative. Thus, $Y_0 > 0$ and $Y_0' < 0$. The solution is, therefore, bounded and exists for all t. It follows from (27) that $t^2 Y_0'$ tends to a limit L as $t \to \infty$ with $L < 0$, finite or infinite. We assume L is finite; a similar argument applies for the case $L = \infty$. Thus, $Y_0' \sim L/t^2$, $Y_0 \to M$, a constant which, however, must be zero in view of

(24). Assuming $Y_0 \sim -L/t$, then (24) gives

$$(t^2 Y_0')' \sim -L^2 \quad \text{or} \quad Y_0' \sim -\frac{L^2}{t},$$

which contradicts $Y_0' \sim -L/t$. Therefore, Y_0 must meet $z = 2/t^2$ a second time, and at the point of intersection we have

$$Y_0 = \frac{2}{t^2}, \qquad Y_0' \leq -\frac{4}{t^3}. \tag{29}$$

The solution $z = 2/t^2$ satisfies the DE

$$z'' + \frac{2}{t}z' - \frac{2}{t^2}z = 0. \tag{30}$$

The initial conditions for this equation are taken to be the point of intersection with the solution curve

$$z(t_0) = \frac{2}{t_0^2}, \qquad z'(t_0) = -\frac{4}{t_0^3} \tag{31}$$

(mark the equality sign for $z'(t_0)$).

Equations (30)–(31) form a convenient comparison system for (24) and (29). The solution $z = 2/t^2$ remains positive and tends to zero as $t \to \infty$. Now we form the DE for the difference function $Y_0 - z$:

$$(Y_0 - z)'' + \frac{2}{t(Y_0 - z)'} - \frac{2}{t^2(Y_0 - z)} = -Y_0^2. \tag{32}$$

We can solve (32) as an inhomogeneous DE (cf. Lemma 5.5.1) so that

$$Y_0 - z = At + \frac{B}{t^2} - \frac{1}{3}\int_{t_0}^{t}\left(t - \frac{u^3}{t^2}\right) Y_0^2(u)\, du. \tag{33}$$

The initial conditions for $Y_0 - z$ follow from (29) and (31):

$$(Y_0 - z)(t_0) = 0, \qquad (Y_0 - z_0)'(t_0) \leq 0. \tag{34}$$

By substituting (34) into (33) we easily verify that $A \leq 0$, $B \geq 0$, and so as $t \to \infty$,

$$Y_0 - z = At + \frac{B}{t^2} - L(t), \text{ say}, \tag{35}$$

where $L(t)$ stands for the integrals term in (33) and is positive increasing. Since $A \leq 0$, (35) implies that $Y_0 - z = Y_0 - 2/t^2$ becomes negative as t increases. Hence the lemma.

LEMMA 5.5.6 If α (> 0) is sufficiently small, then the solution given in Lemma 5.5.1 has property (ii) of Lemma 5.5.2 so that $\alpha \in S_2$.

Existence Theory for BVP via Shooting Techniques

Proof. Here the proof is similar to that for Lemma 5.5.4. We have merely to show that, for sufficiently small α, there is some r_0, possibly depending on α, for which the function (see (17))

$$G(r_0, \alpha) < 0.$$

to establish the existence of such an r_0, we write (1) in the form

$$y'' + \frac{2}{r}y' - \left(1 + \frac{2}{r^2}\right)y = -y^2, \tag{36}$$

and treat it again as an "inhomogeneous" equation. The solutions of the homogeneous equation

$$y'' + \frac{2}{r}y' - \left(1 + \frac{2}{r^2}\right)y = 0 \tag{37}$$

are

$$f(r) = \frac{e^r}{r}\left(1 - \frac{1}{r}\right), \quad g(r) = \frac{e^{-r}}{r}\left(1 + \frac{1}{r}\right). \tag{38}$$

Using variation of parameters, we can write the solution of (36) as

$$y = \frac{3}{2}\alpha(f + g) - \frac{1}{2}\int_0^r \{f(r)g(t) - f(t)g(r)\}t^2 y^2(t)\,dt. \tag{39}$$

By iteration, (39) yields

$$y - \frac{3}{2}\alpha(f + g) = O(\alpha^2) \quad \text{as } \alpha \longrightarrow 0, \tag{40}$$

the estimates being uniform in any fixed interval $[0, R]$. To evaluate the function G, it suffices to set $y = 3\alpha(f + g)/2$ and to use $f + g \sim f = e^r(1 - 1/r)/r$, since, for large r, we ignore in the process at worst polynomials in $1/r$. Referring, therefore, to (17) we have

$$\frac{1}{2}y'^2 - \frac{1}{2}y^2 - \frac{y^2}{r^2} + \frac{2}{r}\int_0^r y'^2(t)\,dt \sim -\frac{3}{2}\frac{e^{2r}}{r^4}, \tag{41}$$

where we have used the approximation $y \sim e^r(1 - 1/r)/r$. We can therefore, find, for α sufficiently small, r_0 sufficiently large such that

$$G(r_0, \alpha) \sim -\frac{3}{2}\left(\frac{3\alpha}{2}\right)^2 \frac{e^{2r_0}}{r_0^4} < -K_0\alpha^2, \tag{42}$$

where K_0 is a positive constant independent of α. Therefore, the important first step is proved, and the rest of the argument is the same as in Lemma 5.5.4.

From Lemma 5.5.1–5.5.6, we easily deduce the following existence theorem.

THEOREM 5.5.1 There exists a nonnegative solution to the boundary value problem (1)–(2), the solution being strictly positive except at $r = 0$.

Proof. We find from Lemmas 5.5.3–5.5.6 that the sets S_1, S_2 are open and nonempty. Moreover, since the positive semiaxis is connected, not all $\alpha > 0$ belong to S_1, S_2. It follows that there exists at least one value of α which belongs neither to S_1 nor to S_2 and for which the corresponding solution as given by Lemma 5.5.1 has property (iii) of Lemma 5.5.2. We further adduce that the solution is strictly positive for $r > 0$, since, otherwise, it would have to touch the value 0 at $r = r_0$ where $y(r_0) = y'(r_0) = 0$. Equation (1) then shows that $y(r) \equiv 0$, contradicting $\alpha > 0$. The theorem is proved.

5.6 A BOUNDARY VALUE PROBLEM ASSOCIATED WITH AN EQUATION OF PLASMA PHYSICS

We consider the boundary value problem

$$\frac{d^2y}{dx^2} - xy = 2y|y|^\alpha, \qquad -\infty < x < \infty, \tag{1}$$

with the conditions

$$y(x) \sim \left(-\frac{1}{2}x\right)^{1/\alpha} \quad \text{as } x \longrightarrow -\infty, \tag{2}$$

$$y(x) \longrightarrow 0 \quad \text{as } x \longrightarrow +\infty, \tag{3}$$

where the number α is strictly positive. This equation arose in plasma physics in the work of De Boer and Ludford (1975) and for $\alpha = 2$ coincides with the second Painlevé transcendent (see Section 8.6). We follow the work of Hastings and McLeod (1980), who used a shooting argument very similar to that in Section 5.5. They also studied asymptotic and other properties of the solutions of (1)–(3) as $x \to -\infty$ and their dependence on the amplitude parameter k, appearing in the asymptotic form of the solution in the limit $x \to \infty$, namely $y \sim kAi(x)$. In particular, they proved that, for $\alpha = 2$, the second Painlevé equation, the amplitude parameter k equals 1 and the solution of the problem (1)–(3) exists for all x. This constitutes a nonlinear connection problem for which the asymptotic behavior of the solution of (1) as $x \to \infty$ is related to that for $x \to -\infty$.

We prove the following existence theorem through Lemmas 5.6.1–5.6.5.

THEOREM 5.6.1 For each $\alpha > 0$, the problem (1)–(3) has a solution.

As a matter of fact, this solution is also unique, but we do not concern ourselves with that aspect here.

Existence Theory for BVP via Shooting Techniques

LEMMA 5.6.1 There exists a unique solution of (1) which is asymptotic to $kAi(x)$ as $x \to \infty$, k being any given positive number. This solution may not exist for all x as x decreases to $-\infty$, but at each x for which it continues to exist, the solution and its derivatives are continuous functions of k. This solution is denoted $y_k(x)$.

Proof. Treating (1) as an inhomogeneous equation, with the Airy equation as the homogeneous counterpart having two solutions $Ai(x)$ and $Bi(x)$, we can write its "solution," with appropriate behavior as $x \to +\infty$, in the form

$$y_k(x) = kAi(x)$$
$$+ 2 \int_x^\infty \{Ai(x)Bi(t) - Ai(t)Bi(x)\} y_k(t) |y_k(t)|^\alpha \, dt. \quad (4)$$

The asymptotic behavior of Ai and Bi is

$$Ai \sim \frac{1}{2} \pi^{-1/2} x^{-1/4} \exp\left(-\frac{2}{3} x^{3/2}\right),$$
$$Bi \sim \pi^{-1/2} x^{-1/4} \exp\left(\frac{2}{3} x^{3/2}\right), \quad \text{as } x \to +\infty. \quad (5)$$

In view of (5) the integral in (4) exists, and the integral equation (4) can be solved uniquely. This then gives the solution $y_k(x)$ as well as its continuous dependence on k.

LEMMA 5.6.2 The set k (> 0), for which $y_k(x)$ remains positive as x decreases and becomes infinite for some finite value of x, is an open set, denoted by S_1.

Proof. Suppose that $y_{k_0}(x)$, for a given $k = k_0$, blows up at $x = x_0$. Then we can find x_1 near x_0 such that

$$y_{k_0}^\alpha(x_1) > |x_1| + 1, \qquad y_k'(x_1) < 0. \quad (6)$$

Since y is continuous with respect to k, we can find k sufficiently close to k_0 such that

$$y_k^\alpha(x_1) > |x_1| + 1. \quad (7)$$

Furthermore, if $1 + |x_1| > |x|$ (i.e., $-1 \le x - x_1 \le 0$), we have

$$y_k^\alpha(x) > |x|, \qquad y_k'(x) < 0. \quad (8)$$

Equation (1) shows that

$$y_k'' = xy_k + 2y_k|y_k|^\alpha$$
$$= y_k(x + |y_k|^\alpha) + y_k|y_k|^\alpha$$
$$> y_k|y_k|^\alpha$$

or
$$y_k'' > y_k^{\alpha+1}, \tag{9}$$
since $y_k > 0$. Integrating inequality (9) twice, we find that
$$y_k^{\alpha/2} > \text{constant} \times \frac{1}{c-x}, \tag{10}$$
where c is another constant. Inequality (10) shows that the solution blows up at a finite point. Indeed, this point will be in the interval $x - x_1 \geq -1$ if
$$y_k(x_1) > K, \qquad y_k'(x_1) < -K, \tag{11}$$
where K is some sufficiently large positive constant independent of x_1. The point $x = x_1$ can be chosen such that (6) is satisfied together with (11). Hence the lemma is proved.

LEMMA 5.6.3 The set k (> 0) for which $y_k(x)$ takes negative values before it ceases to exist is open. It is denoted by S_2.

Proof. This follows from the continuity of y_k with respect to k. If $y_k(x_0) < 0$ for some k_0 and x_0, then $y_k(x_0) < 0$ for all k sufficiently close to k_0.

LEMMA 5.6.4 The set S_1 (the set of $k > 0$ for which $y_k(x)$ remains positive as x decreases and becomes infinite for some finite value of x) is nonempty.

Proof. In view of the asymptotic behavior of Ai and Bi and of Ai' and Bi' (see (5)), the integral term in (4) may be shown to be positive; therefore $y_k(x) > kAi(x)$ for sufficiently large x. In a similar manner, we can show, by differentiating (4), that $y_k'(x) < kAi'(x) < 0$. We can now choose x_1 and k such that the inequalities (6) and (11) are satisfied. Then, proceeding as in Lemma 5.6.2, we can show that y_k blows up at a finite point. Hence the lemma.

LEMMA 5.6.5 The set S_2 (the set of $k > 0$ for which y becomes negative before it ceases to exist) is nonempty.

Proof. For $k = 0$, $y_k = 0$ is a solution of (4). Making use of the continuity of y_k with respect to k, y_k remains small for k sufficiently small. We write (1) in the form
$$y'' - [x + 2|y|^\alpha]y = 0 \tag{12}$$
and choose k and in turn $y_k(-1)$ and $y_k'(-1)$ so small that, for $-1 \geq x \geq -(1 + \pi/\sqrt{2})$,
$$x - 2|y_k(x)|^\alpha \leq -\frac{1}{2}. \tag{13}$$

Existence Theory for BVP via Shooting Techniques

Now comparing (12) with

$$y'' + \frac{1}{2}y = 0, \tag{14}$$

we note that since the solution of the latter has a zero at 0 and $-\pi\sqrt{2}$, y_k must vanish somewhere in $[-1-\pi\sqrt{2}, -1]$ (see Birkhoff and Rota 1978, chapter 10). Hence the lemma.

Proof of Theorem 5.6.1. The positive semiaxis ($k > 0$) is a connected set; therefore, it cannot be divided into two nonempty disjoint open sets. We have proved that the sets S_1 and S_2 of $k > 0$ (see Lemmas 5.6.2 and 5.6.3) are nonempty, open, and, therefore, disjoint. Therefore, there exists at least one positive value of k which lies neither in S_1 nor in S_2. For such a value of $k = k_*$, say, the solution $y_{k_*}(x)$ exists for all x and is always positive. This solution cannot vanish anywhere, for if $y = 0$ for some x, it would be a minimum; that is, $y = y' = 0$ at this point. Equation (1) then would give $y = 0$ for all x. Thus, we have proved the theorem.

5.7 SOME OTHER BOUNDARY VALUE PROBLEMS

We now summarize a few other boundary value problems which have been treated by shooting methods. Zhidkov and Shirikov (1964) considered the problem

$$y'' + \frac{2}{x}y' - y + y^n = 0, \qquad n > 0, \; x \geq 0, \tag{1}$$

$$y(0) = y_0 < \infty, \quad y'(0) = 0, \; y(\infty) = 0, \tag{2}$$

where y_0 is an unknown positive parameter. Equation (1) is of Thomas–Fermi type, and problem (1)–(2) arises in the statistical theory of the nucleus.

The axisymmetric deformation of a circular membrane subjected to normal pressure is given by

$$(x^3 y')' + \frac{Q}{y^2} = 0, \tag{3}$$

where y is the radial membrane stress and the function $Q(x)$ has a piecewise continuous derivative on $[0, 1]$. The boundary conditions at $x = 0$ arise from the symmetry and boundedness of stresses and displacements. Thus, we require that $y(0)$ be bounded and that

$$y'(0) = 0. \tag{4}$$

At the edge one of the following two conditions are satisfied:

$$y(1) > 0 \tag{5}$$

or

$$y'(1) + ay(1) = 0, \qquad a > \frac{1}{2}. \tag{6}$$

This problem was analyzed by Callegari and Reiss (1968), using a shooting argument.

The boundary value problems for the equation

$$y'' = y^2 - x \tag{7}$$

requiring either

$$y(0) = 0, \quad y(x) \sim +\sqrt{x} \qquad \text{as } x \longrightarrow \infty, \tag{8}$$

or

$$y(0) = 0, \quad y(x) \sim -\sqrt{x} \qquad \text{as } x \longrightarrow \infty, \tag{9}$$

were considered by Holmes and Spence (1984). These problems arise in connection with studies of natural convective flows with viscous dissipation. Equation (7) is closely related to the first Painlevé transcendent

$$w'' = 6w^2 + x \tag{10}$$

or more closely to

$$w'' = 6w^2 - 6x \tag{11}$$

(see Section 8.5). The boundary value problems (7), (8) and (7), (9), however, have not been covered by earlier investigations on the first Painlevé transcendent.

A higher-order boundary value problem

$$f''' + ff'' + \frac{2}{3}(h - f'^2) = 0, \tag{12}$$

$$h'' + \sigma f h' = 0, \tag{13}$$

$$f = f' = 0, \quad h = 1 \quad \text{at } x = 0, \tag{14}$$

$$f' = h = 0 \quad \text{at } x = \infty, \tag{15}$$

describes some laminar boundary layer flows (McLeod and Serrin 1968). Here, the prime denotes differentiation with respect to x, and the functions $f'(x)$, $h(x)$, $0 \leq x \leq \infty$, represent, respectively, the vertical component of velocity and the temperature in the fluid. The parameter σ is the Prandtl number of the fluid and may assume any positive constant value. The existence of the solutions of (12)–(15) was proved using shooting methods.

6
Phase Space Study of Autonomous Systems

6.1 INTRODUCTION

A system of n first-order ordinary differential equations may be written in the "normal" form

$$\frac{dx_1}{dt} = X_1(x_1, x_2, \ldots, x_n, t),$$
$$\frac{dx_2}{dt} = X_2(x_1, x_2, \ldots, x_n, t), \qquad (1)$$
$$\vdots$$
$$\frac{dx_n}{dt} = X_n(x_1, x_2, \ldots, x_n, t),$$

where X_i are given functions of $n + 1$ real variables $x_1, x_2, \ldots, x_n, t$. We call a set of n functions $x_1(t), x_2(t), \ldots, x_n(t)$ (which should be once continuously differentiable) a *solution* of (1) if it satisfies (1). The conditions under which system (1) has a unique solution passing through a given point $(x_{10}, x_{20}, \ldots, x_{n0})$ are standard and may be found, for example, in Chapter 6 of Birkhoff and Rota (1978). This solution may be visualized as a curve in a given region R of the $(n + 1)$-dimensional space of the variables $x_1(t)$, $x_2(t), \ldots, x_n(t)$ and t, over which the functions $X_1, X_2, \ldots, X_n$ are assumed to be continuous and real-valued. If we specialize (1) to $n = 1$, we have a region R in the two-dimensional (x_1, t) plane, and the solution curve lies in this planar region. Since the right-hand sides of (1) contain t explicitly, this system is nonautonomous. However, if t does not appear explicitly in (1), so that it reads as

$$\frac{dx_i}{dt} = X_i(x_1, x_2, \ldots, x_n), \qquad i = 1, 2, \ldots, n, \tag{2}$$

we obtain an autonomous system. We shall assume that the functions X_i are continuously differentiable with respect to each of the arguments, so by the existence and uniqueness theorem any set of initial conditions $x_1(0) = a_1$, $x_2(0) = a_2, \ldots, x_n(0) = a_n$ determines a unique trajectory through the point $(a_1, \ldots, a_n)$. Moreover, the solution depends continuously on the initial conditions. The uniqueness of the solution implies that at every point of the trajectory $[x_1(t), x_2(t), \ldots, x_n(t)]$ (or the solution curve as it is equivalently called) equation (2) defines a unique velocity vector $\dot{x}_1, \dot{x}_2, \ldots, \dot{x}_n$ which is tangential to the trajectory at that point. This, in turn, implies that no two trajectories can cross, since this would lead to two values of the tangent vector at the point of crossing.

The nonautonomous system (1) can be changed into an autonomous system of order $n + 1$ by solving one of the equations (1) for t explicitly in terms of $x_1, x_2, \ldots, x_n$ and dx_i/dt, say, and then differentiating it with respect to t. This leads to a second-order equation in the variable x_i, which is equivalent to two first-order equations. The independent variable t can be eliminated from the other $n - 1$ equations.

The solutions of the autonomous system (2) have the following important property. If $x_i(t, \mathbf{c})$, $i = 1, 2, \ldots, n$, is a solution of (2), then so is $x_i(t + a, \mathbf{c})$, $i = 1, 2, \ldots, n$, for any constant a. Here, the vector $\mathbf{c}$ is the initial position of the vector $\mathbf{x}$ with components x_i. The solution $\mathbf{x}(t + a, \mathbf{c})$ can be interpreted as the solution of (2) which passes through $\mathbf{c}$ at $t = -a$. We may infer that there is always a one-parameter family of solutions of (2) describing the same solution curve in the n-dimensional $(x_1, x_2, \ldots, x_n)$ Euclidean space. The curve is traced in the direction of increasing value of the independent variable t and is indicated by an arrow in the phase space. For the autonomous system (2), if the algebraic system obtained by equating the right-hand sides to zero, namely,

$$\begin{aligned} X_1(x_1, x_2, \ldots, x_n) &= 0, \\ X_2(x_1, x_2, \ldots, x_n) &= 0, \\ &\vdots \\ X_n(x_1, x_2, \ldots, x_n) &= 0, \end{aligned} \tag{3}$$

has a solution $(\bar{x}_1, \bar{x}_2, \ldots, \bar{x}_n)$, it is called a *critical*, *singular*, or *fixed point* of the system (2). The velocity at this point is zero, so it is also called

Phase Space Study of Autonomous Systems

an *equilibrium* or *rest point*. We shall use these terms interchangeably. It is the purpose of phase space analysis to investigate the solution about such critical points and to connect them by appropriate trajectories. It is remarkable that in the case of several types of critical points (which we shall classify in Section 6.3), local (linear) analysis contributes so significantly to the understanding of the global properties of the full nonlinear solution, which, in general, is difficult to obtain in a closed form.

The autonomous system (2) has been studied in great detail for $n = 2$ for two reasons: (i) it is relatively simpler; (ii) as we have said earlier, second-order DE appear most frequently in applications. The (x_1, x_2) plane is referred to as the *phase plane*. The cases $n = 1$ and $n = 3$ lead to phase line and phase space, respectively. We have seen in Sections 2.9–2.10 that some nonautonomous second-order systems can either be transformed to second-order autonomous ones or to a system of two equations, one of which is the principal equation holding in a Lie or phase plane, while the other is an auxiliary equation relating the new system to the original one. Thus, the phase plane study is of paramount importance in the analysis of even some nonautonomous second-order systems.

We shall first briefly discuss the phase line, which will illustrate the elegance of phase space analysis. We shall outline the types of "simple" singularities which can appear in the phase plane and illustrate them with a host of examples. We shall discuss many other examples to bring out more complicated cases when either the singular points are not simple or the linear analysis about them does not lead to global understanding of the solution.

We shall refer to the literature where complicated phase plane studies have been rigorously carried out. We shall treat several examples arising from physical phenomena such as gas dynamics and heat diffusion to show how the phase plane study actually solves in a neat qualitative way some boundary value problems. The phase space interpretation in more than two dimensions is relatively more difficult, and we shall content ourselves with a brief discussion of this aspect. We shall, however, list for ready reference all the cases that arise in the linear phase space study. This categorization itself turns out to be elaborate.

6.2 ONE-DIMENSIONAL PHASE SPACE OR THE PHASE LINE

For the autonomous scalar equation

$$\dot{x} = X(x), \tag{1}$$

Figure 6.2.1a Phase line—a stable node or attractor.

Figure 6.2.1b Phase line—a saddle point or shunt.

where x belongs to a subset of the real line and the dot denotes differentiation with respect to t, we may write the solution in the form of a quadrature. Alternatively, we may graphically draw the solution curve in the (x,t) plane. The third possibility which we pursue here is that the solution is represented on the x-line. The point c for which $X(c) = 0$ is a fixed point of (1) since $x(t) \equiv c$ remains there for all time. There may be more than one fixed point if $X(x)$ is a nonlinear function of x. If $X(x) \neq 0$, then $\dot{x} = X(x)$ is either less than zero or greater than zero so that the trajectory on the x-line either approaches the fixed point as t increases or recedes from it. The phase portrait of the solution of (1) is drawn on a line passing through the fixed points.

If we consider a single fixed point, the trajectories in the neighborhood of this point can display just three types of behavior:

1. All trajectories approach the critical point as $t \to \infty$. Such a critical point is called a *stable node* or an *attractor* (see Figure 6.2.1a). The point $x = 0$ for $\dot{x} = -x^2$ is an attractor.
2. All trajectories move into the critical point from one side of it while they move away from it on the other side, as $t \to \infty$. This behavior may appropriately be called a *shunt* or *saddle point* (this latter name will become more meaningful when we discuss phase plane in the next section). The shunt is shown in Figure 6.2.1b. For example, the fixed point $x = 0$ is a shunt or saddle point for the equation $\dot{x} = x^2$.
3. In this case all trajectories move away from the critical point as $t \to +\infty$. This point is called a *repeller* or an *unstable node* (see Figure 6.2.1c). A simple example of this type is $x = 0$ for $\dot{x} = x^3$.

Figure 6.2.1c Phase line—an unstable node or repeller.

Phase Space Study of Autonomous Systems

We consider two other interesting examples.

EXAMPLE 1 $\dot{x} = x^2 - 1$

This equation has two fixed points, $x = 1$ and $x = -1$. Since $x^2 > 1$ for $|x| > 1$, and $x^2 < 1$ for $|x| < 1$, the solutions approach $x = 0$ from both sides while they move away from $x = -1$ and $x = +1$. The points $x = \pm 1$ are therefore unstable nodes. The general solution (apart from the fixed point solutions $x = \pm 1$) can easily be found by separation of variables to be

$$x = \frac{1 \pm e^{2(t-c)}}{1 \mp e^{2(t-c)}} = \begin{Bmatrix} \tanh \\ \coth \end{Bmatrix} (c-t),$$

where c is a constant of integration.

EXAMPLE 2 $\dot{x} = x^3 - x$

This equation has three fixed points $x = 0$, $x = 1$, and $x = -1$. Since $x^3 > x$ in the strips $-1 < x < 0$ and $x > 1$, while $x^3 < x$ in the complementary strips $0 < x < 1$ and $x < -1$, it is easily checked that the point $x = 0$ is a stable node while the points $x = \pm 1$ are unstable nodes. The explicit general solution, apart from the fixed ones, namely $x = 0, x = 1, x = -1$, may easily be found, by separating the variables:

$$x = \pm [1 \mp \exp(2t - k)]^{-1/2},$$

where k is the constant of integration. The above solution is real only for values of t for which the quantity in brackets is positive.

6.3 NATURE OF SINGULAR POINTS IN THE PHASE PLANE

Now we consider a system of two first-order autonomous equations:

$$\frac{dx}{dt} = P(x,y),$$
$$\frac{dy}{dt} = Q(x,y). \tag{1}$$

The points of equilibrium for the system are given by intersection of the curves

$$P(x,y) = 0, \quad Q(x,y) = 0, \tag{2}$$

in the finite or infinite part of the (x,y) plane.

An equilibrium point is simple if the curves (2) have no singularities there and the tangential directions at the point are distinct. For example, if

$$P(x,y) = x^2 + y^2 - 1, \qquad Q(x,y) = y, \tag{3}$$

the intersection consists of two points $(\pm 1, 0)$ in the finite part of the (x,y) plane, and the tangents to the curves $P(x,y) = 0$ and $Q(x,y) = 0$ at these points are distinct.

In contrast, the point of intersection (2) is a multiple equilibrium state if it is either a point of contact of the two curves or a point at which one or both curves have a singularity.

For example, if

$$P(x,y) = x^2 + y^2 - 1, \qquad Q(x,y) = y - 1, \tag{4}$$

the intersection $(0, 1)$ is a double point where the curves $P(x,y) = 0$, $Q(x,y) = 0$ have a common tangent.

An equilibrium point of (1) is isolated if it has a neighborhood containing no other equilibrium point; on the other hand, any neighborhood, however small, of a nonisolated equilibrium point contains other such point(s).

The system (1) may be written in the equivalent form

$$\frac{dy}{dx} = \frac{Q(x,y)}{P(x,y)}, \tag{5}$$

where $P(x,y) \neq 0$. Now, equation (5) may be treated in the (x,y) "phase" plane without reference to the parameter t. This equation has a unique solution in the neighborhood of an "ordinary" point, i.e., a point at which $Q(x,y)/P(x,y)$ has a continuous derivative (see Section 1.3). If, on the other hand, a point (x,y) is singular (this is so, for example, when $Q(x,y)/P(x,y)$ is not defined at the point), then the existence and uniqueness theorem does not apply and the solution at this point is discontinuous, nonunique, or does not exist. The equilibrium points (2) where $P(x,y) = 0$ and $Q(x,y) = 0$ will now be referred to as *singular points* for the system (5).

The advantage of the form (1) is that the points of vertical tangency $P(x,y) = 0$ of (5) are nonsingular for the former. The solutions of (5) terminate where $P(x,y) = 0$, while those of system (1) do so at its equilibrium points (2). As we noted earlier, the integral curves of (1) are directed; in contrast, those of (5) have no directions associated with them. We pointed out in the introduction that many second-order autonomous or nonautonomous systems of nonlinear DE may be treated in the phase plane. For example, dynamical systems with one degree of freedom lead to plane autonomous systems.

Phase Space Study of Autonomous Systems

The book by Andronov et al. (1973) is devoted entirely to the qualitative theory of second-order dynamical systems. These may be represented by

$$\ddot{x} = F(x, \dot{x}). \tag{6}$$

This equation arises from Newton's second law of motion and may be interpreted as describing the motion of a particle on a straight line such that its acceleration is a function of its instantaneous position x and velocity $\dot{x}$. Indeed, we can write (6) as the system

$$v = \frac{dx}{dt}, \qquad \frac{dv}{dt} = v \frac{dv}{dx} = F(x, v), \tag{7}$$

which, in turn, may be changed by division into the single equation of form (5):

$$\frac{dv}{dx} = \frac{F(x, v)}{v}. \tag{8}$$

The (x, v) plane is referred to as the (Poincaré) phase plane. The equilibrium states are given by the simultaneous solution of the equations $v = 0$ and $F(x, v) = 0$.

The first step in the analysis of (1) is the determination of equilibrium points by the solution of the system (2): this may itself present a nontrivial problem. It can be proved (see Andronov et al. 1973) that the system (1) for which $P(x, y)$ and $Q(x, y)$ are analytic can have either only finitely many equilibrium states in any bounded plane region or singular curves all of whose points are equilibrium states. Moreover, an infinite set of isolated equilibrium states with a point of accumulation, which may occur in nonanalytic systems, is impossible in an analytic system. In any case, we assume that (x_0, y_0) is a simple rest point, which requires that the tangential directions of the curves $P(x, y) = 0$ and $Q(x, y) = 0$ at (x_0, y_0) are not the same. Thus, the Jacobian

$$\Delta(x_0, y_0) = \begin{vmatrix} P_x(x_0, y_0) & P_y(x_0, y_0) \\ Q_x(x_0, y_0) & Q_y(x_0, y_0) \end{vmatrix} \neq 0. \tag{9}$$

Subject to (9), a simple rest point satisfies (2); that is,

$$P(x_0, y_0) = 0, \qquad Q(x_0, y_0) = 0.$$

At such a point, by the uniqueness theorem, there is only one solution of (1), namely the constant solution $x = x_0$, $y = y_0$. A constant solution does not define a trajectory; therefore, no trajectory passes through a critical point. If $C = [x(t), y(t)]$ is a trajectory of (1), then we may say that C approaches (x_0, y_0) as $t \to \infty$ if $\lim_{t \to \infty} x(t) = x_0$ and $\lim_{t \to \infty} y(t) = y_0$. Geometrically, if $P = (x, y)$ traces out C according to the solution $x = x(t)$

and $y = y(t)$, then $P \to (x_0, y_0)$ as $t \to \infty$. If, in addition,

$$\lim_{t \to \infty} \frac{y(t) - y_0}{x(t) - x_0}$$

exists or if this quotient becomes positively or negatively infinite as $t \to \infty$, then we say that the trajectory C enters the critical point (x_0, y_0) as $t \to \infty$. The quotient $(y(t) - y_0)/(x(t) - x_0)$ is the slope of the line joining (x_0, y_0) and the point P with the coordinate $x(t)$ and $y(t)$, so the additional requirement means that this line approaches a definite direction as $t \to \infty$. The above argument may also apply to the limit $t \to -\infty$; moreover, the above properties belong to the trajectory C and do not depend on which solution is used to represent it.

We may shift the origin to the singular point (x_0, y_0). Thus, the singular point may hereafter be assumed to be located at $(0, 0)$ without any loss of generality. We further assume that the functions $P(x,y)$ and $Q(x,y)$ are single-valued and analytic in the neighborhood of $(0, 0)$ so that they may be expanded in a Taylor series. The system (1) therefore becomes

$$\begin{aligned} \frac{dx}{dt} &= P_x(0,0)x + P_y(0,0)y + \phi(x,y) \\ &= ax + by + \phi(x,y), \\ \frac{dy}{dt} &= Q_x(0,0)x + Q_y(0,0)y + \psi(x,y) \\ &= cx + dy + \psi(x,y), \end{aligned} \qquad (10)$$

where the functions $\phi(x,y)$ and $\psi(x,y)$ are the remainder terms satisfying the conditions

$$\begin{aligned} \phi(0,0) &= \psi(0,0) = 0, \\ \phi_x(0,0) &= \phi_y(0,0) = \psi_x(0,0) = \psi_y(0,0) = 0, \end{aligned} \qquad (11)$$

so

$$\phi(x,y) = O(\rho^2), \qquad \psi(x,y) = O(\rho^2), \qquad (12)$$

where $\rho = (x^2 + y^2)^{1/2}$. Condition (9) for a critical point to be simple now becomes

$$\Delta(0,0) = \left. \frac{\partial(P,Q)}{\partial(x,y)} \right|_{\substack{x=0 \\ y=0}} = \begin{vmatrix} a & b \\ c & d \end{vmatrix} = ad - bc \neq 0. \qquad (13)$$

Two questions in respect of the legitimacy of the process of linearization of (1) about the critical point $(0,0)$, say, arise. Is the topological structure or the phase portrait of (1) completely determined by the linear terms? The answer to this question is that this is the case for three of the four

Phase Space Study of Autonomous Systems

major types of simple critical points that we shall presently classify. In the exceptional case, the critical point is the so-called center. For nonlinear systems such as (1), is the global phase portrait determined by the local phase portrait? The answer to this question is in the negative. Nonlinear systems, in general, will have more than one critical point. Even though the local phase portrait at the critical points is the same for two systems, the global portraits can show qualitative variation. We shall illustrate these points subsequently with the help of examples.

Now we study in detail the nature of the critical point $(0,0)$ of the linearized form

$$\frac{dx}{dt} = ax + by, \tag{14a}$$

$$\frac{dy}{dt} = cx + dy \tag{14b}$$

of (1). The coefficient matrix of system (14), namely,

$$A = \begin{bmatrix} a & b \\ c & d \end{bmatrix}, \tag{15}$$

is nonsingular if $ad - bc \neq 0$. In this case (see (13)) the point $(0,0)$ is a simple critical point. When $ad - bc = 0$, we obtain a degenerate case for which the integral curves of the equivalent equation

$$\frac{dy}{dx} = \frac{cx + dy}{ax + by} \tag{16}$$

do not form a regular family. We shall consider both the circumstances and the subcases arising from them.

First we note that if $x(t)$ and $y(t)$ are solutions of (14), they are also solutions of the "secular" equation

$$\ddot{u} - (a + d)\dot{u} + (ad - bc)u = 0, \tag{17}$$

This is easily seen by eliminating y and $\dot{y}$ from (14), where

$$y = \frac{1}{b}(\dot{x} - ax),$$

$$\dot{y} = \frac{1}{b}(\ddot{x} - a\dot{x}), \quad b \neq 0.$$

(For $b = 0$, system (14) becomes decoupled and is immediately integrated.) Thus, equation (14b) becomes (17) with u equal to x. A similar elimination of x and $\dot{x}$ gives (17) again for y. The general solution for x and y can

therefore be written as
$$x = x_1 e^{\lambda_1 t} + x_2 e^{\lambda_2 t},$$
$$y = y_1 e^{\lambda_1 t} + y_2 e^{\lambda_2 t}, \tag{18}$$

where x_1, x_2, y_1, and y_2 are arbitrary constants, and λ_1 and λ_2 are the roots of the characteristic polynomial for (17), namely

$$\lambda^2 - (a + d)\lambda + (ad - bc) = 0. \tag{19}$$

It is also evident that (19) is the characteristic polynomial for the coefficient matrix (15) of the system (14), with λ_1 and λ_2 as its eigenvalues.

The form (18) of the solution is not particularly illuminating. It gives the false impression that there are four arbitrary constants for the second-order linear system (14). We must transform (14) in such a manner that the equations become either fully or partially decoupled with their properties rendered more transparent. This can be effected if the matrix coefficient A assumes one of the Jordan canonical forms. For the latter we have the well-known result from linear algebra (see Shilov 1977): for real A, there exists a nonsingular matrix M such that $\mathcal{J} = M^{-1}AM$ is one of the following types:

$$\text{(i)} \begin{bmatrix} \lambda_1 & 0 \\ 0 & \lambda_2 \end{bmatrix}, \quad \lambda_1 > \lambda_2; \quad \text{(ii)} \begin{bmatrix} \lambda_0 & 0 \\ 0 & \lambda_0 \end{bmatrix};$$
$$\text{(iii)} \begin{bmatrix} \lambda_0 & 1 \\ 0 & \lambda_0 \end{bmatrix}; \quad \text{(iv)} \begin{bmatrix} \alpha & -\beta \\ \beta & \alpha \end{bmatrix}, \quad \beta > 0, \tag{20}$$

where $\lambda_0, \lambda_1, \lambda_2, \alpha$, and β are real constants.

The matrix $\mathcal{J}$ is called the *Jordan form* of A. It is known from linear algebra that the eigenvalues of $\mathcal{J}$ and A are the same and are given by (19). If we write the system (14) in vector notation,

$$\frac{d\mathbf{x}}{dt} = A\mathbf{x}, \tag{21}$$

and a linear transformation by the nonsingular matrix M as

$$\mathbf{u} = M\mathbf{x}, \tag{22}$$

then (21) becomes

$$\frac{d\mathbf{u}}{dt} = M\frac{d\mathbf{x}}{dt} = MA\mathbf{x} = (MAM^{-1})\mathbf{u}. \tag{23}$$

Thus, M can be chosen such that MAM^{-1} assumes one of the canonical forms (20). The systems of linear ODE (21) and (23) have matrix coefficients A and MAM^{-1} which, in the language of linear algebra, are similar and have the same eigenvalues. These autonomous systems are linearly equivalent in the sense that the nonsingular (linear) transformation (22) exists, which carries the former into the latter. The regions over which (21) and (23) are defined correspond according to (22).

We shall follow Birkhoff and Rota (1978) to classify the nature of the solution of (14) in the neighborhood of the orbital point (0, 0). For this purpose, two results are used. The first is that linearly equivalent plane autonomous systems have the same secular equation. This follows from the similarity of matrices A and MAM^{-1} of the systems (21) and (23), as we have noted earlier; the roots λ_1 and λ_2 of (19) are the same for both, and hence the solution is in the form (18). The second result is that unless $a = d$ and $b = c = 0$, the (linear) plane autonomous system (14) is linearly equivalent to the Poincaré plane representation

$$\frac{du}{dt} = v,$$

$$\frac{dv}{dt} = -qu - pv, \qquad (24)$$

$$p = -(a + d), \qquad q = ad - bc,$$

of its secular equation (17).

To prove this, we assume that $b \neq 0$ and set $u = x$ and $v = ax + by$. The transformation matrix $M = \begin{bmatrix} 1 & 0 \\ a & b \end{bmatrix}$ is therefore nonsingular. The system (14) becomes

$$\dot{u} = v,$$

$$\dot{v} = a\dot{x} + b\dot{y} = a(ax + by) + b(cx + dy)$$
$$= (a^2 + bc)x + (ab + bd)y. \qquad (25)$$

We write the right-hand side of $\dot{v}$ as $(a^2 + ad)x + (ab + bd)y - (ad - bc)x = (a + d)v - (ad - bc)u$. Therefore, the system (14) becomes

$$\dot{u} = v,$$
$$\dot{v} = (a + d)v - (ad - bc)u = -pv - qu, \qquad (26)$$

with the coefficient matrix $\begin{bmatrix} 0 & 1 \\ -q & -p \end{bmatrix}$, which may be verified to be equal to MAM^{-1}. The case $c \neq 0$ can be treated similarly by letting $u = y$, $v = cx + dy$, etc.

The case $b = c = 0$, $a \neq d$ for which the matrix (15) of (14) is already diagonal (see 20(i)) can be handled by writing

$$x = \frac{du - v}{d - a}, \qquad y = \frac{au - v}{a - d}.$$

Equations (24) follow again.

Now we turn to the exceptional case $b = c = 0$ and $a = d$. The system (14) becomes

$$\dot{x} = ax, \qquad \dot{y} = ay. \tag{27}$$

Its coefficient matrix $\begin{bmatrix} a & 0 \\ 0 & a \end{bmatrix}$ is diagonal, and we have case (ii) of (20). The secular equation for (27) is $\ddot{x} - 2a\dot{x} + a^2 x = 0$, and its characteristic roots are equal. In this case the discriminant of the characteristic equation (19), $\Delta = p^2 - 4q = (a-d)^2 + 4bc$, vanishes.

To summarize, the nature of the solution of (14) near $(0,0)$, for the nonexceptional case, is determined by the roots of the characteristic polynomial (19). The nature of the roots decides which of the four canonical forms (20) is chosen to represent (14), using linear equivalence. The canonical form then determines the nature of the solution of (14). The roots of the characteristic polynomial

$$\lambda^2 - (a+d)\lambda + ad - bc \equiv \lambda^2 + p\lambda + q, \text{ say,} \tag{19}$$

are

$$\lambda_{1,2} = \frac{-p \pm (p^2 - 4q)^{1/2}}{2};$$

their nature depends principally on the sign of the discriminant

$$\Delta = p^2 - 4q = (a-d)^2 + 4bc. \tag{28}$$

The following cases arise:

CASE I If $\Delta > 0$ and $q > 0$, then the roots $\lambda = \mu_1, \mu_2$ of (19) are real, distinct, and of the same sign. The critical point is called a *node*. The linearly equivalent canonical form corresponds to case (i) of (20) and may be chosen to be

$$\frac{dx}{dt} = \mu_1 x, \qquad \frac{dy}{dt} = \mu_2 y, \qquad 0 < |\mu_1| < |\mu_2|. \tag{29}$$

The general solution of (29) is $(c_1 e^{\mu_1 t}, c_2 e^{\mu_2 t})$. The critical point is stable when μ_1 and μ_2 are negative. In this case, the trajectories $(x(t), y(t))$ move into the origin as t increases from $-\infty$ to $+\infty$. The solution in the (x,y) plane may be written as $y = cx^{\mu_2/\mu_1}$, $c = c_2/c_1^{\mu_2/\mu_1}$, which look like parabolas tangential to a line at the origin (see Figure 6.3.1). Each value of c determines a distinct curve. If μ_1 and μ_2 are positive, the critical point is unstable, the orientation of the trajectories reverses, and they diverge from the origin as t increases from $-\infty$ to $+\infty$. The form of the curves is, however, the same, irrespective of whether μ_1 and μ_2 are both negative or both positive.

Phase Space Study of Autonomous Systems

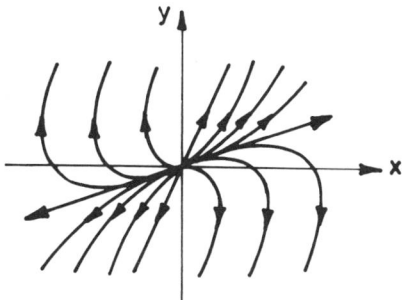

Figure 6.3.1 Phase portrait of system (14): case I—nodal point.

CASE II   If $\Delta > 0$ and $q < 0$, the roots of (19) are real and of opposite sign. The critical point is called a *saddle*. The canonical form of the solution (case (i) of (20)) is again chosen to be (29), but the solution $(ae^{\mu_1 t}, be^{\mu_2 t})$ shows that either $x(t)$ or $y(t)$ tends to infinity as t increases from $-\infty$ to $+\infty$. The critical point is unstable. The solution in the (x,y) plane now becomes $x^m y = c$, $m = -\mu_2/\mu_1 > 0$. The integral curves for different values of c now appear like similar hyperbolas (Figure 6.3.2).

CASE III   If $\Delta < 0$, the roots $\lambda_{1,2} = \mu \pm i\nu$ ($\nu \neq 0$) of the characteristic equation (19) are complex conjugate. The critical point is called a *focal* or *spiral point*. The canonical form corresponds to case (iv) of (20), and we may choose the linearly equivalent system

$$\frac{dx}{dt} = \mu x - \nu y, \qquad \frac{dy}{dt} = \nu x + \mu y. \tag{30}$$

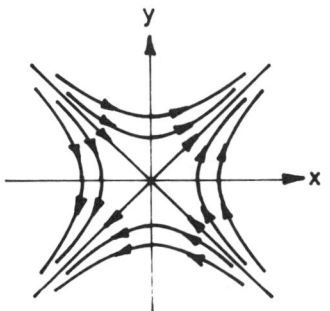

Figure 6.3.2 Phase portrait of system (14): case II—saddle point.

The solution of this system is conveniently found in polar form by writing $x = r\cos\theta$, $y = r\sin\theta$, etc., in (30). The transformed system is

$$\frac{dr}{dt} = \mu r, \qquad \frac{d\theta}{dt} = \nu, \tag{31}$$

which integrates to give

$$r = \rho e^{\mu t}, \qquad \theta = \nu t + \tau, \tag{32}$$

where $\rho \geq 0$ and τ are arbitrary constants. Here, the sum of the roots is $\lambda_1 + \lambda_2 = 2\mu = -p$. Therefore, the spirals (32) are stable if $p > 0$; i.e., $\mu < 0$. The trajectories approach the origin as t varies from $-\infty$ to $+\infty$ (Figure 6.3.3). In the opposite circumstance, $p < 0$, $\mu > 0$, the trajectories diverge from the origin as t increases from $-\infty$ to $+\infty$ and the spiral point is unstable. When $p = 0$, $\mu = 0$, the roots λ_1 and λ_2 are pure imaginary, and the trajectories are given by $r = \rho$, a constant. These are closed curves and represent periodic oscillations. The critical point is now referred to as a *center* or (neutrally stable) *vortex point* (Figure 6.3.4).

Now we discuss the exceptional and degenerate cases when either $\Delta = p^2 - 4q = (a-d)^2 + 4bc = 0$ or $q = 0$.

CASE IV If $\Delta = 0$, $q = a^2 > 0$, the roots of (19) are equal and have the same sign. The critical point is a node and is sometimes called a *star* (Figure 6.3.5). The trajectories are given by $(c_1 e^{at}, c_2 e^{at})$. In the (x,y) plane, they are straight lines $y = mx$, $m = c_2/c_1$. They converge or diverge depending on the sign of a. Correspondingly, the point is stable or unstable. The canonical form corresponds to case (ii) of (20).

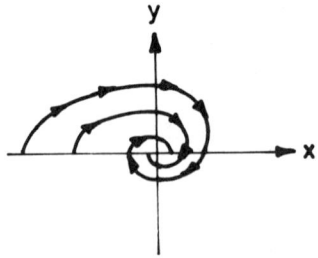

Figure 6.3.3 Phase portrait of system (14): case III—stable focal or spiral point.

Phase Space Study of Autonomous Systems

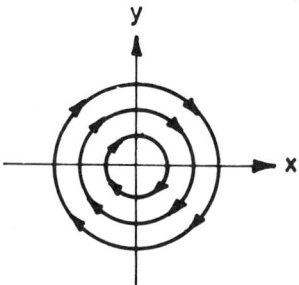

Figure 6.3.4 Phase portrait of system (14): case III—center or vortex point.

CASE V If $\Delta = 0$, $a = d$, but b and c are not both zero, and the system (14) is already in the canonical form (iii) of (20). We have, with $c = 1$, $b = 0$,

$$\frac{dx}{dt} = ax, \qquad \frac{dy}{dt} = x + ay. \tag{33}$$

The equations now are partially separated and can be easily integrated sequentially. The solution is $x = c_1 e^{at}$, $y = (c_1 t + c_2) e^{at}$. The critical point $(0, 0)$ is again a node, stable or unstable, according as $a < 0$ or > 0. The trajectories have the form shown in Figure 6.3.6.

CASE VI In the case $q = ad - bc = 0$, $\Delta \neq 0$, one of the roots of the characteristic equation (19) is zero. The secular equation (17) becomes $\ddot{u} - (a + d)\dot{u} = \ddot{u} + p\dot{u} = 0$, $p \neq 0$. Its solution is $u = c_0/p + c_1 e^{-pt}$. Both x and y have this form of solution, as functions of t. The solution in the (x, y) plane consists of parallel straight lines. This becomes clear from the first integral $\dot{u} + pu = $ constant. The critical point $(0, 0)$ is stable if $p > 0$ and unstable if $p < 0$. The trajectories are shown in Figure 6.3.7.

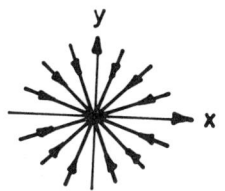

Figure 6.3.5 Phase portrait of system (14): case IV—stable nodal star.

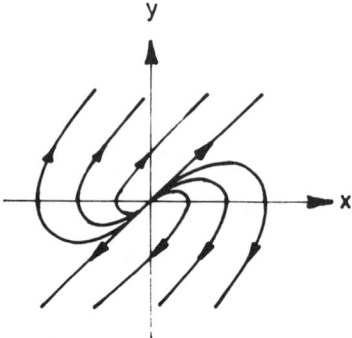

Figure 6.3.6 Phase portrait of system (14): case V—unstable node.

CASE VII The last case pertains to $\Delta = 0$, $q = 0$, and $a = d = b = c = 0$. This exceptional case of (14) refers simply to $\dot{x} = 0$, $\dot{y} = 0$. The solution is constant and neutrally stable.

Thus, to analyze the nature of the solution in the neighborhood of a critical point of (1), we first check whether this point is simple. We then linearize (1) about it. We determine the type of the critical point from the classification detailed above. The trajectories, with the appropriate arrow denoting the direction with increasing time, can be easily drawn by numerical integration. The graphical method of isoclines (see Birkhoff and Rota 1978) is occasionally resorted to, but numerical integration may be preferred. The orientation of the curves is easily fixed by considering the points on the x- or y-axis. For example, for the former, one substitutes $y = 0$ in the given system and verifies whether $\dot{y} > 0$ or $\dot{y} < 0$ for $x > 0$.

It may be verified that the trajectories approach and enter the critical point with a definite slope in cases I, IV, and V as $t \to \infty$, if it is stable.

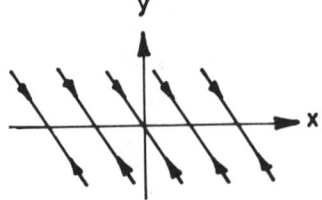

Figure 6.3.7 Phase portrait of system (14): case VI—degenerate case.

Phase Space Study of Autonomous Systems

In case II of a saddle point, only two half-lines, which are asymptotes of the hyperbolas, approach and enter the origin as $t \to \infty$. For case III, the trajectories wind round the (stable) spiral point and approach it as $t \to \infty$, but do not enter it. Indeed, it may be inferred from (32) that x and y change sign infinitely often as $t \to \infty$, and so the trajectories do not approach the origin at a specific angle in this limit. Similar statements may be made if the critical point is unstable.

We again emphasize that the above analysis holds strictly only when the system (1) is exactly linear. When (1) is nonlinear, the local analysis about the critical point still gives the correct type provided the point (i) is simple and (ii) not a center. The critical point for the nonlinear case need not be a center when the linearized system indicates it to be so with the characteristic polynomial possessing pure imaginary eigenvalues. A center is a rather special kind of critical point for which the orbits or trajectories exactly close. Any small perturbation or distortion of a closed orbit may give a nonclosed orbit. Other critical points such as nodes, saddle points, and spiral points are not as sensitive to small changes; their qualitative behavior does not change with small perturbations in the linearized form of the system (1).

6.4 EXAMPLES OF SIMPLE CRITICAL POINTS

EXAMPLE 1

$$\dot{x} = x - y, \qquad \dot{y} = 4x^2 + 2y^2 - 6. \tag{1}$$

The critical points are

$$x - y = 0, \qquad 4x^2 + 2y^2 - 6 = 0.$$

These are found to be $(1, 1)$ and $(-1, -1)$. The determinant

$$\begin{vmatrix} P_x & P_y \\ Q_x & Q_y \end{vmatrix} = \begin{vmatrix} 1 & -1 \\ 8x & 4y \end{vmatrix} = 4y + 8x$$

is not zero at either of these points. Therefore, these points are simple. Shifting the origin to $(1, 1)$, we write

$$X = x - 1, \qquad Y = y - 1,$$

so (1) becomes

$$\dot{X} = X - Y, \qquad \dot{Y} = 4X^2 + 2Y^2 + 8X + 4Y. \tag{2}$$

The linearized form of (2) is

$$\dot{X} = X - Y, \qquad \dot{Y} = 8X + 4Y,$$

for which $a = 1$, $b = -1$, $c = 8$, and $d = 4$. The characteristic equation
$$\lambda^2 - 5\lambda + 12 = 0$$
has roots $(5 \pm 23i)/2$, which are complex conjugate with a positive real part. Therefore $(1, 1)$ is an unstable spiral point.

Changing the origin to the other singular point $(-1, -1)$, we put
$$X = x + 1, \quad Y = y + 1,$$
so equations (1) become
$$\dot{X} = X - Y, \quad \dot{Y} = 4X^2 + 2Y^2 - 8X - 4Y.$$
With $a = 1$, $b = -1$, $c = -8$, and $d = -4$, the characteristic equation
$$\lambda^2 + 3\lambda - 12 = 0$$
has roots $(-3 \pm (57)^{1/2})/2$, which have opposite sign. Therefore $(-1, -1)$ is a saddle point and so is unstable.

EXAMPLE 2

$$\dot{x} = e^{x+y} - y, \quad \dot{y} = -x + xy. \tag{3}$$

The critical points are given by the intersection of the curves
$$x(y - 1) = 0, \quad e^{x+y} - y = 0.$$
The only point of intersection of these curves is $(-1, 1)$. The determinant
$$\begin{vmatrix} P_x & P_y \\ Q_x & Q_y \end{vmatrix} = \begin{vmatrix} e^{x+y} & e^{x+y} - 1 \\ -1 + y & x \end{vmatrix}$$
is evidently not zero at $(-1, 1)$; therefore, this point is simple. Shifting the origin to $(-1, 1)$ by setting
$$X = x + 1, \quad Y = y - 1$$
changes (3) to the form
$$\dot{X} = e^{X+Y} - Y - 1, \quad \dot{Y} = -Y + XY. \tag{4}$$
Expanding (4) about $(0, 0)$ and retaining only the linear terms give
$$\dot{X} = X, \quad \dot{Y} = -Y. \tag{5}$$
The system (5) is already separated and therefore integrable: $X = X_0 e^t$, $Y = Y_0 e^{-t}$. The point $x = -1$, $y = 1$ is therefore a saddle point. In the notation of Section 6.3, $a = 1$, $b = 0$, $c = 0$, and $d = -1$ in (5), so the characteristic equation is
$$\lambda^2 - 1 = 0$$

with roots ± 1. These are real and have opposite sign. Hence the critical point is a saddle.

EXAMPLE 3

$$\dot{x} = 2y, \quad \dot{y} = 12x - 3x^2. \tag{6}$$

The critical points of this system are $O(0,0)$ and $A(4,0)$. Linearizing about $(0,0)$ we have

$$\dot{x} = 2y, \quad \dot{y} = 12x. \tag{7}$$

Here we have $a = 0$, $b = 2$, $c = 12$, and $d = 0$. The characteristic equation is

$$\lambda^2 - 24 = 0$$

with roots $\pm(24)^{1/2}$. Therefore, $(0,0)$ is a saddle point.

Shifting the origin to $(4,0)$, we write

$$X = x - 4, \quad Y = y,$$

so (6) becomes

$$\begin{aligned} \frac{dX}{dY} &= 2Y, \\ \frac{dY}{dt} &= -12X - 3X^2. \end{aligned} \tag{8}$$

The linearized form of (8) is

$$\frac{dX}{dt} = 2Y, \quad \frac{dY}{dt} = -12X. \tag{9}$$

so $a = 0$, $b = 2$, $c = -12$, and $d = 0$. The characteristic equation of (9) is

$$\lambda^2 + 24 = 0;$$

the two roots are $\pm i\sqrt{24}$. The point $(4,0)$ is therefore a center.

Equations (6) can easily be shown to possess a first integral

$$y^2 - 6x^2 + x^3 = C. \tag{10}$$

The integral curve which passes through the origin corresponds to $C = 0$ in (10). This curve is the union of a critical point and three paths: the critical point $(0,0)$, two nonclosed paths, one of which tends to $(0,0)$ as $t \to -\infty$ while the other tends to it as $t \to +\infty$, and a "loop" which tends to $(0,0)$ as $t \to -\infty$ and as $t \to +\infty$. The curve (10) passes through the other critical point $(4,0)$ if $C = -32$, and is described by

$$y^2 - 6x^2 + x^3 = -32. \tag{11}$$

This curve passes through the isolated critical point $(4,0)$ and describes one trajectory. Apart from these two curves, no other integral curve has a critical point lying on it. It may be easily verified that when $C < -32$, the curve (10) has one branch, situated to the left of the infinite branch of the curve (10) with $C = 0$, namely $y^2 - 6x^2 + x^3 = 0$. When $-32 < C < 0$, (10) is the union of two branches, one of which is a closed oval curve enclosing the point $(4,0)$. When $C > 0$, the integral curve (10) has a single branch situated to the right of $y^2 - 6x^2 + x^3 = 0$. To find the direction of the trajectory, we consider $y > 0$ in the first of (6):

$$\dot{x} = 2y > 0,$$

so x increases at points on the positive y-axis, determining the direction of the trajectories.

The special curves which pass through the saddle point $(0,0)$ are *separatrices*. There are four such curves, two moving into $(0,0)$ and two moving out of it. Two of these separatrices lie on the loop $y^2 - 6x^2 + x^3 = 0, x > 0$ (see Figure 6.4.1).

EXAMPLE 4

$$\frac{dx}{dt} = -y - x(x^2 + y^2)^{1/2}, \tag{12a}$$

$$\frac{dy}{dt} = x - y(x^2 + y^2)^{1/2}. \tag{12b}$$

We show that even when the linear system has pure imaginary characteristic roots for a singular point—and therefore the latter is classified as a center—it may actually not be one. It is easily checked that $(0,0)$ is the

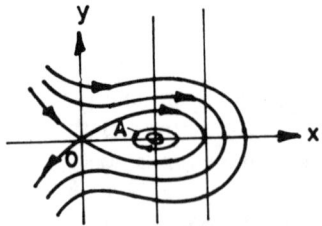

Figure 6.4.1 Phase portrait of system (6). (From Andronov et al. 1973.)

only singular point of (12). Its linearized form about $(0,0)$ is

$$\frac{dx}{dt} = -y,$$
$$\frac{dy}{dt} = x. \tag{13}$$

The system (13) has pure imaginary eigenvalues $\pm i$. The point $(0,0)$, accordingly, is a center. Indeed the solution of (13) can be written as

$$x = x_0 \cos(t - t_0) - y_0 \sin(t - t_0),$$
$$y = x_0 \sin(t - t_0) + y_0 \cos(t - t_0), \tag{14}$$

with initial conditions t_0, x_0, y_0. Moreover,

$$x^2 + y^2 = C \tag{15}$$

is a first integral of the system. The trajectories of the system are therefore the equilibrium state $(0,0)$ and closed concentric circles about the origin. The solution (14) is 2π-periodic.

Now, we turn to the full system (12). It seems convenient to introduce the polar coordinates

$$x = \rho \cos\theta, \qquad y = \rho \sin\theta.$$

Then (12) transforms to

$$\frac{d\rho}{dt} = -\rho^2,$$
$$\frac{d\theta}{dt} = 1. \tag{16}$$

Multiplying (12a) by x and (12b) by y and adding leads to (16a); (16b) is similarly obtained. In the (ρ, θ) plane, (16) reduces to

$$\frac{d\rho}{d\theta} = -\rho^2, \tag{17}$$

which can be integrated, with initial conditions $\rho = \rho_0$, $\theta = \theta_0$ at $t = 0$, to yield

$$\rho = \frac{1}{\theta + 1/\rho_0 - \theta_0}. \tag{18}$$

Equation (18) represents the polar form for the trajectories. Since $\rho > 0$, $\theta_0 - 1/\rho_0 < \theta < \infty$, giving the range of ρ as $(0, \infty)$. The integrated form of (16) is

$$\rho = \frac{1}{t + 1/\rho_0},$$
$$\theta = t + \theta_0.$$
(19)

The solution (18) in the (ρ, θ) plane represents a hyperbolic spiral. The radius ρ tends to 0 as θ tends to $+\infty$ and tends to ∞ as θ tends to $\theta_0 - 1/\rho_0$. Thus, all hyperbolic paths (18) tend to the origin as t tends to $+\infty$, and escape to infinity with t decreasing to a finite value $-1/\rho_0$. Thus, $(0,0)$ is a stable focus (see Figure 6.4.2).

EXAMPLE 5

$$\frac{dx}{dt} = -y + x(x^2 + y^2)\sin\frac{\pi}{(x^2 + y^2)^{1/2}},$$
$$\frac{dy}{dt} = x + y(x^2 + y^2)\sin\frac{\pi}{(x^2 + y^2)^{1/2}},$$
$(x,y) \neq (0,0)$
(20)

the right-hand sides being defined as zero when $(x,y) = (0,0)$. The right-hand sides of (20) are once continuously differentiable, and $(0,0)$ is the only equilibrium point. The linearized form is the same as in example 4 and gives $(0,0)$ as center. The form (20) again suggests introduction of polar coordinates. It then assumes the form

$$\frac{d\rho}{dt} = \rho^3 \sin\frac{\pi}{\rho},$$
(21a)

$$\frac{d\theta}{dt} = 1,$$
(21b)

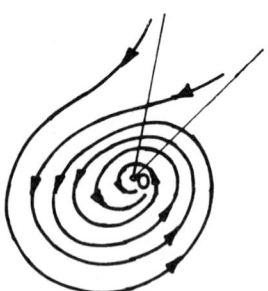

Figure 6.4.2 Phase portrait of system (16)—the polar form of (12). (From Andronov et al. 1973.)

Phase Space Study of Autonomous Systems

the reduction being similar to that for Example 4. In the (ρ, θ) plane, (21) becomes

$$\frac{d\rho}{d\theta} = \rho^3 \sin \frac{\pi}{\rho}. \tag{22}$$

Evidently, $\rho = 1/n$, $n = 1, 2, 3, \ldots$, are closed trajectories of (22) and divide the (x,y) plane into countably infinite annuli. It is easily seen from (22) that

$$\frac{d\rho}{d\theta} > 0 \quad \text{if } \rho > 1, \tag{23a}$$

$$\frac{d\rho}{d\theta} < 0 \quad \text{if } \frac{1}{2k} < \rho < \frac{1}{2k-1}, \tag{23b}$$

$$k = 1, 2, \ldots.$$

$$\frac{d\rho}{d\theta} > 0 \quad \text{if } \frac{1}{2k+1} < \rho < \frac{1}{2k}, \tag{23c}$$

Now we show that no trajectory passing through an interior point of any of the annuli can be closed. Suppose, for contradiction, that there is a trajectory passing through such a point (θ_0, ρ_0) at $t = t_0$. Since this trajectory lies entirely in one annulus, it follows from (21a) that $d\rho/dt$ has the same sign at all points, $d\rho/dt > 0$, say. If this trajectory is closed and has a period $T > 0$, then the point corresponding to $t_0 + T$ coincides with (θ_0, ρ_0). This contradicts $d\rho/dt > 0$. Hence the trajectories passing through a point of an annulus cannot be closed. Thus, all the closed trajectories are given by the circles $\rho = 1/n$, $(n = 1, 2, 3, \ldots)$.

It may be checked numerically and follows more rigorously from general theorems (see §4 of Andronov et al. 1973) that trajectories lying in the interior of all annuli, with the exception of the region $\rho > 1$, tend to one of the bounding circles as $t \to -\infty$ and to the other as $t \to +\infty$. Indeed, (23b) and (21) show clearly that all the trajectories in the annulus $1/2k < \rho < 1/(2k-1)$ $(k = 1, 2, 3, \ldots)$ are spirals tending to the closed integral curve $\rho = 1/(2k-1)$ as $t \to -\infty$ and to $\rho = 1/2k$ as $t \to +\infty$. On the other hand, the trajectories inside the annulus $1/(2k+1) < \rho < 1/2k$ are spirals which wind out from the circle $\rho = 1/(2k+1)$ and finally merge into the circle $\rho = 1/2k$. Equations (23a) and (21a) show that trajectories starting from any point in the finite annulus $\rho > 1$ are spirals, tending to the circle $\rho = 1$ as $t \to -\infty$ and to infinity as $t \to +\infty$. We thus conclude that the circles $\rho = 1/n$ are stable (limit) cycles of the system (20) when n is even and are unstable (limit) cycles when n is odd.

6.5 SINGULAR POINTS AT INFINITY

So far we have restricted our attention to the nature of the singular points in the finite part of the (x,y) plane. We must enquire whether there are singular points at infinity, and hence derive the results concerning their nature. This would complete the study of the totality of singular points and also prove useful in determining the asymptotic behavior of the solution.

We again study the system

$$\frac{dx}{dt} = P(x,y),$$
$$\frac{dy}{dt} = Q(x,y), \tag{1}$$

where we consider P and Q to be polynomials in x and y. To study the points at infinity in the (x,y) plane, it is convenient to project the latter onto a sphere, often referred to as the *Poincaré sphere* (see Figure 6.5.1). We take a unit sphere with center $C(0,0,1)$ tangent to the (x,y) plane at the origin $O(0,0)$. The orientation of the z-axis is as shown in the figure. Any (finite) point P is represented on the sphere by a pair of points P_1 and P_2 where the straight line CP cuts it. It is clear that as this point moves to infinity, it is represented by a pair of points on the great circle on the

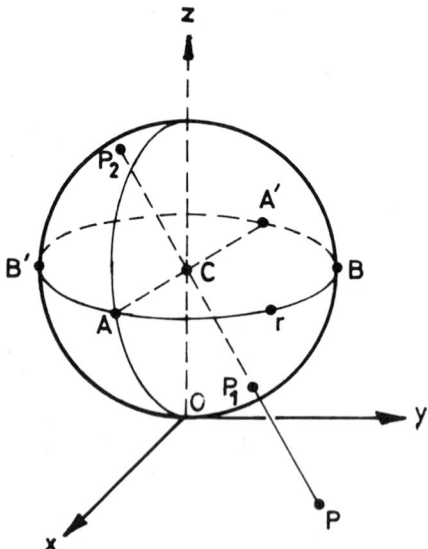

Figure 6.5.1 The Poincaré sphere.

plane $z = 1$. This is shown as joining the points $ABA'B'A$. $z = 1$ is called the *equatorial plane*.

From the projection described above, it is clear that the points on the equator correspond to all the points at infinity in the (x,y) plane. It is also evident that the mapping of the (x,y) plane to the Poincaré sphere takes a single family of integral curves on the (x,y) plane to two families on the sphere, the straight lines on the (x,y) plane onto great circles of the sphere, and the straight lines through the origin O into great circles perpendicular to the equator. It, therefore, follows that tangents to integral curves in the (x,y) plane transform into great circles touching the corresponding curves on the sphere, and an angle between a straight line in the (x,y) plane and the positive x-axis goes into an arc of the equatorial diameter. The hemisphere $ABA'B'O$ is referred to as the *lower* hemisphere, while the other half is called the *upper* hemisphere.

Actually, it is more convenient to consider another projection—that of the sphere on a plane drawn through the point A tangential to it (see Figure 6.5.1). Then one draws perpendiculars from P_1 and P_2 on this plane. Thus, all directions at infinity, except one along Oy which is parallel to the plane of projection, can be easily treated. For the latter, one draws a plane through the point B tangent to the sphere at that point, and the points on the sphere are again projected by drawing lines perpendicular to this plane.

To obtain the analytic transformations for projections on the plane through A, we consider a coordinate system (x',y',z') with origin at the center C of the sphere. The axes Cx' and Cy' of this right-handed system are taken parallel to Ox and Oy and Cz' is taken positive in the upward direction. The coordinates of the points P and P_1 in the present system are $(x,y,-1)$ and $(x',y',-z')$, respectively. Since P_1 lies on the unit sphere with center C, we have

$$x'^2 + y'^2 + z'^2 = 1. \tag{2}$$

The straight line PP_1 passes through the origin C. Therefore, the coordinates (x',y',z') of P_1 are given by

$$\frac{x'}{x} = \frac{y'}{y} = \frac{z'}{1} = \frac{1}{(x^2+y^2+1)^{1/2}}. \tag{3}$$

The coordinates of P_2 are

$$x' = -\frac{x}{(x^2+y^2+1)^{1/2}}, \quad y' = -\frac{y}{(x^2+y^2+1)^{1/2}},$$

$$z' = \frac{-1}{(x^2+y^2+1)^{1/2}}.$$

The perpendicular projection on $x' = 1$ relates the points $(x, y, -1)$ and $(1, y', -z')$ through

$$y' = \frac{y}{x}, \qquad z' = \frac{1}{x}. \tag{4}$$

Equations (4) give the inverse transformation

$$x = \frac{1}{z'}, \qquad y = \frac{y'}{z'}. \tag{5}$$

It is convenient to write (5) as

$$x = \frac{1}{z}, \qquad y = \frac{u}{z}. \tag{6}$$

In this form, the directions at infinity can be easily identified. For example, $z = 0$, $u = \alpha$ corresponds to the point at infinity in the direction $y = \alpha x$ of the (x, y) plane. The point $z = 0$, $u = 0$ corresponds to the point A. If we consider projections on the plane tangent to the sphere at B, the transformations similar to (6) are

$$x = \frac{v}{z}, \qquad y = \frac{1}{z}. \tag{7}$$

The system (1), in terms of the variables z and u of (6), is

$$\begin{aligned}
\frac{du}{dt} &= -uzP\left(\frac{1}{z}, \frac{u}{z}\right) + zQ\left(\frac{1}{z}, \frac{u}{z}\right), \\
\frac{dz}{dt} &= -z^2 P\left(\frac{1}{z}, \frac{u}{z}\right).
\end{aligned} \tag{8}$$

For $z \neq 0$, the trajectories of this system are obviously projections of those on the sphere or, what is the same thing, projections of the trajectories of the system (1) on the (x, y) plane.

We have assumed that the functions $P(x, y)$ and $Q(x, y)$ in (1) are polynomials. If the highest degree of terms in $P(x, y)$ and $Q(x, y)$ is n, we can write (8) in the form

$$\begin{aligned}
\frac{du}{dt} &= \frac{P^*(u, z)}{z^n}, \\
\frac{dz}{dt} &= \frac{Q^*(u, z)}{z^n},
\end{aligned} \tag{9}$$

where $P^*(u, z)$ and $Q^*(u, z)$ are polynomials in u and z. Since the system (9) is undefined for $z = 0$, we may introduce a new independent variable τ via

$$d\tau = \frac{dt}{z^n}$$

Phase Space Study of Autonomous Systems

and write (9) as

$$\frac{du}{d\tau} = P^*(u,z), \qquad \frac{dz}{d\tau} = Q^*(u,z), \tag{10}$$

where

$$P^*(u,z) = z^n \left\{ -uzP\left(\frac{1}{z}, \frac{u}{z}\right) + zQ\left(\frac{1}{z}, \frac{u}{z}\right) \right\},$$

$$Q^*\left(\frac{1}{z}, \frac{t}{z}\right) = z^n \left\{ -z^2 P\left(\frac{1}{z}, \frac{u}{z}\right) \right\}. \tag{11}$$

To get a working rule, we note that the singular points of (1) at infinity are the projections of singular points of the system (10) on $z = 0$. (The point A and its antipode are excluded.) Thus, the singular points on the equator $z = 0$ satisfy the equations

$$P^*(u,0) = 0, \qquad Q^*(u,0) = 0. \tag{12}$$

The roots of (12) are finite in number. One investigates the local phase portrait of (10) in the neighborhood of each root of (12). The two points on the equator, corresponding to each root (12), are obtained by the intersection with it of the line $y/x = u$.

By way of the other Poincaré transformation (7), system (1) becomes

$$\frac{dv}{dt} = zP\left(\frac{v}{z}, \frac{1}{z}\right) - zvQ\left(\frac{v}{z}, \frac{1}{z}\right) = \frac{\hat{P}(v,z)}{z^m},$$

$$\frac{dz}{dt} = -z^2 Q\left(\frac{v}{z}, \frac{1}{z}\right) = \frac{\hat{Q}(v,z)}{z^m}, \tag{13}$$

where m is the smallest integer such that $\hat{P}(v,z)$ and $\hat{Q}(v,z)$ are polynomials in v and z. One then considers the system

$$\frac{dv}{d\tau} = \hat{P}(u,z), \qquad \frac{dz}{d\tau} = \hat{Q}(v,z) \quad d\tau = z^{-m} dt, \tag{14}$$

and the specific points on the plane $z = 0$.

We note, in conclusion, that the nature of a singular point at infinity will be the same if one investigates the system

$$\frac{du}{dt} = P^*(u,z),$$

$$\frac{dz}{dt} = Q^*(u,z), \tag{15}$$

instead of (10).

EXAMPLE 1

$$\frac{dx}{dt} = x(3 - x - ny), \qquad n > 3,$$
$$\frac{dy}{dt} = y(-1 + x + y). \tag{16}$$

Applying the Poincaré transformation $x = 1/z$, $y = u/z$ to (16), we get

$$\frac{du}{dt} = \frac{2u - 4uz + (n + 1)u^2}{z},$$
$$\frac{dz}{dt} = 1 + nu - 3z. \tag{17}$$

Multiplying the right-hand sides of (17) by z, we have

$$\frac{du}{dt} = 2u - 4uz + (n + 1)u^2,$$
$$\frac{dz}{dt} = z + nuz - 3z^2. \tag{18}$$

The only singular points of (18) are $C(0, 0)$ and $B(-2/(n + 1), 0)$. By simple linearization and integration of the resulting system $du/dt = 2u$, $dz/dt = z$, the point $(0, 0)$ is seen to be an unstable node. A simple calculation shows that the other point is a stable node. The line $z = 0$ is a union of trajectories.

To investigate the point corresponding to D and its antipode, in Figure 6.5.2 we write $x = v/z$, $y = 1/z$ in (16). We have

$$\frac{dv}{dt} = \frac{4vz - 2v^2 - (n + 1)v}{z},$$
$$\frac{dz}{dt} = z - v - 1. \tag{19}$$

Again multiplying the right-hand sides by z, we obtain

$$\frac{dv}{dt} = 4vz - 2v^2 - (n + 1)v,$$
$$\frac{dz}{dt} = z(z - v - 1). \tag{20}$$

By simple linearization and integration, $(0, 0)$ corresponding to D is seen to be a stable node. We have multiplied (17) and (19) by z, an odd function, so that the nature of trajectories in $z < 0$ near the equator will change to the extent of arrows being reversed (a stable point becomes unstable, and conversely). Hence the phase portrait near the equator is as shown in Figure 6.5.2.

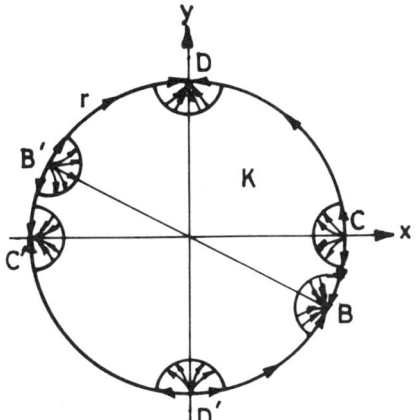

Figure 6.5.2 Phase portrait at infinity (on the equator of the Poincaré sphere) of system (16). (From Andronov et al. 1973.)

6.6 NONSIMPLE SINGULAR POINTS

We consider the system

$$\frac{dx}{dt} = ax + by + P_2(x,y),$$
$$\frac{dy}{dt} = cx + dy + Q_2(x,y), \tag{1}$$

such that the equilibrium state is an isolated nonsimple (or multiple) point. Here $P_2(x,y)$ and $Q_2(x,y)$ are analytic in the neighborhood of the origin, and their series expansions involve terms of second degree at least. *Nonsimple* singular points occur when $\Delta = ad - bc = 0$ and at least one of the characteristic roots is zero (see Section 6.3). When this is the case, it is possible to reduce the system (1) such that one of the right-hand sides in the neighborhood of the equilibrium state involves one first-order term. Two cases will be treated: $\Delta = 0$, $\sigma = a + d \neq 0$, so that one of the characteristic roots is nonzero; (ii) $\Delta = 0$, $\sigma = a + d = 0$ implying that both characteristic roots are zero. (Note a change of notation here.)

We shall transform system (1) in each case, using a nonsingular linear transformation, so that three important theorems from Andronov et al. (1973), due originally to Bendixon, can be applied and the phase plane configuration determined. We shall omit the proofs of these theorems (which are rather lengthy), but shall illustrate them with the help of examples.

Case (i)

The system (1) for $\sigma \neq 0$ can be changed to the form

$$\frac{d\bar{x}}{d\bar{t}} = \overline{P}_2(\bar{x},\bar{y}),$$
$$\frac{d\bar{y}}{d\bar{t}} = \bar{y} + \overline{Q}_2(\bar{x},\bar{y}), \qquad (2)$$

by the transformations

$$\bar{x} = -dx + by, \quad \bar{y} = ax + by, \quad \text{if } b \neq 0, \qquad (3)$$

or

$$\bar{x} = x, \quad \bar{y} = \frac{c}{d}x + y, \quad \text{if } b = a = 0, \qquad (4)$$

or

$$\bar{x} = -\frac{c}{a}x + y, \quad \bar{y} = x, \quad \text{if } b = d = 0, \qquad (5)$$

Since $\sigma \neq 0$, a and d cannot both be zero. The variable $\bar{t}$ in (2) equals kt, where k is a constant. The functions $\overline{P}_2(\bar{x},\bar{y})$ and $\overline{Q}_2(\bar{x},\bar{y})$ satisfy the same conditions as $P_2(x,y)$ and $Q_2(x,y)$. Thus, without loss of generality, we may consider the system (2) instead of (1) and drop bars therein:

$$\frac{dx}{dt} = P_2(x,y),$$
$$\frac{dy}{dt} = y + Q_2(x,y). \qquad (6)$$

Denoting the right-hand sides of (6) by $P(x,y)$ and $Q(x,y)$, respectively, we find that

$$\sigma = \frac{\partial P(x,y)}{\partial x} + \frac{\partial Q}{\partial y}(x,y) = 1 \qquad \text{at } (0,0).$$

Since $\sigma(x,y)$ is continuous, $\sigma > 0$ in a small neighborhood of the origin; therefore, by Bendixon's test (Andronov et al., §12) there are no closed paths or loops in the neighborhood of $(0,0)$. Thus, the origin is not a center and does not have elliptic sectors. An *elliptic sector* is defined as the region through all of whose points pass paths that tend to an equilibrium state 0 as $t \to +\infty$ and as $t \to -\infty$; some of these paths may be arbitrarily stable, while others may be arbitrarily unstable.

System (6) may be further transformed. Using the implicit function theorem, we write the solution of

$$y + Q_2(x,y) = 0 \qquad (7)$$

Phase Space Study of Autonomous Systems

in a small neighborhood of the origin as $y = \phi(x)$ such that $\phi(0) = 0$, $\phi'(0) = 0$. Defining a function

$$\psi(x) = P_2(x, \phi(x)), \tag{8}$$

we see that it cannot vanish identically since, otherwise, it would imply (by definition of $\phi(x)$) that all points of the curve $y = \phi(x)$ are equilibrium states of system (6), contradicting the assumption that the origin is an isolated state. Therefore, the function $\psi(x)$ has the form

$$\psi(x) = \Delta_m x^m + \cdots, \tag{9}$$

where $m \geq 2$, $\Delta_m \neq 0$.

We now quote the first theorem (number 65 in Andronov et al. 1973).

THEOREM 6.6.1 Let $O(0,0)$ be an isolated equilibrium state of system (6). Let $y = \phi(x)$ be a solution of the equation $y + Q_2(x,y) = 0$ in the neighborhood of $O(0,0)$. Assume that the series expansion of the function $\psi(x) = P_2(x, \phi(x))$ has the form (9), where $m \geq 2$, $\Delta_m \neq 0$. Then

1. If m is odd and $\Delta_m > 0$, $O(0,0)$ is a topological node.
2. If m is odd and $\Delta_m < 0$, $O(0,0)$ is a topological saddle point, two of whose separatrices tend to 0 in the directions 0 and π, the other two in the directions $\pi/2$ and $3\pi/2$.
3. If m is even, $O(0,0)$ is a saddle node—i.e., an equilibrium state whose canonical neighborhood is the union of one parabolic and two hyperbolic sectors. If $\Delta_m < 0$, the hyperbolic sectors contain a segment of the x-axis bordering on the origin O (Figure 6.6.1); if $\Delta_m > 0$, they contain a segment of the negative x-axis (Figure 6.6.2).

Case (ii)

Now we consider system (1) when the origin is an isolated equilibrium point and the following conditions hold:

$$|a| + |b| + |c| + |d| \neq 0, \tag{10}$$
$$\sigma = a + d = 0, \tag{11}$$
$$\Delta = ad - bc = 0. \tag{12}$$

Under these conditions, the origin is an isolated multiple point with zero characteristic roots, and the right-hand sides in (1) have nontrivial linear terms. As in case (i), it is possible to change (1) to the form (6) by the transformation

$$\bar{x} = -y, \quad \bar{y} = -cx + ay, \quad \text{if } a \neq 0 \text{ (so that } b \neq 0, c \neq 0\text{)}, \tag{13}$$

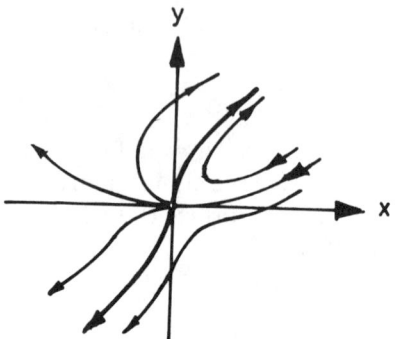

Figure 6.6.1 Phase portrait of system (6) with m even, $\Delta_m < 0$ (Theorem 6.6.1). (From Andronov et al. 1973.)

or

$$\bar{t} = bt, \quad \text{if } a = 0,\ b \neq 0, \tag{14}$$

or

$$\bar{x} = y, \quad \bar{y} = cx \quad \text{if } a = b = 0. \tag{15}$$

As before, the functions $P_2(x,y)$ and $Q_2(x,y)$ are analytic in the neighborhood of the origin, and their series expansions involve terms of at least second degree. System (6) can be further simplified by the transformation

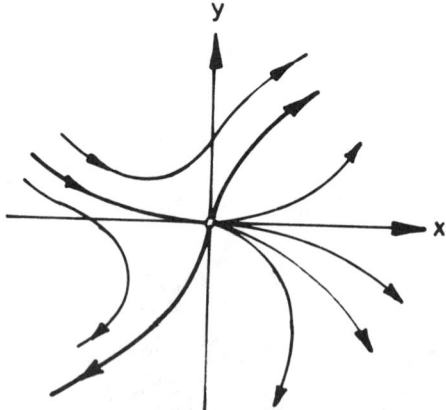

Figure 6.6.2 Phase portrait of system (6) with m even, $\Delta_m > 0$ (Theorem 6.6.1). (From Andronov et al. 1973)

Phase Space Study of Autonomous Systems

$$\xi = x, \quad \eta = y + P_2(x,y). \tag{16}$$

Since the Jacobian of the transformation (16) at $O(0,0)$ is unity, it maps a certain neighborhood of O in the (x,y) plane one-to-one onto a neighborhood of $\tilde{O}(0,0)$ in the (ξ,η) plane such that O is mapped onto $\tilde{O}$. The transformation inverse to (16) is

$$x = \xi, \quad y = f(\xi,\eta), \tag{17}$$

where f is analytic and $f(0,0) = 0$. System (6) via (16) becomes

$$\frac{d\xi}{dt} = \eta, \tag{18a}$$

$$\frac{d\eta}{dt} = Q_2(\xi,f(\xi,\eta)) + P_{2x}(\xi,f(\xi,\eta))\eta + P_{2y}(\xi,f(\xi,\eta))Q_2(\xi,f(\xi,\eta)). \tag{18b}$$

The function on the right side of (18b) is analytic, and its expansion in powers of ξ and η involves terms of second degree at least. If we write this function as $\tilde{Q}_2(\xi,\eta)$, system (6) reduces to the form

$$\frac{d\xi}{dt} = \eta, \quad \frac{d\eta}{dt} = \tilde{Q}_2(\xi,\eta). \tag{19}$$

The integral curves of system (6) passing through a sufficiently small neighborhood of O correspond in a one-to-one way to the paths of system (19) near $\tilde{O}$. Moreover, $\tilde{O}$, like O for (16), is clearly an isolated equilibrium point of (19). Replacing ξ and η in (19) by x and y, respectively, we have

$$\frac{dx}{dt} = y, \tag{20a}$$

$$\frac{dy}{dt} = \tilde{Q}_2(x,y) = a_k x^k[1 + h(x)] + b_n x^n y[1 + g(x)] + y^2 f(x,y), \text{ say,} \tag{20b}$$

where $h(x), g(x)$, and $f(x,y)$ are analytic in the neighborhood of the origin, $h(0) = g(0) = 0, k \geq 2, a_k \neq 0$. The coefficient b_n may vanish; if $b_n \neq 0$, then $n \geq 1$. The possible topological structure of the equilibrium state $O(0,0)$ of system (20) is given in Theorems 6.6.2 and 6.6.3 (see Theorems (66) and (67) in Andronov et al. 1973).

THEOREM 6.6.2 Let the number k in (20) be odd, $k = 2m + 1$ $(m \geq 1)$, and $\lambda = b_n^2 + 4(m+1)a_{2m+1}$. Then if $a_{2m+1} = a_k > 0$, the equilibrium state O of the system (20) is a topological saddle point (Figure 6.6.3). However, if $a_k < 0$, the point O is (1) a focus or center if $b_n = 0$ or if $b_n \neq 0, n = m$, and $\lambda < 0$; (2) a topological node if $b_n \neq 0$, n is even, and $n < m$, or if

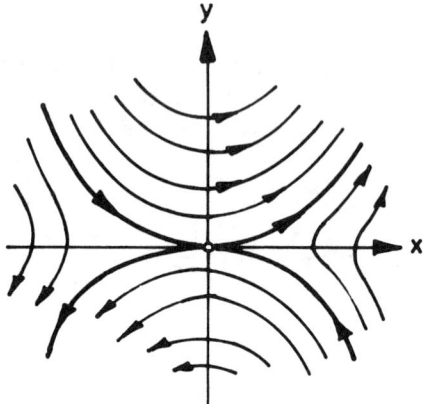

Figure 6.6.3 Phase portrait of system (20) with $k = 2m + 1$, $m \geq 1$, and $a_k > 0$ (Theorem 6.6.2). (From Andronov et al. 1973.)

$b_n = 0$, n is even, $n = m$, and $\lambda \geq 0$; (3) an equilibrium state with elliptic region if $b_n \neq 0$, n is odd, and $n < m$, or if $b_n \neq 0$, n is odd, $n = m$, and $\lambda \geq 0$ (see Figure 6.6.4).

Figure 6.6.3 refers to the case $b_n > 0$; the phase portrait for $b_n < 0$ is obtained by reflection in the x-axis.

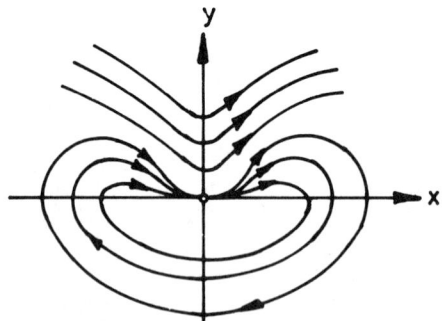

Figure 6.6.4 Phase portrait of system (20) with $a_k < 0$ and either $b_n \neq 0$, n is odd, and $n < m$, or $b_n \neq 0$, n is odd, $n = m$, and $\lambda \geq 0$ (Theorem 6.6.2). (From Andronov et al. 1973.)

THEOREM 6.6.3 Let the number k in system (20) be even ($k = 2m$, $m \geq 1$). Then the equilibrium state $O(0,0)$ is (1) a degenerate equilibrium state if $b_n = 0$ or $b_n \neq 0$ and $n \geq m$ (a *degenerate* equilibrium state is defined as one whose canonical neighborhood is the union of two hyperbolic sectors; see Figure 6.6.5 for $a_{2m} > 0$); (2) a saddle node if $b_n \neq 0$ and $n < m$ (see Figure 6.6.6 for $b_n > 0$, $a_{2m} < 0$).

To be able to apply Theorems 6.6.2 and 6.6.3, we need to determine the lowest-order terms $a_k x^k$ and $b_n x^n$ in (20b). The former is the first term in the expansion of the function $\tilde{Q}_2(x, 0)$ in powers of x, and the latter is the first term in the similar expansion of $\partial \tilde{Q}_2(x, 0)/\partial y$. If ξ and η in (18b) are replaced by x and y, respectively, and $f(x, 0) = \phi(x)$, $a_k x^k$ is the first term in the expansion of

$$Q_2(x, \phi(x))[1 + P_{2y}(x, \phi(x))], \tag{21}$$

while $b_n x^n$ is the first term in the expansion of

$$P_{2x}(x, \phi(x)) + Q_{2y}(x, \phi(x))[1 + P_{2y}(x, \phi(x))]\frac{\partial}{\partial y} f(x, 0)$$

$$+ P_{2yy}(x, \phi(x)) Q_2(x, \phi(x)) \frac{\partial}{\partial y} f(x, 0). \tag{22}$$

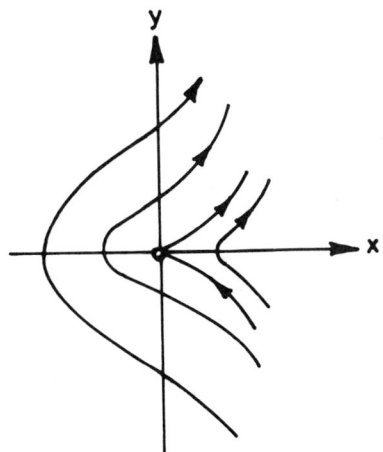

Figure 6.6.5 Phase portrait of system (20); with $k = 2m$ ($m \geq 1$), $a_{2m} > 0$, and either $b_n = 0$ or $b_n \neq 0$ and $n \geq m$ (Theorem 6.6.3). (From Andronov et al. 1973.)

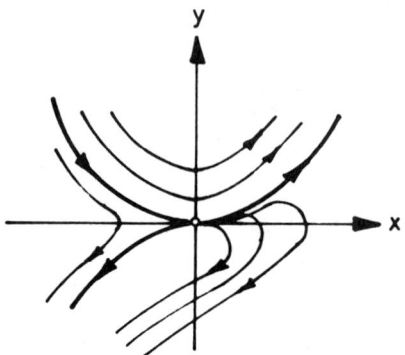

Figure 6.6.6 Phase portrait of system (20) with $k = 2m$ ($m \geq 1$), $b_n > 0$, $a_{2m} < 0$, and $n < m$ (Theorem 6.6.3). (From Andronov et al. 1973.)

It follows from the identity
$$y = f(x,y) + P_2(x, f(x,y)), \qquad (22)$$
that
$$\frac{\partial}{\partial y} f(x,0) = \frac{1}{1 + P_{2y}(x, \phi(x))}. \qquad (23)$$
We may therefore write (22) in the form
$$P_{2x}(x, \phi(x)) + Q_{2y}(x, \phi(x)) + \frac{P_{2yy}(x, \phi(x)) Q_2(x, \phi(x))}{1 + P_{2y}(x, \phi(x))}. \qquad (24)$$

It follows from (21) that the first term in its expansion, namely $a_k x^k$, also serves as the first term in the expansion $Q_2(x, \phi(x))$.

We thus arrive at the following simple rule to find the topological structure of the equilibrium state of system (6):

(a) Determine the solution $y = \phi(x)$ of the equation
$$y + P_2(x,y) = 0 \qquad (25)$$
in the form
$$\phi(x) = \alpha_1 x + \alpha_2 x^2 + \cdots + \alpha_k x^k + \cdots \qquad (26)$$
by substituting it in (25) and equating the coefficients of x^k, etc., to zero.

(b) After finding the coefficients α_i in (26), substitute the truncated series
$$\phi(x) = \alpha_1 x + \alpha_2 x^2 + \cdots + \alpha_n x^n$$

Phase Space Study of Autonomous Systems

in the functions $Q(x,y)$ and $\sigma(x,y)$ to obtain

$$\psi(x) \equiv Q_2(x, \phi(x)), \tag{27}$$
$$\sigma(x) = P_{2x}(x, \phi(x)) + Q_{2y}(x, \phi(x)) \tag{28}$$

and determine the first nonvanishing coefficient in the expansion of $\psi(x)$, namely $a_k x^k$ ($a_k \neq 0$, $a_i = 0$, $i < k$), and the first nonvanishing coefficient in the expansion of $\sigma(x)$, namely $b_n x^n$ ($b_n \neq 0$, $b_j = 0$, $j < n$). If $\sigma(x) \equiv 0$, then $b_n = 0$.

(c) If k is odd (even), apply Theorem 6.6.2 (6.6.3), thus characterizing the state $O(0,0)$ according to the values of k, n, a_n, and b_n in the respective theorems.

We now illustrate Theorems 6.6.1–6.6.3 with the help of examples, Andronov et al. (1973).

EXAMPLE 1

$$\frac{dx}{dt} = x(\beta x - y), \quad \beta > 0,$$
$$\frac{dy}{dt} = -\frac{1}{\alpha} y - y^2 + \alpha x^2, \quad \alpha > 0. \tag{29}$$

To reduce (29) to the form (6), we have to simply introduce $\bar{t} = -(1/\alpha)t$. We then have

$$\frac{dx}{d\bar{t}} = -\alpha \beta x^2 + \alpha xy,$$
$$\frac{dy}{d\bar{t}} = y + \alpha y^2 - \alpha^2 x^2. \tag{30}$$

In our notation, we have $a = b = c = 0$, $d = 1$. Therefore, $\Delta = 0$ and $\sigma = 1$, and Theorem 6.6.1 is applicable. Thus, $P_2(x,y) = -\alpha\beta x^2 + \alpha xy$ and $Q_2(x,y) = \alpha y^2 - \alpha^2 x^2$. Assuming that

$$y = \phi(x) = c_1 x + c_2 x^2 + c_3 x^3 + \cdots \tag{31}$$

and substituting in $y + \alpha y^2 - \alpha^2 x^2 = 0$, we readily find that $c_1 = c_3 = 0$, $c_2 = \alpha^2$, $c_4 = -\alpha^5$, etc. Thus,

$$\phi(x) = \alpha^2 x^2 - \alpha^5 x^4 + \cdots, \tag{32}$$
$$\psi(y) = P_2(x, \phi(x)) = -\alpha\beta x^2 + \alpha^3 x^3 + \cdots. \tag{23}$$

Therefore, $m = 2$ and $\Delta_m = -\alpha\beta$ in Theorem 6.6.1, and the equilibrium point $(0,0)$ of (30) and hence of (29) is a saddle node whose hyperbolic sectors enclose a segment of the positive x-axis.

To construct the phase portrait, we note the following points. The y-axis is a union of trajectories of the system (29), and all other trajectories tending

EXAMPLE 2

$$\frac{dx}{dt} = y - \frac{1}{2}xy - 3x^2,$$
$$\frac{dy}{dt} = -yx - \frac{3}{2}y^2. \tag{34}$$

Here we have $a = c = d = 0$, $b = 1$, so $\Delta = 0$ and $\sigma = 0$. Either Theorem 6.6.2 or 6.6.3 is applicable. System (34) is already in the form (6) with

$$P_2(x,y) = -\frac{1}{2}xy - 3x^2, \tag{35}$$

$$Q_2(x,y) = -xy - \frac{3}{2}y^2. \tag{36}$$

The solution of the equation $y + P_2(x,y) = 0$ is

$$y = \phi(x) = 3x^2 + \frac{3}{2}x^3 + \cdots, \tag{37}$$

so

$$\psi(x) = Q_2(x, \phi(x)) = -3x^3 - 15x^4 - \cdots, \tag{38}$$

$$\sigma(x) = P_{2x}(x, \phi(x)) + Q_{2y}(x, \phi(x)) = -7x - \frac{21}{2}x^2 - \cdots. \tag{39}$$

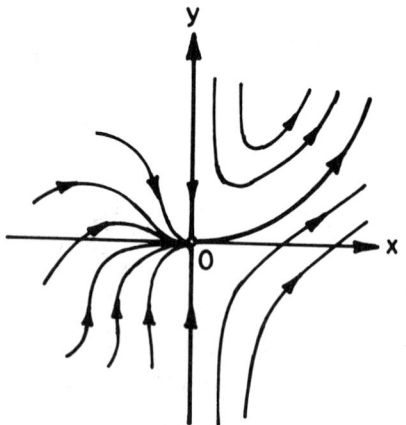

Figure 6.6.7 Phase portrait of system (29). (From Andronov et al. 1973.)

Therefore, in the notation of Theorem 6.6.2, $k = 2m + 1 = 3$; i.e., $m = 1$, $a_k = -3 < 0$, $n = 1$, $b_n = -7 < 0$, and $\lambda = b_n^2 + 4(m+1)a_{2m+1} = 25 > 0$. Further, since k is odd, $a_k < 0$, $m = n$, $\lambda > 0$, and n is odd, Theorem 6.6.2 is applicable and the equilibrium state O of (34) is one with an elliptic region. Moreover, it is evident that the x-axis is an integral curve of system (34). It is easily seen from (34), by putting $y = 0$, etc., that the two halves of the x-axis are separatrices.

The phase portrait of (34) is as shown in Figure 6.6.8.

EXAMPLE 3

$$\frac{dx}{dt} = y + y^2 - x^3,$$
$$\frac{dy}{dt} = 3x^2y + y^3 - 3x^5. \tag{40}$$

Here $a = 0$, $b = 1$, $c = 0$, $d = 0$, so $\Delta = 0$ and $\sigma = 0$. Hence Theorem 6.6.2 or 6.6.3 is applicable. System (40) is already in the form (6). As in the previous examples, we may find the solution of $y + P_2(x,y) = y + y^2 - x^3 = 0$ as

$$y = \phi(x) = x^3 - x^6 + 2x^9 + \cdots.$$

Here $Q_2(x,y) = 3x^2y - y^3 - 3x^5$; hence,

$$\psi(x) = Q_2(x, \phi(x)) = -3x^8 + x^9 + \cdots, \tag{41}$$
$$\sigma(x) = P_{2x}(x, \phi(x)) + Q_{2y}(x, \phi(y)) = 3x^6 - 18x^9 + \cdots. \tag{42}$$

In the notation used in Theorem 6.6.3, $k = 2m = 8$ is an even integer; this theorem is therefore applicable. Since, here, $m = 4$, $n = 6$, we have $n > m$ and $O(0,0)$ is a degenerate equilibrium state, whose canonical

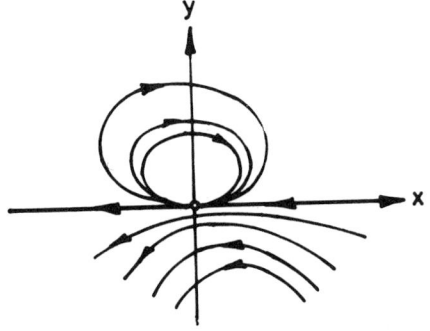

Figure 6.6.8 Phase portrait of system (34). (From Andronov et al. 1973.)

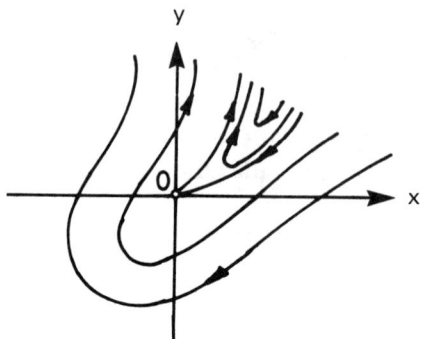

Figure 6.6.9 Phase portrait of system (40). (From Andronov et al. 1973.)

neighborhood is the union of two hyperbolic sectors. The phase portrait is as shown in Figure 6.6.9.

This example incidentally shows that knowledge of the lowest-order term in the series expansion of the function $\phi(x)$ is not always sufficient to determine the lowest-order terms in the expansions of $\psi(x)$ and $\sigma(x)$; higher-order terms may have to be included.

See Section 6.11 for a physical application of Theorem 6.6.2.

6.7 QUADRATIC SYSTEMS

Now we consider the quadratic system

$$\frac{dy}{dt} = P(x,y),$$
$$\frac{dx}{dt} = Q(x,y), \tag{1}$$

where

$$P(x,y) = \sum_{i+k=0}^{2} a_{ik}x^i y^k,$$
$$Q(x,y) = \sum_{i+k=0}^{2} b_{ik}x^i y^k. \tag{2}$$

Numerous problems in mathematical biology, nonlinear mechanics, and compressible flow are described by special cases of (1). There is a subclass

Phase Space Study of Autonomous Systems

of (1), called *normal quadratic form*, which occurs still more frequently in applications. We shall devote the next section to this subclass. Here we reduce (1) to a certain canonical form and give conditions under which it may possess a center. First of all, by a translation in x and y we can remove the constant terms from the polynomials $P(x,y)$ and $Q(x,y)$. The latter may then be written as

$$P(x,y) = Ax + By + Lx^2 + Mxy + Ny^2,$$
$$Q(x,y) = Cx + Dy + Gx^2 + Hxy + Ky^2. \tag{3}$$

We now write

$$x = -\frac{1}{A}X - \frac{C}{kA}Y, \quad y = -\frac{1}{k}Y, \tag{4}$$

where

$$k^2 = -AD - C^2.$$

so that (1) with P and Q given by (3) assumes the form

$$\frac{dY}{dt} = kX + kp(X,Y),$$
$$\frac{dX}{dt} = -kY + kq(X,Y), \tag{5}$$

where

$$p(X,Y) = L'X^2 + H'XY + N'Y^2,$$
$$q(X,Y) = G'X^2 + M'XY + K'Y^2.$$

Dividing each equation of the system (5) by k and writing $\tau = kt$, we obtain

$$\frac{dY}{d\tau} = X + p(X,Y),$$
$$\frac{dX}{d\tau} = -Y + q(X,Y), \tag{6}$$

where p and q can be put in "standard" form (see Davis 1962, Davies and James 1966):

$$p(x,y) = ax^2 + (2b + \alpha)xy + cy^2,$$
$$q(x,y) = -[bx^2 - (2a - \beta)xy + dy^2]. \tag{7}$$

Thus, the desired canonical system is

$$\frac{dy}{dt} = x + ax^2 + (2b + \alpha)xy + cy^2,$$
$$\frac{dx}{dt} = -[y + bx^2 + (2c + \beta)xy + dy^2], \tag{8}$$

which, in the phase plane, becomes
$$\frac{dy}{dx} = -\frac{x + ax^2 + (2b + \alpha)xy + cy^2}{y + bx^2 + (2c + \beta)xy + dy^2}. \tag{9}$$

Coppel (1966) has summarized all the previous work relating to (9), with special reference to the question of the existence of a center at $(0,0)$.

In a review of quadratic systems, Chicone and Jinghuang (1982) summarize recent results, again with particular attention to the questions of existence of center and limit cycles. A most comprehensive treatment of the topological structure of (9) in the phase plane does not seem to be available. Dickson and Perko (1970) considered those two-dimensional autonomous systems with quadratic right-hand sides, which have all their trajectories bounded for $t > 0$. These were referred to as *bounded quadratic forms*. After obtaining further "canonical" forms for such bounded systems, phase plane portraits were drawn.

Lukashevich (1965) considered system (9) in the large and gave "rough" phase portraits under varying conditions on the coefficients. Coppel (1966) summarized the previous results as follows: Equation (9) has a center at the origin if and only if one of the following conditions is satisfied:

$$a + c = b + d = 0. \tag{10}$$

$$\alpha(a + c) = \beta(b + d),$$
$$a\alpha^3 - (3b + \alpha)\alpha^2\beta + (3c + \beta)\alpha\beta^2 - d\beta^3 = 0. \tag{11}$$

$$\alpha + 5(b + d) = \beta + 5(a + c) = ac + bd + 2(a^2 + d^2) = 0. \tag{12}$$

In each case, equation (9) can be integrated in terms of elementary functions. For example, for (12) there is an integral of the form

$$\frac{[f(x,y)]^3}{[g(x,y)]^2} = \text{constant}, \tag{13}$$

where f is a quadratic and g is a cubic polynomial. The details of these results are in Lukashevich (1965).

The homogeneous quadratic system

$$\frac{dx}{dt} = a_1 x^2 + b_1 xy + c_1 y^2,$$
$$\frac{dy}{dt} = a_2 x^2 + b_2 xy + c_2 y^2, \tag{14}$$

treated in a cursory manner by Davies and James (1966), has been thoroughly discussed in the phase plane by Newton (1978). Though the system (14) appears rather simple, it gives rise to quite a diversity of solutions with varied phase portraits.

Phase Space Study of Autonomous Systems

Finally, we note that the book by Andronov et al. (1973) is a rich source of varied phase plane studies in the large (see Chapter XII). We list the quadratic (and higher-order) systems, describing actual physical situations, which have been treated by Andronov et al., for ready reference of the reader:

1. $\dot{x} = 2xy$, $\dot{y} = 1 + y - x^2 + y^2$.
2. $\dot{x} = 2x(1 + x^2 - 2y^2)$, $\dot{y} = -y(1 - 4x^2 + 3y^2)$.
3. $\dot{x} = x[(x^2+y^2+1)(x^2+y^2-1)-4y^2]$, $\dot{y} = y[(x^2+y^2+1)(x^2+y^2-1)+4x^2]$.
4. $\dot{x} = x(3 - x - ny)$, $\dot{y} = y(-1 + x + y)$.
5. $\dot{x} = x(3 - x - y)$, $\dot{y} = y(x - 1)$.
6. $\dot{x} = x(y - \beta)$, $\dot{y} = \beta(\alpha - y) - kxy$.
7. $\dot{x} = x(y + 3/2)$, $\dot{y} = x + y - 2y^2$.
8. $\dot{x} = -x(2 + y)$, $\dot{y} = \alpha x + \beta y$.
9. $\dot{x} = x(y + \lambda/(n - 1))$, $\dot{y} = y^2/(1 - n)$.

6.8 NORMAL QUADRATIC SYSTEMS

We shall now discuss in detail a special form of the quadratic systems referred to in Section 6.7, which Hille (1976) refers to as *normal*. This form appears very frequently in applications; Coppel (1966) enumerates many such applications. This normal form is

$$\frac{dx}{dt} = x(a'x + b'y + c),$$
$$\frac{dy}{dt} = y(ax + by + c), \tag{1}$$

or, in the phase plane,

$$\frac{dy}{dx} = \frac{y}{x} \frac{ax + by + c}{a'x + b'y + c'}, \tag{2}$$

where the coefficients (and the variables) are assumed to be real. System (2) displays a very rich variety of solutions. We first give two examples. The Emden–Fowler equation

$$(\xi^2 \eta')' + \xi^\lambda \eta^n = 0 \tag{3}$$

(where a prime denotes derivative with respect to ξ) is transformed by the change of variables

$$x = \frac{\xi \eta'}{\eta}, \quad y = \frac{\xi^{\lambda-1} \eta^n}{\eta'}, \quad t = \ln|\xi|, \tag{4}$$

to the system
$$\frac{dx}{dt} = -x(1+x+y),$$
$$\frac{dy}{dt} = y(\lambda + 1 + nx + y). \tag{5}$$

Similarly, the Blasius equation
$$\eta''' + \eta\eta'' = 0 \tag{6}$$
via the change of variables
$$x = \frac{\eta\eta'}{\eta''}, \quad y = \frac{\eta'^2}{\eta\eta''}, \quad t = \ln|\eta'| \tag{7}$$
becomes
$$\frac{dx}{dt} = x(1+x+y),$$
$$\frac{dy}{dt} = y(2+x-y). \tag{8}$$

Hille (1976) has studied the existence and nature of the movable singularities of (1) and the expansions for solutions with simple movable singularities. Here, we carry out a detailed study of the structure of all the singularities of (1) in the phase plane. It is instructive first to study geometrically the qualitative nature of singularities of the more complex systems which generalize the simple system
$$\frac{dy}{dx} = \frac{ax+by}{cx+dy}. \tag{9}$$

We have already discussed in detail the analytic conditions under which a variety of simple singularities of (9) exist. (Note, however, a slight change of notation in (9).) We first summarize these, and give qualitative geometrical configurations covering systems which lie between (9) and (2) and are defined by the number of special lines on which the slope of the integral curves is either zero or infinity. We follow the work of Jones (1953). This approach also helps determine the index of a singularity, which we define presently.

We recall from Section 6.3 that, in the present notation, the type of singularity of (9) at the origin depends on the nature of roots of the characteristic equation
$$\lambda^2 - 2h\lambda - k = 0, \tag{10}$$
where
$$2h = b+c, \quad k = ad - bc. \tag{11}$$

Phase Space Study of Autonomous Systems

The criteria for different singularities are

$$
\begin{array}{lll}
h^2 + k > 0, & k > 0, & \text{saddle,} \\
h^2 + k > 0, & k < 0, & \text{node (1).} \\
h^2 + k = 0, & & \text{node (2),} \\
h^2 + k < 0, & h = 0, & \text{center,} \\
h^2 + k < 0, & h \neq 0, & \text{focus.}
\end{array} \tag{12}
$$

The above classification is based on the assumption that $k = ad - bc \neq 0$. When this condition is not satisfied, the singularities at the origin are nonsimple; they are a mixture of two or more of the elementary types enumerated. In this case, if (9) is the linearized form of a quadratic system, it does not describe the nature of the singularities for the latter. We would have to consider the nonlinear terms as well (see Section 6.6).

For equation (9), there are two special lines: on $y/x = -a/b$, the slope of an integral line is zero, while on $y/x = -c/d$ it is infinite. Either or both of these lines may, in special circumstances, coincide with a coordinate axis, but generally they would not but would divide the (x,y) plane into four sectors. The slope dy/dx of an integral curve has a constant sign throughout any sector. It changes sign as it crosses a special line. The simple classification of (9) enumerated above is shown geometrically in a qualitative manner in Figure 6.8.1 as a preliminary to more complex singularities for the more general equations. The geometry does not replace the analytic criteria that fix the nature of singularities; the geometrical construction and the drawing of special lines help us to write the value of the index of a singularity. The index may be defined as follows: draw a small closed curve C around a singular point, and let a point on it make a complete circuit counterclockwise. Suppose the number of times the slope dy/dx of an integral curve (surrounding the singular point) changes from $+\infty$ to $-\infty$ is m and from $-\infty$ to $+\infty$ is n. Then the index of the singular point is

$$i = \frac{1}{2}(m - n). \tag{13}$$

For (9) it is given analytically by

$$i = \frac{1}{2\pi} \int_C d \arctan \frac{ax + by}{cx + dy} \tag{14}$$

(see Figure 6.8.1).

Now we generalize (9) and consider the equation

$$\frac{dy}{dx} = \frac{(ax + by)(ex + fy)}{cx + dy}, \tag{15}$$

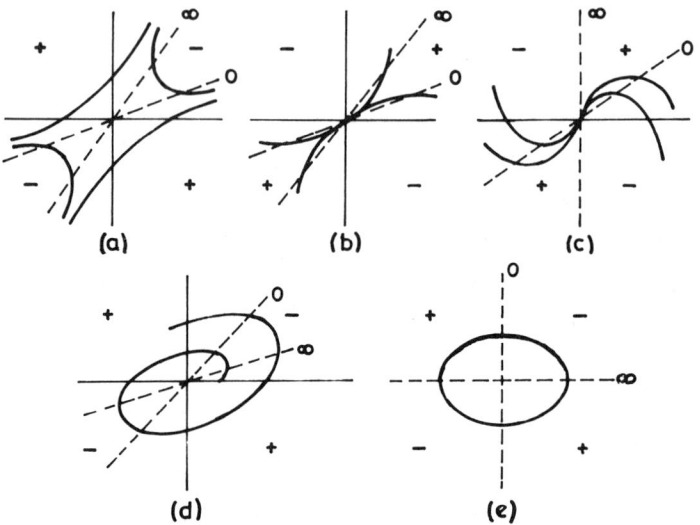

Figure 6.8.1 Phase plane for equation (9): (a) saddle, index -1, (b) node (1), index $+1$; (c) node (2), index $+1$; (d) focus, index $+1$; (e) center, index $+1$. (From Jones 1953.)

which has three special lines. We assume that

$$(ad - bc)(af - be)(cf - de) \neq 0.$$

For Equation (15) there are just two basic types of singularities, each with index zero; combinations of a saddle with a focus and a node, respectively. These are illustrated in Figure 6.8.2, where again the two varieties of nodes are clearly distinguished.

We carry the discussion of the singular points further by considering an equation with four special lines. By a suitable choice of variables, it may be put in the form

$$\frac{dy}{dx} = \frac{y}{x} \frac{ax + by}{cx + dy}, \qquad (16)$$

where $ad - bc \neq 0$. The following types of singular points may be distinguished (see Figure 6.8.3): (i) a double saddle with index -2, (ii) a node focus with index $+2$, (iii) a saddle node with index 0, which is different in structure from that in Figure 6.8.2c.

System (1) is a generalization of (16) and forms the major point of discussion of this section. While Jones (1953) has dealt with (1) to some extent, a more thorough treatment is due to Serebriakova (1963), which we follow

Phase Space Study of Autonomous Systems

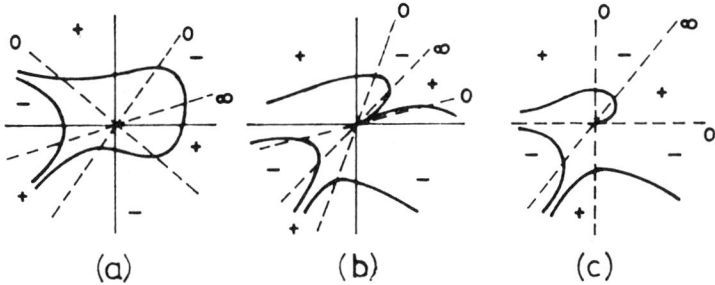

Figure 6.8.2 Phase portrait of equation (15): (a) saddle focus, index 0; (b), (c) saddle nodes, index 0. (From Jones 1953.)

in the sequel. Actually, it is possible to reduce system (1) to a simpler form by writing

$$x = \frac{c'}{a'}x_1, \quad y = \frac{c'}{b'}y_1, \quad t = \frac{t_1}{c'}, \tag{17}$$

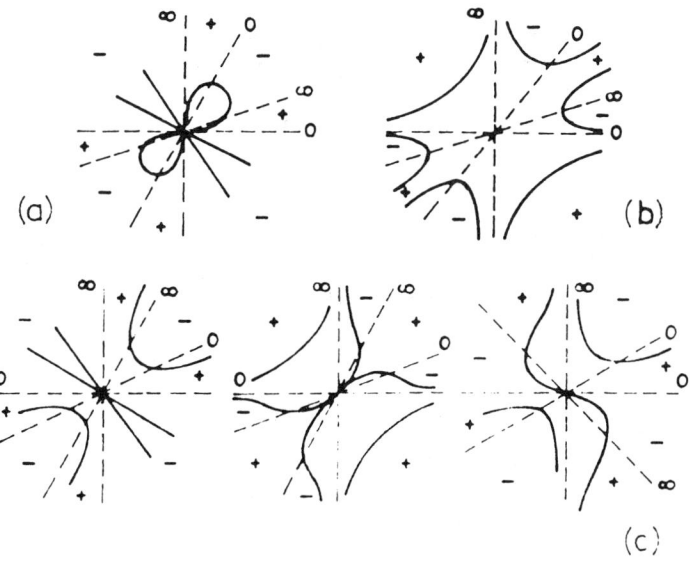

Figure 6.8.3 Phase portraits of equation (16): (a) node focus, index $+2$; (b) double saddle, index -2, (c) saddle nodes, index 0. (From Jones 1953.)

provided a', b', or c' do not vanish. System (1) becomes

$$\frac{dx_1}{dt_1} = x_1(x_1 + y_1 + 1),$$
$$\frac{dy_1}{dt_1} = y_1(a_1 x_1 + b_1 y_1 + c_1),$$
(18)

where $a_1 = a/a'$, $b_1 = b/b'$, $c_1 = c/c'$.

We rewrite (18) in the original notation as

$$\frac{dx}{dt} = x(x + y + 1),$$
(19a)

$$\frac{dy}{dt} = y(ax + by + c),$$
(19b)

This is the system we shall discuss in detail under varying conditions. In the phase plane, it becomes

$$\frac{dy}{dx} = \frac{y(ax + by + c)}{x(x + y + 1)}.$$
(20)

Its equilibrium points in the finite plane are

$$P_1(0,0), \quad P_2\left(0, -\frac{c}{b}\right), \quad P_3(-1, 0), \quad P_4\left(\frac{c-b}{b-a}, \frac{a-c}{b-a}\right),$$
(21)

provided

$$\Delta = \begin{vmatrix} 1 & 1 \\ a & b \end{vmatrix} \neq 0.$$
(22)

The roots λ_1 and λ_2 of the corresponding characteristic equations are

$$\lambda_1 = c, \quad \lambda_2 = 1 \qquad \text{for } P_1,$$

$$\lambda_1 = -c, \quad \lambda_2 = \frac{b-c}{b} \qquad \text{for } P_2,$$
(23)

$$\lambda_1 = -1, \quad \lambda_2 = c - a \qquad \text{for } P_3,$$

$$\lambda_{1,2} = \frac{ab - bc + c - b \pm \{(ab - bc + c - b)^2 - 4(b-a)(c-b)(a-c)\}^{1/2}}{2(b-a)} \qquad \text{for } P_4.$$

Points P_1, P_2, and P_3 can be either nodes or saddle points, depending on whether λ_1 and λ_2 (which are real) have the same or opposite signs. P_4 will be a saddle point if $(b-a)(c-b)(a-c) < 0$. It will be a center if $(b-a)(c-b)(a-c) > 0$ and $ab - bc + c - b = 0$; in this case the roots are pure imaginary.

To study the points at infinity, we introduce the Poincaré transformation (see Section 6.5)

$$x = \frac{1}{z}, \quad y = \frac{\tau}{z}. \tag{24}$$

System (19) becomes

$$\frac{dz}{dt} = -(\tau + z + 1), \tag{25}$$

$$\frac{d\tau}{dt} = \frac{\tau}{z}[(b-1)\tau + (c-1)z + (a-1)]$$

or

$$\frac{dz}{d\tau} = -\frac{z}{\tau} \frac{\tau + z + 1}{(b-1)\tau + (c-1)z + (a-1)}. \tag{26}$$

Counting both diametrically opposite points on the Poincaré sphere as distinct, we find that (26) has four singular points: P_5 and P_5', corresponding to $(z = 0, \tau = 0)$, lie on the positive and negative ends of the x-axis, while P_6 and P_6' are located at the ends of the diameter whose direction makes an angle $\tan^{-1}[(a-1)/(1-b)]$ with the x-axis. Points without primes are assumed to lie on the right half-plane. The characteristic roots are

$$\begin{aligned}\lambda_1 &= -1, \quad \lambda_2 = a - 1 \quad \text{for } P_5, \\ \lambda_1 &= 1 - a, \quad \lambda_2 = \frac{a-b}{b-1} \quad \text{for } P_6.\end{aligned} \tag{27}$$

To cover possible singular points in the direction of the y-axis, we use the transformation

$$x = \frac{\tau}{z}, \quad y = \frac{1}{z}. \tag{28}$$

The equation corresponding to (26) is

$$\frac{dz}{d\tau} = -\frac{z}{\tau} \frac{a\tau + cz + b}{(a-1)\tau + (c-1)z + (b-1)}. \tag{29}$$

Thus, in the direction of the y-axis, we have two singular points P_7 and P_7' corresponding to the singular point $(z = 0, \tau = 0)$ of (29). The roots of the characteristic equation for P_7 are $-b$ and $b - 1$.

To consider the qualitative features of the phase trajectories for system (20), we have to take into account the variation of the parameters a, b, and c. The eigenvalues (23) for P_1 and P_2 suggest that the parameter c has three distinct ranges: (i) $-\infty < c < 0$, (ii) $0 < c < 1$, (iii) $1 < c < \infty$. For a given value of c in each of these intervals, a and b may be considered to lie in several regions which are formed by the lines

$$a - b = 0, \quad a - 1 = 0, \quad a - c = 0,$$
$$b - 1 = 0, \quad b - c = 0, \quad ab - bc + c - b = 0, \tag{30}$$

where $(b - a)(c - b)(a - c) > 0$. This division is again suggested by (23).

The singular points corresponding to each of these parametric regions possess a definite qualitative picture about them in the phase plane. The various regions are shown in Figure 6.8.4. Table 6.8.1 details the nature of the singular points.

The point marked with a single asterisk is a stable or unstable node depending upon whether $b - c - ab + bc < 0$ or > 0. They are centers if $b - c - ab + bc = 0$. The double asterisk refers to a stable or unstable focus, depending on whether $b - c - ab + bc > 0$ or < 0. For the range $0 < c < 1$, the qualitative picture of the phase trajectories in the regions

$\{1\} \ -\infty < a < c, \ 1 < b < \infty,$
$\{3\} \ 1 < a < b, \ 1 < b < \infty,$
$\{4\} \ b < a < \infty, \ 1 < b < \infty,$
$\{3a\} \ -\infty < a < b, \ 0 < b < c,$
$\{4a\} \ b < a < c, \ 0 < b < c,$
$\{10\} \ 1 < a < \infty, \ 0 < b < c,$
$\{3b\} \ 1 < a < \infty, \ -\infty < b < 0,$
$\{1b\} \ b < a < c, \ -\infty < b < 0,$
$\{11\} \ -\infty < a < b, \ -\infty < b < 0$

come out to be the same as in the corresponding regions for $1 < c < \infty$.

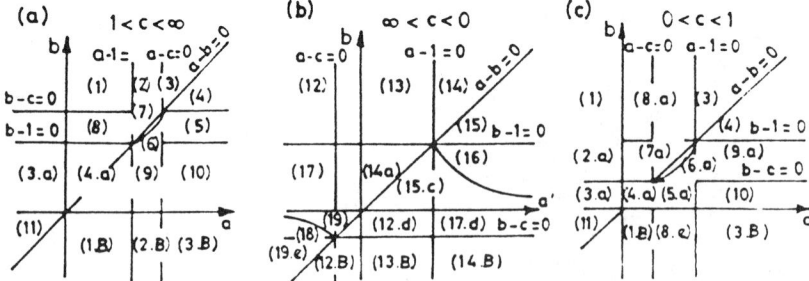

Figure 6.8.4 Schematic division of parametric ranges defining distinct cases for equation (20). (From Serebriakova 1963.)

Phase Space Study of Autonomous Systems

Table 6.8.1 Nature of Singular Points of Equation (20) for Various Cases

No	Regions	P_1	P_2	P_3	P_4	P_5	P_5'	P_6	P_6'	P_7	P_7'
					$1 < c < \infty$						
1	$-\infty < a < 1,\ c < b < \infty$	α_2	γ	γ	$\alpha_1(\beta_1)$	α_1	α_2	γ	γ	α_1	α_2
2	$1 < a < c,\ c < b < \infty$	α_2	γ	γ	$\alpha_1(\beta_1)$	γ	γ	α_1	α_2	α_1	α_2
3	$c < a < b,\ c < b < \infty$	α_2	γ	α_1	γ	γ	γ	α_1	α_2	α_1	α_2
4	$b < a < \infty,\ c < b < \infty$	α_2	γ	α_1	$\alpha_1(\beta_1)$	γ	γ	γ	γ	α_1	α_2
5	$c < a < \infty,\ 1 < b < c$	α_2	α_1	α_1	γ	γ	γ	γ	γ	α_1	α_2
6	$b < a < c,\ 1 < b < c$	α_2	α_1	γ	*1,2,3	γ	γ	γ	γ	α_1	α_2
7	$1 < a < b,\ 1 < b < c$	α_2	α_1	γ	γ	γ	γ	α_1	α_2	α_1	α_2
8	$-\infty < a < 1,\ 1 < b < c$	α_2	α_1	γ	γ	α_1	α_2	γ	γ	α_1	α_2
9	$1 < a < c,\ 0 < b < 1$	α_2	α_1	γ	$\alpha_1(\beta_1)$	γ	γ	α_1	α_2	γ	γ
10	$c < a < \infty,\ 0 < b < 1$	α_2	α_1	α_1	γ	γ	γ	α_1	α_2	γ	γ
11	$-\infty < a < b,\ -\infty < b < 0$	α_2	γ	γ	γ	α_1	α_2	α_2	α_1	α_2	α_1
1.b	$b < a < 1,\ -\infty < b < 0$	α_2	γ	γ	$\alpha_1(\beta_1)$	α_1	α_2	γ	γ	α_2	α_1
2.b	$1 < a < c,\ -\infty < b < 0$	α_2	γ	γ	$\alpha_1(\beta_1)$	γ	γ	α_1	α_2	α_2	α_1
3.a	$-\infty < a < b,\ 0 < b < 1$	α_2	α_1	γ	γ	α_1	α_2	α_2	α_1	γ	γ
3.b	$c < a < \infty,\ -\infty < b < 0$	α_2	γ	α_1	γ	γ	γ	α_1	α_2	α_2	α_1
4.a	$b < a < 1,\ 0 < b < 1$	α_2	α_1	γ	$\alpha_1(\beta_1)$	α_1	α_2	γ	γ	γ	γ
					$-\infty < c < 0$						
12	$-\infty < a < c,\ 1 < b < \infty$	γ	α_2	γ	$\alpha_1(\beta_1)$	α_1	α_2	γ	γ	α_1	α_2
13	$c < a < 1,\ 1 < b < \infty$	γ	α_2	α_1	γ	α_1	α_2	γ	γ	α_1	α_2
14	$1 < a < b,\ 1 < b < \infty$	γ	α_2	α_1	γ	γ	γ	α_1	α_2	α_1	α_2
15	$b < a < \infty,\ 1 < b < \infty$	γ	α_2	α_1	$\alpha_1(\beta_1)$	γ	γ	γ	γ	α_1	α_2
16	$1 < a < \infty,\ 0 < b < 1$	γ	α_2	α_1	*1,2,3	γ	γ	α_1	α_2	γ	γ
17	$-\infty < a < c,\ 0 < b < 1$	γ	α_2	γ	$\alpha_1(\beta_1)$	α_1	α_2	α_2	α_1	γ	γ
18	$-\infty < a < c,\ c < b < 0$	γ	γ	γ	**1,2,3	α_1	α_2	α_2	α_1	α_2	α_1
19	$c < a < b,\ c < b < 0$	γ	γ	α_1	γ	α_1	α_2	α_2	α_1	α_2	α_1
12.b	$b < a < c,\ -\infty < b < c$	γ	α_2	γ	$\alpha_1(\beta_1)$	α_1	α_2	γ	γ	α_2	α_1
12.d	$b < a < 1,\ c < b < 0$	γ	γ	α_1	$\alpha_2(\beta_2)$	α_1	α_2	γ	γ	α_2	α_1
13.b	$c < a < 1,\ -\infty < b < c$	γ	α_2	α_1	γ	α_1	α_2	γ	γ	α_2	α_1
14.b	$1 < a < \infty,\ -\infty < b < c$	γ	α_2	α_1	γ	γ	γ	α_1	α_2	α_2	α_1
14.c	$c < a < b,\ 0 < b < 1$	γ	α_2	α_1	γ	α_1	α_2	α_2	α_1	γ	γ
15.c	$b < a < 1,\ 0 < b < 1$	γ	α_2	α_1	$\alpha_2(\beta_2)$	α_1	α_2	γ	γ	γ	γ
17.d	$1 < a < \infty,\ c < b < 0$	γ	γ	α_1	$\alpha_2(\beta_2)$	γ	γ	α_1	α_2	α_2	α_1
19.e	$-\infty < a < b,\ -\infty < b < c$	γ	α_2	γ	γ	α_1	α_2	α_2	α_1	α_2	α_1

Table 6.8.1 (*Continued*)

No	Regions	P_1	P_2	P_3	P_4	P_5	P_5'	P_6	P_6'	P_7	P_7'
					$0 < c < 1$						
2.a	$-\infty < a < c, c < b < 1$	α_2	γ	γ	$\alpha_1(\beta_1)$	α_1	α_2	α_2	α_1	γ	γ
5.a	$c < a < 1, 0 < b < c$	α_2	α_1	α_1	γ	α_1	α_2	γ	γ	γ	γ
6.a	$b < a < 1, c < b < 1$	α_2	γ	α_1	*1,2,3	α_1	α_2	γ	γ	γ	γ
7.a	$c < a < b, c < b < 1$	α_2	γ	α_1	γ	α_1	α_2	α_2	α_1	γ	γ
8.a	$c < a < 1, 1 < b < \infty$	α_2	γ	α_1	γ	α_1	α_2	γ	γ	α_1	α_2
8.e	$c < a < 1, -\infty < b < 0$	α_2	γ	α_1	γ	α_1	α_2	γ	γ	α_2	α_1
9.a	$1 < a < \infty, c < b < 1$	α_2	γ	α_1	$\alpha_1(\beta_1)$	γ	γ	α_1	α_2	γ	γ

α_1 = stable node; α_2 = unstable node; β_1 = stable focus; β_2 = unstable focus; γ = saddle; $\gamma\alpha$ = saddle node; δ = a complicated singular point obtained when four simple singular points coalesce.
Source: From Serebriakova (1963).

Now, we consider the exceptional case

$$\Delta = \begin{vmatrix} 1 & 1 \\ a & b \end{vmatrix} = 0, \quad \text{or } a = b \tag{32}$$

(see (22)). If $a = b \neq 1$, then it is clear from (21) that P_4 goes into the equator and can be shown to be a saddle node there.

The case $a = b = 1$ has three singular points in the finite plane: $P_1(0,0)$, $P_2(0,-c)$, $P_3(-1,0)$. Here, the equator $z = 0$ is not an integral curve (unlike the general case, see equations (26) and (29)). Table 6.8.2 summarizes the results of the investigation of the singular points of (20) in this special case.

Now we consider the special cases of (1) when some of the coefficients vanish. Suppose $c' = a = 0$. Then changing the variables to

$$x = \frac{c}{a'}x_1, \quad y = \frac{c}{b'}y_1, \quad t = \frac{1}{c}t_1, \tag{33}$$

and returning to unsubscripted variables, we get the form

$$\frac{dy}{dx} = \frac{y(by+1)}{x(x+y)} \tag{34}$$

in the phase plane. Equation (34) has three singularities in the finite (x,y) plane: $P_1(0,0)$, $P_2(0,-1/b)$, $P_3(1/b,-1/b)$. There exist six (pairwise) diametrically opposite points on the equator: P_4 and P_4' coincide with the positive

Phase Space Study of Autonomous Systems

Table 6.8.2 Nature of Singular Points for Equation (20) with $a = b = 1$ (see Eq.(32))

		Points		
No.	Regions	P_1	P_2	P_3
20	$-\infty < c < 0$	γ	α_2	α_1
21	$0 < c < 1$	α_2	γ	α_1
21.a	$1 < c < \infty$	α_2	α_1	γ

Source: From Serebriakova (1963).

and negative ends, respectively, of the x-axis, P_5 and P_5' are located at the ends of the diameter making an angle $\tan^{-1}[1/(b-1)]$ with the x-axis, while P_6 and P_6' are situated at the positive and negative ends, respectively, of the y-axis. The orientation of the axes is such that P_5 is located in the right half-plane. The nature of these singularities is summarized in Table 6.8.3.

Another interesting case arises when $c' = b = 0$ in (1). Then the scaling (33) leads to the form

$$\frac{dy}{dx} = \frac{y(ax + 1)}{x(x + y)} \tag{35}$$

in the phase plane. Equation (35) has two singular points in the finite (x, y) plane: $P_1(0, 0)$ and $P_2(-1/a, 1/a)$. The points at infinity (on the equator of the Poincaré sphere) have the following locations: P_3 and P_3' at the positive and negative ends, respectively, of the x-axis, P_4 and P_4' at the corresponding ends of a diameter inclined at an angle $\tan^{-1}(a - 1)$ with the x-axis, and

Table 6.8.3 Nature of Singular Points of Equation (34)

		Points								
No.	Regions	P_1	P_2	P_3	P_4	P_4'	P_5	P_5'	P_6	P_6'
22	$-\infty < b < 0$	$\gamma\alpha$	γ	α_1	α_1	α_2	γ	γ	α_2	α_1
23	$0 < b < 1$	$\gamma\alpha$	α_1	γ	α_1	α_2	α_2	α_1	γ	γ
24	$1 < b < \infty$	$\gamma\alpha$	α_1	γ	α_1	α_2	γ	γ	α_1	α_2

Source: From Serebriakova (1963).

Table 6.8.4 Nature of Singular Points of Equation (35)

No.	Regions	\multicolumn{7}{c}{Points}

No.	Regions	P_1	P_2	P_3	P_3'	P_4	P_4'	P_5	P_5'
25	$-\infty < a < 0$	$\gamma\alpha$	γ	α_1	α_2	α_2	α_1	γ	α_1
26	$0 < a < 1$	$\gamma\alpha$	$\alpha_1(\beta_1)$	α_1	α_2	γ	γ	γ	α_1
27	$1 < a < \infty$	$\gamma\alpha$	$\alpha_1(\beta_1)$	γ	γ	α_1	α_2	γ	α_1

Source: From Serebriakova (1963).

P_5 and P_5' at the positive and negative ends, respectively, of the y-axis. The results of this case are summarized in Table 6.8.4.

The special case $c' = c = 0$ of (1) after a simple scaling $x = x_1/a'$, $y = y_1/b'$, etc., leads to an equation of the form

$$\frac{dy}{dx} = \frac{y(ax + by)}{x(x + y)}. \tag{36}$$

The right-hand side of (36) is "homogeneous" of degree 2 and has been discussed in detail by Davies and James (1966, p. 39). This has a complicated (δ in our notation) singular point (see insert figures (28)–(30) in Figure 6.8.7). Again there are six singular points on the equator: P_2 and P_2' at the ends of the x-axis, P_3 and P_3' at the ends of the diameter with an angle $\tan^{-1}[(a-1)/(1-b)]$ with the x-axis, and P_4 and P_4' at the ends of the positive and negative y-axes. The orientation is such that the points P_2 and P_3 are located on the right half-plane. Figure 6.8.5 gives the various regions in the (a,b) plane with distinct singularity structure. The nature of the singular points in the present case is described summarily in Table 6.8.5. Equation (36) is the only one displaying the most complicated singular point structure denoted by δ. Other cases show only combinations of two singularities (see Section 6.6).

All the cases discussed in Tables 6.8.1 to 6.8.5 are neatly drawn in Figures 6.8.6 and 6.8.7. They represent the trajectories in the lower hemisphere including the equator. There are some subcases tagged as a, b, or c to the original cases, which may be obtained by some rotation or reflection, as indicated by Serebriakova (1963)

a. The pictures for {2a}, {3a}, {4a}, {5a}, {6a}, {7a}, {8a}, {9a}, {21a}, {29a}, and {30a} can be obtained by a clockwise 90° rotation and a reflection with respect to the x-axis of the pictures for {2}, {3}, {4}, {5}, {6}, {7}, {8}, {9}, {21}, {29}, and {30}, respectively.

Phase Space Study of Autonomous Systems

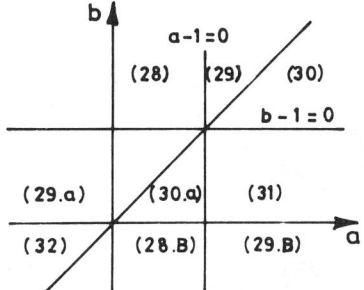

Figure 6.8.5 Schematic division of parametric ranges defining distinct cases for equation (36). (From Serebriakova 1963.)

b. The qualitative pictures for cases {1b}, {2b}, {3b}, {12b}, {13b}, {14b}, {28b}, and {29b} can be obtained by a mere reflection with respect to the x-axis of the phase portraits for {1}, {2}, {3}, {12}, {13}, {14}, {28}, and {29}, respectively.
c. The pictures {14c} and {15c} are obtained from {14} and {15} by three transformations: a counterclockwise 90° rotation, a reflection with respect to the x-axis, and a reversal of the direction along the trajectories.

Table 6.8.5 Nature of Singular Points for Equation (36)

		Points						
No.	Regions	P_1	P_2	P_2'	P_3	P_3'	P_4	P_4'
28	$-\infty < a < 1$, $1 < b < \infty$	δ	α_1	α_2	γ	γ	α_1	α_2
29	$1 < a < b$, $1 < b < \infty$	δ	γ	γ	α_1	α_2	α_1	α_2
30	$b < a < \infty$, $1 < b < \infty$	δ	γ	γ	γ	γ	α_1	α_2
31	$1 < a < \infty$, $0 < b < 1$	δ	γ	γ	α_1	α_2	γ	γ
32	$-\infty < a < b$, $-\infty < b < 0$	δ	α_1	α_2	α_2	α_1	α_2	α_1
28.b	$b < a < 1$, $-\infty < b < 0$	δ	α_1	α_2	γ	γ	α_2	α_1
29.a	$-\infty < a < b$, $0 < b < 1$	δ	α_1	α_2	α_2	α_1	γ	γ
29.b	$1 < a < \infty$, $-\infty < b < 0$	δ	γ	γ	α_1	α_2	α_2	α_1
30.a	$b < a < 1$, $0 < b < 1$	δ	α_1	α_2	γ	γ	γ	γ

See Table 6.8.1 for notation
Source: From Serebriakova (1963).

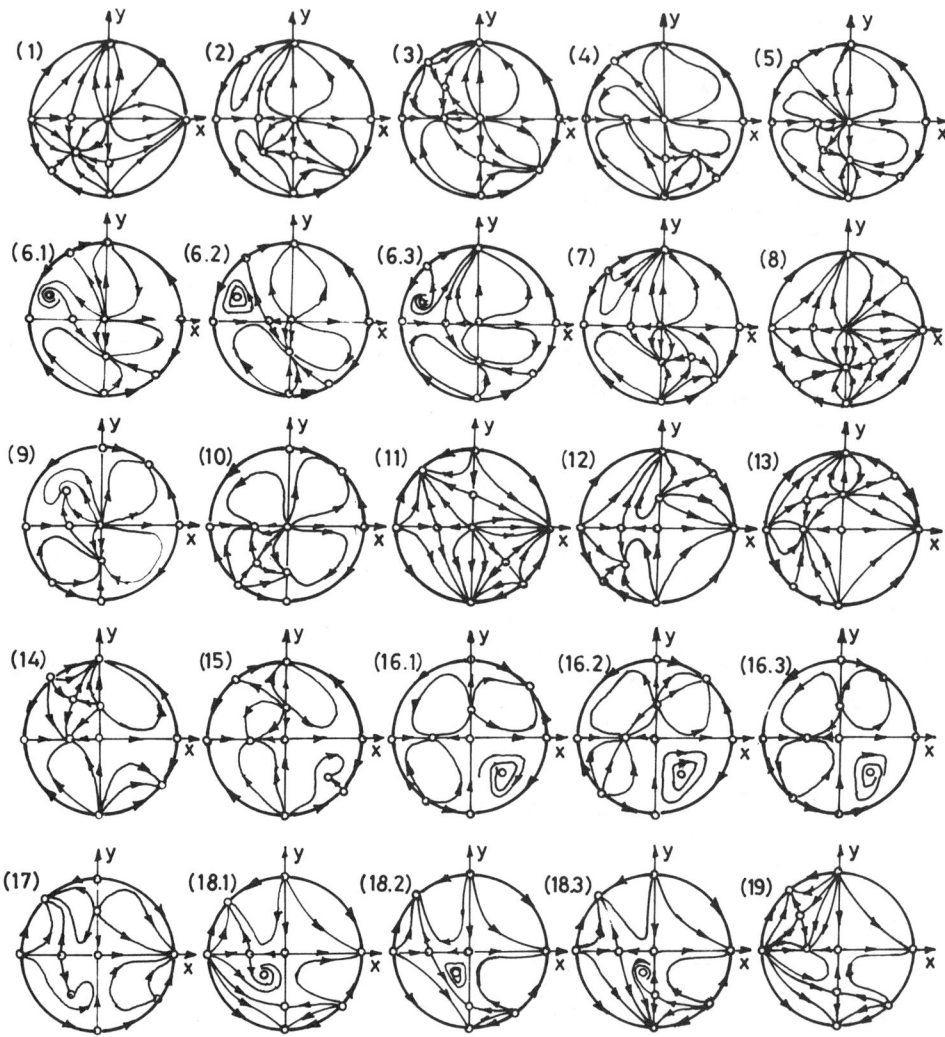

Figure 6.8.6 Phase portraits for cases listed in Tables 6.8.1– 6.8.5 (From Serebriakova 1963.)

Phase Space Study of Autonomous Systems

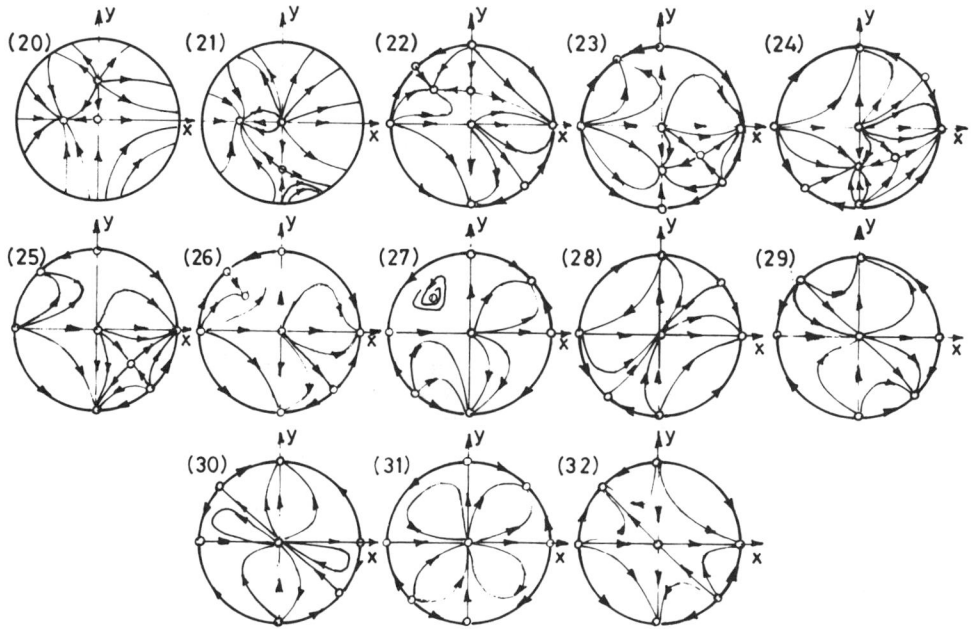

Figure 6.8.7 Phase portraits for cases listed in Tables 6.8.1– 6.8.5 (From Serebriakova 1963.)

d. The pictures {12d} and {17d} can be obtained from {12} and {17} by a counterclockwise 90° rotation and reversal of the orientation of the trajectories.
e. The case {8e} is obtained from {8} by a clockwise 90° rotation of the phase portrait of the latter.
f. Finally, {19e} is obtained from {19} by a reflection with respect to the x-axis, a clockwise 90° rotation, and a change in the orientation of the trajectories.

Emden's Equation

We discuss in detail Emden's equation

$$y'' + \frac{2}{x}y' + y^n = 0, \tag{37}$$

which arises in the theories of the internal structure of stars. This is a special case of the Emden–Fowler equation (3). This equation has been well studied, and tables of its solution have been prepared. For different positive values of the constant n, solutions have been found satisfying initial conditions $x = 0$, $y = 1$, $y' = 0$, from which a class of similar solutions can be constructed through scaling.

Equation (37) can be changed into the normal form (19) via the transformations

$$X = \frac{xy'}{y}, \qquad Y = \frac{xy^n}{y'} \tag{38}$$

so that

$$XY = x^2 y^{n-1}$$

(see also Section 2.9) and

$$\frac{dX}{dx} = \frac{y'}{y} - \frac{xy'^2}{y^2} + \frac{xy''}{y}$$

$$= \frac{1}{x}\left\{X - X^2 - \frac{x^2}{y}\left(\frac{2}{x}y' + y^n\right)\right\} = \frac{1}{x}\{-X - X^2 - XY\},$$

$$\frac{dY}{dx} = \frac{y^n}{y'} + nxy^{n-1} + \frac{xy^n}{y'^2}\left(\frac{2}{x}y' + y^n\right) = \frac{1}{x}\{3Y + nXY + Y^2\}. \tag{39}$$

Introducing $\ln|x| = t$ in (39), we have

$$\frac{dX}{dt} = -X(1 + X + Y),$$
$$\frac{dY}{dt} = Y(3 + nX + Y); \tag{40}$$

$$\frac{dY}{dX} = \frac{Y}{X}\frac{-nX - Y - 3}{X + Y + 1}. \tag{41}$$

In the notation of equation (20), we have $a = -n$, $b = -1$, and $c = -3$. The nature of the singularities depends on the constant n. Referring to (21) and (23), we have the following singularities (we call these points A, B, C, D, etc.) with their respective nature (see equation (23)):

A. $X = 0$, $Y = 0$; saddle.
B. $X = 0$, $Y = -3$; saddle.
C. $X = -1$, $Y = 0$; since the characteristics are $\lambda_1 = -1$, $\lambda_2 = n - 3$, the point is a node if $n < 3$, it is a saddle node for $n = 3$, and a saddle if $n > 3$.
D. $X = -2/(n-1)$, $Y = -(n-3)/(n-1)$ ($n \neq 1, 3$). This point is denoted as P_4 in (23), and the nature of this singularity depends on the characteristic roots given there. The expression under the radical is $-7n^2 + 22n + 1$. It vanishes for $n_1 = -0.0448$ and $n_2 = +3.1877$. So

Phase Space Study of Autonomous Systems

the following cases arise: $n < n_1$, a focus; $n_1 < n < 1$, node; $1 < n < 3$, saddle; $3 < n < n_2$, node; $n_2 < n < 5$, focus; $n > 5$, focus.
E. $X = \infty$, $Y/X = 0$; $n < -1$, saddle; $n = -1$, saddle node; $n > -1$, node.
F. $X = \infty$, $Y/X = -(n+1)/2$ $(n \neq -1)$; $n < -1$, node; $-1 < n < 1$, saddle; $n = 1$, saddle node; $n > 1$, node.
G. $X/Y = 0$, $Y = \infty$, node.

In the scheme of Serebriakova which we discussed earlier, this case belongs to {18} if $n > 3$ and to {19} if $3 > n > 1$. Actually, there are three cases included in {18}, depending on whether $5 > n$, $5 = n$, or $5 < n$ (see $\lambda_{1,2}$ for P_4 in (23)); point D is, respectively, a focus, a center, and a focus. Thus $n = 2$ and $n = 5$ correspond to the cases {19} and {18.2} (see Figure 6.8.6), respectively, while $n = 3$ corresponds to {19} provided D moves down to coincide with C, the rest of the configuration remaining the same. One, however, has to reverse the orientation of the trajectories in these figures since dX/dt in (40) has a sign opposite to that in (19a).

The integral curves are divided into various families by the separatrices; these are the particular integral curves which run through the saddle points. These curves correspond to the special one-parameter family solutions of equation (37). Thus, for example, the separatrix from $B(0, -3)$ into the third quadrant corresponds to the Emden solution found in astrophysics, namely,

$$y = a\left(1 - \frac{1}{6}a^{n-1}x^2 + \cdots\right),$$

where a is an arbitrary parameter. Similarly, for $n = 2$, the pair of separatrices through $D(-2, 1)$ corresponds to the two singular solutions

$$y = x^{-2}\left\{-2 + bx^k\left(1 - \frac{b}{6}\frac{x^k}{1+k} + \cdots\right)\right\},$$

where $k = (3 \pm \sqrt{17})/2$ and b is arbitrary.

In the rest of this chapter we present applications.

6.9 NONLINEAR DIFFUSION EQUATION OF POPULATION GROWTH—FISHER'S EQUATION

The nonlinear PDE

$$u_\tau = \nu u_{yy} + ku(1-u), \tag{1}$$

first studied by Fisher (1936) and later by Kolmogoroff et al. (1937), describes the propagation of a virile mutant in an infinitely long habitat. The mutant population grows due to diffusion and nonlinear local multiplication. The parameters ν and k in (1) are diffusion coefficients and a positive multiplication factor, respectively, while τ and y are time and distance. With minor changes (1) also describes the evolution of the neutron population in a nuclear reactor, where the domain is obviously finite (see Canosa 1969). The time and space variables can be rendered nondimensional by writing

$$t = k\tau, \qquad x = \left(\frac{k}{\nu}\right)^{1/2} y \qquad (2)$$

so that (1) transforms to

$$u_t = u_{xx} + u(1-u). \qquad (3)$$

Fisher (1936) postulated that (3) describes the nonlinear evolution of a population in a one-dimensional habitat which can support only a certain maximum population per unit length. This maximum, for convenience, may be chosen to be unity. The initial conditions for (3), therefore, must satisfy the restriction

$$0 \leq u(x,0) \leq 1, \qquad -\infty < x < \infty. \qquad (4)$$

The solutions of (3) are sought such that the initial condition satisfies (4), while the boundary conditions are

$$\lim_{x \to -\infty} u(x,t) = 1, \quad \lim_{x \to +\infty} u(x,t) = 0, \qquad t > 0. \qquad (5)$$

All spatial derivatives of u are assumed to vanish at $x = \pm\infty$. Kolmogorov et al. (1937) proved that, for each condition of the form (4), equation (3) has a unique solution that is bounded for all time, as the initial distribution is; that is,

$$0 \leq u(x,t) \leq 1, \qquad -\infty < x < \infty, \, t > 0. \qquad (6)$$

Further, they showed that, for the following forms of initial conditions,

$$u(x,0) = \begin{cases} 1, & x < 0, \\ 0, & x > 0, \end{cases} \qquad (7a)$$

or

$$u(x,0) = \begin{cases} 1, & x < a, \\ f(x), & a < x < b, \\ 0, & x > b, \end{cases} \qquad (7b)$$

where $f(x)$ is an arbitrary function subject however to (4), the solution in the limit $t \to \infty$ becomes a shocklike traveling wave that satisfies (5)

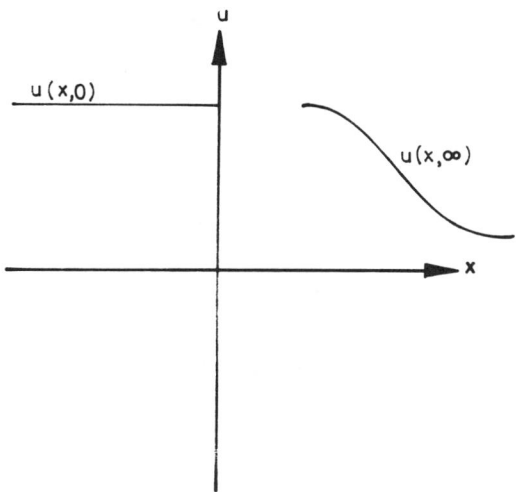

Figure 6.9.1 Qualitative plot of the profile of the minimum-speed traveling wave.

and propagates to the right with minimum allowable characteristic speed $c_{min} = 2$ (see Figure 6.9.1).

We shall obtain two results from the phase plane analysis of ODE derived from (1) describing the traveling wave solutions:

1. There are an infinite number of traveling wave solutions of (1) with characteristic speeds $c > 2$.
2. A traveling wave propagates with a speed linearly proportional to its thickness.

Writing

$$u = u(x - ct) \equiv u(s), \tag{8}$$

where c is the speed of the wave or propagation front, and substituting it into (3), we get

$$\frac{d^2u}{ds^2} + c\frac{du}{ds} + u - u^2 = 0. \tag{9}$$

We look for shocklike solutions of (9) satisfying

$$u(-\infty) = 1, \quad u(+\infty) = 0. \tag{10}$$

System (9)–(10) defines a nonlinear eigenvalue problem over an infinite domain, with the propagation speed c as the eigenvalue. Introducing the variable

$$\frac{du}{ds} = y,$$

we write equation (9) in the (y, u) plane as

$$\frac{dy}{du} = \frac{u^2 - u - cy}{y}. \tag{11}$$

Equation (11) has two singular points: $(0, 0)$ and $(0, 1)$. Its linearized form about $(0, 0)$ is

$$\frac{dy}{du} = \frac{-u - cy}{y}. \tag{12}$$

In the notation of Section 6.3, we have, for (12), $\Delta = c^2 - 4$ and $q = 1$. Since u, being the population of the habitat, is always positive, we would want $(0, 0)$ to be a stable node; if it were a center or a focus, the trajectory would enclose the origin and thus lead to u assuming negative values. Thus, for $(0, 0)$ to be a node, the discriminant Δ should be positive so that $c > 2$. The point $(1, 0)$ is easily checked to be a saddle point. We therefore have a continuous infinite spectrum of eigenvalues of problem (9)–(10). The trajectories for $c > 2$ are shown in Figure 6.9.2.

It is possible to carry out a simple perturbation analysis in the phase plane, with $\epsilon = 1/c^2$ as a convenient perturbation parameter. Since $c^2 \geq 4$,

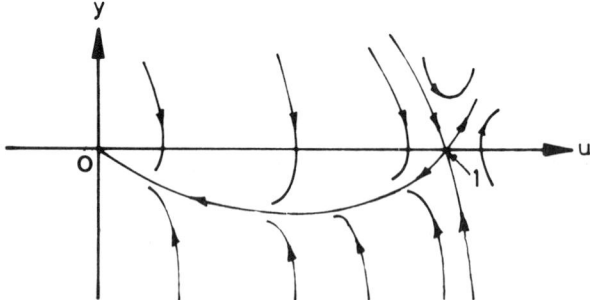

Figure 6.9.2 Qualitative plot of the trajectories of equation (11). The solution to the nonlinear eigenvalue problem defined by equations (9) and (10) is given by the trajectory that intercepts the critical points: saddle point $(1, 0)$ and stable node $(0, 0)$. (From Canosa 1973.)

$\epsilon \leq 0.25$. Introducing the dependent variable $\bar{y} = cy$ into (11), we get

$$\epsilon \frac{d\bar{y}}{du} = \frac{u^2 - u - \bar{y}}{\bar{y}}. \tag{13}$$

Equation (13) appears to be singular, since ϵ multiplies the highest-order derivative, but a straightforward perturbation expansion

$$\bar{y}(u, \epsilon) = g_0(u) + \epsilon g_1(u) + \epsilon^2 g_2(u) + \cdots, \tag{14}$$

on substitution into (13), yields the equations

$$\begin{aligned} g_0 &= u^2 - u, \\ g_1 &= -g_0 g_0', \\ g_2 &= -g_0 g_1' - g_1 g_0', \text{ etc.} \end{aligned} \tag{15}$$

The solution of the system (15) passing through the singular points $u = 0$ and $u = 1$ comes through by mere substitution; there is no need to solve the coupled system of ODE. The boundary conditions are appropriately satisfied by terms of each order, as may easily be verified by inspection. The solution of (13) is thus

$$y(u, \epsilon) = \epsilon^{1/2}(u^2 - u) - \epsilon^{3/2}(2u^3 - 3u^2 + u) \\ + 2\epsilon^{5/2}(5u^4 - 10u^3 + 6u^2 - u) + O(\epsilon^{7/2}). \tag{16}$$

The series (16) is an accurate asymptotic series in the limit $\epsilon \to 0$ ($c^2 \to \infty$). It is a good approximation even for $\epsilon = 0.25$, which corresponds to the slowest wave with $c^2 = 4$.

The solution (16) of (9) enables us to obtain a relation between shock wave thickness and its speed. The wave front is steepest at the point of inflection where $d^2y/ds^2 = 0$; that is, $y(dy/du) = 0$. Substituting $dy/du = 0$ in (11) and assuming that $u = 1/2 + a\epsilon$ at the point of inflection, we easily obtain $a = -1/4$. Inserting $u = 1/2 - \epsilon/4$ into (16) then gives the coordinates of the point of inflection as

$$(u, y) \equiv \left(u, \frac{du}{ds} \right) = \left(\frac{1}{2} - \frac{\epsilon}{4}, \frac{-(1 - \epsilon^2/4)}{4c} \right). \tag{17}$$

If the steepness S of the wave profile is defined as the magnitude of the slope at its point of inflection, then (17) shows that, to an error $O(\epsilon^2)$,

$$\left| \frac{du}{ds} \right| = S = \frac{1}{4c}.$$

Further, if L denotes the thickness of the profile whose total height is unity, it is easily verified geometrically that

$$L = \frac{1}{S} = 4c$$

or

$$c = \frac{L}{4}. \tag{18}$$

We arrive at the interesting result that the propagation speed of the wave is linearly proportional to its thickness, and the time L/c for all the waves to pass a stationary observer is constant.

6.10 BOUNDARY VALUE PROBLEMS FOR A NONLINEAR DIFFUSION EQUATION

The nonlinear diffusion equation

$$u_t = r^{1-\lambda} \{r^{\lambda-1} u^\beta u_r\}_r, \qquad u \geq 0, \tag{1}$$

has been studied by many investigators, particularly in the context of their similarity solutions. Here r is the spatial variable, t is time, β is a constant index, and λ is a geometry index which takes values 1, 2, or 3 for plane, cylindrical, or spherical geometry. Two classes of similarity solutions of (1) exist. The first is of the form

$$u = t^\alpha f(\eta), \qquad \eta = rt^{-\delta} \tag{2}$$

with $\delta > 0$, and is governed by the ODE

$$\{\alpha f - \delta \eta f'\} \eta^{\lambda-1} = \{\eta^{\lambda-1} f^\beta f'\}', \qquad f > 0, \tag{3}$$

where the prime denotes differentiation with respect to η. The form (2) is valid only when

$$\alpha \beta = 2\delta - 1. \tag{4}$$

The second class of solutions can be written as

$$u = \exp(\alpha t) f(\eta), \qquad \eta = r \exp(-\delta t), \ \delta > 0.$$

and are governed by (3) again, provided

$$\alpha \beta = 2\delta.$$

We shall restrict our attention to the case $\lambda = 1$ of the first class, and so deal exclusively with the special case

$$(\alpha f - \delta \eta f') = (f^\beta f')' \tag{5}$$

Phase Space Study of Autonomous Systems

of (3). We follow Grundy (1979) in the sequel. The form (3) of (1) obtained via (2) has been discussed in great detail by several authors (see, for example, Gilding and Peletier 1976) for special values of the parameters α, δ, β, and λ. Grundy has used phase plane analysis to investigate the existence, uniqueness, and nature of all solutions of (3) for every parameter range. Although he has studied in detail the case $\lambda = 1$, corresponding to plane symmetry, the results can be extended in a straightforward manner to $\lambda = 2, 3$. The method is to reduce (3) to a system of two first-order ODE (see Section 2.9), one of which is the "main" equation whose phase plane analysis gives all the details while the other relates the independent variable with the dependent variable along the integral curve.

Thus, we introduce the variables

$$X = \frac{\eta f'}{f}, \qquad Y = \frac{\eta^2}{2f^\beta} > 0; \tag{6a, b}$$

then (5) becomes

$$\frac{dY}{dX} = \frac{Y(2 - \beta X)}{2(\alpha - \delta X)Y - (\beta + 1)X^2 + X}. \tag{7}$$

By eliminating f from (6a,b), we get an equation for η:

$$\frac{1}{\eta}\frac{d\eta}{dY} = \frac{1}{(2 - \beta X)Y}. \tag{8}$$

The following procedure is adopted. First, all of the singular points of (7) are enumerated. Using standard linearization techniques, the nature of the integral curves in the neighborhood of the singular points is determined. Indeed, the explicit form of the solution in the close vicinity of the singular point is written out. Then, (8) and (6b) give η and f in the vicinity of the singular point. With this information, the integral curves representing the solution of (5) can be drawn in the (X, Y) plane. From this analysis we may identify all the solutions of (5), as well as the number of parameters needed to specify them. We may then choose the solution with the appropriate parameters, which satisfies a given set of boundary conditions. We shall here solve a specific boundary value problem over the half-line $x > 0$.

We easily check that (7) has the following singular points:

1. $X = 0, Y = 0$.
2. $X = 2/\beta, Y = -(\beta + 2)/\beta$.
3. $X = 1/(\beta + 1), Y = 0$.
4. $X = \alpha/\delta, Y = \infty$.
5. $X = \pm\infty, Y = 0$.
6. $X = \pm\infty, Y = \infty$.

It is useful to note (see Section 6.8) that the integral curves of (7) have zero slope along $X = 2/\beta$ and infinite slope along $Y = X\{(\beta + 1)X - 1\}/2(\alpha - \delta X)$.

We discuss the nature of each of the singular points, the integral curves (or solutions) passing through each point, and the arbitrary sets of parameters associated with a solution.

1. The point $X = 0$, $Y = 0$ is a node; linearizing (7) about it, we have

$$\frac{dY}{dX} = \frac{2Y}{X + 2\alpha Y}. \tag{9}$$

The equation integrates to give

$$Y \sim A_1 \{X - 2\alpha Y\}^2, \tag{10}$$

where A is an arbitrary parameter. Linearizing (8) about $X = 0$, $Y = 0$ and integrating, we have

$$\eta \sim B_1 Y^{1/2}, \tag{11}$$

where B_1 is another arbitrary parameter. It follows then that $Y = X = 0$ when $\eta = 0$. Substituting (11) into (6b), we get

$$f = \left(\frac{B_1^2}{2}\right)^{1/\beta} \quad \text{at } \eta = 0. \tag{12}$$

This local analysis shows that a two-parameter family of solutions $f(\eta; A_1, B_1)$ is associated with the singular point 1 with $f = (B_1^2/2)^{1/\beta} > 0$ at $\eta = 0$.

2. The singular point $X = 2/\beta$, $Y = -(2+\beta)/\beta$ exists when $-2 \le \beta \le 0$, since $Y \ge 0$ (see (6b)). Then, by shifting the origin to this point, etc., we easily verify that it is a saddle point and the equations of the two separatrices passing through it are (approximately)

$$Y + \frac{\beta + 2}{\beta} \sim \frac{(B - A)(X - 2/\beta)}{4}, \tag{13a}$$

$$Y + \frac{\beta + 2}{\beta} \sim \frac{(B + A)(X - 2/\beta)}{4}, \tag{13b}$$

where $B = \alpha\beta(\beta + 2) - 2(\beta + 1)$, $A^2 = B^2 - 8\beta(\beta + 2)$. Since $A^2 - B^2 = -8\beta(\beta + 2) > 0$ when $-2 < \beta < 0$, $B - A$ and $B + A$ have opposite signs. Substituting for Y from (13a) or (13b) into (8) and integrating, we get $\eta \sim B_2(X - 2/\beta)^{\mu_1}$ along the curves of positive slope while $\eta \sim B_2|X - 2/\beta|^{-\mu_2}$ along those with negative slope, where μ_1 and μ_2 are positive constants (we may choose $B - A > 0$ or $B + A > 0$). In the former case, $\eta \to 0$ at the singular point while in the latter $\eta \to \infty$ there. Substituting these expressions in (6b) we find that

$$f \sim \frac{\beta \eta^{2/\beta}}{2(\beta + 2)} \longrightarrow \infty \quad \text{as } \eta \longrightarrow 0$$

for the first case, while

$$f \sim \frac{\beta \eta^{2/\beta}}{2(\beta + 2)} \longrightarrow 0 \quad \text{as } \eta \longrightarrow \infty$$

for the second.

We recall that $-2 < \beta < 0$. We thus conclude that there are two one-parameter families of solutions $f(\eta; B_2)$ associated with the singular point 2. For one family $f \to \infty$ as $\eta \to 0$, while for the other $f \to 0$ as $\eta \to \infty$.

3. The point $X = 1/(\beta + 1)$, $Y = 0$ can be shown to be a saddle point for $\beta > -1$ or $\beta < -2$ and a node for $-2 < \beta < -1$. The integral curves have the behavior

$$Y \sim A_3 \left\{ (2\beta + 3) \left(X - \frac{1}{(\beta + 1)} \right) - 2(\delta - 1 + \alpha) Y \right\}^{-(\beta+2)/(\beta+1)}. \quad (14)$$

When the singular point is a node, each member of the family, with A_3 as a parameter, passes through the point. Moreover, according to (8) and (6b),

$$\eta \sim B_3 Y^{(\beta+1)(\beta+2)} \longrightarrow \infty \quad \text{as } Y \longrightarrow 0$$

and

$$f \sim \frac{B_3^{(\beta+2)/\beta(\beta+1)}}{2^{1/\beta}} \eta^{1/(\beta+1)} \longrightarrow 0.$$

When the singular point under discussion is a saddle, the single curve of the form

$$Y \sim \frac{2\beta + 3}{2(\delta - 1 + \alpha)} \left(X - \frac{1}{(\beta + 1)} \right)$$

passes through it. Equations (8) and (6b) now give

$$\eta \sim B_3 Y^{(\beta+1)/(\beta+2)} \longrightarrow 0 \quad \text{as } Y \longrightarrow 0,$$

$$f \sim \frac{B_3^{(\beta+2)/\beta(\beta+1)}}{2^{1/\beta}} \eta^{1/(\beta+1)}.$$

In this case, $f \to 0$ if $\beta > -1$ and $f \to \infty$ if $\beta < -2$ as $Y \to 0$.

We therefore arrive at the result that for $\beta > -1$ or $\beta < -2$, the one-parameter family of solutions $f(\eta; B_3)$ is associated with the point 3 where $\eta = 0$ and $f = 0$ if $\beta > -1$, while $f \to \infty$ if $\beta < -2$. For $-2 < \beta < -1$, we have the two-parameter family $f(\eta; A_3, B_3)$, for which $f = 0$ when $\eta = \infty$.

4. To discuss the point $X = \alpha/\delta$, $Y = \infty$, it is convenient to write $Y = 1/y$, $X = x$ so that (7) becomes

$$\frac{dy}{dx} = \frac{y^2(\beta x - 2)}{2(\alpha - \delta x) - (\beta + 1)yx^2 + xy}.$$

The corresponding singular point here is $x = \alpha/\delta$, $y = 0$, which is more complex, the dominant term being a quadratic in the numerator and linear in the denominator (see Davies and James 1966). The solution in the present case passing through $X = \alpha/\delta$, $Y = \infty$ is rather complicated. It consists of the one-parameter family

$$\frac{1}{Y} = \frac{2\delta^3}{\alpha\{\delta - \alpha(\beta+1)\}}\left[X - \frac{\alpha}{\delta} + \frac{\delta(X - \alpha/\delta)^2}{\alpha\{(\beta+1)\alpha - \delta\}} + \cdots \right.$$
$$\left. + A_4 \exp\left(\frac{\alpha\{(\beta+1)\alpha - \delta\}}{\delta(X - \alpha/\delta)}\right)\{1 + \cdots\}\right], \quad (15)$$

where, since $Y \to +\infty$ at the singular point, $X \gtrless \alpha/\delta$ for $\{(\beta+1)\alpha - \delta\}\alpha \lessgtr 0$. For each of these curves we get, from (8) and (6b), by integration, etc.,

$$\eta \sim B_4 Y^\delta \longrightarrow \infty$$

and

$$f \sim \frac{B_4^{1/\delta\beta}\eta^{\alpha/\delta}}{2^{1/\beta}},$$

so $f \to \infty$ if $\alpha > 0$ and $f \to 0$ if $\alpha < 0$ as $Y \to \infty$. We thus have a two-parameter family of solutions $f(\eta; A_4, B_4)$ associated with the point 4, where $\eta \to \infty$. Moreover, $f \to \infty$ if $\alpha > 0$, and $f \to 0$ if $\alpha < 0$.

5. To discuss this case with $X = \pm\infty$, $Y = 0$, it is helpful to change the variables to $x = 1/X$, $y = Y$ so that (7) becomes

$$\frac{dx}{dy} = -\frac{x[2y(\alpha x - \delta) - (\beta+1) + x]}{(2x - \beta)y}.$$

Its linearized form, after integration and reversion to original variables, gives

$$X = A_5 Y^{(\beta+1)/\beta}, \quad X \gtrless 0, A_5 \gtrless 0. \quad (16)$$

For $\beta(\beta+1) < 0$ (i.e., $-1 < \beta < 0$) equation (8) leads, for each A_5, to the integral

$$\eta = \eta_5 \exp\left(\frac{-1}{(\beta+1)X}\right), \quad X \to \pm\infty, \; 0 < \eta_5 < \infty,$$

where $\eta = \eta_5$ at the singular point 5. Equation (6b) then shows that

$$f \sim |\eta - \eta_5|^{1/(\beta+1)} \longrightarrow 0, \quad f' \sim |\eta - \eta_5|^{-\beta/(\beta+1)} \longrightarrow 0, \text{ as } \eta \longrightarrow \eta_5.$$

We conclude that, for $-1 < \beta < 0$, a two-parameter family of solutions $f(\eta; A_5, \eta_5)$ is associated with 5. These solutions pass through $\eta = \eta_5$ for $\eta \lessgtr \eta_5$ where $f(\eta_5) = f'(\eta_5) = 0$.

6. The points $X = \pm\infty$, $Y = \infty$ can be conveniently handled by the substitution

$$X = \frac{1}{x}, \quad Y = \frac{1}{y}$$

so that (7) near $x = 0$, $y = 0$ is approximated by

$$\frac{dy}{dx} = \frac{\beta y^2}{x[2\delta x + (\beta + 1)y]}.$$

Since the right-hand side of this equation is homogeneous in x and y, one may put $y = vx$, etc., and integrate. The result, in terms of the variables X and Y, is

$$Y^{(\beta+1)/\beta} \sim A_6(2\delta Y + X). \tag{17}$$

Here A_6 is an arbitrary parameter.

We examine those curves (17) which actually reach the singular points under discussion. Several cases arise.

1. For all β there is a single integral curve, which we call class I, for which

$$Y = -\frac{X}{2\delta} + o(X), \quad X \longrightarrow -\infty, \; Y \longrightarrow \infty.$$

2. For $\beta > 0$ or $\beta < -1$, the left-hand side of (17) is more dominant than $A_6 \delta Y$, so we have

$$X \sim A_{62} Y^{(\beta+1)/\beta},$$

where $A_{62} \gtrless 0$ depending on whether $X \gtrless 0$. This family is called class II.

3. For $\beta < 0$, there is a one-parameter family, class III, which may be obtained from (17) by a suitable expansion as

$$Y \sim -\frac{X}{2\delta} + (-X)^{(\beta+1)/\beta} A_{63} + \cdots, \quad X \longrightarrow -\infty, \; Y \longrightarrow \infty,$$

which evidently is asymptotic to class I curves.

To get the connection with η and f, we again use (8) and (6b). For class I, we easily find from (8) that

$$\eta \sim \eta_6 \exp(-2\delta Y \beta), \quad Y \longrightarrow \infty,$$

where $\eta = \eta_6$ at point 6, $0 < \eta_6 < \infty$. This result implies that

$$Y \sim \frac{\eta_6}{2\delta |\beta(\eta_6 - \eta)|}$$

since $Y \to +\infty$, $\eta_6 \gtrless \eta$ for $\beta \gtrless 0$ as point 6 is approached along the class I curves. Now, (6b) gives

$$f \sim (\eta_6 \delta \beta)^{1/\beta} |\eta_6 - \eta|^{1/\beta};$$

therefore, $f \to 0$ for $\beta > 0$ and $f \to \infty$ for $\beta < 0$. We also check that $(f^{\beta+1})' \to 0$ as $\eta \to \eta_6$ for $\beta > 0$.

A similar analysis for class II curves shows that along each such curve

$$\eta_6 - \eta \sim \frac{Y^{-(\beta+1)/\beta}}{(\beta+1)A_{62}};$$

hence, when $(\beta+1)A_{62} \gtrless 0$, $\eta \gtrless \eta_6$. Equation (6b) then gives

$$f \sim |\eta_6 - \eta|^{1/(\beta+1)},$$

so $f \to 0$ for $\beta > 0$ with $f'(\eta_6) = \infty$, while for $\beta < -1$, $f \to \infty$ as $\eta \to \eta_6$.

For curves of class III, one may verify that

$$f \sim \{\delta \beta \eta_6 (\eta - \eta_6)\}^{1/\beta}$$

with $\eta > \eta_6$; since $\beta < 0$, $f(\eta_6) = \infty$.

We summarize the results for case 6. The integral curves through point 6, where $\eta = \eta_6$, are associated with the following solutions for f. For $\beta > 0$: a one-parameter family of solutions $f(\eta; \eta_6)$ of class I with $\eta < \eta_6$ for which $(f^{\beta+1})' = 0$ at $\eta = \eta_6$; also a two-parameter family, $f(\eta; A_{62}, \eta_6)$ of class II where $f(\eta_6) = 0$, $f'(\eta_6) = \infty$, and $\eta \gtrless \eta_6$. For $-1 < \beta < 0$: a two-parameter family $f(\eta; A_{63}, \eta_6)$ of class III with $\eta > \eta_6$ and $f(\eta_6) = \infty$; also a one-parameter family $f(\eta; \eta_6)$ of class I with $\eta > \eta_6$ and $f(\eta_6) = \infty$. For $\beta < -1$: again a one-parameter family $f(\eta; \eta_6)$ of class I with $\eta > \eta_6$ and $f(\eta_6) = \infty$; also the two-parameter family $f(\eta; A_{62}, \eta_6)$ of class II with $\eta \gtrless \eta_6$, $f(\eta_6)$ and $f'(\eta_6)$ equal to ∞; and finally, the two-parameter family $f(\eta; A_6, \eta_6)$ of class III with $\eta > \eta_6$ and $f(\eta_6) = \infty$.

Grundy (1979) has given detailed phase portraits for each of the above cases and extensive tables to explain which boundary value problems are solved by a given phase curve for a certain range of the parameters. Here, we just consider a particular example. The problem is to find the solutions of

$$\alpha f - \delta \eta f' = (f^\beta f')', \tag{18}$$

subject to

$$f(0) = U \geq 0, \quad f(\infty) = 0. \tag{19}$$

Grundy has identified all the solutions $f(\eta)$ of (18) and (19) which are continuous on $[0, \infty)$. We restrict ourselves to the case $\beta > 0$, $\alpha > 0$. We claim that Figure 6.10.1 contains the solution of this problem. The singular

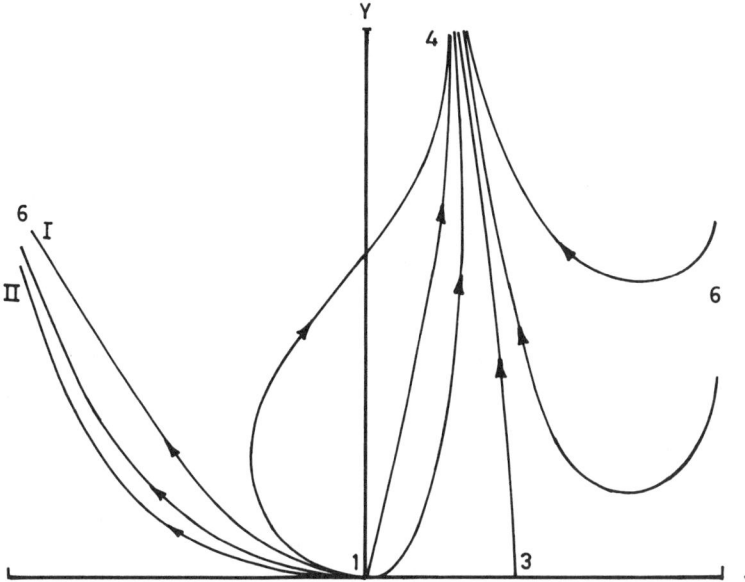

Figure 6.10.1 The solution of problem (18)–(19) in the (X,Y) plane. (From Grundy 1979.)

points corresponding to $\eta = 0$ are 1 and 3. Three types of curves issue from the point 1: (a) 1 to 6 class I, (b) 1 to 6 class II, (c) 1 to 4.

(a) Choosing the value $A_1 = A_1^*$ in the discussion of point 1, and noting that $X < 0$ on the line joining 1 and 6_1, we find from (6a) that $f' < 0$. Also, it follows from (12) that

$$f(0) = \left(\frac{B_1^2}{2}\right)^{1/\beta} = U$$

for each curve passing through $\eta = 0$. Hence, the integral curves 1–6 class I represent a one-parameter family of solutions $f(\eta; A_1^*, B_1)$. We know the behavior of the solution near the point 6 where $\eta = \eta_6$; we join this solution at $\eta = \eta_6$ with the zero solution $f = 0$ for $\eta > \eta_6$. Thus, we have a weak solution, discussed earlier by Gilding and Peletier (1976) by the shooting technique (see Section 5.4).

(b) The solutions joining the point 1 with 6 class II are a two-parameter family $f(\eta; A_1, B_1)$ for which $f'(\eta) < 0$. Again, we take $f = 0$ for $\eta > \eta_6$.

(c) The curves 1–4 do not furnish solutions to our problem since at 4, $\eta \to \infty$ and so $f \sim \eta^{\alpha/\delta} \to \infty$.

We conclude that for $\beta > 0$ and $\alpha > 0$, we have two types of possible solutions, referred to as (a) and (b).

Once $f(\eta)$ has been identified, one may return to the actual similarity solution $u(r,t)$ given by (2). The finite values of η where $u = 0$ represent "wave fronts" in the (r,t) plane. These fronts move according to the law $r = \eta t^\delta$.

The above discussion shows how one may discover the local behavior of all types of possible solutions by writing the asymptotic solutions in the neighborhood of singularities. This helps in excluding the solutions not relevant to a given problem. One may draw the actual integral curves joining the singular points by any numerical technique, such as Runge–Kutta. Local analysis is an important element for understanding global behavior of the solutions.

We also refer to a similar analysis by Kochina and Mel'nikova (1958a,b) in a gas dynamics context. They treated two problems: (i) strong blasts in a compressible medium, and (ii) unsteady motion of gas driven outward by a piston. The self-similar problems were reduced to a phase plane analysis. Local analysis about the singular points aided the shifting of the solutions relevant to given boundary value problems.

6.11 COLLAPSE OF A SPHERICAL CAVITY IN A PERFECT GAS

Now we describe similarity solutions that may govern the symmetric flow of a fluid into an empty spherical cavity. The present class of solutions, which could explain cavitation and implosion phenomena, describes the geometric convergence of the flow. They apply to the flow near the cavity as the cavity radius becomes small. However, numerical solutions of the basic PDE with suitable initial conditions show that the flow near the cavity does not always tend to the similarity form as the cavity radius becomes small (Hunter 1960, Thomas et al. 1986; see Section 9.3 for further details).

From dimensional analysis one may arrive at the conclusion that the flow near the cavity tends to a self-similar form as the cavity collapses. When the cavity becomes small, the only relevant scale for the flow is the cavity radius $R = R(t)$; any other length scale would be too large to be relevant near the collapse. Similarly, since the inward flow velocities tend to become large, the only relevant velocity scale would be $\dot{R}$, the velocity of the cavity. Here, the dot denotes differentiation with respect to t. Thus, we may write the radial flow velocity as

$$u = \dot{R} f_1 \left(\frac{r}{R}, \frac{t}{T} \right), \tag{1}$$

where r is the radial distance from the center of the cavity, T is some dimensional time scale, and f_1 is an unknown function of its arguments. Supposing that the time is measured from the instant at which $R = 0$, we may, for sufficiently small time before R becomes zero, approximate (1) by

$$u = \dot{R} f_1 \left(\frac{r}{R}, 0 \right) = \dot{R} f \left(\frac{r}{R} \right), \quad \text{say.} \tag{2}$$

The kinematic condition that the velocity at the cavity wall is equal to the particle velocity there leads to

$$f(1) = 1. \tag{3}$$

Assuming the gas to be perfect and the flow to be homenotropic, we have

$$p = \kappa \rho^\gamma, \tag{4}$$

where κ and $\gamma = c_p/c_v$ are constants. Using the expression for the square of the sound speed, $c^2 = dp/d\rho$, the equations of continuity and momentum can be written as

$$u_t + u u_r + \frac{1}{\gamma - 1} (c^2)_r = 0, \tag{5}$$

$$(c^2)_t + u(c^2)_r + (\gamma - 1)c^2 \left(u_r + \frac{2u}{r} \right) = 0; \tag{6}$$

the subscripts denote partial derivatives.

We may seek c^2 in the form

$$c^2 = \dot{R}^2 g \left(\frac{r}{R} \right), \tag{7}$$

where g is another unknown function. Inside the cavity, there is a vacuum; therefore, $c^2 = 0$ at $r = R$. This leads to the condition

$$g(1) = 0. \tag{8}$$

Substituting (2) and (7) into (5)–(6) shows that the similarity form exists only if

$$\frac{R\ddot{R}}{\dot{R}^2} = \text{const.} \tag{9}$$

or

$$R = A(-t)^n, \tag{10}$$

where A and n are constants. (The only other possible form, namely, the exponential dependence of R on t, is ignored because that would require an infinite time for the cavity to collapse.) If we further insist that the cavity collapses with a nonzero velocity, then $0 < n \leq 1$.

The case $n = 1$ corresponds to the collapse with a constant velocity; for $0 < n < 1$, the cavity accelerates inward and has infinite velocity at $t = 0$. The two first-order ODE that result from the substitution of (2) and (7) into (5)–(6) are to be solved subject to the boundary conditions (3) and (8). For fixed γ and n, they fail, in general, to provide a solution free from singularities. These singularities occur for some value of r/R which corresponds to the so-called limiting characteristic curve. These singularities are rather unrealistic since they must exist for all time and must be fed into the flow initially at precisely the right moment to ensure their propagation to $r = 0$ at exactly the instant at which R becomes 0.

Hunter (1962) showed that both types of solutions, describing the collapse of the cavity either with uniform speed if $1 < \gamma < 7$ or with an acceleration if $\gamma > 3/2$. Therefore, for $\gamma > 3/2$, assuming that the similarity solutions are the correct asymptotic form near the collapse point, we have both types of solutions possible. However, the relevant type will depend on the initial conditions which give rise to the limiting flow. Only a certain class of initial conditions will lead to each of the asymptotic forms. Energy considerations suggest that strong pressure gradients pushing the fluid would favor accelerating solutions. Moreover, stability arguments may also rule out some of the solutions as unrealistic.

For mathematical convenience, we introduce the similarity variable in the form

$$\xi = -\left(\frac{R}{r}\right)^{1/n} = \frac{A^{1/n}t}{r^{1/n}}, \tag{11}$$

in accordance with (10). The cavity velocity from (10) is

$$\dot{R} = -nA^{1/n}R^{1-(1/n)}. \tag{12}$$

The similarity form of the solution, in view of (2) and (7), can therefore be written as

$$u = -nA^{1/n}r^{1-1/n}F(\xi), \qquad c^2 = n^2 A^{2/n} r^{2-2/n} G(\xi), \tag{13}$$

where F and G are functions to be determined. At the cavity wall, $\xi = -1$, and the boundary conditions (3) and (8) via (13) become

$$F(-1) = 1, \qquad G(-1) = 0. \tag{14}$$

Substituting (13) into (5) and (6) yields the ODE

$$(\gamma - 1)(1 + \xi F)F' + \xi G' + (1 - n)[(\gamma - 1)F^2 + 2G] = 0, \tag{15}$$

Phase Space Study of Autonomous Systems

$$(\gamma - 1)\xi G F' + (1 + \xi F)G' + (1 + n + \gamma - 3n\gamma)FG = 0. \tag{16}$$

We seek a solution of the problem outside the cavity before the collapse, so $r > R$. This region corresponds to $-1 < \xi \leq 0$. The two boundary conditions (14) fix, for given γ and n, the solutions of (15) and (16), but, as we remarked earlier, not all these solutions are acceptable; two further conditions must be imposed. The first condition arises from the fact that, far from the cavity, as $r \to \infty$, the particle velocity and sound speed must tend to some finite values. The similarity solution is expected to apply only in a limited region near the cavity, but at the outer edge of this region it must be matched to a solution in which u and c remain finite. From (13), we have

$$u = -n \frac{A(-\xi)^{(1-n)}}{(-t)^{1-n}} F(\xi), \qquad c^2 = \frac{n^2 A^2 (-\xi)^{2-2n}}{(-t)^{2-2n}} G(\xi). \tag{17}$$

Thus, if u and c^2 are to remain finite for finite r as $(-t) \to 0$, we must have

$$(-\xi)^{1-n} F(\xi) = (-\xi)^{2-2n} G(\xi) = 0 \qquad \text{at } \xi = 0. \tag{18}$$

The case $0 < n < 1$ of accelerating cavities has been discussed in great detail by Hunter (1962). Here we study the special case $n = 1$ corresponding to the uniform velocity of the cavity. It is for this special case that we also attempt to find the global solution of the cavity collapse problem in Section 9.3. Since the basic equations (5)–(6) are invariant with respect to translation in t, we may redefine ξ in (11) as $(\beta t - 1)/r$ with $\beta = 2/(\gamma - 1)$. With this definition of ξ, $A = \beta$ in (11) and the cavity surface $\xi = -1$, moving with constant speed, would reach the focusing point $r = 0$ at $t = \beta^{-1}$. The conditions (18) in the present case are replaced by the requirement that

$$F(\xi), G(\xi) \text{ are bounded as } \xi \longrightarrow 0, \tag{19}$$

corresponding to r tending to infinity for $t \neq 1/\beta$.

The ODE (15)–(16) for $n = 1$ may be rewritten as

$$[(1 + \xi F)^2 - G\xi^2] F' = 2\xi GF, \tag{20}$$

$$[(1 + \xi F)^2 - G\xi^2] G' = \frac{4}{\beta} GF(1 + \xi F). \tag{21}$$

In view of the boundary conditions (14), equations (20)–(21) are singular at $\xi = -1$. To analyze the present problem and obtain its appropriate solution, we first reduce system (20)–(21) to a phase plane and hence study its singularities. We write

$$Y = -\xi F, \qquad Z = \xi^2 G \tag{22}$$

so that (20)–(21) become

$$\xi \frac{dY}{d\xi} = \frac{Y[(1-Y)^2 - 3Z]}{(1-Y)^2 - Z}, \tag{23}$$

$$\xi \frac{dZ}{d\xi} = \frac{2Z[-\beta Z + (1-Y)(\beta(1-Y) - 2Y)]}{\beta[(1-Y)^2 - Z]}. \tag{24}$$

The single equation in the (Y, Z) plane follows immediately:

$$\frac{dY}{dZ} = \frac{\beta Y}{2Z} \frac{(1-Y)^2 - 3Z}{-\beta Z + (1-Y)(\beta(1-Y) - 2Y)}. \tag{25}$$

The cavity conditions (14), in the light of (22), require that the solution of (25) must pass through the point C with coordinates $Y = 1$, $Z = 0$. The boundary conditions (19) demand that the solution of (25) starting from C must join with the point O with coordinates $Y = 0$, $Z = 0$ (see (22)). Moreover, since the density behind the cavity is always positive, we restrict ourselves to the half-plane $Z > 0$.

Equation (25) has the following singular points:

$O: Y = 0, Z = 0; \quad B: Y = 0, Z = 1; \quad C: Y = 1, Z = 0;$

$D: Y = \dfrac{\beta}{3+\beta}, \; Z = \dfrac{3}{(3+\beta)^2}.$

Points O, B, and D are easily checked to be nodes. Point C is not elementary, but a multiple singular point. Its topological structure may be investigated by following the general theory developed by Andronov et al. (1973) and summarized in Section 6.6. We write $Y = 1 + y$ and approximate (25) as

$$\frac{dy}{dZ} = \frac{\beta[y^2 - 3Z(1+y)]}{2Z[-\beta Z + 2y]}. \tag{26a}$$

Equation (26a) may be rewritten as

$$\frac{dy}{dz_1} = \frac{Z_1(1+y) + \beta y^2}{2Z_1(2y + Z_1/3)}, \quad Z_1 = -3\beta Z. \tag{26b}$$

In the notation of Andronov et al., we have $a = b = d = 0$, $c = 1$, $\sigma = a + d = 0$, $\Delta = 0$. The characteristic equation is $\lambda^2 = 0$, with $0,0$ as its roots. Referring to Section 6.6, we also have

$$P_2(y, Z_1) = y[\beta y + Z_1], \tag{27}$$

$$Q_2(y, Z_1) = 2Z_1[2y + Z_1/3]. \tag{28}$$

Phase Space Study of Autonomous Systems

Assuming that $Z_1 = \phi_1(y)$ is the solution of the equation $P_2(y, Z_1) + Z_1 = 0$, we easily find that

$$\phi_1(y) = \beta[-y^2 + y^3 + \cdots]. \tag{29}$$

It therefore follows that

$$\psi_1(y) = Q_2(y, \phi_1(y)) = -4\beta[y^3 + \cdots] \tag{30}$$

and

$$\begin{aligned}\delta_1(y) &= P_{2y}(y, \phi_1(y)) + Q_{2y}(y, \phi_1(y)) \\ &= 2\beta y + \cdots.\end{aligned} \tag{31}$$

Now, if we denote the first nonvanishing coefficient in the series expansion of $\psi_1(y)$ and $\delta_1(y)$ about zero, respectively, by a_k and b_n, then $k = 3$, $n = 1$, so in the notation of Section 6.6, $k = 2m + 1 = 3$; thus $m = 1$, $a_3 = -4\beta < 0$, $n = 1$, $b_1 = 2\beta > 0$, and $\lambda_1 = b_1^2 + 8a_3 = 4\beta(\beta - 8)$. Since k is odd, $m = n$, $\lambda = 0$, and n is odd, Theorem 6.6.2 is applicable and the singular point $y = 0$, $Z_1 = 0$ of (26b) is an equilibrium state with an elliptic region if $\gamma \leq 5/4$ and center if $\gamma > 5/4$. The integral curves in the neighborhood of point C are as shown in Figure 6.11.1.

It is possible to obtain an approximate analytic representation of the solution of (26a). First we note that the integral curves near the point C ($y = 0, Z = 0$) lie below the parabola $y^2 = 3Z$. In the region between this

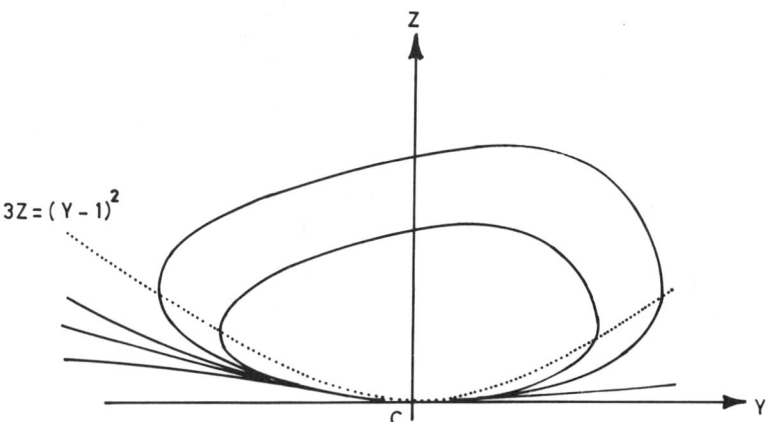

Figure 6.11.1 The integral curves in the (y, z) plane in the neighborhood of C. (From Hunter 1962.)

parabola and the y-axis, equation (26a) may be further approximated by

$$\frac{dy}{dZ} = \frac{\beta[y^2 - 3Z]}{4yZ}. \tag{32}$$

Equation (32), being linear in y^2, immediately integrates to give

$$y^2 = \frac{3\beta Z}{\beta - 2} + KZ^{\beta/2}, \quad \gamma \neq 2; \tag{33}$$

$$y^2 = -3Z \ln Z + KZ, \quad \gamma = 2, \tag{34}$$

where K is constant of integration. Thus, for $1 < \gamma < 2$, $\beta > 2$, and the integral curves (33) consist of the parabolas $y^2 = 3\beta Z/(\beta - 2)$, corresponding to the constant $K = 0$, and a family of curves which asymptote to this parabola as point C is approached. For $\gamma > 2$, the approximate form of the

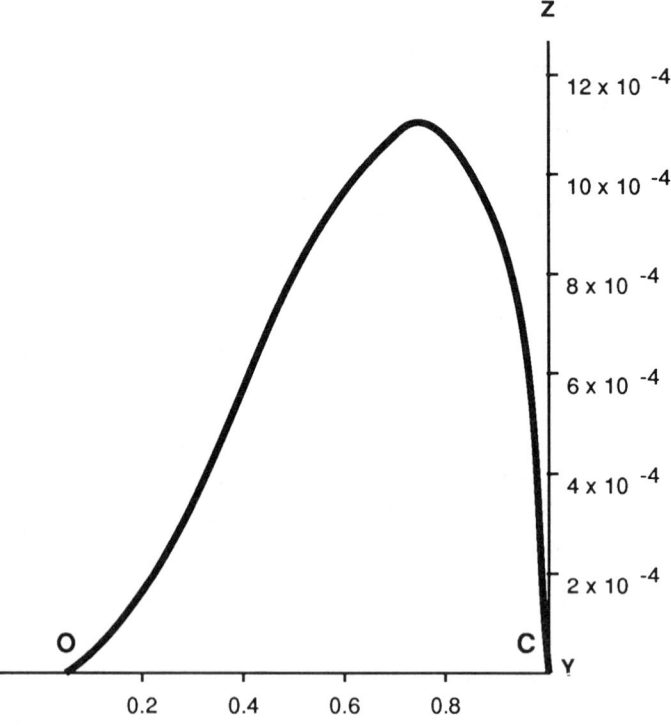

Figure 6.11.2 Solution of the system (23)–(24), joining the singular points C and 0, for $\gamma = 7/5$.

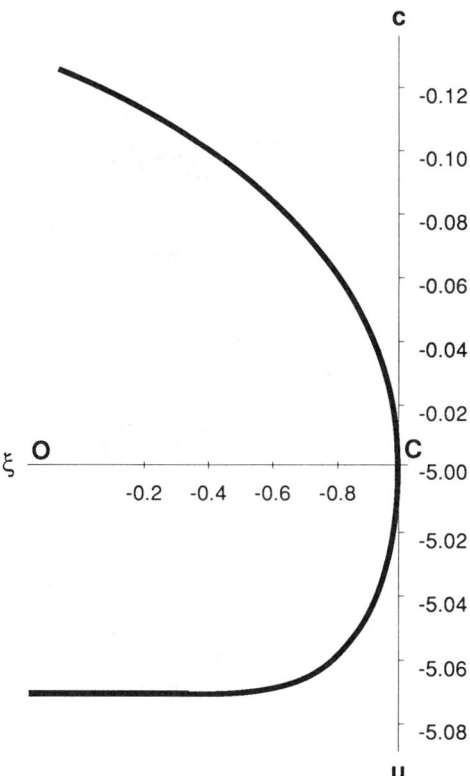

Figure 6.11.3 Solution u and c versus ξ, corresponding to that shown in Figure 6.11.2, for $\gamma = 7/5$.

solution curve near C is $Z = K^{-2/\beta} y^{4/\beta}$, so all integral curves are asymptotic to the curve $y^2 = -3Z \ln Z$.

By balancing of the dominant terms and simple Painlevé analysis, it is also possible to write the solution of the full equation (25) in the neighborhood of point C. Writing the solution in the form

$$Z = \frac{\beta - 2}{3\beta} y^2 + a_3 y^3 + a_4 y^4 + \cdots, \tag{35}$$

we find that

$$a_3 = \frac{(2-\alpha)(4+\alpha)}{9(3\alpha - 4)}, \tag{36}$$

where $\alpha = 4/\beta = 2(\gamma - 1)$. The general coefficient a_k is obtained from

$$(-4 + k\alpha)a_k = F(a_1, \ldots, a_{k-1}). \tag{37}$$

The coefficient of a_k in (37) vanishes when $k = 3, 4, 5, \ldots$, corresponding to $\gamma = 5/3, 3/2, 7/5, \ldots$, respectively. The solution is therefore not regular for these values of γ, and a logarithmic term has to be introduced (see Section 4.2). Thus, for $\gamma = 5/3$, the solution may be found to be

$$Z = \frac{1}{9}y^2 + a_3 y^3 + \frac{2}{5}\left(\frac{4}{27} - 3a_3\right) y^3 \ln y + \cdots, \tag{38}$$

where a_3 is arbitrary.

The solution of the system (23)–(24), and hence u and c, for $\gamma = 7/5$, are depicted in Figures 6.11.2 and 6.11.3. A more detailed analysis and numerical results for this problem may be found in Sachdev, Gupta, and Ahluwalia (1990).

6.12 NONLINEAR TRAVELING WAVES IN AN ISOTHERMAL ATMOSPHERE

Here we give an example of exact periodic solutions for a system of (inhomogeneous) nonlinear PDE governing the propagation of internal gravity waves in an isothermal atmosphere. If we denote by u and w the fluid velocity components in the horizontal and vertical directions, respectively, and by p and ρ the pressure and density, the PDE governing the propagation of internal gravity waves in an inviscid compressible medium are

$$u_t + uu_x + wu_z + \frac{1}{\rho}p_x = 0, \tag{1}$$

$$w_t + uw_x + ww_z + \frac{1}{\rho}p_z = -g, \tag{2}$$

$$p_t + up_x + wp_z + \gamma p u_x + \gamma p w_z = 0, \tag{3}$$

$$\rho_t + u\rho_x + w\rho_z + \rho u_x + \rho w_z = 0, \tag{4}$$

where γ is the ratio of specific heats and g is the acceleration due to gravity.

We introduce the nondimensional variables

$$X = \frac{x}{H}, \quad Z = \frac{z}{H}, \quad T = \left(\frac{g}{H}\right)^{1/2} t,$$

$$U = \frac{u}{(gH)^{1/2}}, \quad W = \frac{w}{(gH)^{1/2}}, \quad P = \frac{p}{p_0 \exp(-z/H)}, \tag{5}$$

$$R = \frac{\rho}{\rho_0 \exp(-z/H)}, \quad \frac{\gamma p_0}{\rho_0} = C_0^2 = \gamma g H,$$

Phase Space Study of Autonomous Systems

where H is the vertical scale height of the isothermal atmosphere, and 0 refers to constant reference conditions. The exponential functions scale out the undisturbed stratification. We seek for the transformed nondimensionalized system traveling wave solutions in the form

$$U = U(\phi), \quad W = W(\phi), \quad P = P(\phi),$$
$$R = R(\phi), \quad \phi = \phi(X, Z, T). \tag{6}$$

A simple argument (see Seshadri and Sachdev 1977 and Seshadri 1977) shows that the phase function ϕ in the present case is linear in X, Z, and T:

$$\phi = \frac{X}{\lambda_1} + \frac{Z}{\lambda_2} - T, \tag{7}$$

where λ_1, λ_2 are arbitrary constants. Substituting (6) into the normalized PDE and solving for the derivatives with respect to ϕ, we have

$$U_\phi = -\frac{E}{D} \frac{P}{\lambda_1 R} \left[\frac{\gamma(1 - P/R)}{\lambda_2} + WE \right], \tag{8}$$

$$W_\phi = \frac{E}{D} \left[\left(1 - \frac{P}{R}\right) \left(\frac{\gamma P}{\lambda_1^2 R} - E^2\right) - \frac{WEP}{\lambda_2 R} \right], \tag{9}$$

$$(\ln P)_\phi = \frac{E^2}{D} \left[WE + \frac{\gamma}{\lambda_2} \left(1 - \frac{P}{R}\right) \right], \tag{10}$$

$$(\ln R)_\phi = \frac{E}{D} \left[WE^2 + \frac{E}{\lambda_2} \left(1 - \frac{P}{R}\right) - (\gamma - 1)nW \frac{P}{R} \right], \tag{11}$$

where

$$E = -1 + \frac{U}{\lambda_1} + \frac{W}{\lambda_2}, \tag{12}$$

$$D = E^2 \left(E^2 - \gamma n \frac{P}{R} \right), \tag{13}$$

$$n = \frac{1}{\lambda_1^2} + \frac{1}{\lambda_2^2}. \tag{14}$$

The form (8)–(11) suggests the introduction of the variable $K = P/R$, the "sound speed" square. We combine (10) and (11) suitably so that

$$(\ln K)_\phi = \frac{(\gamma - 1)E}{D} \left[\frac{E(1 - K)}{\lambda_2} + nKW \right]. \tag{15}$$

When (15) is suitably manipulated with (8) and (9), we obtain an intermediate integral

$$EK^{1/(\gamma-1)} = \text{constant} = -B, \quad B > 0. \tag{16}$$

Using (15) and (16), we can eliminate U and write P and R in terms of K. Thus, we reduce the discussion of system (8)–(11) to the (W, K) phase plane:

$$\frac{dW}{dK} = \left[\frac{BWK^{1-1/(\gamma-1)}}{\lambda_2} + (1-K)\left(\frac{\gamma K}{\lambda_1^2} - B^2 K^{-2/(\gamma-1)} \right) \right]$$

$$\times \left\{ (\gamma - 1)K \left[nWK + \frac{K-1}{\lambda_2} BK^{-1/(\gamma-1)} \right] \right\}^{-1}. \quad (17)$$

After equation (17) has been analyzed, the other flow quantities can be obtained in terms of the phase function ϕ by simple quadrature, using (8)–(16).

The singular points of (17) are

$$\alpha: W = 0, \qquad K = 1, \quad (18)$$

$$\beta: K = K_\beta \equiv \left[\frac{\beta^2}{\gamma n} \right]^{(\gamma-1)/(\gamma+1)}, \quad (19)$$

$$W = W_\beta \equiv (1 - K_\beta)(BK_\beta^{-1/(\gamma-1)})(n\lambda_2 K_\beta)^{-1},$$

where n is given by (14). Points α and β coincide when $K_\beta = 1$.

The parameter $B^2/\gamma n$ plays a crucial role in the discussion of the solutions of (17). A thorough phase plane analysis has been carried out by Seshadri and Sachdev (1977) to show that there are no solutions with shock discontinuities. This conclusion could perhaps have been arrived at by noting the dispersive nature of the system (1)–(4). Here we content ourselves with a study of the periodic solutions of (17).

We note that the singular point α is a saddle. The point β is a center only if $\lambda_2 = \infty$. In this case, equation (17) reduces to

$$\frac{dW}{dK} = \frac{\gamma}{\gamma - 1} \frac{(1-K)[1 - (B^2 \lambda_1^2/\gamma) K^{-(\gamma+1)/(\gamma-1)}]}{WK}. \quad (20)$$

The singular points now are α as in (18) and

$$\beta: W = 0, \qquad K = \left(\frac{B^2 \lambda_1^2}{\gamma} \right)^{(\gamma-1)/(\gamma+1)} \quad (21)$$

Indeed, (20) can be integrated exactly in this case to give

$$W^2 = (2\gamma/(\gamma - 1))(\ln K - K)$$

$$+ 2B^2 \lambda_1^2 K^{-2/(\gamma-1)} \left(\frac{1}{(\gamma+1)K} - \frac{1}{2} \right) + C, \quad (22)$$

Phase Space Study of Autonomous Systems

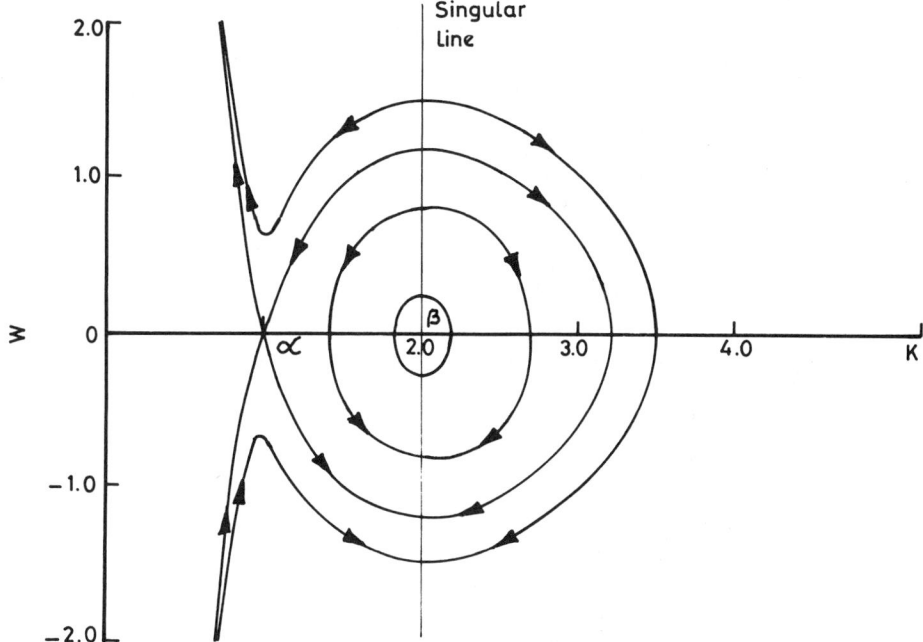

Figure 6.12.1 Phase portrait of equation (20): $B_1 > 1$.

where C is an arbitrary constant.

Three cases arise, depending on the magnitude of $B^2 \lambda_1^2 / \gamma \equiv B_1$, say. If $B_1 > 1$, the singular point α is a saddle while β is a center. There are closed orbits around point β; the solution for a particular choice of parameters is shown in Figure 6.12.1. These closed trajectories change sign as they cross the singular line $K = K_\beta$. Therefore, the solutions are unphysical. If $B_1 = 1$, the singular points α and β coincide. The nature of the singularity for a particular choice of parameters is depicted in Figure 6.12.2; this singularity is not elementary. In the third case $B_1 < 1$, the singular point α is a center while β is a saddle point. We now have periodic solutions with correct orientation. A specific case belonging to this class is shown in Figure 6.12.3. These solutions are physically meaningful and represent periodic waves propagating in the horizontal direction with arbitrary speed λ_1. These waves, in the atmospheric context, are referred to as *surface* or *evanescent waves*.

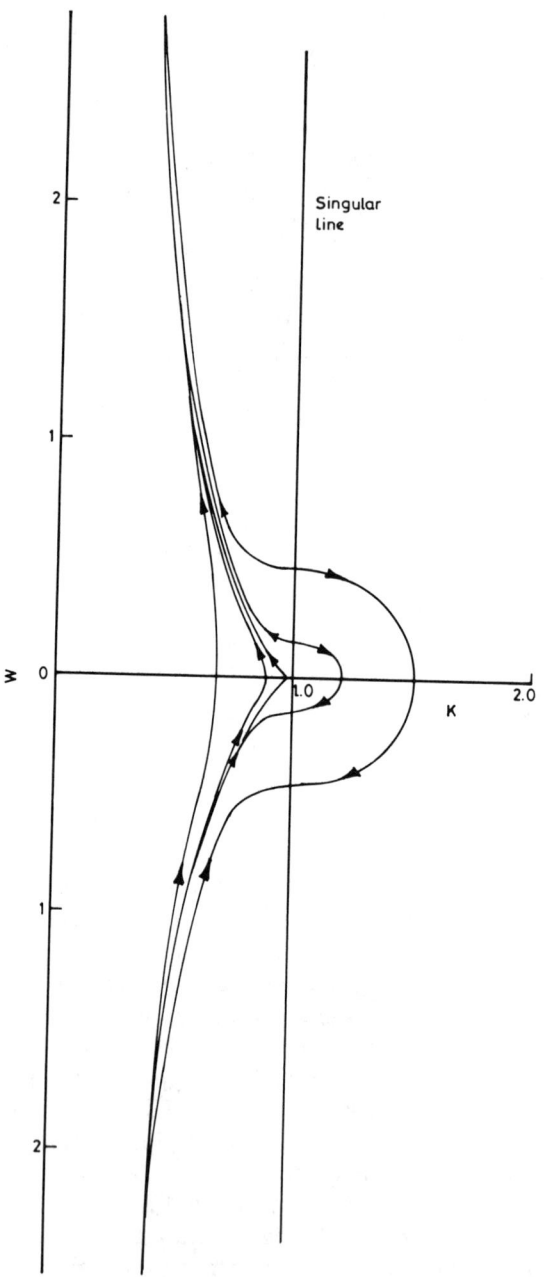

Figure 6.12.2 Phase portrait of equation (20): $B_1 = 1$.

Phase Space Study of Autonomous Systems

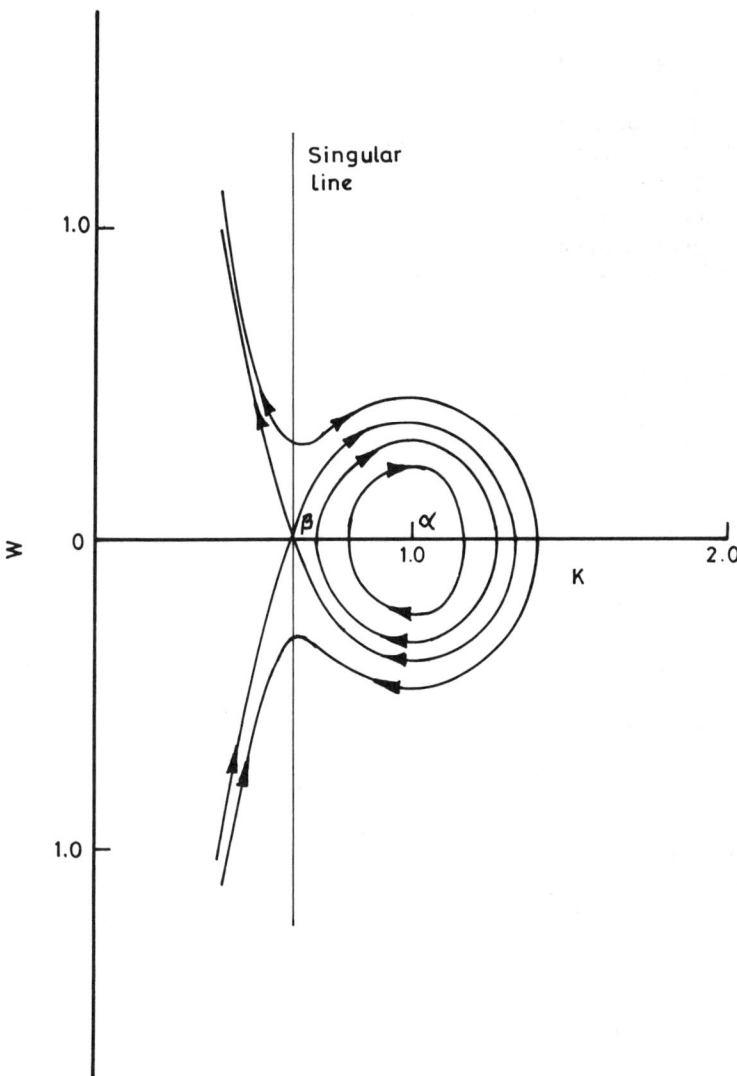

Figure 6.12.3 Phase portrait of equation (20): $B_1 < 1$.

6.13 SINGULAR POINTS OF A SYSTEM OF THREE LINEAR DIFFERENTIAL EQUATIONS

Often it is not possible to reduce the discussion of systems of ODE to that in a plane. We have discussed some cases of the reduction either by some transformations as in the problem of cavity collapse or by the discovery of an intermediate integral as for traveling waves in an isothermal atmosphere. There is, however, a need to have a classification for a system of three equations, which occurs frequently in applications. This would naturally be much more varied and complex, since the integral curves now have greater freedom to turn and twist in a three-dimensional space. Reyn (1964) has given a detailed classification for a system of three linear differential equations. We shall discuss in detail the case when the matrix of the system has three real and distinct eigenvalues, and summarize the results for other cases due to space limitations. We observe again that if the linear system is an approximation of a nonlinear system, the qualitative nature of the solution of the nonlinear system particularly away from the singular point need not be the same as for the linear system.

We study the system

$$\frac{dx}{dt} = a_1 x + b_1 y + c_1 z,$$
$$\frac{dy}{dt} = a_2 x + b_2 y + c_2 z, \qquad (1)$$
$$\frac{dz}{dt} = a_3 x + b_3 y + c_3 z,$$

where $x = x(t)$, $y = y(t)$, and $z = z(t)$ are functions of t, and a_i, b_i, and c_i ($i = 1, 2, 3$) are real constants.

If we denote the coefficient matrix of system (1) by

$$A = \begin{bmatrix} a_1 & b_1 & c_1 \\ a_2 & b_2 & c_2 \\ a_3 & b_3 & c_3 \end{bmatrix}, \qquad (2)$$

then (1) may be written as

$$\dot{X} = AX, \qquad (3)$$

where

$$X = \begin{bmatrix} x \\ y \\ z \end{bmatrix} \qquad (4)$$

is a 3×1 matrix and the dot denotes differentiation with respect to t. The eigenvalues of A determined by

$$|A - \lambda I| = 0 \tag{5}$$

may be denoted by λ_1, λ_2, and λ_3. Equation (5) has the explicit form

$$\lambda^3 - S_1\lambda^2 + S_2\lambda - S_3 = 0, \tag{6}$$

where $S_i (i = 1, 2, 3)$ denotes the sum of all principal minors of A of order i:

$$S_1 = a_1 + b_2 + c_3 = \lambda_1 + \lambda_2 + \lambda_3,$$

$$S_2 = \begin{vmatrix} a_1 & b_1 \\ a_2 & b_2 \end{vmatrix} + \begin{vmatrix} b_2 & c_2 \\ b_3 & c_3 \end{vmatrix} + \begin{vmatrix} a_1 & c_1 \\ a_3 & c_3 \end{vmatrix} = \lambda_1\lambda_2 + \lambda_2\lambda_3 + \lambda_3\lambda_1,$$

$$S_3 = \begin{vmatrix} a_1 & b_1 & c_1 \\ a_2 & b_2 & c_2 \\ a_3 & b_3 & c_3 \end{vmatrix} = \lambda_1\lambda_2\lambda_3. \tag{7}$$

The nature of the solution of (3) depends on the nature of the eigenvalues. Indeed, the latter dictates how system (3) may be changed to one of the following canonical forms, which, in turn, facilitate the solution and its discussion (cf. Section 6.3). If the matrix of transformation is T so that $U = TX$, where $U = (u, v, w)$ is a 3×1 matrix, system (3) can be changed to

$$\dot{U} = (TAT^{-1})U \tag{8}$$

with real coefficient matrix $\mathcal{J} = TAT^{-1}$. The latter may have one of the following canonical forms.

1. If the three eigenvalues are real and distinct, then

$$\mathcal{J}_r = \begin{bmatrix} \lambda_1 & 0 & 0 \\ 0 & \lambda_2 & 0 \\ 0 & 0 & \lambda_3 \end{bmatrix}.$$

2. If the three eigenvalues are real and two of them coincide, then

$$\mathcal{J}_{23} = \begin{bmatrix} \lambda_1 & 0 & 0 \\ 0 & \lambda_2 & 0 \\ 0 & 0 & \lambda_2 \end{bmatrix} \quad \text{or} \quad \mathcal{J}_{22} = \begin{bmatrix} \lambda_1 & 0 & 0 \\ 0 & \lambda_2 & 1 \\ 0 & 0 & \lambda_2 \end{bmatrix}.$$

3. If the three eigenvalues are real and coincident, then

$$\mathcal{J}_{33} = \begin{bmatrix} \lambda & 0 & 0 \\ 0 & \lambda & 0 \\ 0 & 0 & \lambda \end{bmatrix} \quad \text{or} \quad \mathcal{J}_{32} = \begin{bmatrix} \lambda & 0 & 0 \\ 0 & \lambda & 1 \\ 0 & 0 & \lambda \end{bmatrix}.$$

or $\quad J_{31} = \begin{bmatrix} \lambda & 1 & 0 \\ 0 & \lambda & 1 \\ 0 & 0 & \lambda \end{bmatrix}.$

4. If two of the three eigenvalues are complex, then

$$J_c = \begin{bmatrix} \lambda_1 & 0 & 0 \\ 0 & \text{Re}\,\lambda_2 & \text{Im}\,\lambda_2 \\ 0 & -\text{Im}\,\lambda_2 & \text{Re}\,\lambda_2 \end{bmatrix}.$$

The various cases that arise can be classified conveniently in (S_1, S_2, S_3) space (see Figure 6.13.1). If we write

$$D = \frac{1}{27}\left(S_2 - \frac{1}{3}S_1^2\right)^3 + \frac{1}{4}\left(-\frac{2}{27}S_1^3 + \frac{1}{3}S_1 S_2 - S_3\right)^2, \qquad (9)$$

then three cases can be distinguished.

i. $D < 0$; three eigenvalues are distinct.
ii. $D = 0$; at least two eigenvalues coincide.
iii. $D > 0$; one eigenvalue is real, and the other two are complex conjugate.

On the surface $D = 0$ there exists a special curve

$$S_2 = \frac{1}{3}S_1^2, \quad S_3 = \frac{1}{27}S_1^3 \qquad (10)$$

on which all the three eigenvalues coincide.

The following cases and subcases arise and are shown in Figure 6.13.1. A dash between two cases, for example, I–II, will denote a curve or a surface which represents a transition between two distinct regions.

1. If the three eigenvalues are real and distinct ($D < 0$), then

    I. $\lambda_1 > 0, \lambda_2 > 0, \lambda_3 > 0; S_1 > 0, S_2 > 0, S_3 > 0.$
    II. $\lambda_1 < 0, \lambda_2 > 0, \lambda_3 > 0;$ either $S_1 > 0, S_3 < 0$ or $S_1 < 0, S_2 < 0, S_3 < 0.$
    III. $\lambda_1 > 0, \lambda_2 < 0, \lambda_3 > 0;$ either $S_1 > 0, S_2 < 0, S_3 > 0$ or $S_1 < 0, S_3 > 0.$
    IV. $\lambda_1 < 0, \lambda_2 < 0, \lambda_3 < 0; S_1 < 0, S_2 < 0, S_3 < 0.$
    I–II. $\lambda_1 > 0, \lambda_2 > 0, \lambda_3 = 0; S_1 > 0, S_2 > 0, S_3 = 0.$ (11)
    II–III. $\lambda_1 < 0, \lambda_2 > 0, \lambda_3 = 0; S_2 < 0, S_3 = 0.$
    III–IV. $\lambda_1 < 0, \lambda_2 < 0, \lambda_3 = 0; S_1 < 0, S_2 > 0, S_3 = 0.$

2. If the three eigenvalues are real and two of them coincide ($D = 0$), then

    I–V. $\lambda_1 > 0, \lambda_2 > 0; S_1 > 0, S_2 > 0, S_3 > 0.$

Phase Space Study of Autonomous Systems

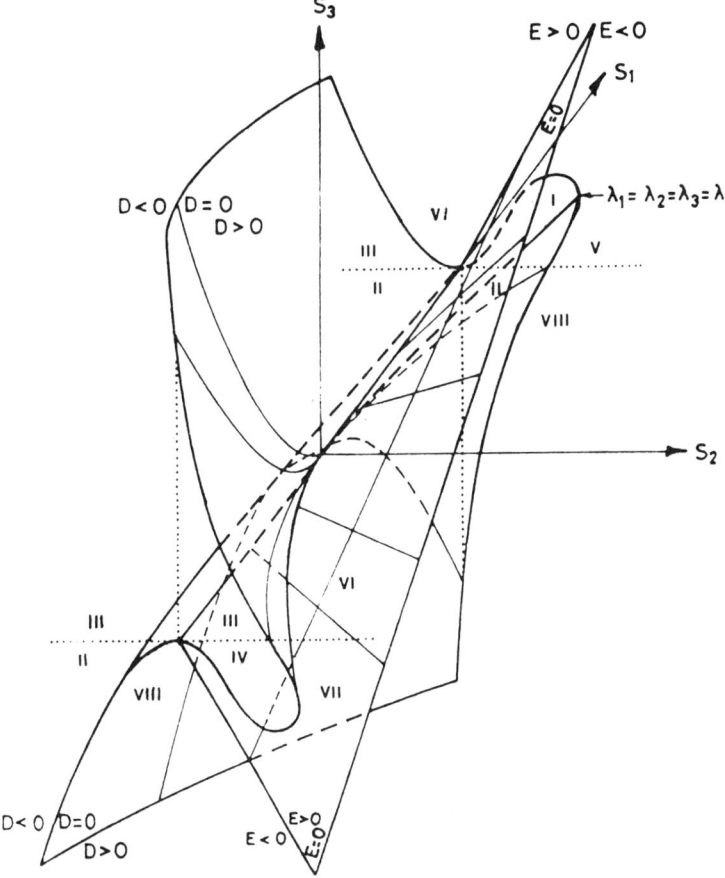

Figure 6.13.1 Classification of the eigenvalues of matrix (2) (see equations (7), (9), (10), and (14)). (From Reyn 1964.)

$$
\begin{aligned}
&\text{II–VIII.} && \lambda_1 < 0,\ \lambda_2 > 0;\ \text{either } S_1 > 0,\ S_3 < 0 \text{ or } S_1 < 0,\ S_2 < 0,\\
&&& S_3 < 0.\\
&\text{III–VI.} && \lambda_1 > 0,\ \lambda_2 < 0;\ \text{either } S_1 < 0,\ S_2 < 0,\ S_3 > 0 \text{ or } S_1 < 0,\\
&&& S_3 > 0.\\
&\text{IV–VII.} && \lambda_1 < 0,\ \lambda_2 < 0;\ S_1 < 0,\ S_2 < 0,\ S_3 < 0.\\
&S_1\text{-axis} && \lambda_1 > 0,\ \lambda_2 = 0;\ S_1 > 0,\ S_2 = 0,\ S_3 = 0.\\
&S_1\text{-axis} && \lambda_1 < 0,\ \lambda_2 = 0;\ S_1 < 0,\ S_2 = 0,\ S_3 = 0.\\
&S_2 = \tfrac{1}{4}S_1^2\text{:} && \lambda_1 = 0,\ \lambda_2 > 0;\ S_1 > 0,\ S_2 < 0,\ S_3 = 0.
\end{aligned}
\qquad (12)
$$

$S_2 = \frac{1}{4}S_1^2$: $\lambda_1 = 0$, $\lambda_2 < 0$; $S_1 < 0$, $S_2 > 0$, $S_3 = 0$.

3. If the three eigenvalues are real and coincident ($D = 0$), then

$$S_2 = \frac{1}{3}S_1^2, \quad S_3 = \frac{1}{27}S_1^3: \quad \lambda > 0; \ S_1 > 0.$$

$$S_2 = \frac{1}{3}S_1^2, \quad S_3 = \frac{1}{27}S_1^3: \quad \lambda = 0; \ S_1 = 0, \ S_2 = 0, \ S_3 = 0. \quad (13)$$

$$S_2 = \frac{1}{3}S_1^2, \quad S_3 = \frac{1}{27}S_1^3: \quad \lambda < 0; \ S_1 < 0.$$

4. If two of the three eigenvalues are complex ($D > 0$), then there exists a surface $E = 0$, where

$$E = S_3 - S_1 S_2, \tag{14}$$

on which the complex eigenvalues are pure imaginary, separating a region ($E < 0$) with complex eigenvalues having a positive real part from a region ($E > 0$) with complex eigenvalues having a negative real part. In this situation, the following cases arise:

V.	$\lambda_1 > 0$, Re $\lambda_2 > 0$; $E < 0$, $S_1 > 0$, $S_2 > 0$, $S_3 > 0$.
VI.	$\lambda_1 > 0$, Re $\lambda_2 < 0$; $E > 0$, $S_3 > 0$.
VII.	$\lambda_1 < 0$, Re $\lambda_2 < 0$; $E > 0$, $S_1 < 0$, $S_2 > 0$, $S_3 < 0$.
VIII.	$\lambda_1 < 0$, Re $\lambda_2 > 0$; $E < 0$, $S_3 < 0$.
V–VIII.	$\lambda_1 = 0$, Re $\lambda_2 > 0$; $E < 0$, $S_1 > 0$, $S_2 > 0$, $S_3 = 0$.
VI–VII.	$\lambda_1 = 0$, Re $\lambda_2 < 0$; $E > 0$, $S_1 < 0$, $S_2 > 0$, $S_3 = 0$.
V–VI.	$\lambda_1 > 0$, Re $\lambda_2 = 0$; $E = 0$, $S_1 > 0$, $S_2 > 0$, $S_3 > 0$.
VII–VIII.	$\lambda_1 < 0$, Re $\lambda_2 = 0$; $E = 0$, $S_1 < 0$, $S_2 > 0$, $S_3 < 0$.
S_2-axis:	$\lambda_1 = 0$, Re $\lambda_2 = 0$; $E = 0$, $S_1 = 0$, $S_2 > 0$, $S_3 = 0$.

We consider the case of three real and distinct eigenvalues in some detail. The normal form is $\mathcal{J}_r$ so system (1) can be written as

$$\frac{du}{dt} = \lambda_1 u, \quad \frac{dv}{dt} = \lambda_2 v, \quad \frac{dw}{dt} = \lambda_3 w \tag{15}$$

with the solution

$$u = c_1 e^{\lambda_1 t}, \quad v = c_2 e^{\lambda_2 t}, \quad w = c_3 e^{\lambda_3 t}, \tag{16}$$

where c_i ($i = 1, 2, 3$) are real constants. The following cases arise.

Case I. Unstable Three-Branched Node. Here all the roots λ_i are positive. The transformation T can be so arranged that $0 < \lambda_1 < \lambda_2 < \lambda_3$. The behavior of the integral curves in the neighborhood of the origin is shown in Figure 6.13.2a. The constants c_1, c_2, and c_3 can be so chosen that through

Phase Space Study of Autonomous Systems

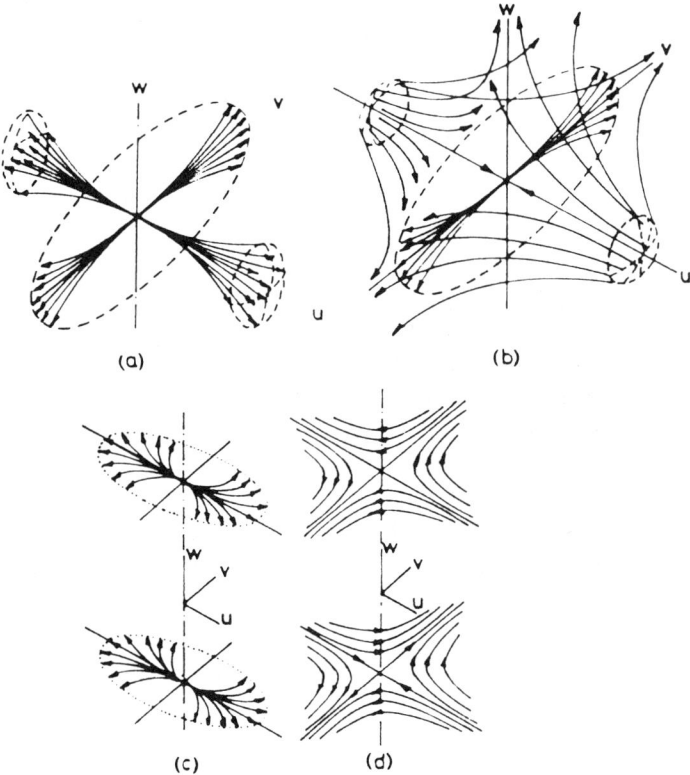

Figure 6.13.2 Phase portrait of system (1): the case of three real and distinct eigenvalues ($D < 0$). (From Reyn 1964.)

every point of (u,v,w) space, one integral curve passes, which tends to the origin as $t \to -\infty$. The orientation in the figures shows the curve in the direction of increasing t. By writing the solution in terms of u, v, and w alone from (16) by elimination of t, one can verify that all curves except those in the (v,w) plane are tangent to the u-axis at the origin; those in the (v,w) plane except the w-axis itself are tangent to the v-axis at the origin. This follows from the ordering of λ_i. In the (v,w) plane, therefore, we have an unstable two-branched plane node. In general, the projection of an integral curve on a coordinate plane coincides with an integral curve in that pane. It is easy to conclude from Figure 6.13.2a that the integral curves in the (u,v) plane and (u,w) plane also form an unstable two-branched plane node. In general, the integral curves are three-dimensional, the exceptions being those in the coordinate planes which are planar.

Case II. One of the characteristic roots is negative while the other two are positive: $\lambda_1 < 0 < \lambda_2 < \lambda_3$. The integral curves are shown in Figure 6.13.2b. The curves in the (v,w) plane are

$$\left(\frac{v}{c_2}\right)^{\lambda_3} = \left(\frac{w}{c_3}\right)^{\lambda_2},$$

and since $\lambda_2 < \lambda_3$ the curves in the (v,w) plane are tangent to the v-axis at the origin. The only exception is the w-axis, which is also an integral curve. We thus have planar curves forming an unstable two-branched node in the (v,w) plane. As shown in the figure, there are just two other curves which pass through the origin: these are the positive and negative u-axes, which approach the origin as $t \to \infty$. It is also seen that all other curves are tangent to the u-axis for $t \to -\infty$, so $u \to +\infty$ or $u \to -\infty$. They approach the (v,w) plane such that $v^2 + w^2 \to \infty$ as $t \to \infty$. The integral curves in the (u,v) and (u,w) planes form a plane saddle point pattern about the singular point. The curves in the coordinate planes are the only plane curves of the system.

Case III. The integral curves with $\lambda_3 < \lambda_2 < 0 < \lambda_1$ have the same behavior as those for case II (Figure 6.13.2b); only the direction of the arrows has to be reversed.

Case IV. With all negative characteristic roots $\lambda_3 < \lambda_2 < \lambda_1 < 0$, the behaviour is the same as for case I, except that the orientation of the curves has to be reversed. We have a stable three-branched node at the origin.

Case I–II. This case concerns the situation when one root, say λ_3, is zero and two other roots are positive, $0 < \lambda_1 < \lambda_2$. It follows from (15) that the w-axis ($u = 0$, $v = 0$) is a line of singular points and the integral curves lie on the plane w = const. The discussion reduces essentially to plane flows. The integral curves form an unstable two-branched plane node with $u \to 0$, $v \to 0$ as $t \to -\infty$. The phase portrait is shown in Figure 6.13.2c. We may refer to the w-axis as the line of unstable two-branched plane nodes.

Case II–III. This case refers again to the case when one root, say λ_3, is zero, while the other two have opposite signs $\lambda_1 < 0 < \lambda_2$. Again the w-axis is a line of singular points and the integral curves lie in planes w = const. as shown in Figure 6.13.2d. Since the signs of λ_1 and λ_2 are opposite, the singular points on the w-axis are saddles. The case $\lambda_2 < 0 < \lambda_1$ is similar;

the orientation of the curves in the configuration, however, is opposite to that in Figure 6.13.2d. We may refer to the w-axis as a line of saddle points.

Case III–IV. This case, with $\lambda_2 < \lambda_1 < 0$, $\lambda_3 = 0$, may be obtained from Figure 6.13.2c by simply reversing the direction of arrows.

For the other cases, we list salient features of the solutions and the configurations for ready reference. The projections of the space curves on the coordinate planes can easily be determined from the solutions and their tangency, etc., to various coordinates axes identified.

Case	Eigenvalues	Figure	Nature of the singular point	Remarks
2A.	Three eigenvalues are real, and two of them coincide ($D = 0$) (see (12)). Coefficient matrix of the normal form is $\mathcal{J}_{23}$, with the corresponding system $du/dt = \lambda_1 u$, $dv/dt = \lambda_2 v$, $dw/dt = \lambda_2 w$ and solutions $u = c_1 e^{\lambda_1 t}$, $v = c_2 e^{\lambda_2 t}$, $w = c_3 e^{\lambda_2 t}$, where c_i ($i = 1, 2, 3$) are real constants.			
(i) I–V	$\lambda_2/\lambda_1 > 1$	6.13.3Aa	Unstable pointed star node	Each plane through the u-axis is filled with integral curves forming an unstable two-branched plane node in that plane. All integral curves are planar.
(ii) I–V	$\lambda_2/\lambda_1 < 1$	6.13.3Ab	Unstable blunt star node	Each plane through the u-axis is filled with integral curves forming an unstable two-branched plane node. The u-axis is the only integral curve tangent to the u-axis at the origin. All integral curves are plane curves.
(iii) II–VIII		6.13.3Ac	Saddle star with unstable plane star	The positive u-axis, the negative u-axis, and the integral curves in the (v, w) plane pass through the origin. The latter form an unstable plane star there. All other curves are planar and lie in planes through the u-axis. They form a saddle structure there.
(iv) III–VI		6.13.3Ac	Saddle star with stable plane star	Reverse the arrows in 6.13.3Ac

Phase Space Study of Autonomous Systems

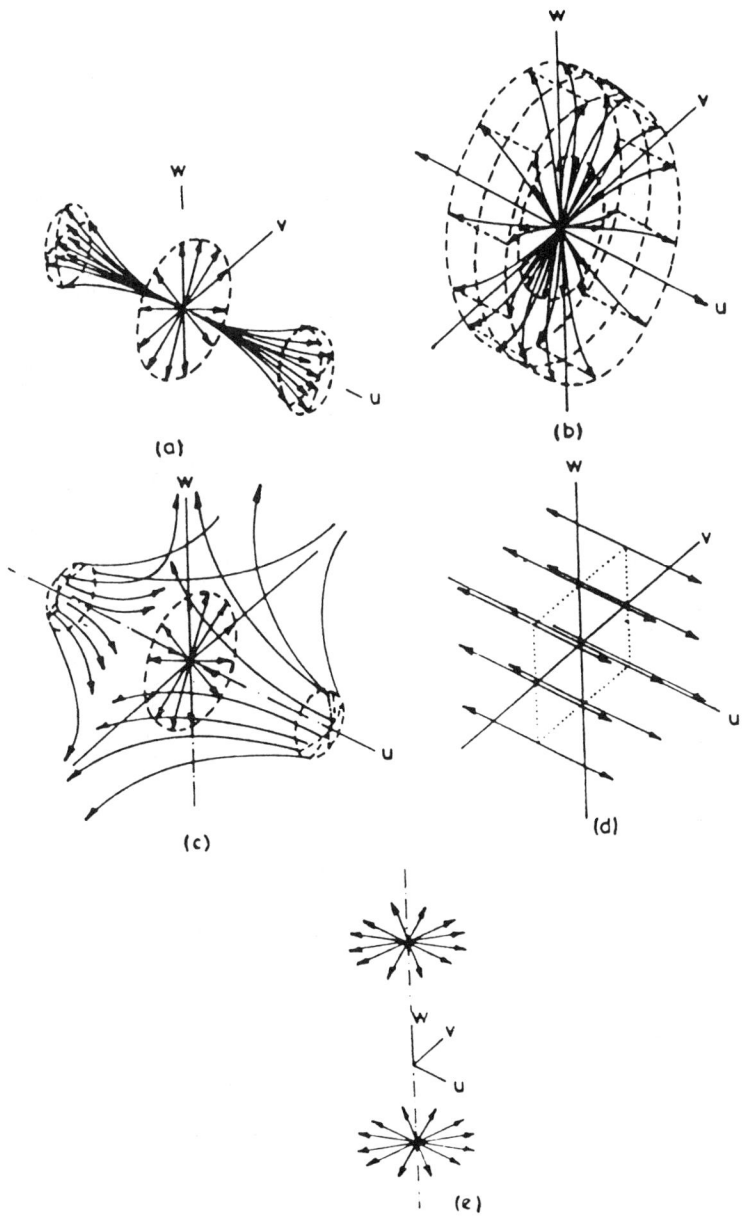

Figure 6.13.3A Phase portrait of system (1): the case of three real eigenvalues, two being coincident ($D = 0$), normal form $\mathcal{J}_{23}$. (From Reyn 1964.)

Case	Eigenvalues	Figure	Nature of the singular point	Remarks
(v) IV–VII	$\lambda_2/\lambda_1 > 1$	6.13.3Aa	Stable pointed star node	Reverse the arrows in 6.13.3Aa
(vi) IV–VII	$\lambda_2/\lambda_1 < 1$	6.13.3Ab	Stable blunt star node	Reverse the arrows in 6.13.3Ab
(vii) S_1-axis	$\lambda_1 > 0$	6.13.3Ad	Unstable normal plane	The whole (v,w) plane is singular, and all integral curves are lines parallel to the u-axis. $u \to 0$ as $t \to -\infty$.
(viii) S_1-axis	$\lambda_1 < 0$	6.13.3Ad	Stable normal plane	Reverse the arrows in 6.13.3Ad
(ix) $S_2 = \frac{1}{4}S_1^2$	$\lambda_2 > 0$	6.13.3Ae	Line of unstable plane stars	The whole u-axis is singular. In each plane $u = $ constant, there is an unstable plane star.
(x) $S_2 = \frac{1}{4}S_1^2$	$\lambda_2 < 0$	6.13.3Ae	Line of stable plane stars	Reverse the arrows in 6.13.3Ae.

2B. When the coefficient matrix is of the normal form $\mathcal{J}_{22}$, the corresponding system of ODE is

$$\frac{du}{dt} = \lambda_1 u, \quad \frac{dv}{dt} = \lambda_2 v + w, \quad \frac{dw}{dt} = \lambda_2 w$$

with the solution $u = c_1 e^{\lambda_1 t}$, $v = (c_2 + c_3 t)e^{\lambda_2 t}$, $w = c_3 e^{\lambda_2 t}$, where c_i ($i = 1, 2, 3$) are real constants.

Case	Eigenvalues	Figure	Nature of the singular point	Remarks
(i) I–V	$\lambda_2/\lambda_1 > 1$	6.13.3Ba	Unstable wide two-branched node	In the (v,w) plane all curves are tangent to v-axis and form an unstable one-branched plane node there. In the (u,w) plane the projections of the integral curves form an unstable two-branched plane node. In the (u,v) plane, the

Phase Space Study of Autonomous Systems 341

			integral curves form an unstable two-branched plane node. The only plane integral curves lie in the (u,v) and (v,w) planes.
(ii) I–V	$\lambda_2/\lambda_1 < 1$	6.13.3Bb	Unstable slender two-branched node	The integral curves form an unstable two-branched plane node in the (u,v) plane and an unstable one-branched plane node in the (v,w) plane. There are no integral curves in the (u,w) plane, except those coinciding with the u-axis. The only plane curves lie in the (u,v) and (v,w) planes.
(iii) II–VIII		6.13.3Bc	Two-branched saddle node with unstable one-branched plane node	The integral curves in the (v,w) plane are tangent to the v-axis at the origin and form an unstable one-branched plane node. The plane integral curves form a saddle point configuration in the (u,v) plane and do not coincide with the projections of the space curves on the (u,v) plane. The projections on (u,w) plane also form a saddle point.
(iv) III–VI		6.13.3Bc	Two-branched saddle node with stable one-branched plane node	Reverse the arrows in 6.13.3Bc and rotate this figure 180° around the w-axis.

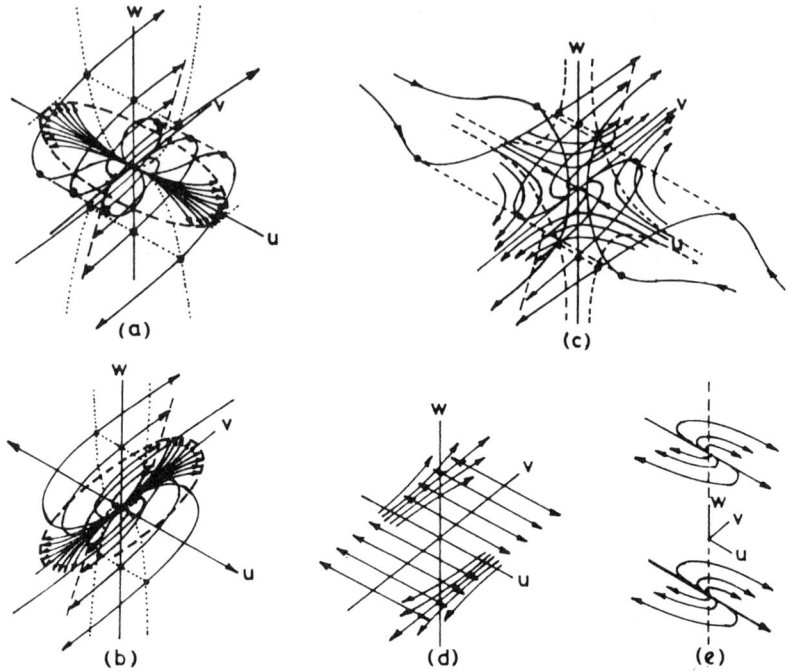

Figure 6.13.3B Same as in Figure 6.13.3A, normal form $\mathcal{J}_{22}$. (From Reyn 1964.)

Phase Space Study of Autonomous Systems

Case	Eigenvalues	Figure	Nature of the singular point	Remarks
(v) IV–VII	$\lambda_2/\lambda_1 > 1$	6.13.3Ba	Stable wide two-branched node	Reverse the arrows in 6.13.3Bc and rotate the figure 180° around the w-axis.
(vi) IV–VII	$\lambda_2/\lambda_1 < 1$	6.13.3Bb	Stable slender two-branched node	Reverse the arrows in 6.13.3Bb after rotating the figure 180° around the w-axis.
(vii) S_1-axis	$\lambda_1 > 0$	6.13.3Bd	Unstable normal line	All integral curves lie in $w = $ const. planes. v-axis is a singular line. In the (u,v) plane, integral curves are parallel to the u-axis such that $u \to 0$ as $t \to -\infty$. Integral curves in $w = $ const. planes diverge in the direction of increasing (decreasing) v for $w > (<) 0$.
(viii) S_1-axis	$\lambda_1 < 0$		Stable normal line	Reverse the arrows in 6.13.3Bd after rotating the figure about the w-axis by 180°.
(ix) $S_2 = \frac{1}{4}S_1^2$	$\lambda_2 > 0$	6.13.3Be	Line of unstable one-branched plane nodes	The u-axis is a line of singular points, and in each plane $u = $ const there is an unstable one-branched plane node.
(x) $S_2 = \frac{1}{4}S_1^2$	$\lambda_2 < 0$	6.13.3Be	Line of stable one-branched plane nodes	Reverse the arrows in 6.13.3Be after rotating the figure by 180° around the w-axis.

Case	Eigenvalues	Figure	Nature of the singular point	Remarks

3. When the three eigenvalues are real and coincident ($D = 0$).

3A. The coefficient matrix is of the normal form $\mathcal{J}_{33}$; the system of ODE has the form

$$\frac{du}{dt} = \lambda u, \quad \frac{dv}{dt} = \lambda v, \quad \frac{dw}{dt} = \lambda w$$

with solutions: $u = c_1 e^{\lambda t}, v = c_2 e^{\lambda t}, w = c_3 e^{\lambda t}$, where c_i ($i = 1, 2, 3$) are real constants.

Case	Eigenvalues	Figure	Nature of the singular point	Remarks
(i)	$\lambda > 0$	6.13.4Aa	Unstable three-dimensional star	Each ray through the origin is an integral curve, and the distance $r \to 0$ as $t \to -\infty$.
(ii)	$\lambda = 0$	6.13.4Ab	Space with equilibrium points	Each point in (u, v, w) space is an equilibrium point.
(iii)	$\lambda < 0$	6.13.4Aa	Stable three-dimensional star	Reverse the arrows in 6.13.4Aa.

3B. The coefficient matrix is of the normal form $\mathcal{J}_{32}$, and the corresponding ODE system is

$$\frac{du}{dt} = \lambda u, \quad \frac{dv}{dt} = \lambda v + w, \quad \frac{dw}{dt} = \lambda w$$

with solutions $u = c_1 e^{\lambda t}, v = (c_2 + c_3 t)e^{\lambda t}, w = c_3 e^{\lambda t}$, where c_i ($i = 1, 2, 3$) are real constants.

Case	Eigenvalues	Figure	Nature of the singular point	Remarks
(i)	$\lambda > 0$	6.13.4Ba	Unstable antisymmetric node star	In the (u, v) plane, integral curves form an unstable plane star. All curves are planar, the planes passing through the v-axis. They form an unstable one-branched node at the origin.

Phase Space Study of Autonomous Systems

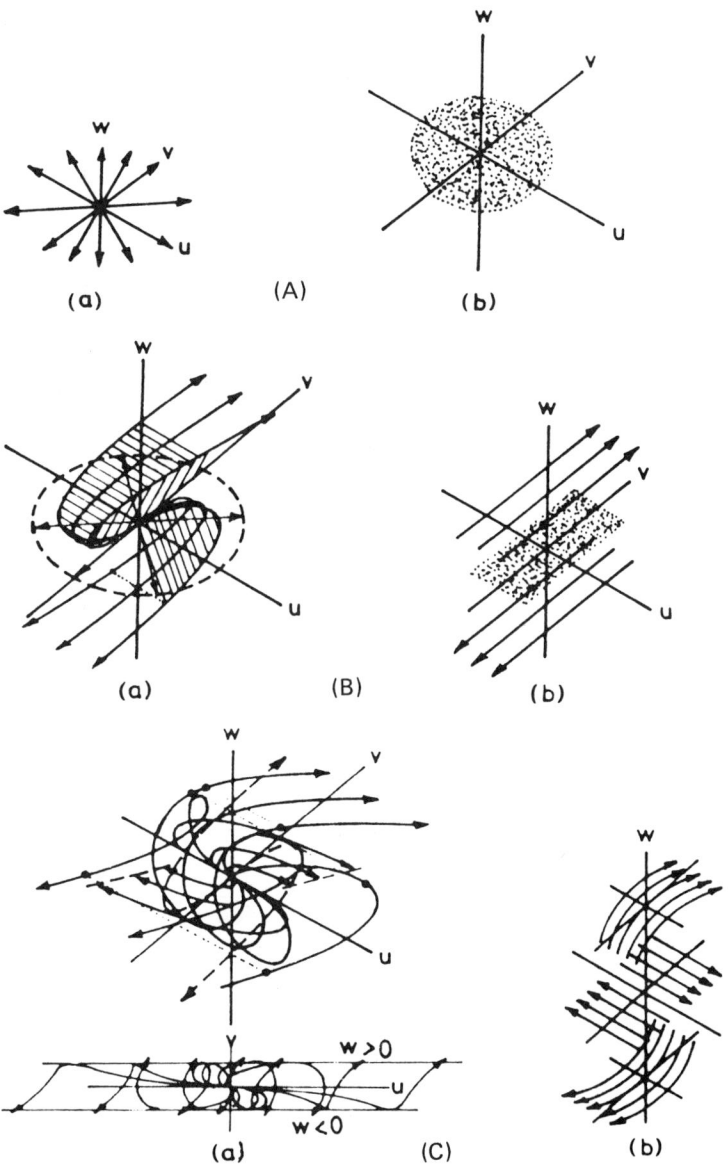

Figure 6.13.4A, 6.13.4B, and 6.13.4C Phase portrait of system (1): the case of three real coincident eigenvalues ($D = 0$), normal form $\mathcal{J}_{33}$, $\mathcal{J}_{32}$, and $\mathcal{J}_{31}$. (From Reyn 1964.)

Case	Eigenvalues	Figure	Nature of the singular point	Remarks
(ii)	$\lambda = 0$	6.13.4Bb	Shear plane with equilibrium points	The (u,v) plane consists of singular points. For $w > 0 \,(< 0)$, the integral curves are straight lines parallel to the v-axis; v increases (decreases) as t increases.
(iii)	$\lambda < 0$	6.13.4Ba	Stable antisymmetric node star	Reverse the arrows after rotating 6.13.4Ba by 180° around the w-axis.

3C. The coefficient matrix is of the normal form $\mathcal{J}_{31}$ with the corresponding ODE system

$$\frac{du}{dt} = \lambda u + v, \quad \frac{dv}{dt} = \lambda v + w, \quad \frac{dw}{dt} = \lambda w$$

having solutions $u = (c_1 + c_2 t + \tfrac{1}{2} c_3 t^2) e^{\lambda t}$, $v = (c_2 + c_3 t) e^{\lambda t}$, $w = c_3 e^{\lambda t}$, where $c_i \; (i = 1, 2, 3)$ are real constants.

Case	Eigenvalues	Figure	Nature of the singular point	Remarks
(i)	$\lambda > 0$	6.13.4Ca	Unstable one-branched node	The only plane curves are in the (u,v) plane and form an unstable one-branched plane node at the origin. The figure also contains projections of other integral curves on the (u,v) plane for $w > 0$ and $w < 0$.
(ii)	$\lambda = 0$	6.13.4Cb	Shear line with equilibrium points	The u-axis consists of the singularities. The plane $w = 0$ contains integral curves which are straight lines parallel to the u-axis. In the planes $w = $ const. the

Phase Space Study of Autonomous Systems

(iii) $\lambda < 0$ 6.13.4Ca   Stable one-branched node

integral curves are parabolas with axes $v = 0$, $w = $ const. > 0.

Reverse the arrows in 6.13.4Ca after rotating the figure by 180° around the w-axis.

4. Two of the three eigenvalues are complex ($D > 0$). The system of ODE can be changed, in terms of polar coordinates, $v = r \sin\phi$, $w = r \cos\phi$, into

$$\frac{du}{dt} = \lambda_1 u, \quad \frac{dr}{dt} = (\operatorname{Re}\lambda_2)r, \quad \frac{d\phi}{dt} = \operatorname{Im}\lambda_2.$$

The solutions of this system are $u = c_1 e^{\lambda_1 t}$, $r = c_2 \exp(\operatorname{Re}\lambda_2 t)$, $\phi = \operatorname{Im}\lambda_2(t + c_3)$, where c_i ($i = 1, 2, 3$) are real constants.

(i) V $\lambda_1^{-1}\operatorname{Re}\lambda_2 > 1$ 6.13.5a   Unstable pointed spiral

The integral curves form an unstable plane focus in the (r, ϕ) plane. In general, the integral curves are situated on surfaces of revolution with the u-axis as the axis of revolution. These surfaces are cusped at the origin. Change the conical surface in 6.13.5a into a surface of revolution having a cusp at the origin to get the case with $\operatorname{Im}\lambda_2 > 0$.

(ii) V $\lambda_1^{-1}\operatorname{Re}\lambda_2 = 1$ 6.13.5a   Unstable conical spiral

The spatial integral curves lie on circular cones with the u-axis as the axis. The rest is as in the unstable pointed spiral.

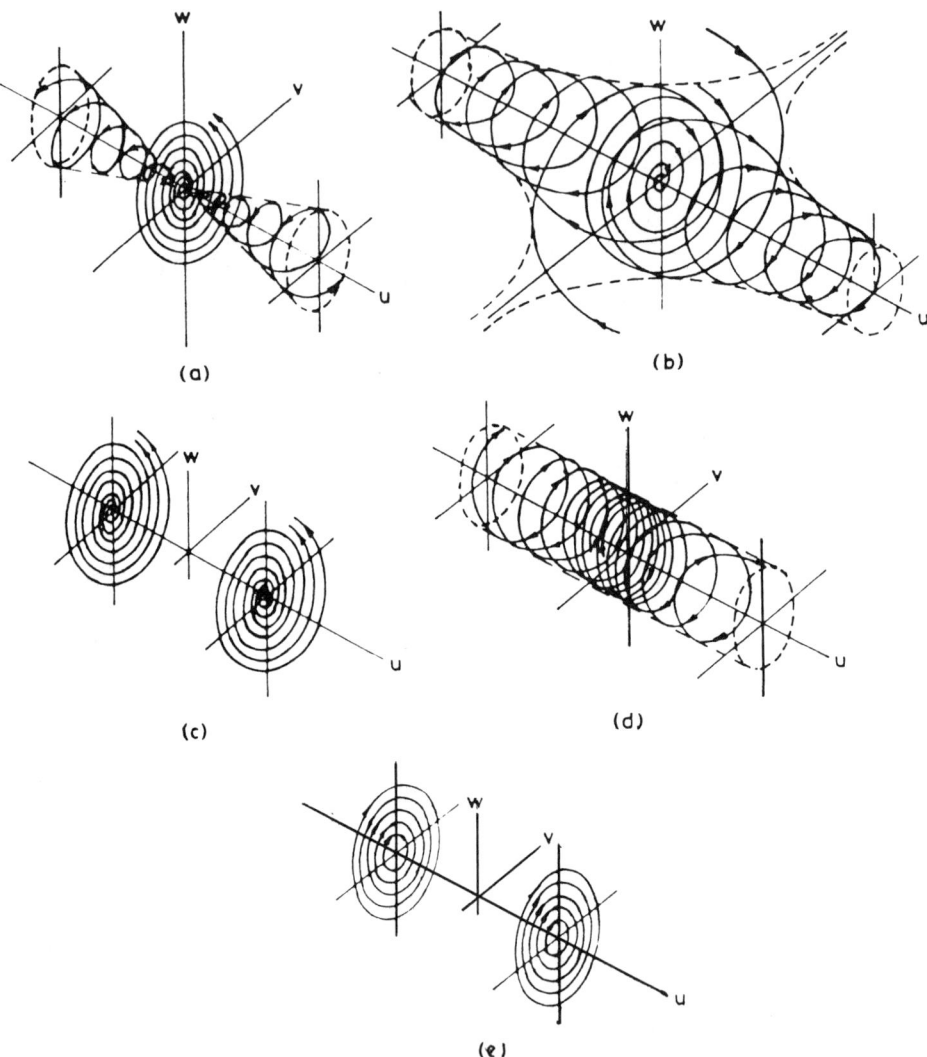

Figure 6.13.5 Phase portrait of system (1): the case of two complex eigenvalues ($D > 0$), normal form $\mathcal{J}_c$. (From Reyn 1964.)

Phase Space Study of Autonomous Systems

Case	Eigenvalues	Figure	Nature of the singular point	Remarks
(iii) V	$\lambda_1^{-1}\operatorname{Re}\lambda_2 < 1$	6.13.5a	Unstable blunt spiral	The description of the unstable pointed spiral applies except that the surfaces of revolution on which the integral curves lie are perpendicular to the u-axis at the origin.
(iv) VI	$\operatorname{Im}\lambda_2 < 0$	6.13.5b	Saddle spiral with stable plane focus	The integral curves form a stable plane focus in the (r, ϕ) plane. The spatial integral curves lie on surfaces of revolution with the u-axis as the axis of revolution; for $t \to -\infty$ they come from $r \to \infty$, $u = 0$ and approach the u-axis for $t \to \infty$.
(v) VII	$\lambda_1^{-1}\operatorname{Re}\lambda_2 > 1$	6.13.5a	Stable pointed spiral	Reverse the arrow in the unstable pointed spiral, case V, with $\lambda_1^{-1}\operatorname{Re}\lambda_2 > 1$.
(vi) VII	$\lambda_1^{-1}\operatorname{Re}\lambda_2 = 1$	6.13.5a	Stable conical spiral	Reverse the arrows in 6.13.5a.
(vii) VII	$\lambda_1^{-1}\operatorname{Re}\lambda_2 < 1$	6.13.5a	Stable blunt spiral	Reverse the arrows in the integral curves of the unstable blunt spiral, case V, $\lambda_1^{-1}\operatorname{Re}\lambda_2 < 1$.
(viii) VIII		6.13.5b	Saddle spiral with unstable plane focus.	Reverse the arrows in the curves of 6.13.5b.

Case	Eigenvalues	Figure	Nature of the singular point	Remarks
(ix) V–VIII		6.13.5c	Line of unstable plane foci	The u-axis is a line of singular points, and there is an unstable plane focus in each plane $u =$ const.
(x) VII–VIII		6.13.5c	Line of stable plane foci	The u-axis is a line of singular points, and in each plane $u =$ const. there is a stable plane focus. Reverse the arrows in Figure 6.13.5c.
(xi) V–VI	$\operatorname{Im} \lambda_2 < 0$	6.13.5d	Divergent vortex spiral	The integral curves form a plane center in the (v,w) plane. The spatial integral curves lie on circular cylinders with their axes coinciding with the u-axis, and approach the (v,w) plane as $t \to -\infty$; they wind away from this plane as $t \to \infty$. For $\operatorname{Im} \lambda_2 > 0$, rotate 6.13.5d around the w-axis by $180°$.
(xiii) VII–VIII		6.13.5d	Converging vortex spiral	Reverse the arrows in 6.13.5d.
(xiv) S_2-axis	$\operatorname{Im} \lambda_2 < 0$	6.13.5e	Line of plane centers	The u-axis is a line of singular points, and there is a plane center in each plane $u =$ constant.
(xv)	$\operatorname{Im} \lambda_2 > 0$	6.13.5e	Line of plane centers	Reverse arrows in 6.13.5e.

7
Singularity Structure and Chaotic Behavior of Nonlinear Ordinary Differential Equations

7.1 INTRODUCTION

Nonlinear ODE are still far from being completely understood. In many physical problems, one is interested in the ultimate behavior of the solution, as opposed to transient behavior, associated with arbitrary initial conditions. And it is here that the problems of predictability arise. We have already referred in Section 4.2 to the class of nonlinear evolutionary partial differential equations, which are exactly integrable by an inverse scattering transform (IST). Predictability is their hallmark. That is, (i) these equations are computationally deterministic for all time, (ii) their solutions are neutrally stable with respect to perturbation in the initial data, and (iii) these solutions are predictable. In short, the initial value problems are well-posed for these equations in the Hadamard sense, and thus they represent real-world models as we normally understand them. Because these equations are completely integrable, they represent models of deterministic problems.

We also discussed in Section 4.2 the conjecture of Ablowitz et al. concerning the integrability of a PDE via IST. This involves checking whether the nonlinear ODE which result from the PDE by similarity and other transformations enjoy the Painlevé property in the strong or weak sense. The strong P-property requires that the only movable singularities of the ODE are poles. The weak property, which has been introduced in recent years, permits logarithmic (or sometimes branch point) singularities as well.

In sharp contrast to the class of predictable or completely integrable evolution equations, there are models of physical systems which manifest a certain amount of unpredictability. This unpredictability may arise from a variety of sources. Typically, such a system depends on an external controllable parameter, and for some value of the parameter its dynamical behavior is well understood. However, as the parameter is changed from this value, the qualitative behavior of the system may change, too. After a finite or infinite succession of such changes, the system may present erratic behavior in the sense that its time evolution may be quite unpredictable on large time scales, or it may show broadband spectrum, or it may not be periodic any more. This behavior is called *irregular, aperiodic, erratic, chaotic,* or *(weakly) turbulent.*

One of the best-known such systems is called the *Lorenz model,* which we shall describe in great detail in the next section. This model has the form

$$\dot{x} = \sigma(y - x),$$
$$\dot{y} = -x(z - r) - y, \qquad (0)$$
$$\dot{z} = xy - bz.$$

Here σ, r, and b are some parameters, which are defined in Section 7.2. This system of equations is deterministic and contains no random, noisy, or stochastic terms. Although the time evolution of (0) obeys strict deterministic laws, it seems to behave according to its own free will. To quote Ruelle (1980), "the solutions of equations of type (0) are such that these systems of curves, these clouds of points suggest sometimes fireworks or galaxies, sometimes strange and disquieting vegetal proliferations. A realm lies there of forms to explore, and harmonies to discover." We may regard the Lorenz model with the special parameters chosen by Lorenz as a prototype of a chaotic model. Its main features may be described as follows (see Segur 1982):

1. System (0) is deterministic for a finite time; that is, given finite initial data, a unique solution exists for a finite time. However, unless the initial data are known with infinite accuracy, the solution becomes less and less determined by the initial data as time increases without bound.
2. The solution is unstable with respect to perturbations of the initial data; it is extremely sensitive to initial conditions.
3. Given any initial data, the solution is unpredictable over a long time scale in any practical sense.

Singularity Structure and Chaotic Behavior

This unpredictability over a long time is not a consequence of errors that may be introduced in the numerical solution, or of the type of numerical method used. An Euler method with a small mesh size or a sophisticated Runge–Kutta or a predictor-corrector scheme will give the same numbers. Moreover, the insensitivity is with respect to the general pattern of the solution. The details of the solution, however, depend crucially on the choice of the initial conditions and the integrating routine employed.

The question arises, how do we recognize analytically whether a given dynamical system will display chaotic behavior? Is there a means of doing it apart from carrying out expensive (and necessarily limited) computations in some ranges of the parameters? The computations may indeed be unavoidable, but there must be some alternative analytic approach to uncover the underlying structures of the ODE. It turns out that the singularity structure of the ODE determines to a fair degree the chaotic or regular behavior of the solutions in a given range of parameters. Closely connected with this is the question of integrability (or otherwise) of the given system, a concept which we presently elucidate.

We first explain and define some terms occurring in the chaotic phenomenon (Eckmann 1981), by referring to the system

$$\frac{d}{dt}\mathbf{x}(t) = \mathbf{F}(\mathbf{x}(t)), \tag{1}$$

where $\mathbf{x}$ is a vector in R^m, $m \geq 1$; each of its components describes a mode or a coordinate. In general F will depend on a parameter μ. Typical examples of systems of the form (1) may be found in Hamiltonian mechanics, particle accelerators, fluid mechanics, chemical reactions, and electrical circuits. In Hamiltonian dynamics, the Liouville theorem asserts that the flow $t \to t(x)$ preserves volumes: if $\mathbf{x}(\mathbf{y},t)$ is the solution of (1) with initial condition $\mathbf{x}(\mathbf{y}, t = 0) = \mathbf{y}$, and if

$$\sum_{i=1}^{m} \frac{\partial F_i}{\partial x_i}(\mathbf{x}) = 0, \tag{2}$$

then the flow preserves volumes locally. On the other hand, for dissipative systems with internal friction, the flow contracts volumes so that

$$\sum_{i=1}^{m} \frac{\partial F_i}{\partial x_i} < 0.$$

Let us assume that there is a finite volume V in state space R^m such that if $\mathbf{y} \in V$ then $T_\mathbf{y}^t = \mathbf{x}(\mathbf{y},t)$ is in V for all $t > 0$. Since, for dissipative cases that we consider, the flow T^t decreases volumes, the set $T^t V$ decreases as $t \to \infty$ to a set $W = \cap_{t>0} T^t V$ of zero volume. Thus, every solution curve starting at some $\mathbf{y} \in V$ approaches W as $t \to \infty$. We may alternatively say that if $\mathbf{y} \in V \setminus W$, then $\mathbf{y}$ is transient and the curve $T_\mathbf{y}^t$ will for some sufficiently large t definitively depart from $\mathbf{y}$ and converge to W. In contrast, in nondissipative conservative systems, almost all curves $T_\mathbf{y}^t$ return infinitely often arbitrarily close to their initial state $\mathbf{y}$. We consider only systems which have assumed some sort of "internal equilibrium"; that is, we analyze the motion on W or on parts of W, assuming the orbits which tend to W but are not in it behave similarly to those in W, at least after a sufficient lapse of time. These parts of W will be called *attractors*, and studying attractors amounts to neglecting transient behavior. Thus, we arrive at the following definition of an attractor.

DEFINITION An *attractor* for the flow T^t is a compact set X satisfying the following conditions:

1. X is invariant under T^t: $T^t X = X$.
2. X has a shrinking neighborhood; i.e., there is an open neighborhood U of X, $U \supset X$, such that $T^t U \subset U$ for $t > 0$ and $X = \cap_{t>0} T^t U$.

While an attractor is a generalization of a stable equilibrium point, a *repeller* corresponds to an unstable equilibrium point or a saddle point.

There is a third important ingredient to the definition of an attractor, which generalizes the description of k separate stable equilibria to k separate attractors. This is contained in the following requirements:

3. The flow T^t on X is recurrent and indecomposable. Here, *recurrent* means T^t is nowhere transient on X: if U is an open set in V and if $U \cap V \neq \emptyset$, then there are arbitrarily large values for t such that $T_\mathbf{x}^t \in X \cap U$ when $\mathbf{x} \in X \cap U$. *Indecomposable* means that X cannot be split into two nontrivial closed invariant pieces.

As an illustration, the simplest dynamical system may be as shown in Figure 7.1.1. There are two attractors x_1 and x_2, which are stable fixed points. Their basins of attraction are, respectively, the left and right sides of line L. The line L is attracted by x_3, which, however, is not an attractor since it also has an unstable direction. It is a saddle point. The set W, in our previous definition, is $\{x_1, x_2, x_3\}$. (If X is an attractor, its basin of attraction is defined to be the set of initial points $\mathbf{x}$ such that $T^t \mathbf{x}$ approaches X as $t \to \infty$.)

For illustration, we consider the following simple examples.

Singularity Structure and Chaotic Behavior

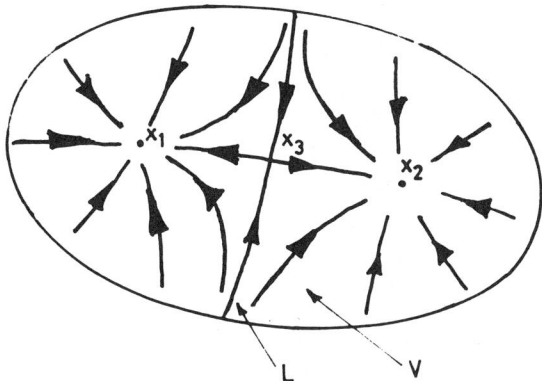

Figure 7.1.1 Phase portrait illustrating two stable (x_1, x_2) and one unstable (x_3) fixed points. (From Eckmann 1981.)

EXAMPLE 1 The damped harmonic oscillator is described by

$$\dot{x} = -y,$$
$$\dot{y} = x - y.$$

Here the set of all initial conditions form a plane, and, as the oscillator ultimately stops, all solutions tend toward the fixed point $(0,0)$, the simplest attractor. The dimensionality of the solutions is reduced from 2 to 0 (see Figure 7.1.2).

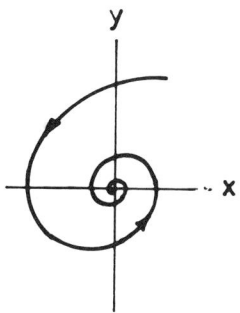

Figure 7.1.2 Phase portrait for the system $\dot{x} = -y$, $\dot{y} = x - y$.

EXAMPLE 2 For the Van der Pol oscillator

$$\dot{x} = -y,$$
$$\dot{y} = x + \mu y(1-x^2),$$

all solutions tend toward a stable orbit, or a limit cycle, another example of an attractor. Here, the dimensionality of the solution set is reduced from two to one (see Figure 7.1.3).

Attractors are called *strange* if they show the following special behavior: even though T^t contracts volumes, it does not contract length. If we take snapshots of T^t at $t = 0, 1, 2$, say, we may have the picture shown in Figure 7.1.4a, but we could also get the one in Figure 7.1.4b or 7.1.4c. In the latter two, contraction of volume in phase space takes place with stretching of length with folding. In particular, even if all points in V converge to a single attractor X, one may still find that points which are arbitrarily close initially may get macroscopically separated on the attractor after sufficiently large time intervals. This property spells sensitive dependence on initial conditions. This situation can and does occur in dissipative dynamic systems. Strange attractors are neither periodic points nor periodic orbits. There exists a large variety of attractors which are neither trivial (i.e., they are neither periodic orbits nor fixed points) nor strange attractors; yet, they seem to predict more or less chaotic features.

We do not discuss chaos in any great detail. A most readable semitechnical account of this fascinating topic, with a comprehensive bibliography, may be found in Gleick (1988). We use some of the terms introduced above in trying to establish the singularity structure of the solutions of the dynamical systems and their possible chaotic behavior. As we noted, the dynamical systems may be conservative or dissipative. For the former, the total energy

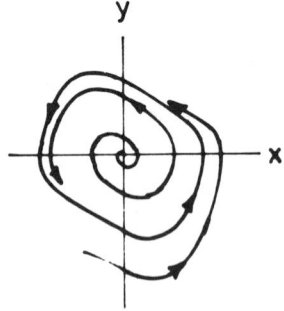

Figure 7.1.3 Phase portrait for the system $\dot{x} = -y$, $\dot{y} = x + y\mu(1-x^2)$.

Singularity Structure and Chaotic Behavior

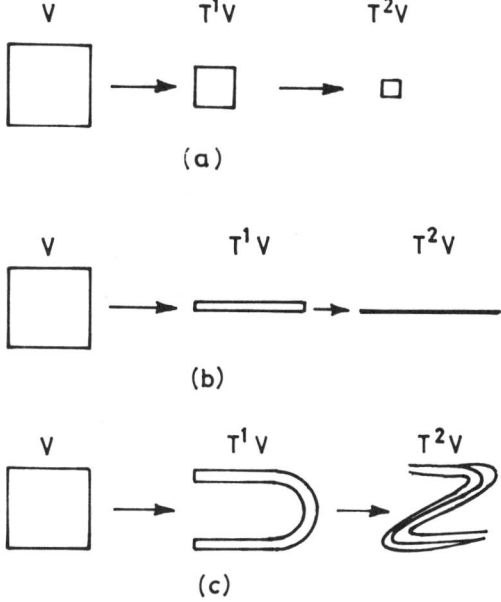

Figure 7.1.4 (a) Contraction of volume in phase space, (b) contraction of volume in phase space, with stretching of length, (c) contraction of volume, stretching of length, and folding.

or some other quantity does not vary with time, while for the latter a "noble" form of energy (for example, mechanical, electrical, or chemical) changes into heat. Conservative systems often show sensitive dependence on initial conditions. Dissipative systems show some (peculiarly) interesting behavior only if they are constantly fed some noble energy; otherwise they go to rest.

What is integrability of a system? The first response is the literal one: the system can be integrated in a closed form in terms of known functions and has a sufficient number of constants of integration. In the present day context of chaos, completely integrable (or integrable for short) systems are defined to be those which are characterized by regular, or predictable, behavior for all initial conditions and all time. By contrast, nonintegrable systems have, in many cases, regions in the phase space of their dependent variables, where the motion is irregular and chaotic in the sense described above (Ramani et al. 1988). In the context of (conservative) Hamiltonian systems with N degrees of freedom, this definition can be made more precise: there exist N analytic, single-valued, global integrals which do not depend explicitly on time t and are in involution (see Gelfand and Fomin

1963, p. 72). For a dissipative system of N first-order ODE, one definition of integrability requires that it possess N analytic, global constants of motion which may explicitly depend on time. It is well known that few nonlinear dynamical systems can be explicitly integrated. For example, in classical mechanics the positions and velocities of a system of mass particles are governed by a set of ODE which are in general coupled and nonlinear. Only a few of these systems can be integrated by known methods of analysis. Many solutions have been derived by heuristic considerations. However, to recognize an integrable dynamical system, a systematic approach, due originally to Kovalevskaya, was given by Ablowitz, Ramani, and Segur, which we have discussed in some detail in Section 4.2. This is to look for all parameter values for which the solutions have the Painlevé property; that is, their only movable singularities (in the complex time plane) are poles; moreover, these solutions involve as many constants as there are initial conditions. It was felt that the equations having this property might be easier to integrate or solve analytically.

It turns out that there is also a close connection between the (complex time) singularities of dynamical systems and their general behavior—regular versus chaotic in real time. As the Painlevé property is partially lifted so that sufficient number of arbitrary constants in the expansions about singularities do not appear and only some classes of particular solutions have poles as the only movable singularities, or it is weakened in the sense that, besides poles, the solution may exhibit logarithmic singularities of the form $[\ln(t - t_0)]^p$ (with p an integer), then most orbits still appear regular and very little chaos is evident. However, if worse singularities of the form $\ln\ln(t - t_0)$ show up in the expansions of the solutions about the singularities, then the solutions undergo a dramatic change in their behavior with respect to real time, manifesting large-scale chaotic regions in phase space.

One must therefore devise a scheme to discover whether a given system of nonlinear ODE is integrable and (hence) shows regular behavior. In the opposite circumstance, the solution would exhibit chaos, its degree depending on the intricacy of the complex-time singularity. The first step is to discover (intuitively or otherwise) whether there are any special values of parameters for which the system is explicitly integrable in terms of known functions or reducible to elliptic equations or to equations of Painlevé type. Even the existence of one integral (for a system of three autonomous DE) spells regularity. The next step is to carry out the local analysis as suggested by Ablowitz et al. to find the dominant behavior and resonances (see Section 4.2) and to check whether the Painlevé property is satisfied. The major step then is to construct psi series—usually a double series—about the singular point, when the latter is not just a (single-valued) pole. Such series do not generally have closed recursive formulas for their coefficients; yet they throw much light on the structure of the singularities, and for special sets of parameters may even have closed recursion formulas for the coefficients. These psi series have different nature for different systems of ODE and have to be individually constructed. The questions of their convergence, etc., are far from simple.

Hille (1976), for example, has quoted some results due to Smith (1973) regarding the second-order normal system

$$\dot{x} = x(a_0 + a_1 x + a_2 y),$$
$$\dot{y} = y(b_0 + b_1 x + b_2 y).$$

This system has solutions of the form

$$x(t) = \alpha t^{-1} + \sum_{n=0}^{\infty} \alpha_n t^n,$$

$$y(t) = \beta t^{-1} + \sum_{n=0}^{\infty} \beta_n t^n, \qquad (3)$$

if and only if

$$q = \frac{(a_1 - b_1)(a_2 - b_2)}{a_1 b_2 - a_2 b_1} \qquad (4)$$

is not a positive integer. In the latter circumstances, the coefficients in (3) are uniquely determined with

$$\alpha = \frac{a_2 - b_2}{a_1 b_2 - a_2 b_1}, \qquad \beta = \frac{b_1 - a_1}{a_1 b_2 - a_2 b_1}. \qquad (5)$$

In addition, there are solutions in the form of psi series

$$x(t,c) = \sum_{m=0}^{\infty} \sum_{n=0}^{\infty} a_{mn} c^n t^{m-1+nq},$$

$$y(t,c)) = \sum_{m=0}^{\infty} \sum_{n=0}^{\infty} b_{mn} c^n t^{m-1+nq}, \qquad (6)$$

where c is an arbitrary parameter. Note that these are double series in t and t^q. Alternatively, if q is a positive integer and if

$$\mathbf{z}(t) = \begin{bmatrix} x(t) \\ y(t) \end{bmatrix}$$

is a solution which becomes infinite when t decreases to zero, then

$$\mathbf{z}(t) = \sum_{n=0}^{\infty} \mathbf{V}_n \left[\ln \frac{1}{t} \right] t^{n-1}, \qquad (7)$$

where $\mathbf{V}_n[s]$ is a vector-valued polynomial in s of degree $< [n/q]$. There exists a number $K > 0$ such that the series

$$\sum_{n=0}^{\infty} \mathbf{V}_n(s) e^{(1-n)s} \tag{8}$$

converges in norm for s real and greater than the positive root of the equation $K(s+z) = e^{qs}$. We refer the reader to Hille (1976) and Smith (1973) for further details. Hille has also considered the more general second-order system

$$\dot{x} = P(x,y), \qquad \dot{y} = Q(x,y), \tag{9}$$

where $P(x,y)$ and $Q(x,y)$ are finite sums of homogeneous polynomials in x and y.

We shall take up a detailed study of several dynamical systems—both dissipative and conservative—from their singularity structure point of view. These include the Lorenz system and the Henon–Heiles Hamiltonian system. While our major concern will be their analytic behavior, we shall also discuss some of their numerical solutions to bring out the relation between their (global) regular or chaotic behavior and the analytic structure of their singularities.

First we take up an elementary discussion of some special cases of the nonlinear autonomous system of DE

$$\dot{x}_i = \sum_{j=1}^{3} a_{ij} x_j + \sum_{j,k=1}^{3} b_{ijk} x_j x_k, \tag{10}$$

where $i = 1, 2, 3$, which occurs frequently in physics, chemistry, and biology (Steeb, Kunick, and Strampp 1983). We list a few of these models.

Rikitake Two-Disk System

The system

$$\begin{aligned}
\dot{x}_1 &= -\mu x_1 + x_2 x_3, \\
\dot{x}_2 &= -\mu x_2 - \alpha x_1 + x_1 x_3, \\
\dot{x}_3 &= 1 - x_1 x_2
\end{aligned} \tag{11}$$

describes earth's magnetohydrodynamic dynamo. It can be changed to the form (10). It shows a chaotic behavior for a wide range of parameters α and μ. We shall see by singularity analysis that for a certain choice of parameters, system (11) admits a first integral and, hence, on reduction to a plane system does not exhibit chaos.

Lorenz Model

The system

$$\dot{x}_1 = \sigma(x_2 - x_1),$$
$$\dot{x}_2 = -x_1 x_3 + r x_1 - x_2, \qquad (12)$$
$$\dot{x}_3 = -b x_3 + x_1 x_2,$$

will be discussed in great detail in the next section.

The Oregonator

The system

$$\dot{x}_1 = s(x_2 - x_1 x_2 + x_1 - q x_1^2),$$
$$\dot{x}_2 = \frac{1}{s}(-x_2 - x_1 x_2 + f x_3), \qquad (13)$$
$$\dot{x}_3 = w(x_1 - x_3),$$

where s, w, f, and q are real parameters, is a simple and idealized model of the Belousov–Zhabotinskii reaction.

Lotka–Volterra Model

The system

$$\dot{x}_1 = \gamma_1 x_1 + \gamma_{12} x_1 x_2 + \gamma_{13} x_1 x_3,$$
$$\dot{x}_2 = \gamma_2 x_2 - \gamma_{12} x_1 x_2 + \gamma_{23} x_2 x_3, \qquad (14)$$
$$\dot{x}_3 = \gamma_3 x_3 - \gamma_{13} x_1 x_3 - \gamma_{23} x_2 x_3,$$

where γ_i and γ_{ij} are real parameters, describes three species in interaction. Here x_1, x_2, and x_3 are positive quantities.

A Model of Plasma Physics

The system

$$\dot{x}_1 = \gamma_1 x_1 + \gamma_{23} x_2 x_3,$$
$$\dot{x}_2 = \gamma_2 x_2 + \gamma_{13} x_1 x_3, \qquad (15)$$
$$\dot{x}_3 = \gamma_3 x_3 + \gamma_{12} x_1 x_2,$$

where γ_i and γ_{ij} are real parameters, appears in plasma physics (Fuchs 1975).

We shall study system (11) in some detail and briefly quote the results for (14) and (15). We first write the former in the complex domain:

$$\frac{dw_1}{dz} = -\mu w_1 + w_2 w_3,$$

$$\frac{dw_2}{dz} = -\mu w_2 - \alpha w_1 + w_1 w_3, \qquad (16)$$

$$\frac{dw_3}{dz} = 1 - w_1 w_2,$$

where $z = t_1 + it_2$, $w_i = u_i + iv_i$, and t_1, t_2, u_i, and v_i are real quantities. We now attempt the solution of (16) (see Section 4.2) in the form

$$w_i \sim A_i(z - z_1)^{k_i}, \qquad k_i < 0, \quad i = 1, 2, 3. \qquad (17)$$

Substitution into (16) shows that the nonlinear terms dominate and balance with the derivative terms on the left, to yield

$$k_1 = k_2 = k_3 = -1 \qquad (18)$$

and either

$$A_1 = A_2 = \pm i, \qquad A_3 = -1 \qquad (19)$$

or

$$A_1 = \pm i, \quad A_2 = \mp i, \quad A_3 = 1. \qquad (20)$$

To obtain the resonances—i.e., the powers in the Laurent series where the arbitrary parameters appear—we retain only the leading (dominant) terms in (16), namely,

$$\frac{dw_1}{dz} = w_2 w_3,$$

$$\frac{dw_2}{dz} = w_1 w_3, \qquad (21)$$

$$\frac{dw_3}{dz} = -w_1 w_2,$$

and substitute

$$w_i = A_i(z - z_1)^{k_i} + B_i(z - z_1)^{k_i + r}. \qquad (22)$$

Here k_i and A_i are given by (18) and (19) or (18) and (20). If we retain leading terms in B_i after substituting (22) into (21), we get a system of three homogeneous equations

$$Q(r)B = 0,$$

Singularity Structure and Chaotic Behavior

where Q is a 3×3 matrix

$$Q = \begin{bmatrix} r-1 & -A_3 & -A_2 \\ -A_3 & r-1 & -A_1 \\ A_2 & A_1 & r-1 \end{bmatrix} \tag{23}$$

whose elements depend on r and A_i. Some of the coefficients B_i are arbitrary provided $\det Q(r) = 0$. The roots of this equation yield the resonances. Explicitly, we find from (23) that

$$\det Q(r) = r^3 - 3r^2 + 4 = 0. \tag{24}$$

Apart from $r = -1$, which refers to the arbitrariness of the location z_1 of the pole, we have $r_{2,3} = 2, 2$. This multiple root, in general, may hint at the appearance of a logarithmic movable branch point with an arbitrary coefficient. In the present case, however, system (21) has no logarithmic term in its solution. Indeed, it is easily seen to be completely integrable and has a Laurent series with three constants of integration.

Now we insert the Laurent series

$$w_i(z) = A_i(z - z_1)^{k_i} + \sum_{j=1}^{\infty} a_{i,k_i+j}(z - z_1)^{k_i+j} \tag{25}$$

into system (16), where k_i and A_i are given by (18) and (19) or (18) and (20). We treat each of these subcases separately. For the values (18) and (19) we find that the first few coefficients in the series in (25) are

$$a_{1,0} = \frac{-i\alpha}{2}, \quad a_{2,0} = \frac{i\alpha}{2}, \quad a_{3,0} = \mu + \frac{\alpha}{2}. \tag{26}$$

The coefficients $a_{1,1}$, $a_{2,1}$, and $a_{3,1}$ are determined by the following three linear (inhomogeneous) equations:

$$a_{1,1} + a_{2,1} - ia_{3,1} = i\alpha\mu + \frac{i\alpha^2}{4}, \tag{27a}$$

$$a_{1,1} + a_{2,1} - ia_{3,1} = -i\alpha\mu + \frac{i\alpha^2}{4}, \tag{27b}$$

$$a_{1,1} + a_{2,1} - ia_{3,1} = -i + \frac{i\alpha^2}{4}, \tag{27c}$$

where we have used the values $A_1 = A_2 = i$ and $A_3 = -1$. We recall that system (27) appears at the resonance $r = 2$, so $k_i + j = -1 + 1 = 0$ in (25). For $A_1 = A_2 = -i$ and $A_3 = -1$ we find similar results.

We notice that the rank of the homogeneous form of system (27) is 1. Moreover, the inhomogeneous system (27) has no solution for arbitrary values of the parameters α and μ. Equations (27a) and (27b) are consistent if and only if $\alpha = 0$ while μ is arbitrary. For these parametric values, it becomes possible to find an intermediate integral or "constant of motion" for (11), namely,

$$(x_1^2 - x_2^2) \exp(2\mu t) = C. \tag{28}$$

In this case the dynamical system (11) becomes planar and does not behave chaotically.

For the other subcase, we consider $A_1 = -i$, $A_2 = i$, and $A_3 = 1$. The coefficients now are

$$a_{1,0} = a_{2,0} = -\frac{i\alpha}{2}, \quad a_{3,0} = -\mu + \frac{\alpha}{2}. \tag{29}$$

The coefficients $a_{1,1}$, $a_{2,1}$ and $a_{3,1}$ are determined from the linear system

$$a_{1,1} - a_{2,1} - ia_{3,1} = i\alpha\mu - \frac{i\alpha^2}{4},$$

$$a_{1,1} - a_{2,1} - ia_{3,1} = -i\alpha\mu - \frac{i\alpha^2}{4}, \tag{30}$$

$$a_{1,1} - a_{2,1} - ia_{3,1} = -i - \frac{i\alpha^2}{4}.$$

Similar results are found for the values $A_1 = i$, $A_2 = -i$, and $A_3 = 1$. As in the first subcase, system (30) has no solutions for arbitrary values of α and μ; but the choice $\alpha = 0$, with arbitrary μ, points to the existence of the intermediate integral (28).

Now we summarize the results for systems (14) and (15). For the Lotka–Volterra system (14), we find that if

$$\gamma_{12} + \gamma_{23} - \gamma_{13} = 0 \tag{31}$$

and

$$\gamma_1 = \gamma_2 = \gamma_3,$$

then a Laurent series solution exists and involves three arbitrary constants. We can also find an intermediate integral in this case so that the dynamical system becomes planar. For system (15), a general solution with three arbitrary constants again exists if $\gamma_1 = \gamma_2 = \gamma_3 = \gamma$, say. Indeed, system (15) now admits two intermediate integrals depending explicitly on time:

$$\left(\frac{\gamma_{13} x_1^2}{2} - \frac{\gamma_{23} x_2^2}{2} \right) \exp(-2\gamma t) = C_1$$

and $\tag{32}$

$$\left(\frac{\gamma_{12} x_1^2}{2} - \frac{\gamma_{23} x_3^2}{2} \right) \exp(-2\gamma t) = C_2.$$

The solution can therefore be found by a simple quadrature. If $\gamma_i \neq \gamma_j$, we are forced to introduce logarithmic terms.

The above examples show that if we can find one intermediate integral for a third-order autonomous system of DE, the system becomes planar and chaotic motion can be excluded. Even when an explicit Laurent series expansion may not be found, the discussion of the algebraic equations determining the coefficients of the series can point to the existence of intermediate integral(s). This aspect of singular point analysis will be brought out more clearly in subsequent sections when we discuss some important physical models—the Lorenz and Henon–Heiles systems—and the structure of their singularities.

In general, therefore, once a system of ODE is shown to possess the Painlevé property, further investigation is needed to prove integrability by (i) actually constructing the full set of integrals, or (ii) linearizing the equations exactly, or (iii) reducing them to one of the Painlevé equations P_I–P_{VI} or to new transcendents that may have to be introduced at higher orders.

When the notion of the weak Painlevé property is introduced, permitting the presence of special types of algebraic or logarithmic branch point singularities, the system of ODE may sometimes be found to be completely integrable. However, complex or irrational powers of the dominant terms can be rigorously connected to the nonintegrability of Hamiltonian systems, as we shall see.

7.2 THE LORENZ SYSTEM

We briefly describe the physical model which gave rise to the Lorenz system

$$\frac{dx}{dt} = \sigma(y - x), \tag{1a}$$

$$\frac{dy}{dt} = rx - y - xz, \tag{1b}$$

$$\frac{dz}{dt} = xy - bz. \tag{1c}$$

A two-dimensional fluid cell is heated from below and cooled from above and the resulting motion is modeled by two coupled nonlinear PDE in the stream function and a temperature function. These two functions were expanded in a double Fourier series in the two spatial variables, with functions of t as the coefficients. After substituting the double series into the PDE and equating the coefficients of trigonometric functions, etc., an infinite system of ODE for the functions of time was obtained. This system was truncated severely by setting all but three of the functions identically equal to zero. These three time functions are governed by (1), where σ is the Prandtl number, r is the Raleigh number, and b represents some physical proportions of the region under consideration. All these constants are obviously positive. The variable t is nondimensional time. The dependent variables x, y, and z have the following interpretation: x is proportional to the intensity of convective motion, while y is proportional to the temperature difference between the ascending and descending currents; identical signs of x and y denotes that warm fluid is rising while cold fluid is descending. The variable z is proportional to the distortion of the vertical temperature profile from a linear one, a positive value indicating that the strongest gradients occur near the boundaries.

The severely truncated finite model (1) is not expected to represent a realistic model of the original fluid dynamical problem (as Lorenz (1963) himself pointed out) when strong convection occurs so that $r \gg 1$. Moreover, the larger truncated systems obtained by retaining more terms in the Fourier series show somewhat different behavior. Yet, model (1) has assumed considerable importance chiefly because (i) several other real-world problems have led exactly to (1) with $r \gg 1$ (see Sparrow 1982 for references to these phenomena), and (ii) this model appears to manifest a greater range of different behaviors than most other (similar) systems.

Now we pick up some simple properties of system (1). If we consider the variation of the volume V of a small region in phase space (x,y,z), it is contracted under (1) into a volume element $Ve^{-(\sigma+b+1)t}$. This is easily seen by writing the divergence of the flow with the help of (1):

$$\frac{\partial}{\partial x}\dot{x} + \frac{\partial}{\partial y}\dot{y} + \frac{\partial}{\partial z}\dot{z} = -(\sigma + b + 1); \tag{2}$$

V now becomes $Ve^{-(\sigma+b+1)t}$ in time t. This implies that each small volume shrinks to zero as t goes to infinity. This does not mean that each small volume shrinks to a point. The volume may simply flatten into a surface. This observation leads to the conclusion that all trajectories initially enclosed by a surface S ultimately become confined to a specific subspace having zero volume. Thus, we already have a distinctive character of flows described by (1). This dissipative nature of the flow rules out unstable periodic orbits or unstable stationary points.

Now we consider the stationary points or steady-state solutions of (1). The origin $x = y = z = 0$ is a stationary point for all values of the parameters and represents the state of no convection. System (1) linearized about $(0,0,0)$ has the characteristic equation

$$[\lambda + b][\lambda^2 + (\sigma + 1)\lambda + \sigma(1 - r)] = 0. \tag{3}$$

This equation has three real roots for $r > 0$. All of them are negative when $r < 1$ so that the origin is stable and attracting for $0 < r < 1$. For $r > 1$, one root is positive and the origin is unstable. Equation (1) has two additional steady-state solutions for $r > 1$; $x = y = \pm\sqrt{b(r-1)}, z = r - 1$. The characteristic equation for each of these solutions can be found to be

$$\lambda^3 + (\sigma + b + 1)\lambda^2 + (r + \sigma)b\lambda + 2\sigma b(r - 1) = 0. \tag{4}$$

This equation has one real negative root and, in general, two complex conjugate roots. All three roots are real when r is very close to one. For the values $\sigma = 10$, $b = 8/3$, chosen by Lorenz and investigated subsequently by many other workers, there is one real root and a pair of complex conjugate roots, provided $r > 1.346$. All three roots have a negative real part if $r < \sigma(\sigma + b + 3)/(\sigma - b - 1)$. Therefore, for $\sigma = 10$ and $b = 8/3$, these additional stationary points are stable in the parameter range $1 < r < 470/19 \approx 24.74$. The complex conjugate roots are pure imaginary if

$$r = \sigma(\sigma + b + 3)(\sigma - b - 1)^{-1}. \tag{5}$$

This is therefore the critical value of r which separates the region of instability of steady convection. If $\sigma < b + 1$, no positive value of r satisfies (5), and steady (state) convection is stable. Conversely, for $\sigma > b + 1$, the convection is unstable for sufficiently high values of r. These statements follow from system (1) only and therefore need not hold for the original PDE.

Before we discuss integrability and the Painlevé property, relating to (1), we give a simple regular perturbation scheme for (1) due to Robbins (1979), which gives periodic solutions for high r-values. To this end we first change variables and notation in (1) to agree with Robbins' notation. First we transform (1) according to

$$w = r - z, \quad x \longrightarrow y, \quad \text{and} \quad y \longrightarrow z \tag{6}$$

and set $b = 1$. Then system (1) becomes

$$\begin{aligned}\frac{dw}{dt} &= r - zy - w, \\ \frac{dz}{dt} &= wy - z, \\ \frac{dy}{dt} &= \sigma(z - y).\end{aligned} \tag{7}$$

We assume that $r \gg \sigma$ and introduce the following scaling in (7):

$$t \longrightarrow \epsilon t, \quad w \longrightarrow \frac{w}{\epsilon^2 \sigma}, \quad z \longrightarrow \frac{z}{\epsilon^2 \sigma}, \quad y \longrightarrow \frac{y}{\epsilon}, \tag{8}$$

where

$$\epsilon = \frac{1}{\sqrt{r\sigma}}. \tag{9}$$

System (7) then becomes

$$\frac{dw}{dt} = -zy + \epsilon(1-w),$$
$$\frac{dz}{dt} = wy - \epsilon z, \tag{10}$$
$$\frac{dy}{dt} = z - \sigma\epsilon y.$$

We look for a (regular) perturbation solution of (10) in the form

$$w(t,\epsilon) = w_0(t) + \epsilon w_1(t) + \cdots,$$
$$z(t,\epsilon) = z_0(t) + \epsilon z_1(t) + \cdots, \tag{11}$$
$$y(t,\epsilon) = y_0(t) + \epsilon y_1(t) + \cdots.$$

Substituting (11) into (10), we have, to orders ϵ^0 and ϵ,

$$\frac{dw_0}{dt} = -z_0 y_0, \tag{12a}$$

$$\frac{dz_0}{dt} = w_0 y_0, \tag{12b}$$

$$\frac{dy_0}{dt} = z_0; \tag{12c}$$

$$\frac{dw_1}{dt} = -z_0 y_1 - z_1 y_0 + 1 - w_0, \tag{13a}$$

$$\frac{dz_1}{dt} = w_0 y_1 + w_1 y_0 - z_0, \tag{13b}$$

$$\frac{dy_1}{dt} = z_1 - \sigma y_0. \tag{13c}$$

System (12) is conservative and has two first integrals,

$$w_0^2 + z_0^2 = B^2,$$
$$\frac{1}{2} y_0^2 + w_0 = D, \tag{14}$$

where the constants of integration B and D define a two-parameter family of periodic solutions. For a given value of B, the solutions lie on a cylinder of radius B along the y-axis. If $D < B$, there is a double-looped periodic solution for each (B,D). At $D = B$, the solution is a homoclinic orbit of the critical point $w_0 = B, z_0 = 0, y_0 = 0$. For $D > B$, there are two single-looped

Singularity Structure and Chaotic Behavior

solutions. The first and third cases are shown in Figure 7.2.1. To get the explicit solution, we find

$$z_0 = \pm \left[B^2 - \left(D - \frac{1}{2} y_0^2 \right)^2 \right]^{1/2} \tag{15}$$

from (14) and substitute it into (12c). Integrating the resulting equation shows that $y_0(t)$ is an elliptic function. Substituting $y_0(t)$ into (14) then gives $w_0(t)$ and $z_0(t)$. We thus have, for $B > D > 0$,

$$w_0(t) = B[1 - 2\,dn^2(\sqrt{B}t + u_0)],$$
$$z_0(t) = -\alpha\sqrt{B}\,sn(\sqrt{B}t + u_0)\,dn(\sqrt{B}t + u_0), \tag{16}$$
$$y_0(t) = \alpha\,cn(\sqrt{B}t + u_0),$$

where $\alpha = \sqrt{2(B+D)}$, $\beta = \sqrt{2(B-D)}$, $k^2 = \alpha^2/4B$, $k'^2 = 1 - k^2$. The constants of integration B, D, and u_0 are fixed by the initial conditions; sn, dn, and cn denote Jacobi elliptic functions (see Byrd and Friedman 1954). The dependence of the elliptic functions on the modulus k^2 has not been explicitly given for the sake of conciseness. The Jacobi elliptic functions are doubly periodic—that is, periodic in both real and imaginary directions—and have singularities which are simple poles arranged on an infinite periodic lattice in the complex t-plane.

To fix the solution, the phase constant u_0 is chosen such that the zeroth-order solution has the simple initial conditions

$$w_0(0) = 0,$$
$$z_0(0) = -B, \tag{17}$$
$$y_0(0) = \sqrt{2D},$$

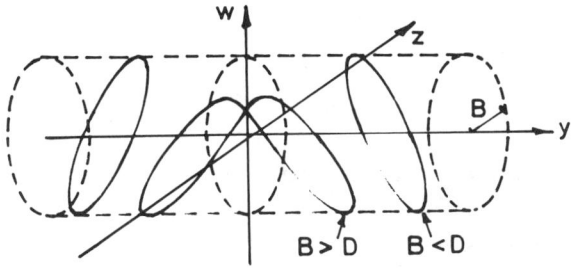

Figure 7.2.1 As $r \to \infty$ the periodic solutions of (7) lie on a cylinder of radius B: Cases $B > D$ and $B < D$. (From Robbins 1979.)

in accordance with (14). Indeed, the full solution is sought with (17) as the initial conditions so that

$$w(0, \epsilon) = w_0(0),$$
$$z(0, \epsilon) = z_0(0), \qquad (18)$$
$$y(0, \epsilon) = y_0(0).$$

Equations (11) and (17) then imply that

$$w_i(0) = z_i(0) = y_i(0) = 0 \qquad \text{for } i > 0. \qquad (19)$$

The first-order system (13) can be rearranged as

$$\frac{d}{dt}(w_0 w_1 + z_0 z_1) = w_0 - z_0^2 - w_0^2,$$
$$\frac{d}{dt}(y_0 y_1 + w_1) = 1 - w_0 - \sigma y_0^2. \qquad (20)$$

The right-hand sides of (20) are known from (16). System (20) can be integrated subject to the initial conditions $w_1(0) = z_1(0) = y_1(0) = 0$. We thus obtain

$$w_0 w_1 + z_0 z_1 = R_0(t),$$
$$y_0 y_1 + w_1 = S_0(t), \qquad (21)$$

where

$$R_0(t) = B(1 - B)t - 2\sqrt{B}\,[E(\sqrt{B}\,t + u_0) - E(u_0)]$$

and $\qquad (22)$

$$S_0(t) = (1 - B + \sigma\beta^2)t - 2\sqrt{B}\,(2\sigma - 1)[E(\sqrt{B}\,t + u_0) - E(u_0)],$$

where E and $E(t)$ are complete and incomplete elliptic integrals of the second kind, respectively. (See Abramowitz and Stegun 1964, pp. 589–590.)

Assuming $z_0(t) \neq 0$, $z_1(t)$ can be eliminated from (13c) with the help of (21) to obtain a single linear equation for $y_1(t)$,

$$\frac{d}{dt} y_1(t) - q(t) y_1(t) = p(t), \qquad (23)$$

where

$$q(t) = \frac{w_0 y_0}{z_0} \qquad \text{and} \qquad p(t) = \frac{1}{z_0}[R_0(t) - w_0 S_0(t)] - \sigma y_0. \qquad (24)$$

The solution of (23) in terms of the variable $x = \sqrt{B}\,t + u_0$ has a complicated form:

$$y_1 = \frac{sn\,x\,dn\,x}{\alpha\sqrt{B}} \left\{ Y_0 + \left(\frac{\alpha^2}{\beta^2}\gamma + \eta \right)(k^2 cn^2 x - E^2(x)) + 2\eta x E(x) \right.$$

$$-\frac{x^2}{2}(2-2B+\sigma\beta^2) - \frac{1}{\sqrt{B}}$$

$$\times \left[(a_1+a_2B)(x-E(x)) + (a_1-a_2B)\frac{\alpha^2}{\beta^2}E(x)\right]\Big\}$$

$$+ \frac{dn^2 x\, cn\, x}{\alpha\sqrt{B}}\{-2\eta(E(x)-E_0) + (2-2B+\sigma\beta^2)(x-u_0)\}$$

$$+ \frac{sn^2 x\, cn\, x}{\alpha\sqrt{B}}\left\{2\gamma k^2\frac{\alpha^2}{\beta^2}(E(x)-E_0) - \sigma\alpha^2 k^2(x-u_0)\right\}, \qquad (25)$$

where $dn^2 u_0 = 1/2$ and $E_0 = E(u_0)$. The other constants are

$$\gamma = (2\sigma-1)B - 1; \qquad \eta = (2\sigma-1)B + 1;$$
$$a_1 = -\sqrt{B}(1-B)u_0 + 2\sqrt{B}E_0;$$
$$a_2 = -(1-B+\sigma\beta^2)\frac{u_0}{\sqrt{B}} + 2\sqrt{B}(2\sigma-1)E_0; \qquad (26)$$
$$Y_0 = -\left(\frac{\alpha^2}{\beta^2}\gamma+\eta\right)(k^2 cn^2 u_0 + E_0^2) + 2\eta u_0 E_0 - (2-2B+\sigma\beta^2)\frac{u_0^2}{2}.$$

The first integrals (21) now provide z_1 and w_1. The zeroth-order solution (16) with $B > D > 0$ is periodic with period $\tau = 4K/\sqrt{B}$, where τ depends on B and D through k^2, the modulus of K. To determine the period T of the solutions for $\epsilon \neq 0$, we seek the constants B, D, and T so that

$$w(T) = w(0) = 0; \quad z(T) = z(0) = -B; \quad y(T) = y(0) = \sqrt{2D}, \qquad (27)$$

the conditions at $t = T$ being the same as those at $t = 0$ (see (17)). Let $T = \kappa + \tau = \kappa + \tau_0 + \epsilon\tau_1$. Then

$$w(T) = w(\kappa+\tau_0+\epsilon\tau_1) = w_0(\kappa+\tau_0+\epsilon\tau_1) + \epsilon w_1(\kappa+\tau_0+\epsilon\tau_1) + O(\epsilon^2)$$

$$= w_0(\kappa) + \tau_0\frac{dw_0}{dt}(\kappa)$$

$$+ \epsilon\left[\tau_1\frac{dw_0}{dt}(\kappa) + \tau_0\frac{dw_1}{dt}(\kappa) + w_1(\kappa)\right] + O(\epsilon^2),$$

$$z(T) = z(\kappa+\tau)$$

$$= z_0(\kappa) + \tau_0\frac{dz_0}{dt}(\kappa) \qquad (28)$$

$$+ \epsilon\left[\tau_1\frac{dz_0}{dt}(\kappa) + \tau_0\frac{dz_1}{dt}(\kappa) + z_1(\kappa)\right] + O(\epsilon^2),$$

$$y(T) = y(\kappa+\tau)$$

$$= y_0(\kappa) + \tau_0\frac{dy_0}{dt}(\kappa)$$

$$+ \epsilon \left[\tau_1 \frac{dy_0}{dt}(\kappa) + \tau_0 \frac{dy_1}{dt}(\kappa) + y_1(\kappa) \right] + O(\epsilon^2).$$

To zeroth order the solution is periodic; therefore $\tau_0 = 0$. It is also periodic to order ϵ provided the bracketed terms in (28) vanish. Using (12) and (13) for the derivatives, the conditions for periodicity to order ϵ become

$$w_1(\kappa) + B(2D)^{1/2}\tau_1 = 0,$$
$$z_1(\kappa) = 0, \tag{29}$$
$$y_1(\kappa) - B\tau_1 = 0.$$

Eliminating τ_1 from these equations, we have

$$\sqrt{2D}\, y_1(\kappa) + w_1(\kappa) = 0,$$
$$z_1(\kappa) = 0. \tag{30}$$

The integrals (21) for w_1, z_1, and y_1 evaluated at $t = \kappa$ give

$$0 - Bz_1(\kappa) = R_0(\kappa),$$
$$\sqrt{2D}\, y_1(\kappa) + w_1(\kappa) = S_0(\kappa). \tag{31}$$

Equations (30) and (31) imply that, for the periodic solutions to exist,

$$R_0(\kappa) = 0, \qquad S_0(\kappa) = 0. \tag{32}$$

Using the definitions (22) for R_0 and S_0 and noting the relation $E(4K + u_0) = E(u_0) + 4E$ for complete elliptic integrals of the second kind (Abramowitz and Stegun 1964, p. 592), we finally obtain the conditions for periodic solutions of (7) to $O(\epsilon)$:

$$(1 - B)K - 2E = 0, \tag{33}$$
$$[(1 - B + \sigma\beta^2)K - 2B(2\sigma - 1)E] = 0.$$

For each $\sigma > 1$, there is a unique pair (B_0, D_0) satisfying (33) (see Figure 7.2.2). We have assumed that $B > D$. Conditions similar to (33) can also be derived for $B < D$.

Figure 7.2.3 shows phase space projections of the stable periodic solution of (7), obtained numerically for $r = 4{,}000$ and $\sigma = 5$. The solution is nearly symmetric about the y- and z-axes.

Now we study the Lorenz system in the form (1). This system has a natural symmetry, $(x, y, z) \to (-x, -y, z)$, for all values of the parameters. Also the z-axis, $x = y = 0$, is an invariant of (1). System (1) easily shows that all trajectories that start on the z-axis remain on it and tend toward the origin $(0, 0, 0)$. This follows from (1c) since $b > 0$. Moreover, all trajectories which rotate around the z-axis do so in a clockwise direction when viewed from above the plane $z = 0$. This is verified by noting that if $x = 0$, then

Singularity Structure and Chaotic Behavior

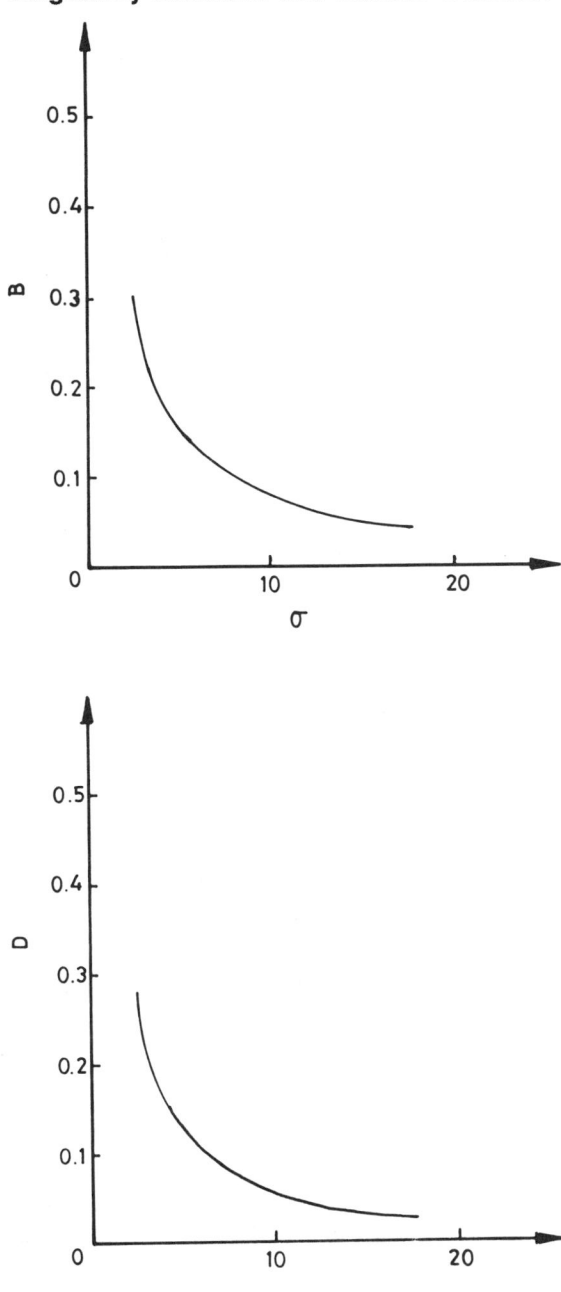

Figure 7.2.2 The values of B and D versus σ for which the stable periodic solution persists at finite r. (From Robbins 1979.)

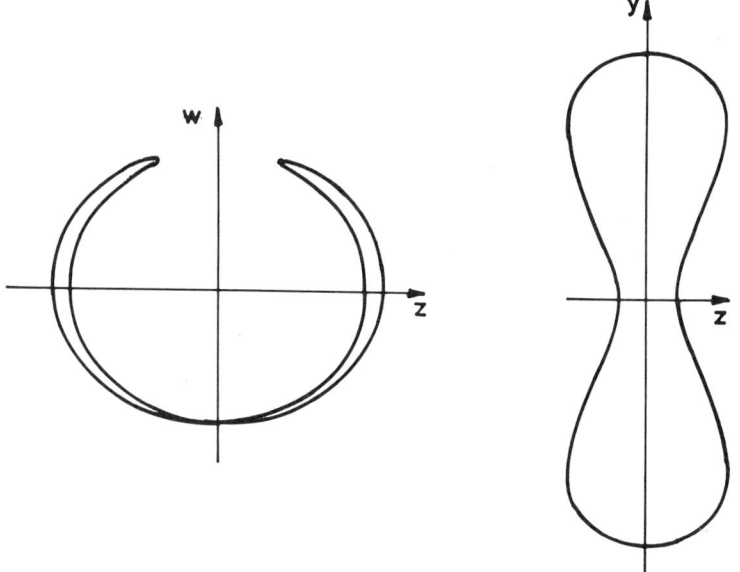

Figure 7.2.3 Phase space projections of the stable periodic solution of (7) for $r = 4{,}000$ and $\sigma = 5$. (From Robbins 1979.)

$dx/dt = \sigma y \gtrless 0$ when $y \gtrless 0$. This helps in the partial description of the periodic orbits of the system by counting the numbers of times they wind around the z-axis. This qualitative feature of the orbits does not depend on the choice of parameters; only we must pursue the same periodic orbit.

It can be easily checked that system (1), by a simple change of variables

$$x' = x, \qquad y' = y, \qquad z' = z - r - \sigma, \tag{34}$$

becomes a special case of forced dissipative systems

$$\frac{dX_i}{dt} = \sum_{j,k} a_{ijk} X_j X_k - \sum_j b_{ij} X_j + c_i, \tag{35}$$

where $\Sigma a_{ijk} X_i X_j X_k$ vanishes identically, $\Sigma b_{ij} X_i X_j$ is positive definite, and $c_1, \ldots, c_M$ are constants. If we write

$$Q = \frac{1}{2} \sum_i X_i^2, \tag{36}$$

an energylike positive definite quantity, and if $e_1, e_2, \ldots, e_M$ are the roots of the equations

$$\sum_j (b_{ij} + b_{ji})e_j = c_i, \tag{37}$$

then differentiating (36) and using (35) and (37) we can write

$$\frac{dQ}{dt} = \sum_{ij} b_{ij} e_i e_j - \sum_{ij} b_{ij}(X_i - e_i)(X_j - e_j). \tag{38}$$

The right-hand side of (38) equated to zero represents the surface of an ellipsoid E and is positive only in the interior of E. According to (36) the surfaces of constant Q are concentric spheres. In view of negative divergence (2), if S denotes one such sphere whose interior R contains the ellipsoid E, it follows that each trajectory must eventually become trapped within R.

We mentioned in Section 7.1 that the Painlevé property is closely related to the complete integrability of a system. We now give four choices of parameters, found first by Segur (1980), for which system (1) is exactly integrable. Subsequently, we shall show how these parameters follow from a detailed analysis of the singularity structure of (1). The following are possibly the only cases for which the Lorenz system is of P type.

1. $\sigma = 0$. In this case, x is a constant and system (1) is linear; therefore, it is of P type.

2. $\sigma = 1/2, b = 1, r = 0$. System (1) becomes

$$\dot{x} = \frac{1}{2}(y - x),$$
$$\dot{y} + y = -xz, \tag{39}$$
$$\dot{z} = xy - z.$$

It is easy to find the combinations

$$2x\dot{x} - \dot{z} = -x^2 + z \quad \text{and} \quad y\dot{y} + z\dot{z} = -y^2 - z^2 \tag{40}$$

of (39) leading to the two integrals

$$x^2 - z = B \exp(-t), \tag{41a}$$
$$y^2 + z^2 = A^2 \exp(-2t), \tag{41b}$$

where B and A are constant. The third integration is effected by a quadrature; the solution may be expressed in terms of elliptic functions.

3. $\sigma = 1, b = 2, r = 1/9$. System (1) becomes

$$\dot{x} = y - x, \tag{42a}$$

$$\dot{y} + y = -x\left(z - \frac{1}{9}\right), \tag{42b}$$

$$\dot{z} = xy - 2z. \tag{42c}$$

Multiplying (42a) by $2x$ and (42c) by 2 and subtracting the latter from the former, we have

$$2x\dot{x} - 2\dot{z} = -2(x^2 - 2z), \tag{43}$$

which integrates to give a first integral

$$x^2 - 2z = C \exp[-2t]. \tag{44}$$

It is now possible to express (42) as a single second-order DE in x

$$\ddot{x} + 2\dot{x} + \frac{8x}{9} + \frac{x}{2}[x - C\exp(-2t)] = 0. \tag{45}$$

This equation, after several changes of variables, may be transformed into P_{II}. The solutions are, therefore, predictable.

4. $\sigma = 1/3$, $b = 0$, r arbitrary. In this case (1a) is written as $y = 3\dot{x} + x$, and the whole system can be transformed to

$$\frac{\ddot{x}x - \ddot{x}\dot{x} + x^3\dot{x}}{\ddot{x}x - \dot{x}^2 + x^4/4} = -\frac{4}{3}, \tag{46}$$

which integrates to

$$x\ddot{x} - \dot{x}^2 + \frac{x^4}{4} = C \exp\left(-\frac{4t}{3}\right). \tag{47}$$

A further change of variables

$$T = \exp\left[\frac{-t}{3}\right], \qquad x(t) = TW(T), \tag{48}$$

transforms (47) into

$$TW\ddot{W} + W\dot{W} - T\dot{W}^2 + \frac{9}{4}TW^4 = 9CT, \tag{49}$$

which is a special case of P_{III} (see Section 8.9).

Thus each of these four cases are of P type and therefore predictable. These isolated points of parameter space are embedded in larger regions in which system (1) has first integrals, although it may not be completely integrable.

5. If $b = 1$, $r = 0$, then (41b) is obtainable for any σ. The existence of this integral rules out chaotic behavior.

6. If $b = 2\sigma$, then for any (r, σ) there exists a first integral

$$x^2 - 2\sigma z = C \exp(-2\sigma t). \tag{50}$$

Singularity Structure and Chaotic Behavior

This again precludes ergodic behavior of the solutions of (1).

These different possibilities of completely integrable or partly integrable cases of the Lorenz system are shown in Figure 7.2.4.

7. $b = 1 - 3\sigma$, r arbitrary. Tabor and Weiss (1981) conjectured from their analysis (which we give below) that a first integral may exist; they were not able to actually find one. However, in this case, (1) does not possess the Painlevé property.

To study the singularity structure of (1), we start from its scaled form (10), but in the original notation, namely,

$$\frac{dx}{dt} = y - \sigma\epsilon x,$$
$$\frac{dy}{dt} = -xz + x - \epsilon y, \qquad (51)$$
$$\frac{dz}{dt} = xy - \epsilon Bz.$$

There is a slight change in notation here; we have replaced b by B. We recall that $\epsilon = (\sigma r)^{-1/2}$.

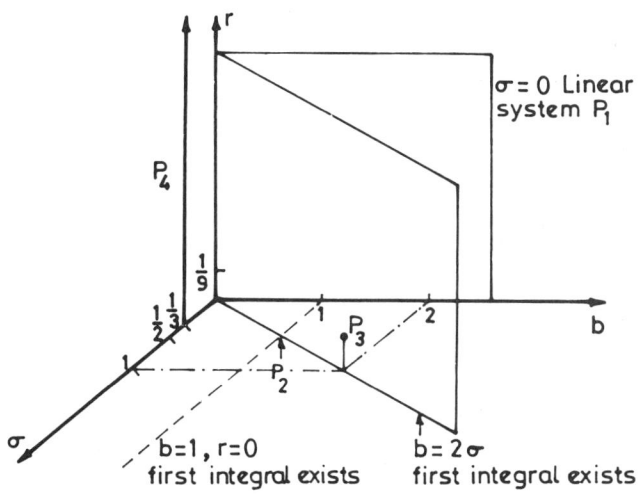

Figure 7.2.4 Map of parameter space for Lorenz model (1), showing where equations have the Painlevé property and where they admit exact or first integrals. (From Segur 1980.)

We consider the leading order behavior of the solution of (51) about the singularity $t = t_*$ in the form

$$x = \frac{a}{(t-t_*)^\alpha}, \quad y = \frac{b}{(t-t_*)^\beta}, \quad z = \frac{c}{(t-t_*)^\gamma}. \tag{52}$$

Substituting (52) into (51), we have

$$\begin{aligned}
-a\alpha\tau^{-(\alpha+1)} &= b\tau^{-\beta} - \sigma\epsilon a\tau^{-\alpha}, \\
-b\beta\tau^{-(\beta+1)} &= -ac\tau^{-(\alpha+\gamma)} + a\tau^{-\alpha} - \epsilon b\tau^{-\beta}, \\
-c\gamma\tau^{-(\gamma+1)} &= ab\tau^{-(\alpha+\beta)} - \epsilon Bc\tau^{-\gamma},
\end{aligned} \tag{53}$$

where $\tau = (t - t_*)$.

A consistent dominant balance takes place if the left-hand sides of (53) balance the first terms on the right-hand sides. This gives

$$\alpha + 1 = \beta,$$
$$\beta + 1 = \alpha + \gamma,$$
$$\gamma + 1 = \alpha + \beta,$$

so

$$\alpha = 1, \quad \beta = 2, \quad \gamma = 2. \tag{54}$$

The corresponding coefficients of the powers of τ also equal if

$$a = \pm 2i, \quad b = \mp 2i, \quad c = -2. \tag{55}$$

The behavior in the neighborhood of the singularity is now sought in detail by making the Ansatz

$$\begin{aligned}
x &= \frac{2i}{t-t_*} \sum_{j=0}^{\infty} a_j (t-t_*)^j, \\
y &= -\frac{2i}{(t-t_*)^2} \sum_{j=0}^{\infty} b_j (t-t_*)^j, \\
z &= -\frac{2}{(t-t_*)^2} \sum_{j=0}^{\infty} c_j (t-t_*)^j.
\end{aligned} \tag{56}$$

On substituting these Laurent series into (51) and equating the coefficients of equal powers, we find that

$$a_0 = b_0 = c_0 = 1 \tag{57}$$

(which agrees with the leading order behavior), and

$$a_1 = \frac{(3\sigma - 2B - 1)\epsilon}{6}, \quad b_1 = -\sigma\epsilon, \quad c_1 = \frac{(B-1-3\sigma)\epsilon}{3}. \tag{58}$$

Singularity Structure and Chaotic Behavior

For $a_j, b_j, c_j, j = 2, 3, 4, \ldots$, we obtain the recurrence relations in matrix form:

$$\begin{bmatrix} j-1 & 1 & 0 \\ 2 & j-2 & 2 \\ 2 & 2 & j-2 \end{bmatrix} \begin{bmatrix} a_j \\ b_j \\ c_j \end{bmatrix} = \begin{bmatrix} -\sigma\epsilon a_{j-1} \\ -2\sum_{k=1}^{j-1} a_{j-k}c_k - a_{j-2} - \epsilon b_{j-1} \\ -2\sum_{k=1}^{j-1} a_{j-k}b_k - \epsilon B c_{j-1} \end{bmatrix}. \tag{59}$$

The coefficient matrix on the left is singular when $j = 2$ or $j = 4$. For the inhomogeneous system (59), written as $PA_j = Q$, to have a (nonunique) solution the matrix P and the augmented matrix $[P \ Q]$ must have the same rank. The corresponding consistency condition for $j = 2$ is

$$\epsilon^2(6\sigma^2 - \sigma B - 2\sigma) = B(B-1)\epsilon^2. \tag{60}$$

The case $j = 4$ leads to two consistency conditions:

$$\frac{\epsilon^2}{9}(B-1)[57(\sigma-1) - 15(B-1) + 24] = \epsilon^2 \sigma(2\sigma - 1) \tag{61}$$

and

$$2(1 + B - \sigma)a_1^2 c_1 \epsilon + \epsilon^2 \left(\frac{5\sigma}{3} - \frac{B}{3} - \frac{5}{3}\right) B a_1 c_1$$
$$+ \epsilon^2 \left(3\sigma - 3B + \frac{5}{3}\right) 2\sigma a_1^2$$
$$- \frac{\epsilon^4 \sigma}{18} \left(\sigma^2(15B - 24) - \sigma B(7B - 3) - B^2 \left(2B - \frac{22}{3}\right)\right.$$
$$\left. + \left(16\sigma - \frac{44B}{3} - \frac{8}{3}\right)\right) + \frac{\epsilon^2 B}{6}(3\sigma - B - 5) = 0. \tag{62}$$

The coefficients matrices (a_2, b_2, c_2) ad (a_4, b_4, c_4) will each be determined up to an arbitrary constant. Thus, subject to (60)–(62), the general solution at the singularity will depend on three arbitrary parameters, one each appearing in the solution of (59) at $j = 2$ and $j = 4$, and the arbitrary position of the pole t_*. Equations (60)–(62) imply restrictions on the parameters (σ, ϵ, B). The other coefficients $(a_j, b_j, c_j), j = 3, j \geq 5$ can now be found recursively.

Conditions (60)–(61) for $\epsilon \neq 0$ specify σ and B, while (62) determines ϵ. For these values of (σ, ϵ, B), the Laurent expansion (56) is valid, and since under these conditions system (1) has poles as its only movable singularities, the Painlevé condition is satisfied. Case (1) of the existence of the first integral we have noted above (due originally to Segur 1982) makes system (51) linear and Ansatz (56) does not apply. It is easy to check that the values of (σ, ϵ, B) corresponding to cases (2)–(4) satisfy (60)–(62). The values of (σ, ϵ, B) for cases (5) and (6) do not satisfy (60)–(62), so the Painlevé property does not hold, even though one first integral exists. Case 7 corresponds to $B = 1 - 3\sigma$, with $r(\epsilon)$ arbitrary, and satisfies the consistency condition (60), but not all the conditions for the Painlevé property are satisfied. Yet, Tabor and Weiss (1981) conjecture that a first integral might exist.

We have thus seen that the Lorenz system does not, in general, satisfy the Painlevé condition. We must weaken the latter and permit logarithmic movable singularities in the solution. In other words, we seek a psi series (Hille 1976) for (51) in the form (we set $t_* = 0$ for convenience in writing)

$$x = \frac{2i}{t} \sum_{k=0}^{\infty} \sum_{j=0}^{\infty} a_{kj} t^j (t^2 \ln t)^k$$

$$y = \frac{-2i}{t} \sum_{k=0}^{\infty} \sum_{j=0}^{\infty} b_{kj} t^j (t^2 \ln t)^k, \qquad (63)$$

$$z = \frac{-2}{t^2} \sum_{k=0}^{\infty} \sum_{j=0}^{\infty} c_{kj} t^j (t^2 \ln t)^k.$$

Note that this double series is in t and $t^2 \ln t$, and not in t and $\ln t$, in view of resonances appearing first at $j = 2$ and $j = 4$. Substituting (63) into (51) leads to rather cumbersome recursion relations:

$$\begin{bmatrix} 2k+j-1 & 1 & 0 \\ 2 & 2k+j-2 & 2 \\ 2 & 2 & 2k+j-2 \end{bmatrix} \begin{bmatrix} a_{k,j} \\ b_{k,j} \\ c_{k,j} \end{bmatrix} \qquad (64)$$

$$= \begin{bmatrix} -(k+1)a_{k+1,j-2} - \sigma\epsilon a_{k,j-1} \\ -(k+1)b_{k+1,j-2} - a_{k,j-2} - \epsilon b_{k,j-1} - 2\sum_{m=1}^{j} a_{k,j-m} c_{0,m} - 2\sum_{m=0}^{j-1} a_{0,j-m} c_{k,m} \\ \qquad\qquad -2\sum_{l=1}^{k-1}\sum_{m=0}^{j} a_{k-l,j-m} c_{l,m} \\ -(k+1)c_{k+1,j-2} - \epsilon B c_{k,j-2} - 2\sum_{m=1}^{j} a_{k,j-m} b_{0,m} - 2\sum_{m=0}^{j-1} a_{0,j-m} b_{k,m} \\ \qquad\qquad -2\sum_{l=1}^{k-1} a_{k-l,j-m} b_{l,m} \end{bmatrix}.$$

The coefficients a_{kj}, b_{kj}, and c_{kj} are obtained by substituting (63) into (51) and equating equal powers of $t^{j-2}(t^2 \ln t)^k$ on both sides. It is clear

Singularity Structure and Chaotic Behavior

from the summations in (64) that the recursion relations are not closed for (a_{kj}, b_{kj}, c_{kj}), $k, j \neq 0$, since to find the latter one must, in general, know the coefficients $(a_{\alpha\beta}, b_{\alpha\beta}, c_{\alpha,\beta})$ for all (α, β) in the range

$$0 \leq \beta \leq j,$$
$$0 \leq 2\alpha + \beta \leq 2k + j. \tag{65}$$

Again, the coefficient matrix in (64) is singular for the following cases:

$$k = j = 0; \tag{66a}$$

$$2k + j = 2, \text{ i.e., } k = 1, \quad j = 0,$$
$$k = 0, \quad j = 2; \tag{66b}$$

$$2k + j = 4, \text{ i.e., } k = 2, \quad j = 0,$$
$$k = 1, \quad j = 2, \tag{66c}$$
$$k = 0, \quad j = 4.$$

At each singular level, consistency conditions are imposed and the arbitrary parameters introduced in the solutions of (64). Of the cases in (66), two corresponding to $k = 0, j = 2$ and $k = 0, j = 4$, which do not involve logarithmic terms in (63), have already been discussed in the expansions (56).

It turns out that the recursion relations for leading order terms involving the expansions in $t^2 \ln t$ only corresponding to $j = 0$ and arbitrary k are closed. In this case, (64) assumes the form

$$\begin{bmatrix} 2k-1 & 1 & 0 \\ 2 & 2k-2 & 2 \\ 2 & 2 & 2k-2 \end{bmatrix} \begin{bmatrix} a_{k,0} \\ b_{k,0} \\ c_{k,0} \end{bmatrix} = \begin{bmatrix} 0 \\ -2\sum_{l=1}^{k-1} a_{k-l,0} c_{l,0} \\ -2\sum_{l=1}^{k-1} a_{k-l,0} b_{l,0} \end{bmatrix}, \quad k \geq 2. \tag{67}$$

The consistency conditions for $k = 0, j = 0$; $k = 1, j = 0$; $k = 2, j = 0$, three of the cases enumerated in (66), are identically satisfied without restrictions on the parameters (σ, ϵ, B). Indeed, one can easily check that

$$(a_{00}, b_{00}, c_{00}) = (1, 1, 1), \tag{68a}$$
$$(a_{10}, b_{10}, c_{10}) = (\lambda, -\lambda, -\lambda), \tag{68b}$$
$$(a_{20}, b_{20}, c_{20}) = (\gamma, -3\gamma, 2\gamma) + (0, 0, \lambda^2). \tag{68c}$$

For (68a), one may refer to (57); (68b), (68c) involve two arbitrary parameters λ and γ, required by the solutions at the respective levels. These parameters are fixed by reference to other levels in (66). For example, the coefficient λ is determined by the consistency conditions at $k = 0, j = 2$. Writing (64) explicitly for this set of k, j we have

$$\begin{bmatrix} 1 & 1 & 0 \\ 2 & 0 & 2 \\ 2 & 2 & 0 \end{bmatrix} \begin{bmatrix} a_{02} \\ b_{02} \\ c_{02} \end{bmatrix} = -\begin{bmatrix} \lambda \\ -\lambda \\ -\lambda \end{bmatrix} + \begin{bmatrix} -\sigma \epsilon a_{01} \\ -a_{00} - \epsilon b_{01} - 2a_{01}c_{01} \\ -\epsilon B c_{01} - 2a_{01}b_{01} \end{bmatrix}. \tag{69}$$

This system of linear inhomogeneous equations for a_{02}, b_{02}, and c_{02} is consistent, provided

$$3\lambda = 2a_{01}(b_{01} - \sigma\epsilon) + \epsilon B c_{01}. \tag{70}$$

Using the values of $a_{01} = a_1$, $b_{01} = b_1$, $c_{01} = c_1$ from (58), we get

$$\lambda = \frac{\epsilon^2}{9}[B(B-1) - 6\sigma^2 + \sigma B + 2\sigma]. \tag{71}$$

Equation (71) shows that if the consistency condition (60) is satisfied (for the expansions (56), which involve a_{0j}, b_{0j}, and c_{0j}), then $\lambda = 0$ and (68b) gives $a_{10} = b_{10} = c_{10} = 0$, so the logarithmic terms in the psi series enter at powers of $t^4 \ln t$. We shall briefly discuss this important special case.

The parameter γ in (68) is fixed by considering the consistency condition at another point listed in (66), namely $k = 1, j = 2$. It can be shown that the consistency condition at this level yields the unknown parameter γ:

$$\gamma = \frac{\lambda \epsilon^2}{5}\left[\frac{\sigma}{3}(2\sigma - 1) + \left(\frac{5 - 22\sigma}{3}\right)\left(\frac{B-1}{3}\right) + 6\left(\frac{B-1}{3}\right)^2\right]. \tag{72}$$

The unknown parameter introduced at $k = 1, j = 2$ is specified by the consistency condition at $k = 0, j = 4$. Finally, we recall that the unknown parameters at $k = 0, j = 2$, and $k = 0, j = 4$ which are the same as enter $(a_j, b_j, c_j), j = 2, j = 4$, in the expressions (56) are two of three (in general, complex) constants of integration in the general solution. These considerations show that the recursion relations (64) are well defined; therefore, (63) represents the general form of the formal expansion for system (51) about a singular point. Even though, in general, the recursion relation (64) is not closed and the analysis is not complete, the above discussion does indicate where the resonances appear and how the various arbitrary parameters are fixed at subsequent stages. The analysis suggests the means by which certain parameter regimes of a system, for which it may have some preferred properties, may be identified.

The leading order logarithmic behavior (for which a closed recursion relation exists) can be studied more closely to throw some light on the possible (partial) integrability of system (51). For this purpose we define

Singularity Structure and Chaotic Behavior

the generating functions

$$\Theta(\hat{x}) = \sum_{k=0}^{\infty} a_{k0} \hat{x}^k,$$

$$\Phi(\hat{x}) = \sum_{k=0}^{\infty} b_{k0} \hat{x}^k, \qquad (73)$$

$$\Psi(\hat{x}) = \sum_{k=0}^{\infty} c_{k0} \hat{x}^k,$$

where

$$\hat{x} = t^2 \ln t. \qquad (74)$$

In view of the recursion relations (67) for the coefficients in the leading logarithmic behavior, we can easily deduce that $\Theta(\hat{x})$, $\Phi(\hat{x})$, and $\Psi(\hat{x})$ satisfy the coupled system of ODE

$$2\hat{x} \frac{d\Theta}{d\hat{x}} - \Theta + \Phi = 0, \qquad (75a)$$

$$2\hat{x} \frac{d\Phi}{d\hat{x}} - 2\Phi + 2\Theta\Psi = 0, \qquad (75b)$$

$$2\hat{x} \frac{d\Psi}{d\hat{x}} - 2\Psi + 2\Theta\Phi = 0. \qquad (75c)$$

Eliminating Φ from (75a) and (75c) and integrating, we get

$$\Psi = \Theta^2 - 3\lambda \hat{x}, \qquad (76)$$

where 3λ is the constant of integration found from using the coefficients (68) of the series solution. Equation (75a) gives

$$\Phi = \Theta - 2\hat{x} \frac{d\Theta}{d\hat{x}}. \qquad (77)$$

Using (76) and (77) in (75b), we get a single second-order equation for Θ,

$$2\hat{x}^2 \frac{d^2\Theta}{d\hat{x}^2} = \hat{x} \frac{d\Theta}{d\hat{x}} - 3\lambda \hat{x} \Theta + \Theta(\Theta^2 - 1), \qquad (78)$$

with initial conditions

$$\Theta = 1, \quad \frac{d\Theta}{d\hat{x}} = \lambda, \quad \frac{d^2\Theta}{d\hat{x}^2} = 2\gamma, \qquad (79)$$

again obtained from (68). The differential operator in (78) is of Euler–Cauchy type. The transformation

$$\Theta = \hat{x}^{1/2} f(\hat{x}^{1/2}) \qquad (80)$$

changes (78) to

$$f'' = 2f^3 - 6\lambda f, \tag{81}$$

where prime denotes differentiation with respect to the argument of f. Multiplying (81) by f', integrating, and using (79), we get

$$f'^2 = f^4 - 6\lambda f^2 + 7\lambda^2 - 10\gamma. \tag{82}$$

The solution of this equation in terms of the Jacobi elliptic function is

$$f(\hat{x}) = \frac{\alpha}{sn(\alpha\hat{x}, k)}, \tag{83}$$

where

$$\alpha = 3\lambda + (2\lambda^2 + 10\gamma)^{1/2}, \tag{84}$$

and the square of the modulus

$$k^2 = 1 - \left(\frac{\beta}{\alpha}\right)^2, \quad \beta = 3\lambda - (2\lambda^2 + 10\gamma)^{\frac{1}{2}}; \tag{85}$$

we assume that $\alpha > 0$, $\beta > 0$. Thus, the solution of (78) subject to (79) is

$$\Theta(\hat{x}) = \frac{\alpha \hat{x}^{1/2}}{sn(\alpha \hat{x}^{1/2}, k)}. \tag{86}$$

Substituting $\Theta(\hat{x})$ as given by (73) and (86) in (63), we have

$$x(t) = \frac{2i}{t} \frac{\alpha t (\ln t)^{1/2}}{sn(\alpha t (\ln t)^{1/2}, k)} + O(t). \tag{87}$$

For different choices of α and β, the solution of (82) can be expressed in terms of combinations of Jacobi elliptic functions. For the special choice

$$2\lambda^2 + 10\gamma = 0, \tag{88}$$

we get $\beta = \alpha = 3\lambda$ and $k^2 = 0$ (see (84)–(85)) and $sn(z, 0) = \sin z$, while

$$7\lambda^2 - 10\gamma = 0 \tag{89}$$

gives $k^2 = 1$ and $sn(z, 1) = \sinh z$. Using (71) and (72) for λ and γ we can simplify conditions (88) and (89). The former leads to

$$2\lambda^2 + 10\gamma = \gamma\epsilon^2 \left[\frac{14}{3}(B-1) + 4 - 14\sigma\right] \left(\frac{B-1}{3}\right) = 0, \tag{90}$$

which, for $\lambda\epsilon^2 \neq 0$, is satisfied by either

$$B = 1 \tag{91}$$

or

$$B = 3\sigma + \frac{1}{7}. \tag{92}$$

Condition (89) becomes

$$7\lambda^2 - 10\gamma = \frac{7\lambda\epsilon^2}{9}\left[(B-1)^2 + (\sigma+1)(B-1) + 3\sigma(1-2\sigma)\right] - 2\lambda\epsilon^2$$

$$\times \left[\frac{\sigma}{3}(2\sigma - 1) + \left(\frac{5-22\sigma}{3}\right)\left(\frac{B-1}{3}\right) + 6\left(\frac{B-1}{3}\right)^2\right]$$

$$= 0, \tag{93}$$

which, for $\lambda\epsilon^2 \neq 0$, is satisfied if either

$$B = 1 + 9\sigma \tag{94}$$

or

$$B = \frac{2}{5}(3\sigma + 1). \tag{95}$$

We recall case (5) given by Segur (1982) with $B = 1, r = 0$, and σ arbitrary, for which a first integral exists. This was not indicated by the structure of the singularities. Equation (91) gives some indication that first integrals may exist when $k^2 = 0$ or $k^2 = 1$, so the logarithmic leading behavior can be expressed in terms of trigonometric or hyperbolic functions. However, this must be treated as conjectural. Equations (91), (92), (94), and (95) give special values of B for which the corresponding solutions may have some preferred property near the singularity. Figure 7.2.5 gives all the special lines in the (B, σ) plane, together with the line $B = 2\sigma$ for which a first integral is known and the line $B = 1 - 3\sigma$ for which Tabor and Weiss (1981) conjecture that an integral exists. Intersection points of these lines, namely $(1/2, 1)$, $(1/7, 4/7)$, and $(-1/7, -2/7)$, are likely to show up some special behavior of the solution near the singularity or point to the existence of a first integral.

Before we relate the above analytic structure of the singularities to the numerical solution, we note that for the special case $\lambda = 0$, for which a_{10}, b_{10}, c_{10} each becomes zero (see ((68))), the psi series have the form

$$x = \frac{2i}{t} \sum_{k=0}^{\infty} \sum_{j=0}^{\infty} a_{kj} t^j (t^4 \ln t)^k,$$

$$y = -\frac{2i}{t} \sum_{k=0}^{\infty} \sum_{j=0}^{\infty} b_{kj} t^j (t^4 \ln t)^k, \tag{96}$$

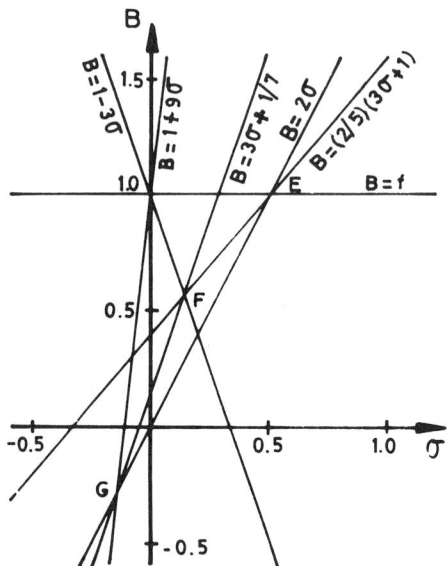

Figure 7.2.5 Special lines in the (σ, B) plane for which system (1) has either a first integral or some other preferred property. (From Tabor and Weiss 1981.)

$$z = -\frac{2}{t^2} \sum_{k=0}^{\infty} \sum_{j=0}^{\infty} c_{kj} t^j (t^4 \ln t)^k.$$

Tabor and Weiss show that, subject to the consistency conditions that arise, the recursion relations for the coefficients in (96) are well defined. Since the analysis for this case is quite similar to one we have carried out for the general case, we omit the details.

Tabor and Weiss found the numerical solution of (51) by two different ways—Taylor series expansions and a finite difference approach. Integration by Taylor series expansion was performed by using an algorithm developed by Chang and Corliss (1980). This algorithm determines accurately the radius of convergence of a Taylor series in the complex plane, locates singularities nearest the real axis, and identifies their leading order behavior. Tabor and Weiss (1981) use a predictor-corrector finite difference scheme to find more accurately the order and multivaluedness of the singularities. For this purpose they write (51) in the complex t-plane by setting $t = u + iv$ and

$$x = x_R(u, v) + i x_I(u, v),$$

Singularity Structure and Chaotic Behavior

$$y = y_R(u,v) + iy_I(u,v), \tag{97}$$

$$z = z_R(u,v) + iz_I(u,v),$$

where R and I stand for real and imaginary parts, respectively. Substituting (97) into (51), we obtain six (real) first-order equations:

$$\frac{dx_R}{du} = y_R - \sigma \epsilon x_R,$$

$$\frac{dx_I}{du} = y_I - \sigma \epsilon x_I,$$

$$\frac{dy_R}{du} = -x_R z_R + x_I z_I + x_R - \epsilon y_R,$$

$$\frac{dy_I}{du} = -x_R z_I + x_I z_R + x_I - \epsilon y_I, \tag{98}$$

$$\frac{dz_R}{du} = x_R y_R - x_I y_I - \epsilon B z_R,$$

$$\frac{dz_I}{du} = x_R y_I - x_I y_R - \epsilon B z_1.$$

To get the corresponding DE with respect to v, we use the Cauchy–Riemann conditions. We do not dwell upon the details of the numerical scheme and the checks which were applied by Tabor and Weiss to ensure the accuracy of the algorithm both on the real t-line and the complex t-plane. We summarize the numerical results. The initial conditions chosen by Tabor and Weiss were $(x_0, y_0, z_0) = (0.0, 0.0602, 0.1626)$; the parametric values were $\sigma = 10$, $B = 8/3$, the same as Lorenz assumed. The parameter r was varied beyond the critical value 24.74 (see discussion above equation (5)). A strange attractor (see Section 7.1) first appears at $r \simeq 25$. As the value of r is increased, the attractor shrinks to a periodic orbit. At a substantially higher value of $r \sim 166$, the periodic orbit starts to exhibit turbulent bursts and undergoes a complicated sequence of bifurcations, and then there appears a new strange attractor. To be specific, we describe what happens in the vicinity of one such burst, say at $r = 166.1$, corresponding to $\epsilon = 0.0245366$. While Figure 7.2.6 shows the real-time behavior of $z(t)$ with a fairly low degree of resolution, Figure 7.2.7 shows the burst region in more detail. Also shown is the corresponding singularity structure *nearest* the real axis in the complex t-plane. The positions of the singularities follow closely the location of the maxima of the real-time solution; moreover, the amplitude of the maxima is a fairly smooth function of the distance of the corresponding singularity from the real axis. This dependence turns out to be inverse quadratic in agreement with the leading order behavior of the solution (63) (see Figure 7.2.8). The correlation between real-time

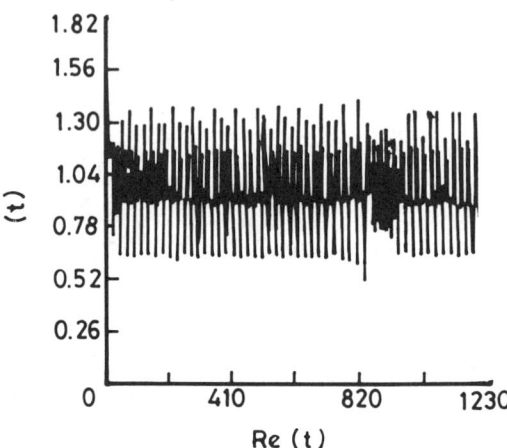

Figure 7.2.6 Real-time behavior of $z(t)$ for $\epsilon = 0.0245366$, showing turbulent burst beginning at about $t = 820$. Irregularity of burst is exaggerated by low degree of resolution. Initial conditions are $(x_0, y_0, z_0) = (0.0, 0.0602, 0.1626)$. (From Tabor and Weiss 1981.)

extrema and the singularity position is of considerable aid in ascertaining the location of the latter. Figure 7.2.7 shows clearly that the lower line of singularities is associated with the large-scale oscillations, while the upper line (of singularities) shows the small-scale oscillations. The frequency of oscillations is determined by the spacing of the singularities. This figure also manifestly shows how the burst is started by a singularity from the lower line moving downward toward the real axis. Then there follows an apparent rearrangement of the double line of poles into a single line. This arrangement unscrambles itself back into the double line structure of the postburst periodic regime. We have discussed the behavior of the solution with references to the arrays of the singularities nearest the real t-axis. There are distant arrays of poles which probably do not affect significantly the real-time behavior of the trajectories.

Figure 7.2.9 shows how the analytic structure of the solution (with regard to singularities) and the nature of the actual solution change (for $\sigma = 10$, $B = 10/3$) as ϵ is increased from zero. Here we show only the nearest two singularities. At $\epsilon = 0$, we observe a symmetric periodic orbit and the corresponding regular lattice of poles. As ϵ is increased, say to 0.020412 ($r = 240$), the orbit distorts in the manner shown, and the singularities (which are no longer simple poles) move closer to the real axis and shift relative to each other. As ϵ moves up to 0.021320, the orbit has already undergone its

Singularity Structure and Chaotic Behavior

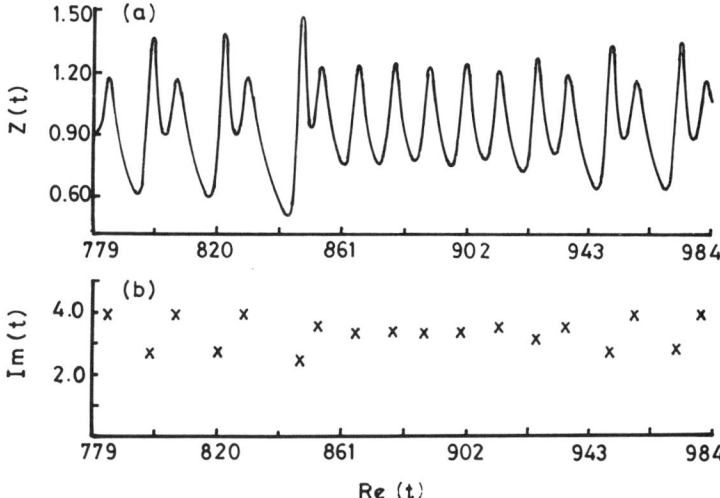

Figure 7.2.7 Turbulent burst in greater detail: (a) real-time solutions, (b) corresponding singularity structure in the complex t-plane. Only singularities (marked ×) nearest real axis are shown. (From Tabor and Weiss 1981.)

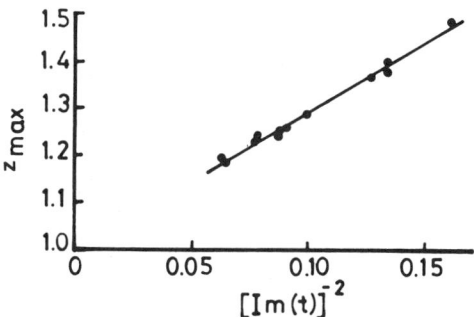

Figure 7.2.8 Amplitude of real-time extrema (z_{max}) versus inverse square of distance of associated singularity from real-time axis. (From Tabor and Weiss 1981.)

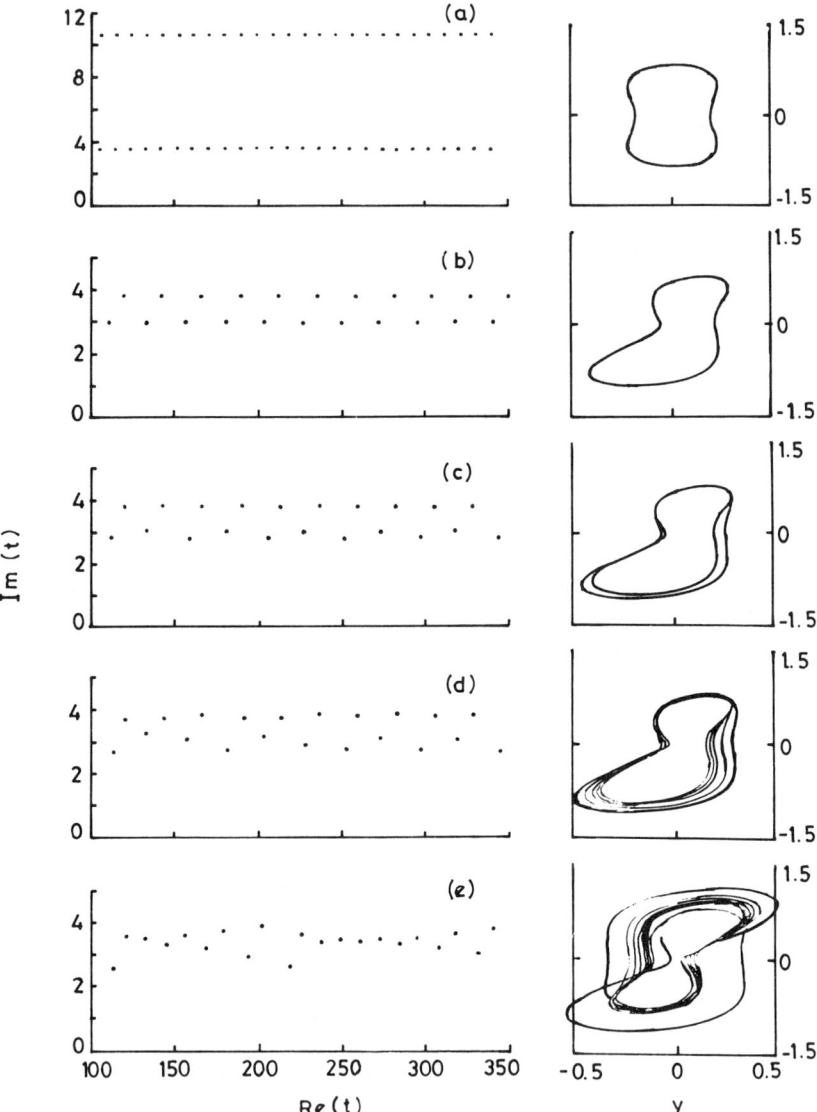

Figure 7.2.9 Real-time solutions (plotted in (x,y) plane) and associated singularity structure for (a) $\epsilon = 0$, (b) $\epsilon = 0.020412$, (c) $\epsilon = 0.021320$, (d) $\epsilon = 0.022086$, and (e) $\epsilon = 0.022361$. (From Tabor and Weiss 1981.)

Singularity Structure and Chaotic Behavior

first period-doubling bifurcation (see Gleick 1988). This becomes apparent from the analytic structure; the lowest line of singularities bifurcates into two distinct levels. As ϵ crosses 0.022086 ($r = 205$), the orbit has undergone further bifurcation, and new levels of poles appear. At a slightly higher value of $\epsilon = 0.022361$ ($r = 200$), the orbit has experienced many more bifurcations and the singularities have lost their ordered positions. They show a random pattern reflecting the randomness of the motion itself. It is quite clear from the above discussion that considerable information can be obtained by knowing the singularity structure of the given system.

The Painlevé condition determines when the system is integrable. When the movable singularities are logarithmic as well as poles, the analysis may hint at the set of parameters for which a first integral exists or point to a set of parameters for which the solutions may have some preferred properties. In the latter cases, the psi series in t and $t^m \ln t$ (m an integer) may not be found in a closed recursive form, yet its leading behavior throws much light on the special form of solutions for certain cases. However, there are many systems which are integrable and yet do not satisfy the Painlevé condition.

We have given some numerical results to show the correlation between singularity structure, as it grows more complex and random, and the evolution of a periodic orbit toward chaos. We end this section by quoting some other numerical results for the Lorenz system, due to Shimizu and Morioka (1978). This work also has some analytic interest. The study concerns the parametric ranges $r > r_T = \sigma(\sigma + b + 3)/(\sigma - b - 1) > 1$ and $\sigma > b + 1$ (see discussion below equation (4)) for which the Lorenz system is unstable about its (nonzero) critical points. First, the form (1) is rewritten as follows:

$$\ddot{X} + (\sigma + 1)\dot{X} = -\sigma(r - 1)X(X^2 - 1 + m), \tag{99a}$$

$$\dot{m} = -b\left[m - \left(\frac{2\sigma}{b} - 1\right)X^2\right], \tag{99b}$$

where

$$X = \frac{x}{\sqrt{2\sigma(r-1)}},$$

$$m = \frac{z}{r-1} - X^2 = \left[\frac{z(0)}{r-1} - X^2(0)\right]e^{-bt}$$

$$+ (2\sigma - b) \int_0^t e^{-b(t-t')} X^2(t') \, dt'. \tag{100}$$

Since $\sigma > b + 1$, it follows from (100) that m is nonnegative for large t. Therefore, (99a) may be regarded as describing a damped motion under the force $f = -\sigma(r-1)X(X^2 - 1 + m)$. This force f has the following simple properties. It depends on the history of the motion via m (see (100)) and is proportional to $\sigma(r-1)$. System (99) shows that f is positive (negative) if $X(X^2 - 1 + m) < 0 \ (> 0)$. The function m increases (decreases) if $m - (2\sigma/b - 1)X^2 < 0 \ (> 0)$. The curves $X^2 - 1 + m = 0$ and $m - (2\sigma/b - 1)X^2 = 0$ intersect at the points $X = \pm\sqrt{b/2a}, m = 1 - b/2\sigma$. These are just the points $x = y = \pm\sqrt{b(r-1)}, z = r - 1$ of the original system (1), which were shown to be unstable. The force f along the X-axis and the velocity component $\dot{m}$ along the m-axis depend only on the coordinates X and m. The velocity component $\dot{m}$ does not depend on the parameter r so that, for fixed b and σ, it does not change with r. The force f, on the other hand, depends linearly on r so that as r is increased the trajectories in the (X,m) plane tend to shrink, particularly in the m-direction.

Shimizu and Morioka carried out a numerical investigation of system (99) with $\sigma = 16, b = 4, \Delta t = 0.01$ using a standard Runge–Kutta–Gill integration routine. The critical value r_T in this case is 33.45. They searched for periodic solutions in the large r regime. The existence of periodic and turbulent regions is shown schematically in Figure 7.2.10: P_j and T_j ($j = 0, 1, 2, 3$) denote the periodic and turbulent regions, respectively. These regions alternate as r varies. The shape of the limit cycles in each periodic region is shown in Figures 7.2.11a–7.2.11f. The following points emerge from a scrutiny of these configurations: (1) As r increases, symmetric and asymmetric limit cycles (with respect to the m-axis) appear alternately. The transformation $(X,m) \to (-X,m)$ leaves (99) invariant so that the asymmetric limit cycles under this configuration map into their mirror images. (2) The limit cycles assume simpler shapes as r increases. (3) The width of the periodic region P_j decreases as r decreases.

Shimizu and Morioka also observe that if r is decreased in each periodic region from its upper end value to the lower end value, the periodic solution bifurcates successively and the symmetric periodic solution gradually becomes asymmetric. Another observation is that the asymmetric limit cycle shown in Figure 7.2.11d seems to change into the symmetric cycle 7.2.11e gradually, without passing through a turbulent region. These results and the schematic Figures 7.2.11 suggest that an infinite number of limit

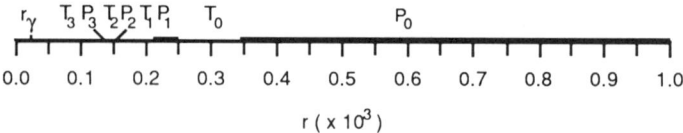

Figure 7.2.10 The r dependence of turbulent and periodic regions. P_j = periodic regions; T_j = turbulent regions. (From Shimizu and Morioka 1978.)

Singularity Structure and Chaotic Behavior

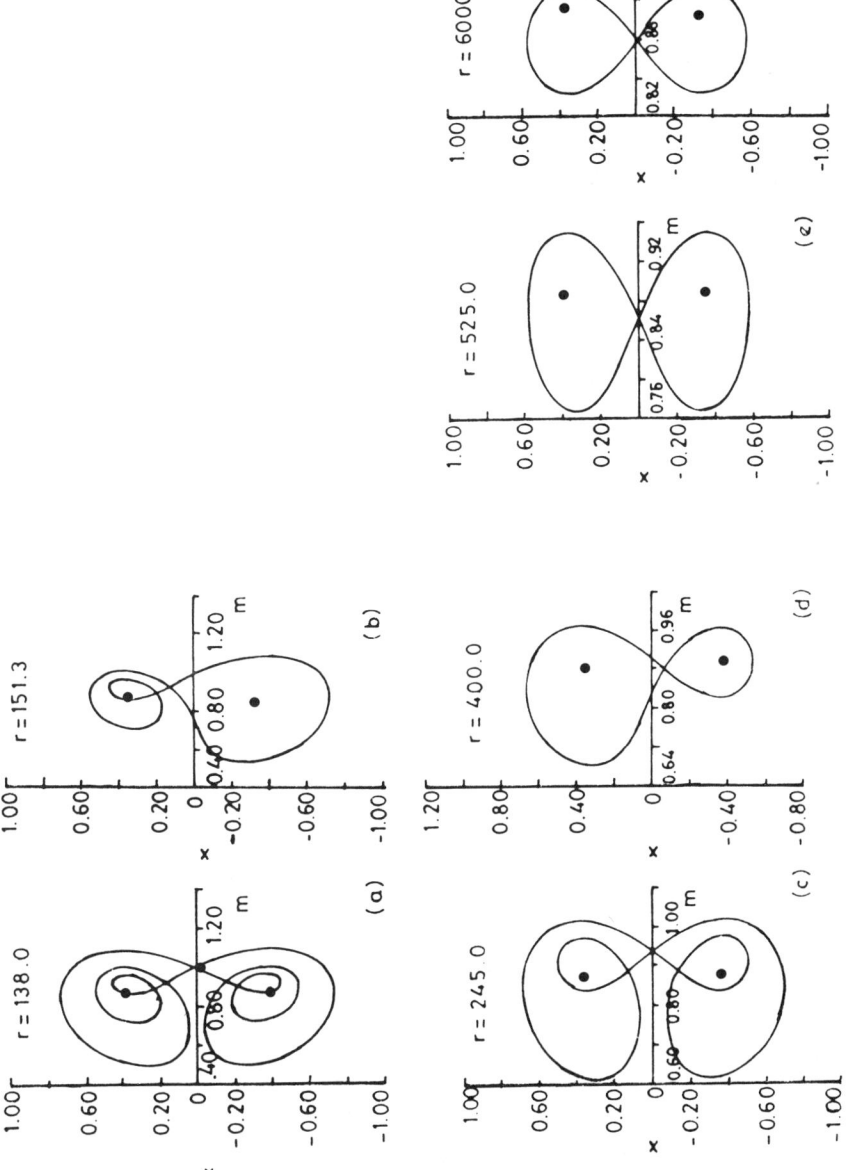

Figure 7.2.11 Limit cycles. The points denote the steady-state solutions ($x = \pm\sqrt{b/2\sigma}$, $m = 1 - b/2\sigma$). (From Shimizu and Morioka 1978.)

cycles exist in the turbulent region T_3, corresponding to a smaller value of r, which are much more complicated than those shown in Figure 7.2.11a. The details of the transitions between turbulent and periodic states may be found in Morioka and Shimizu (1978).

A comprehensive account of the Lorenz system detailing many other features is given by Sparrow (1982).

7.3 HENON–HEILES HAMILTONIAN SYSTEM

Now we take up the study of the singularity structure of a Hamiltonian system and determine how the Painlevé property determines the integrability of the system. This conservative system contrasts with the Lorenz system, which is dissipative. Again, we shall try to correlate the regular and chaotic behaviors of the trajectories with the kind of singularity structure the system possesses for different sets of parameters. Here, also, the numerical integration of the system proves useful. A novel feature here is the existence of a natural boundary in the nonintegrable cases. We shall follow the work of Chang et al. (1982, 1983) and Bountis et al. (1982).

The Henon–Heiles Hamiltonian is written in the general form

$$H = \frac{1}{2}(\dot{x}^2 + \dot{y}^2 + x^2 + y^2) + Dx^2y - \left(\frac{C}{3}\right)y^3, \tag{1}$$

where C and D are constant. The Newtonian equations of motion

$$\frac{d}{dt}\frac{\partial H}{\partial \dot{x}} = -\frac{\partial H}{\partial x}, \qquad \frac{d}{dt}\frac{\partial H}{\partial \dot{y}} = -\frac{\partial H}{\partial y}$$

yield

$$\begin{aligned}\ddot{x} &= -x - 2Dxy, \\ \ddot{y} &= -y - Dx^2 + Cy^2.\end{aligned} \tag{2}$$

To determine the leading order behavior, we write

$$x = a(t - t_*)^\alpha, \qquad y = b(t - t_*)^\beta. \tag{3}$$

Substituting (3) into (2), we have

$$a\alpha(\alpha - 1)\tau^{\alpha-2} = -a\tau^\alpha - 2Dab\tau^{\alpha+\beta}, \tag{4a}$$

$$b\beta(\beta - 1)\tau^{\beta-2} = -b\tau^\beta - Da^2\tau^{2\alpha} + Cb^2\tau^{2\beta}, \tag{4b}$$

where $\tau = t - t_*$. Two cases arise from (4).

Case 1. The dominant terms in (4a) and (4b) are $\tau^{\alpha-2}$ and $\tau^{\alpha+\beta}$, and $\tau^{\beta-2}$, $\tau^{2\alpha}$, and $\tau^{2\beta}$, respectively. The powers balance if $\alpha = -2$, $\beta = -2$.

The coefficients a and b then satisfy the equations

$$6a = -2abD,$$
$$6b = -Da^2 + Cb^2$$

so that we have

$$a = \pm \frac{3}{D}\left(2 + \frac{1}{\lambda}\right)^{1/2}, \qquad b = \frac{-3}{D}, \tag{5}$$

where $\lambda = D/C$.

Case 2. The dominant terms in (4a) and (4b) are $\tau^{\alpha-2}$ and $\tau^{\alpha+\beta}$, and $\tau^{\beta-2}$ and $\tau^{2\beta}$, respectively. These powers balance if $\beta = -2$. The power α is obtained by considering the equations for the coefficients:

$$a\alpha(\alpha - 1) = -2Dab,$$
$$b\beta(\beta - 1) = Cb^2. \tag{6}$$

These equations, for $\beta = -2$, show that $b = 6/C$ and $a \neq 0$ is arbitrary. Moreover,

$$\alpha^2 - \alpha + 12\lambda = 0,$$

so

$$\alpha = \frac{1}{2} \pm \frac{1}{2}(1 - 48\lambda)^{1/2}. \tag{7}$$

For system (2) to enjoy the Painlevé property, the powers α and β in (3) must be negative integers so, from (7),

$$\frac{1}{2} \pm \frac{1}{2}(1 - 48\lambda)^{1/2} = -n, \quad \text{an integer.} \tag{8}$$

This gives

$$\frac{D}{C} = \lambda = -\frac{(2n + 1)^2 - 1}{48}. \tag{9}$$

The first few values of λ which give (negative) integral α are $-1/6, -1/2, -1, -5/3, \ldots$. Equation (7) also shows that α is complex when $\lambda > 1/48$. We note that, for the standard Henon–Heiles Hamiltonian, $C = D = 1$, so that $\lambda = 1$, and the leading behavior for case 2 is given by

$$x = a(t - t_*)^{1/2 \pm i(47)^{1/2}/2}, \qquad y = 6(t - t_*)^{-2}, \tag{10}$$

where a is arbitrary.

System (2) is of order 4, and its general solutions should contain four arbitrary constants. To obtain resonances—that is, the powers at which arbitrary constants appear—we follow Ablowitz et al. (1980) (see Section 4.2). We consider cases 1 and 2 separately.

Case 1. We retain in (2) those dominant terms which lead to this case, namely,

$$\ddot{x} = -2Dxy,$$
$$\ddot{y} = -Dx^2 + Cy^2, \tag{11}$$

and substitute

$$x = \pm \frac{3}{D}\left(2 + \frac{1}{\lambda}\right)^{1/2} \tau^{-2} + p\tau^{-2+r}, \tag{12}$$

$$y = -\left(\frac{3}{D}\right)\tau^{-2} + q\tau^{-2+r}.$$

Here p and q are unknown parameters. The powers r are to be found such that p and q are arbitrary. Thus,

$$\pm \frac{18m}{D}\tau^{-4} + (r-2)(r-3)p\tau^{r-4}$$

$$= 2D\left[\pm \frac{9m}{D^2}\tau^{-4} + \frac{3p}{D}\tau^{r-4} \mp \frac{3m}{D}q\tau^{r-4} - pq\tau^{2r-4}\right],$$

$$-\frac{18}{D}\tau^{-4} + (r-2)(r-3)q\tau^{r-4}$$

$$= -D\left[\frac{9m^2}{D^2}\tau^{-4} + p^2\tau^{-4+2r} \pm \frac{6m}{D}p\tau^{-4+r}\right]$$

$$+ C\left[\frac{9}{D^2}\tau^{-4} + q^2\tau^{2r-4} - \frac{6}{D}q\tau^{r-4}\right], \tag{13}$$

where $m = (2 + 1/\lambda)^{1/2}$. In view of the balance of the highest-order terms, τ^{-4} terms cancel on both sides of equations (13). The next leading order terms are $O(\tau^{-4+r})$. These terms balance if

$$\begin{bmatrix} (r-2)(r-3) - 6 & \pm 6m \\ \pm 6m & (r-2)(r-3) + \frac{6C}{D} \end{bmatrix} \begin{bmatrix} p \\ q \end{bmatrix} = 0. \tag{14}$$

For a nontrivial solution (p,q) of (14), the determinant of the coefficient matrix must vanish. This leads to

$$(\Theta - 12)\left[\Theta + 6\left(1 + \frac{1}{\lambda}\right)\right] = 0, \tag{15}$$

where
$$\Theta = (r-2)(r-3). \tag{16}$$

The roots of (15) are

$$r = -1, 6, \frac{5}{2} \pm \frac{1}{2}\left[1 - 24\left(1 + \frac{1}{\lambda}\right)\right]^{1/2}. \tag{17}$$

The root $r = -1$ corresponds to the arbitrariness of t_* in $\tau = t - t_*$ (see Section 4.2).

Case 2. Writing

$$x = a\tau^\alpha + p\tau^{\alpha+r}, \tag{18}$$

$$y = \frac{6}{C}\tau^{-2} + q\tau^{-2+r},$$

where

$$\alpha = \frac{1}{2} \pm \frac{1}{2}(1 - 48\lambda)^{1/2},$$

and substituting into the dominant terms in (2) for the present case, namely,

$$\ddot{x} = -2Dxy, \tag{19}$$
$$\ddot{y} = Cy^2,$$

we get

$$a\alpha(\alpha - 1)\tau^{\alpha-2} + p(r+\alpha)(r+\alpha-1)\tau^{r+\alpha-2}$$
$$= -2D[a\tau^\alpha + p\tau^{\alpha+r}]\left[\frac{6}{C}\tau^{-2} + q\tau^{-2+r}\right], \tag{20a}$$

$$\frac{36}{C}\tau^{-4} + (r-2)(r-3)q\tau^{r-4} = C\left[\frac{6}{C}\tau^{-2} + q\tau^{-2+r}\right]^2. \tag{20b}$$

In (20a), (20b) the terms lower than those balanced in obtaining the dominant behavior (3) are of order $\tau^{r+\alpha-2}$ and τ^{r-4}, respectively. Their balance requires that

$$\begin{bmatrix} (r+\alpha)(r+\alpha-1) + \dfrac{12D}{C} & 2Da \\ 0 & (r-2)(r-3) - 12 \end{bmatrix} \begin{bmatrix} p \\ q \end{bmatrix} = 0. \tag{21}$$

Again for nontrivial (p, q), the determinant of the coefficient matrix in (21) must vanish. Thus, with $\lambda = D/C$,

$$[(r+\alpha)(r+\alpha-1) + 12\lambda][(r-2)(r-3) - 12] = 0, \tag{22}$$

leading to
$$r = -1, 6 \tag{23}$$
and
$$r = 0, \mp(1 - 48\lambda)^{1/2}. \tag{24}$$

The upper and lower signs in (24) correspond to those of α in the leading order behavior in (18). It follows from (24) that, for $\lambda > 1/48$, the leading orders and resonances are complex. Moreover, the imaginary part becomes infinite as $\lambda \to \infty$. In the range $-1/2 < \lambda < 1/48$, the negative branch $\alpha_- = 1/2 - (1 - 48\lambda)^{1/2}/2$ can define a four-parameter solution, the values of α_- and $r_+ = +(1 - 48\lambda)^{1/2}$ being real in this case. For the special case $\lambda = D/C = 0$, $\alpha_- = 0$, the singularity in the complex t-plane disappears; indeed, equations (11) decouple and are easily integrated.

For the Painlevé property to hold, we require that all leading orders and resonances for cases 1 and 2 be integers. Referring to (9), (17), and (24), we may verify that the only parametric values for which the Painlevé property holds are

$$\lambda = -\frac{1}{6}, -\frac{1}{2}, -1. \tag{25}$$

These values of λ promise a rich variety of possibilities.

The value $\lambda = -1$, according to (17), gives resonances at $r = -1, 2, 3, 6$ for case 1 singularities. To illustrate the expansions and the manner in which the arbitrary constants appear, we give detailed calculations for this case. Since $\alpha = \beta = -2$ in (3) for this case and $\lambda = -1$ ($D = -C$), we seek the solution in the form

$$\begin{aligned} x &= \frac{a_0}{t^2} + \frac{a_1}{t} + a_2 + a_3 t + a_4 t^2 + a_5 t^3 + a_6 t^4 + \cdots, \\ y &= \frac{b_0}{t^2} + \frac{b_1}{t} + b_2 + b_3 t + b_4 t^2 + b_5 t^3 + b_6 t^4 + \cdots \end{aligned} \tag{26}$$

and substitute in (2). Since the resonances appear at t^0, t, and t^4, it suffices to compare the coefficients of t^{-4}, t^{-3}, t^{-2}, t^{-1}, t^0, t, and t^2 in (2).

The relations in the coefficients a_i and b_i are

$$6a_0 = 2Ca_0 b_0,$$
$$2a_1 = 2C(a_0 b_1 + b_0 a_1),$$
$$0 = 2C(a_0 b_2 + a_1 b_1 + a_2 b_0) - a_0,$$
$$0 = 2C(a_0 b_3 + a_1 b_2 + a_2 b_1 + a_3 b_0) - a_1,$$
$$2a_4 = 2C(a_0 b_4 + a_1 b_3 + a_2 b_2 + a_3 b_1 + a_4 b_0) - a_2,$$
$$6a_5 = 2C(a_0 b_5 + a_1 b_4 + a_2 b_3 + a_3 b_2 + a_4 b_1 + a_5 b_0) - a_3,$$

Singularity Structure and Chaotic Behavior

$$12a_6 = 2C(a_0b_6 + a_1b_5 + a_2b_4 + a_3b_3 + a_4b_2 + a_5b_1 + a_6b_0) - a_4$$

and

$$6b_0 = C(a_0^2 + b_0^2),$$
$$2b_1 = 2C(a_0a_1 + b_0b_1),$$
$$0 = C(2a_0a_2 + a_1^2 + 2b_0b_2 + b_1^2) - b_0,$$
$$0 = C(2a_1a_2 + 2a_0a_3 + 2b_1b_2 + 2b_0b_3) - b_1,$$
$$2b_4 = C(2a_0a_4 + 2a_1a_3 + a_2^2 + 2b_0b_4 + 2b_1b_3 + b_2^2) - b_2,$$
$$6b_5 = C(2a_0a_5 + 2a_1a_4 + 2a_2a_3 + 2b_0b_5 + 2b_1b_4 + 2b_2b_3) - b_3,$$
$$12b_6 = C(2a_0a_6 + 2a_1a_5 + 2a_2a_4 + a_3^2 + 2b_0b_6 + 2b_1b_5 + 2b_2b_4 + b_3^2) - b_4.$$

The simultaneous recursive solution of the above systems for a_i and b_i gives

$$a_0 = -\frac{3}{C}, \quad b_0 = \frac{3}{C}, \quad a_1 = 0, \quad b_1 = 0, \quad b_2 = a_2 + \frac{1}{2C}, \quad b_3 = a_3,$$

$$a_4 = a_2^2C - \frac{3}{40C}, \quad b_4 = a_2^2C - \frac{1}{20C}, \quad a_5 = \frac{C}{3}(a_2b_3 + a_3b_2), \quad (27)$$

$$b_5 = a_5, \quad b_6 = \frac{1}{3}(2C^2a_2^3 + Ca_3^2) - \frac{1}{24}a_2 - a_6,$$

where the constants a_2, a_3, and a_6 are arbitrary and appear where $\alpha + r = 0$, 1, and 4, respectively (see (26)). Thus the solution depends on four arbitrary parameters (which include t_*) and has the Painlevé property. The integrability conditions are thus established. The integrals in this case were first found by Aizawa and Saito (1972).

The case $\lambda = -1/2$ is rather peculiar. The coefficient a in (5) for case 1 vanishes. The resonances for case 1 are -1, 0, 5, 6 (see (17)). The root $r = 0$ corresponds to the vanishing of the coefficient a. What happens is that the case 1 singularity merges with the positive branch ($\alpha = 3$) of case 2 (see (23) and (24)). The negative branch corresponding to $\alpha = -2$ in case 2 (see below (18)) is undefined. Thus, according to (18), the leading order terms are $x = at^3$, $y = -3t^{-2}$, where a is arbitrary. The resonance at $r = 6$ introduces another parameter. Thus t_*, a, and the arbitrary parameter at $r = 6$ define a three-parameter family of solutions (as may be checked by writing the appropriate Laurent series), which have poles as the only movable singularities.

For $\lambda = -1/6$, the resonances for the case 1 singularities are $r = -3$, -1, 6, 8. Detailed Laurent series show that for case 1 there is a three-parameter Painlevé form of the solution. On the other hand, for the case 2 singularities, with $\alpha = -1$ (see below (18)), the resonances appear at $r = -1$,

0, 3, 6. The arbitrary parameters appear at $r = 0, 3, 6$. These, together with t_*, make up the four parameters. The Painlevé property holds, and the integrability of the system is therefore indicated. The second integral (besides the Hamiltonian) can be verified to be

$$x^4 + 4x^2y^2 - 4\dot{x}(\dot{x}y - \dot{y}x) + 4x^2y + 3(\dot{x}^2 + x^2) = \text{const.} \tag{28}$$

The numerical investigations of Chang et al. (1982) could identify only the four-parameter family of solutions. Indeed, Chang et al. considered a slightly more general Henon–Heiles system where the coefficients A and B were introduced so that

$$H = \frac{1}{2}(\dot{x}^2 + \dot{y}^2 + Ax^2 + By^2) + Dx^2y - \left(\frac{C}{3}\right)y^3, \tag{29}$$

which leads to the equations of motion

$$\begin{aligned}\ddot{x} &= -Ax - 2Dxy, \\ \ddot{y} &= -By - Dx^2 + Cy^2.\end{aligned} \tag{30}$$

In the parametric range, $-1/2 < \lambda < 0$, identified earlier in the case 2 singularities with negative α (see equation (7)), the solution has a leading order behavior of (30) as

$$\begin{aligned}x &= at^\alpha, \\ y &= bt^{-2},\end{aligned} \tag{31}$$

where $\alpha = 1/2 - (1 - 48\lambda)^{1/2}/2$ and the resonances are

$$r = -1, 0, (1 - 48\lambda)^{1/2}, 6. \tag{32}$$

It defines a four-parameter family of solutions, while the case 1 singularities define, at most, a three-parameter family of solutions. Choosing $D/C = \lambda = -1/16$, we find (see below (31)) that $\alpha = -1/2$; (32) then yields $r = -1, 0, 2, 6$. A special choice of the coefficients such that $B = 16A$ leads to the Laurent series expansion for case 2,

$$\begin{aligned}x(t) &= t^{-1/2} \sum_{j=0}^{\infty} a_j t^j, \\ y(t) &= t^{-2} \sum_{j=0}^{\infty} b_j t^j,\end{aligned} \tag{33}$$

where

$$a_0 = \mu, \quad a_1 = 0, \quad a_2 = \Theta, \quad a_3 = -\frac{\mu^3}{18}, \quad a_4 = \left(\frac{\mu}{5}\right)A^2,$$

Singularity Structure and Chaotic Behavior

$$a_5 = -\frac{\mu^2 \Theta}{18}, \quad a_6 = \frac{1}{24}\left(\frac{\mu^5}{108} - 2\mu\psi\right); \quad b_0 = -\frac{3}{8}, \quad b_1 = 0,$$

$$b_2 = -\frac{A}{2}, \quad b_3 = \frac{\mu^2}{12}, \quad b_4 = -\frac{4}{5}A^2, \quad b_5 = \frac{\mu\Theta}{3}, \quad b_6 = \psi. \tag{34}$$

It is clear from (34) that the arbitrary constants μ, Θ, and ψ appear at $j = 0, 2, 6$ respectively, as predicted earlier. If we define the new variable $\tau = t^{1/2}$, the expansions (33) have poles as the only movable singularities, so the Painlevé condition is satisfied.

Again for $\lambda = -1/16$, the case 1 singularities correspond to leading behavior (see (5))

$$\begin{aligned} x &\simeq \pm 3i(14)^{1/2} t^{-2}, \\ y &\simeq -3t^{-2}. \end{aligned} \tag{35}$$

Here we have chosen $D = 1$ so that $C = -16$. The resonances come about at

$$r = -1, 6, -7, 12 \tag{36}$$

(see equation (17)). This branch in the complex t-plane also has the Painlevé property. It is, however, a (reduced) three-parameter family. The numerical studies of Chang et al. (1982) confirmed that in this case also the singularity structure is a regular lattice, indicative of an integrable system, (see Figure 7.3.1). These results suggest that any case for which the third member of (32) can be expressed as

$$(1 - 48\lambda)^{1/2} = \frac{n}{m},$$

where n and m are relatively prime integers, may yield a four-parameter family of solutions in the variable $\tau = t^{1/2m}$. However, each case must be checked separately for the presence of logarithmic singularities which may betray widespread chaos (see, for example, Figure 7.3.2, where $A = 5$, $B = 16$, $\lambda = -5/16$, and $E = 4/3$).

Chang et al. (1982) introduced the notion of canonical resonance, which their numerical results indicate is associated with a particularly symmetric form of the analytic structure. They mean by canonical resonance the special cases which arise when the power of t at which resonance occurs in one case is identical with the leading order behavior of another case. Referring to equations (7) and (17), the resonance for the first case takes place at the power $r - 2 = 1/2 \pm [1 - 24(1 + 1/\lambda)]^{1/2}/2$; this is equated to the leading order behavior of case 2, namely $1/2 \pm (1 - 48\lambda)^{1/2}/2$. The equality holds for $\lambda = 1$ and $\lambda = -1/2$. The case $\lambda = -1/2$ results in a leading order

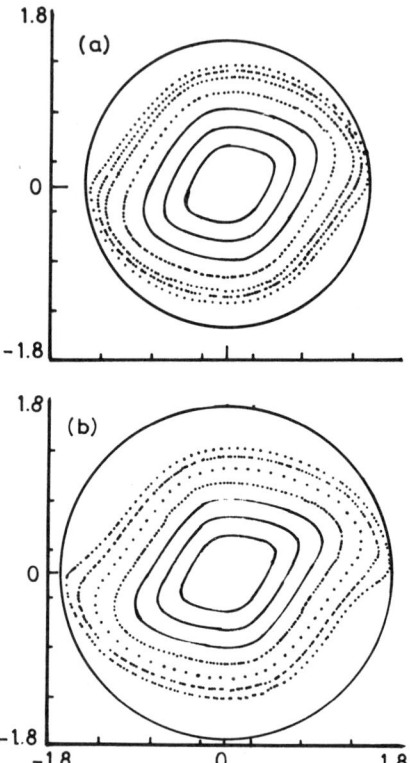

Figure 7.3.1 Surface of section of Henon–Heiles system with $D = 1$, $C = -16$, $A = 1$, $B = 16$ at (a) $E = 1$ and (b) $E = 4/3$. Outermost circle denotes phase space boundary. There were no discernible regions of chaotic motion. (From Chang, Tabor, and Weiss 1982.)

resonance at $\alpha = 3$. The case $\lambda = 1$ corresponds to complex leading order behavior noted in (10).

The full series expansion in the present system, as for the Lorenz system, leads to rather complicated recursion relations, which, in general, are not closed. For the leading order behavior the recursion relations are closed, but the results concerning them can be obtained more easily by studying the asymptotic behavior. We therefore give the general expansion briefly and then pass to the asymptotic behavior. We treat only case 1.

Singularity Structure and Chaotic Behavior

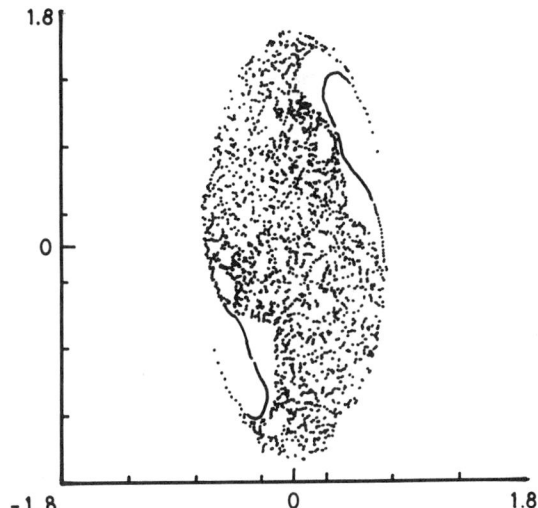

Figure 7.3.2 Surface of section of Henon–Heiles system with $D = 5$, $C = -16$, $A = 5$, $B = 16$ at $E = 4/3$. In contrast with Figure 7.3.1b there are large regions of chaotic motion. (From Chang, Tabor, and Weiss 1982.)

The expansion in this case is sought in the form (we let $t_* = 0$ for convenience)

$$x(t) = t^{-2} \sum_{k=0}^{\infty} \sum_{j=0}^{\infty} a_{kj} \tau^k t^j + t^{-2} \sum_{k=1}^{\infty} \sum_{j=0}^{\infty} \bar{a}_{kj} \bar{\tau}^k t^j, \tag{37a}$$

$$y(t) = t^{-2} \sum_{k=0}^{\infty} \sum_{j=0}^{\infty} b_{kj} \tau^k t^j + t^{-2} \sum_{k=1}^{\infty} \sum_{j=0}^{\infty} \bar{b}_{kj} \bar{\tau}^k t^j, \tag{37b}$$

where

$$\tau = t^\alpha, \quad \alpha = \frac{1}{2} + \frac{1}{2}\left(1 - 24\left(1 + \frac{1}{\lambda}\right)\right)^{1/2},$$
$$\bar{\tau} = t^{\bar{\alpha}}, \quad \bar{\alpha} = \frac{1}{2} - \frac{1}{2}\left(1 - 24\left(1 + \frac{1}{\lambda}\right)\right)^{1/2} \tag{38}$$

and

$$a_{00} = \pm \frac{3}{D}\left(2 + \frac{1}{\lambda}\right)^{1/2}, \quad b_{00} = -\frac{3}{D}. \tag{39}$$

The form (37) derives from the consideration of the leading order behavior, given by (3) and (5) with powers -2 and -2, and the resonances given by (17). The details show that the coefficients a_{12}, $\bar{a}_{12}$, and a_{06} are arbitrary and can be chosen to satisfy the initial conditions. The recursion relations (the details of which may be found in Chang et al. 1983) show that the sets of coefficients $a_{j,2j}$ and $b_{j,2j}$ for $j = 0, 1, 2, \ldots$ or $\bar{a}_{j,2j}$ and $\bar{b}_{j,2j}$ for $j = 0, 1, 2, \ldots$ can be determined in a closed sequential manner. The coefficients can be shown (as for the Lorenz system) to define two functions which are governed by a coupled system of nonlinear DE. These DE yield the asymptotic behavior of the basic system in the neighborhood of the singularity, when $|t| \ll 1$. We derive these DE by an alternative method.

Since the expansions (37) start with t^{-2} as the leading terms, we may ignore the linear terms $-x$ and $-y$ near the singularity in the basic system (2). For the truncated system, which has homogeneous terms of degree 2 on the right-hand sides, we write

$$x(t) = \frac{1}{t^2}\Theta(t), \quad y(t) = \frac{1}{t^2}\psi(t). \tag{40}$$

The (truncated) system (2) now becomes

$$\begin{aligned}t(t\Theta')' - 5t\Theta' + 2\Theta(3 + \lambda\psi) &= 0, \\ t(t\psi')' - 5t\psi' + \psi(6 - \psi) + \lambda\Theta^2 &= 0,\end{aligned} \tag{41}$$

where prime denotes differentiation with respect to t. The equilibrium or steady states of this system are given by

$$2\Theta(3 + \lambda\psi) = 0, \quad \psi(6 - \psi) + \lambda\Theta^2 = 0. \tag{42}$$

We thus have two cases:

$$\psi = -\frac{3}{\lambda}, \quad \Theta = \pm\frac{3}{\lambda}\left(2 + \frac{1}{\lambda}\right)^{1/2}. \tag{43}$$

$$\psi = 6, \quad \Theta = 0. \tag{44}$$

To get the next order terms in the expansions about the steady state, we write for (43),

$$\begin{aligned}\Theta &= \pm\frac{3}{\lambda}\left(2 + \frac{1}{\lambda}\right)^{1/2} + at^s, \\ \psi &= -\frac{3}{\lambda} + bt^s.\end{aligned} \tag{45}$$

Substitution into (41) shows that

$$s = -1, 6, \frac{5}{2} \pm \frac{i}{2}\left(24\left(1 + \frac{1}{\lambda}\right) - 1\right)^{1/2}, \quad \left(\lambda > -\frac{24}{23}\right).$$

Singularity Structure and Chaotic Behavior

Similarly for (44) we find that

$$\Theta = at^s, \qquad \psi = 6 + bt^s, \tag{46}$$

where

$$s = -1, 6, \frac{5}{2} \pm \frac{i}{2}(48\lambda - 1)^{1/2} \qquad \left(\lambda > \frac{1}{48}\right).$$

Thus, we have leading orders and resonances appropriate to both (43) and (44) (cf. (17) and (18)).

Now to proceed to the asymptotic form of the solution near a singularity, we first write the complex time as

$$t = |t|e^{i\theta}$$

and note that

$$|t^s| \gg 1$$

when either $\theta \gg 0$ and $\text{Im}(s) < 0$ or $\theta \ll 0$ and $\text{Im}(s) > 0$. Defining the new independent variable, $X = t^s \equiv t^{2+\alpha}$, where $\alpha \; (= s - 2)$ is chosen to correspond to (43) (see (45)), and substituting in (41), we have

$$(\alpha + 2)^2 X(X\Theta')' - 5(\alpha + 2)X\Theta' + 6\Theta + 2\lambda\Theta\psi = 0,$$
$$(\alpha + 2)^2 X(X\psi')' - 5(\alpha + 2)X\psi' + 6\psi + \lambda\Theta^2 - \psi^2 = 0, \qquad ' \equiv \frac{d}{dx}. \tag{47}$$

System (47) is exactly the one which is obtained by writing

$$\Theta(X) = \sum_{j=0}^{\infty} a_{j,2j} X^j,$$

$$\psi(X) = \sum_{j=0}^{\infty} b_{j,2j} X^j, \tag{48}$$

$$X = t^{\alpha+2}, \qquad \alpha = \frac{1}{2} + \frac{1}{2}\sqrt{1 - 24(1 + 1/\lambda)},$$

and using the closed recursion relations for $a_{j,2j}$ and $b_{j,2j}$ (see (37) and the discussion following it).

The case corresponding to (44) can be treated similarly.

Now to examine the singularities that $\Theta(X)$ and $\psi(X)$ may possess, we apply the standard leading order analysis to (47):

$$\Theta(X) \simeq A(X - X_0)^\gamma, \qquad \psi(X) \simeq B(X - X_0)^\delta, \tag{49}$$

where X_0 is the position of the singularity. Substitution into (47) shows that two possibilities arise:

Case a.

$$\gamma = -2, \quad A = \pm \frac{3}{\lambda}\left(2 + \frac{1}{\lambda}\right)^{1/2} X_0^2 (2+\alpha)^2,$$

$$\delta = -2, \quad B = -\frac{3}{\lambda} X_0^2 (2+\alpha)^2.$$

Case b.

$$\gamma = \frac{1}{2} \pm \frac{1}{2}(1 - 48\lambda)^{1/2},$$

$$\delta = -2, \quad B = 6X_0^2(2+\alpha)^2, \quad A \text{ arbitrary}. \tag{50}$$

Thus $\theta(X)$ and $\psi(x)$ exhibit exactly the same sort of singularities in the X-plane as $x(t)$ and $y(t)$ do in the t-plane. Therefore, we may study the singularities of $\theta(X)$ and $\psi(X)$ to determine those of $x(t)$ and $y(t)$.

Suppose the singularity in the X-plane is situated at

$$X = X_0 \equiv X_0 e^{2\pi i n}, \quad n = 0, 1, 2, \ldots \tag{51}$$

Then the position in the t-plane (see (48)) is given by

$$t_0 = \left[X_0 e^{2\pi i n}\right]^{1/(2+\alpha)}, \tag{52}$$

where, considering (43), we have

$$\alpha = \frac{1}{2} + \frac{i}{2}\left[24\left(1 + \frac{1}{\lambda}\right) - 1\right]^{1/2}.$$

Thus, the singularity t_0, as given by (52), becomes

$$t_0 = X_0^{1/(2+\alpha)} \exp\left\{2n\pi\left[\frac{5i + \sqrt{24(1 + 1/\lambda) - 1}}{2(12\lambda + 6)}\right]\right\}. \tag{53}$$

For the canonical case $\lambda = 1$, we have

$$t_0 = X_0^{1/(2+\alpha)} e^{(5n\pi/18)i} e^{-n\pi(47)^{1/2}/18}, \quad n = 0, 1, 2, \ldots. \tag{54}$$

Thus, each pole in the X-plane yields an equiangular spiral of poles in the t-plane, one pole for each value of n. These poles have an angular displacement about the pole equal to

$$\Delta\theta = \frac{5\pi}{18} \tag{55}$$

and radial decrement

$$\Delta |t| = -\frac{n\pi(47)^{1/2}}{18}, \quad n = 0, 1, 2, \ldots. \tag{56}$$

Similar arguments for the case 2 singularity show that, in general,

$$\Delta \theta = \frac{5\pi}{2(12\lambda + 6)},$$

$$\Delta |t| = \exp\left[\frac{-n\pi(48\lambda - 1)^{1/2}}{2[12\lambda + 6)}\right], \quad n = 0, 1, 2, \ldots. \tag{57}$$

For $\lambda = 1$, these quantities become

$$\Delta \theta = \frac{5\pi}{36} = 25°,$$

$$\Delta |t| = \exp\left[-\frac{n\pi\sqrt{47}}{36}\right], \quad n = 0, 1, 2, \ldots. \tag{58}$$

We have given details of the asymptotic analysis only for case 1 corresponding to (43). The same procedure can be followed for the singularities of case 2 corresponding to (44). The main result from our discussion is the value of the angular displacement, which is 25° for case 2 and 50° for case 1. These values are borne out by the numerical results which we now describe.

The numerical scheme used by Chang et al. (1982) to integrate the basic system (2) is the one due to Chang and Corliss (1980) referred to in Section 7.2; this is a Taylor series approach and yields detailed information concerning the singularity nearest to the initial point. We discuss first the results for $\lambda = 1$, the case of canonical resonance. When the solution is expanded at various points along the real-time axis, one finds a nonuniform row of seemingly isolated singularities. The initial data was so chosen that the motion is bounded for all real time, and the singularities occur at finite distances from the real-time axis. However, when the path of integration is deformed into the complex plane and passes between two singularities, as observed from the real-time axis, there is found a third singularity located at the apex of an approximately isosceles triangle whose base is the line joining the two singularities that are on either side of the path of analytic continuation (see Figure 7.3.3). If the base consists of two order -2 singularities, the singularity at the apex is of order 1/2. If, on the other hand, the base consists of order -2 and order 1/2 singularities (order here refers to the real part of the leading order), there is found an order -2 singularity at the apex. The interesting result is that the base angle of the isosceles triangle is found to be 25° (see (58)).

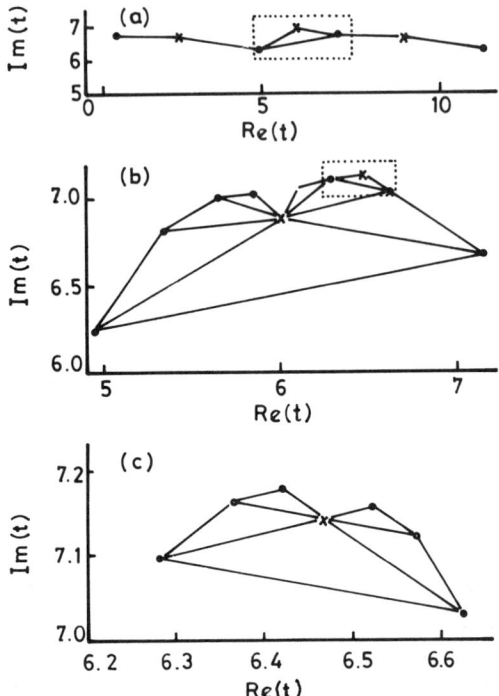

Figure 7.3.3 Analytic continuation of $x(t)$ for $\lambda = 1$. (a) Sequence of singularities found from the real axis and one singularity found at the first stage of analytic continuation. $\cdot$ = singularity of (leading) order -2 and $\times$ = singularity of order $1/2 + (i/2)\sqrt{47}$. (b) Boxed region of (a) in more detail showing double spiral of singularities about apex of "triangle." (c) Boxed region of (b) in more detail showing self-similar nature of the double spiral of singularities. Analytic continuation of $y(t)$ is identical, but all singularities now have order -2. (From Chang, Tabor, and Weiss 1982.)

The case $\lambda = 1$ manifests considerable symmetry in the singularity structure, as we now describe. This is possibly due to the value $\lambda = 1$ being one of the canonical resonances.

If we integrate between any pair of singularities that are observed to be neighboring during the process of analytic continuation (see the box in Figure 7.3.3), the construction described above is repeated. From any singularity there emanates a double spiral one clockwise and another counterclockwise. Moreover, the base between the neighboring singularities contracts geometrically at successive stages of analytic continuation; therefore, starting from a pair of base singularities, the solution can be continued only to a given finite distance in any direction without retracing the original path, provided one remains on the same side of the real t-axis. Thus, any path of analytic continuation between a pair of singularities would appear to be closed in a geometrically converging web of singularities that creates a natural boundary of the solution.

Chang et al. (1982) also describe the singularity structure for other values of λ. The lattice of singularities is very complicated and we refer the reader to their original work for details.

7.4 SOME OTHER HAMILTONIAN SYSTEMS

We now discuss two other nonintegrable systems to further elucidate the connection between the chaotic behavior of Hamiltonian systems and the movable singularities of their solutions in the complex plane. The systems we choose are the quartic lattice and the free-end Toda lattice. We treat the former in some detail and quote the results for the latter (Bountis and Segur 1982).

We first write the equations of motion for the quartic lattice

$$H = \frac{1}{2}(\dot{x}^2 + \dot{y}^2) + \frac{1}{4}[x^4 + y^4 + \eta(x-y)^4], \tag{1}$$

namely,

$$\ddot{x} = -x^3 - \eta(x-y)^3 = -(1+\eta)x^3 + \eta y^3 - 3\eta xy^2 + 3\eta x^2 y, \tag{2}$$

$$\ddot{y} = \eta(x-y)^3 - y^3 = -(1+\eta)y^3 + 3\eta xy^2 - 3\eta x^2 y + \eta x^3, \tag{3}$$

where η is as yet an undetermined parameter. We look for Painlevé solutions of (2) and (3) with the dominant behavior

$$x = a\tau^p, \quad y = b\tau^q, \quad \tau = (t-t_0) \longrightarrow 0, \tag{4}$$

where p and q are negative integers. Inserting (4) into (2) and (3), we easily check that two cases arise:

$$p = -1, \quad q = -1; \tag{5}$$

$$p = q < -1. \tag{6}$$

For (6) it is easy to verify that $a = b = 0$ for all p and q. For (5), a and b are given by

$$2a = -(1+\eta)a^3 + \eta b^3 - 3\eta ab^2 + 3\eta a^2 b, \tag{7}$$

$$2b = -(1+\eta)b^3 + \eta a^3 + 3\eta ab^2 - 3\eta a^2 b. \tag{8}$$

Equations (7) and (8) give rise to three possibilities:

$$a = -b = -i \left(\frac{2}{1 + 8\eta} \right)^{1/2}, \tag{9}$$

$$a = b = i\sqrt{2}, \tag{10}$$

$$a = \left(\frac{2\eta - 1 + \sqrt{1 - 4\eta}}{1 - \eta} \right)^{1/2}, \qquad a = \frac{2\eta}{b(1 - \eta)}. \tag{11}$$

To find the resonances, we write

$$x = a\tau^{-1} + \alpha\tau^{-1+r}, \qquad y = b\tau^{-1} + \beta\tau^{-1+r}, \tag{12}$$

and as usual obtain by substitution into (2) and (3), etc., the following equation for r:

$$s^2 + 3s[a^2 + b^2 + 2\eta(a - b)^2] + 9[a^2 b^2 + \eta(a - b)^2(a^2 + b^2)] = 0, \tag{13}$$

where

$$s = (r - 1)(r - 2). \tag{14}$$

Now, substitution of a and b from (9)–(11) into (13) shows that (10) yields integer r values for all η, while the only common η values for which (9) and (11) give integer r are $\eta = 0$ and $\eta = 1/4$. The value $\eta = 0$ is not very interesting since equations (2) and (3) decouple, integrate, and give rise to four arbitrary constants of integration. The second value $\eta = 1/4$ is more interesting: both (9) and (11) lead to the same resonance equation

$$(r + 1)r(r - 3)(r - 4) = 0, \tag{15}$$

and it may seem that there are four arbitrary constants, entering at $r = -1$ (corresponding to t_0), 0, 3, 4, implying that system (2)–(3) has the Painlevé property and is an integrable Hamiltonian system. This, however, is not the case. The arbitrary constant at $r = 0$ cannot be captured by admitting only poles, since, according to (9) and (11), a and b get fixed. This indicates the need to introduce logarithmic singularities to capture the arbitrary constants. If we seek the solution in the form

$$x = a\tau^{-1} + a_1(\log \tau)^{q_1}, \qquad y = b\tau^{-1} + b_1(\ln \tau)^{q_2}$$

and substitute into (2) and (3), a balance of dominant terms shows that two possibilities arise: (i) $q_1 = q_2 = -1$, (ii) $q_1 = q_2 = -1/2$. Taking these into account, we may finally obtain the logarithmic (psi) series as

$$x = i \left(\frac{2}{3} \right)^{1/2} \frac{1}{\tau} \left[1 + \sum_{k=1}^{\infty} a_k (\ln \tau)^{-k} \right]$$

Singularity Structure and Chaotic Behavior

$$+ \frac{\sqrt{3}}{2\tau} \sum_{k=0}^{\infty} b_k (\ln \tau)^{-k-1/2} + f\tau^2 + g\tau^3 + \cdots, \tag{16a}$$

$$y = -i \left(\frac{2}{3}\right)^{1/2} \frac{1}{\tau} \left[1 + \sum_{k=1}^{\infty} a_k (\ln \tau)^{-k}\right]$$

$$+ \frac{\sqrt{3}}{2\tau} \sum_{k=0}^{\infty} b_k (\ln \tau)^{-k-1/2} + f\tau^2 - g\tau^3 + \cdots, \tag{16b}$$

where $a_1 = 9/16$, $b_0 = 1$, and t_0, b_1, f, and g are the four arbitrary constants.

Figures 7.4.1a–7.4.1b show surfaces of section of solutions of (1). It appears that as $\eta \to 1/4$ large regions of chaotic behavior disappear. This impression is deceptive, however. If more closely spaced sets of initial conditions are chosen, thereby increasing the "resolutions" of the surfaces of section, there appear in between the invariant curves of Figure 7.4.1c chains of islands which are separated by small-scale chaotic regions.

We now summarize the results for the free-end Toda lattice with linear terms:

$$H = \frac{1}{2}(p_1^2 + p_2^2 + p_3^2) + e^{q_1 - q_2} + e^{q_2 - q_3} - q_1 + q_3. \tag{17}$$

Introducing the Flaschka variables

$$a_k = \frac{1}{2} e^{(q_k - q_{k+1})/2}, \quad b_k = \frac{1}{2} p_k, \quad k = 1, 2, \tag{18}$$

and taking zero total momentum, we obtain the equations of motion:

$$\dot{a}_1 = a_1(b_2 - b_1), \quad \dot{a}_2 = -a_2(2b_2 + b_1),$$
$$\dot{b}_1 = 2a_1^2 - \frac{1}{2}, \quad \dot{b}_2 = 2(a_2^2 - a_1^2). \tag{19}$$

The asymptotic expansions with simple poles as $\tau \to 0$ are

$$a_1 = c_1 \tau^{-1} + \cdots, \quad a_2 = c_2 \tau^{-1} + \cdots,$$
$$b_1 = c_3 \tau^{-1} + \cdots, \quad b_2 = c_4 \tau^{-1} + \cdots. \tag{20}$$

Three cases may be distinguished:

$$c_1 = 0, \quad c_2 = \frac{i}{\sqrt{2}}, \quad c_3 = 0, \quad c_4 = \frac{1}{2}, \tag{21}$$

$$c_1 = \frac{i}{2}, \quad c_2 = 0, \quad c_3 = \frac{1}{2}, \quad c_4 = \frac{-1}{2}, \tag{22}$$

$$c_1 = c_2 = \frac{i}{\sqrt{2}}, \quad c_3 = 1, \quad c_4 = 0. \tag{23}$$

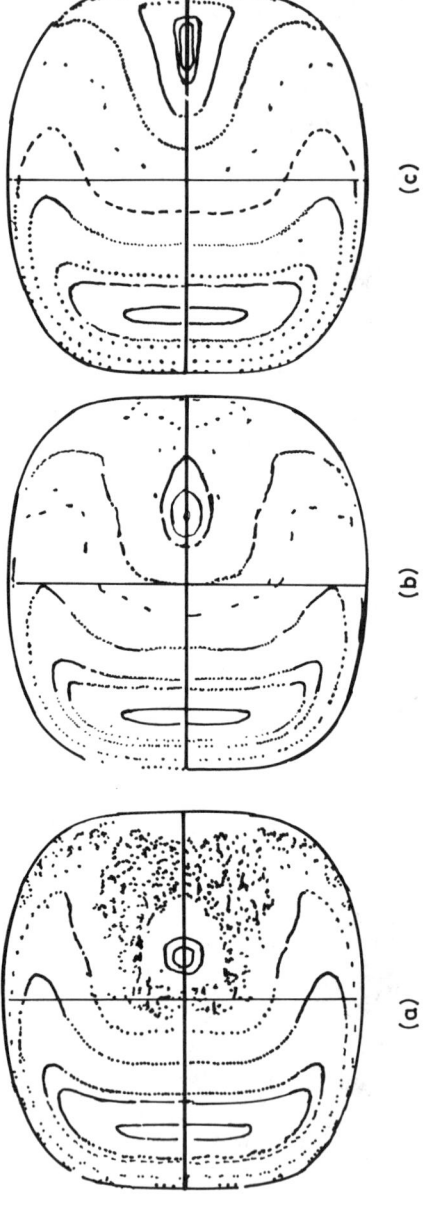

Figure 7.4.1 Surface of section for the quartic lattice (1) at $\eta = 0.15$, 0.2, and 0.25, respectively, and total energy $E = 1000$. (From Bountis and Segur 1982.)

Singularity Structure and Chaotic Behavior

It turns out that it is not possible to complete the series (20) with integral powers of τ only. We must include logarithmic terms of the form $\tau^p \ln t$ as well as τ^q terms, with integer values of p and q. For example, for (22) the asymptotic expansions are

$$a_1 = \frac{i}{2\tau} + \left(\frac{i\tau}{12} + \cdots\right)\ln\tau + h\tau + \cdots, \tag{24a}$$

$$a_2 = g\tau^{1/2}\left(1 + \frac{3}{2}f\tau + \cdots\right), \tag{24b}$$

$$b_1 = \frac{1}{2\tau} + f + \left(-\frac{\tau}{6} + \cdots\right)\ln\tau + \left(2ih - \frac{1}{3}\right)\tau + \cdots, \tag{24c}$$

$$b_2 = \frac{-1}{2\tau} + f + \left(\frac{\tau}{6} + \cdots\right)\ln\tau + \left(-2ih - \frac{1}{6}\right)\tau + \cdots, \tag{24d}$$

where f, g, h, and t_0 are arbitrary constants, and dots represent higher integer powers of τ. Similar expansions with $\tau^p \ln\tau$ terms can be found for (21) and (23), but only with three arbitrary constants. We note that the $\tau^{1/2}$ term in (24b) can be removed by considering the variable $a_2^2 \equiv a$ instead of a_2 in the original system (19). Then, the expansions for a_1, a_2, b_1 and b_2 would involve only $\ln\tau$ and integer powers of τ.

7.5 THE KURAMOTO MODEL

Now we discuss briefly the singularities and natural boundaries of the steadstate form of the Kuramoto equation

$$u_t + uu_x + u_{xx} + u_{xxxx} = 0, \quad x \in [0, L]. \tag{1}$$

Equation (1) with suitable boundary and initial conditions describes the propagation of flame fronts in explosive mixtures or the evolution of concentration fronts in chemical reactions (Kuramoto 1978). This PDE is known to display chaotic behavior (see Frisch 1983). We analyze the singularity structure of the stationary solutions of (1), (Thual and Frisch 1984)—that is, of the equation

$$\left(\frac{u^2}{2} + u_x + u_{xxx}\right)_x = 0,$$

which integrates to yield

$$\frac{u^2}{2} + u_x + u_{xxx} = 0, \tag{2}$$

provided we set the constant of integration equal to zero. It is easy to check that the dominant behavior of (2) is given by $u = u_0(z - z_*)^{-3}$, obtained by the balance of the first and third terms in (2). (We replace x by the complex variable z in (2) to carry out the investigation in the complex plane.) Substitution of the series form

$$u(z) = (z - z_*)^{-3} \sum_{r=0}^{\infty} u_r (z - z_*)^r \qquad (3)$$

into (2) shows that, to order -6, we must have

$$-60 u_0 + \frac{u_0^2}{2} = 0;$$

that is,

$$u_0 = 120, \qquad (4)$$

while, to order $r - 6$, we have

$$[(r-3)(r-4)(r-5) + u_0] u_r = 0. \qquad (5)$$

Putting $u_0 = 120$ in (5), we find that u_r is arbitrary, provided resonances are given by

$$(r-3)(r-4)(r-5) + 120 = 0. \qquad (6)$$

The roots of (6), apart from -1, are

$$r = \frac{13}{2} + i \frac{(71)^{1/2}}{2}, \qquad \bar{r} = \frac{13}{2} - i \frac{(71)^{1/2}}{2}. \qquad (7)$$

Since the resonances occur at complex conjugate values, (2) does not possess the Painlevé property, as predicted by the chaotic behavior of the Kuramoto model (1). The solution of (2) should be sought in the following form as in Sections 7.2 and 7.3:

$$u(z) = (z - z_*)^{-3} \{ 120 + P[A(z - z_*)^r] + Q[B(z - z_*)^{\bar{r}}] \}, \qquad (8)$$

where P and Q are two series without constant terms:

$$P(y) = \sum_{m=1}^{\infty} a_m y^m, \qquad Q(y) = \sum_{m=1}^{\infty} = b_m y^m. \qquad (9)$$

Here, A and B are two arbitrary constants which, together with z_*, constitute three parameters for the general solution of a third-order ODE. To find the natural boundaries, we write (cf. the analysis following (51) of Section 7.3)

$$A(z - z_*)^r = y_0, \qquad (10)$$

Singularity Structure and Chaotic Behavior

where $y = y_0$ is a singularity of one of the series in (9). Every solution of (10) is a singularity of the function $u(z)$. If $z = z_0$ is a solution of (10), then we can construct an infinity of new solutions $z = z_n$, where n is a natural integer, each being a singularity of $u(z)$. This is seen by writing

$$(z_n - z_*) = s^n(z_0 - z_*), \quad n \in \mathbb{Z}, \tag{11}$$

where

$$s = \exp \frac{2\pi i}{r} \tag{12}$$

so that

$$|s| = \exp\left[\frac{2\pi \operatorname{Im}(r)}{|r|^2}\right] \tag{13}$$

and

$$\arg(s) = \frac{2\pi \operatorname{Re}(r)}{|r|^2}. \tag{14}$$

The family (z_n), where $n \in \mathbb{Z}$, forms a discrete spiral converging toward z_*. Using (7) for r, we find that, for the Kuramoto model (2), the poles have

$$\text{Geometric ratio } \rho = \exp\left(-2\pi \frac{\sqrt{71}}{120}\right) = 0.6434 \tag{15}$$

and

$$\text{Rotation angle } \Theta = \pm 2\pi \frac{13}{120} = \pm 0.6807. \tag{16}$$

We also note that the two sets of spirals (clockwise and counterclockwise) coincide if and only if the "triangles" (see the discussion of the numerical results below) are isosceles; that is, $s + \bar{s} = 1$. This, on substituting (12) and (7), requires that

$$2 \exp\left(-\pi \frac{\sqrt{71}}{60}\right) \cos\left(\pi \frac{13}{60}\right) = 1. \tag{17}$$

The numerical results show that this equality is true within an error of less than 10^{-5}.

Now we summarize the numerical results for (2), as reported by Thual and Frisch (1984). The numerical scheme employed by them is again due to Chang and Corliss (1980) (see Section 7.2). The initial conditions for (2) were chosen to be $u(0) = u_x(0) = 0, u_{xx}(0) = 1$. As the trajectory proceeded from these initial conditions, differently placed singularities, each of order

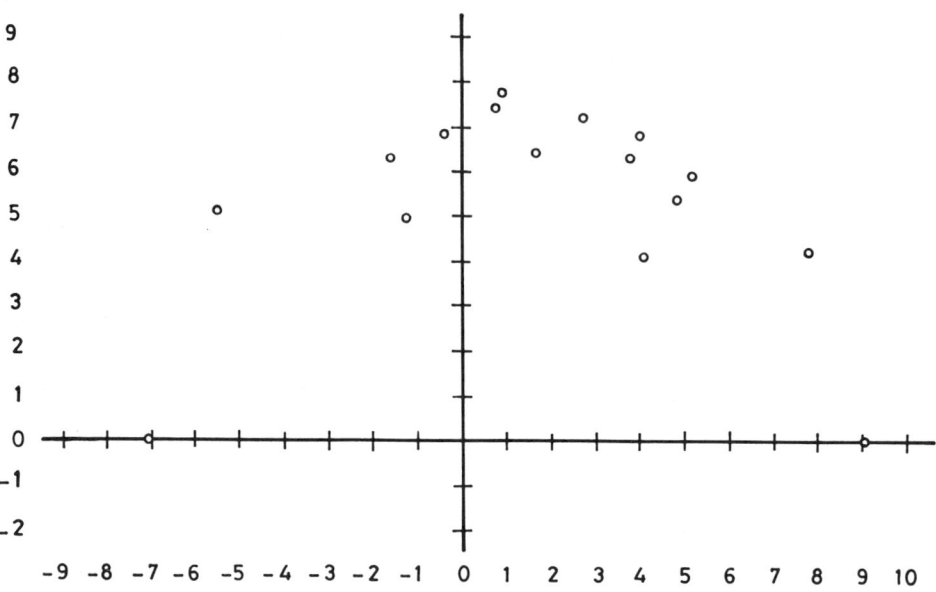

Figure 7.5.1 Singularities in x complex plane of equation (2). (From Frisch 1983.)

$(z - z_*)^{-3}$, were encountered at "reasonable" distances (see Figure 7.5.1). A close scrutiny of the singularities reveals a certain geometrical structure; these are found to be located at the vertices of the isosceles triangles. Given two singularities, one constructs an isosceles triangle with smaller sides 0.6 times the side joining the two singularities. This process is repeated on each of the smaller sides, and a fractal set of singularities is obtained as shown in Figures 7.5.2 and 7.5.3. The sets of singularities for the right-hand and the left-hand sides form clockwise and counterclockwise spirals, respectively. Moreover, each complete turn around the singularity leads to a new Riemann sheet. Numerical results confirm that the ratio of the small sides of the isosceles triangle to the base tends to $\rho = 0.64$, while the two equal angles tend to $\Theta = 0.7$. These compare very well with the analytic values $\rho = 0.6434$ and $\Theta = +0.6807$ in (15) and (16). The pattern depicted in Figure 7.5.4 shows the extent of the natural boundary, as obtained from the two singularities on the real axis and three other main singularities in the complex plane. No open continuation domain can cross this boundary.

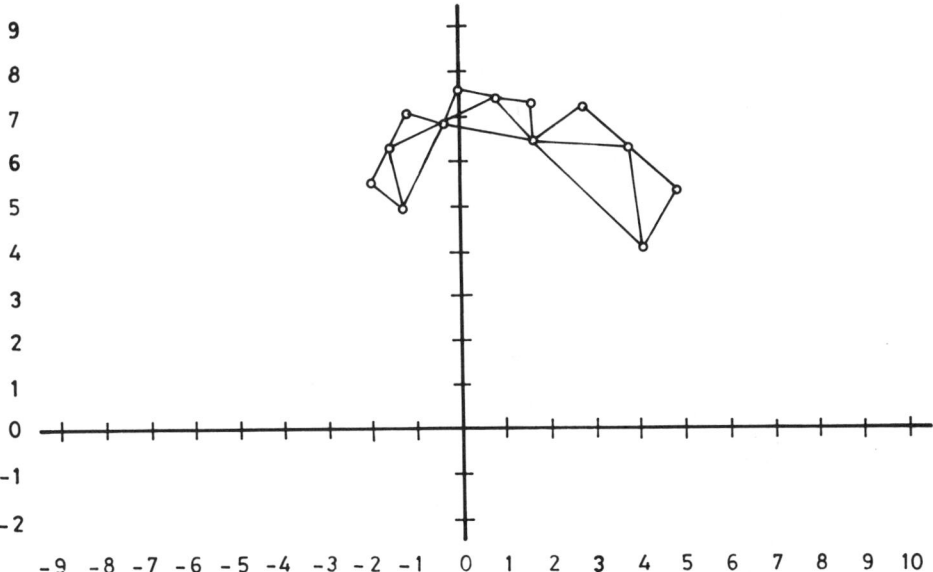

Figure 7.5.2 Singularities in x complex plane of equation (2). (From Frisch 1983.)

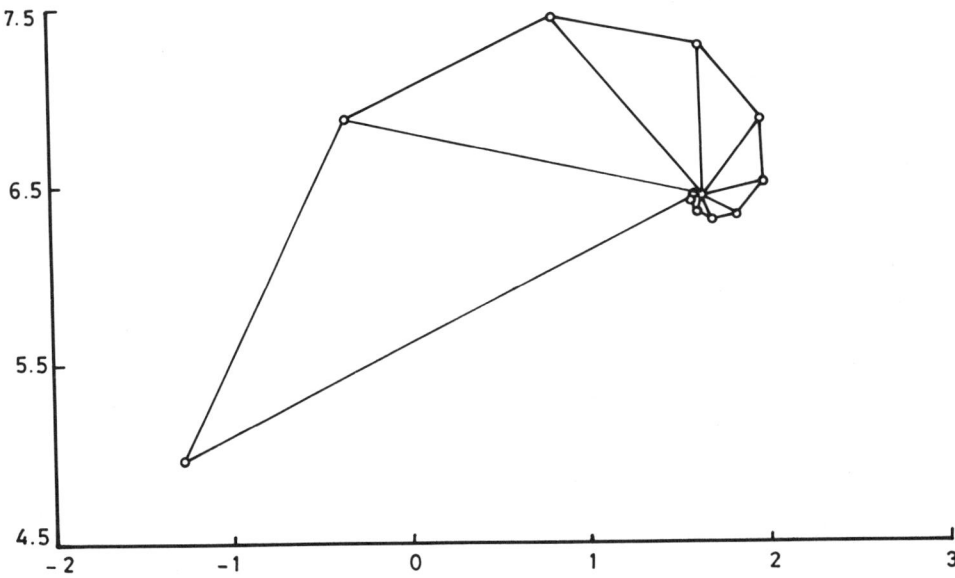

Figure 7.5.3 Spiral of singularities around one singularity for Kuramoto's model (2) (From Frisch 1983.)

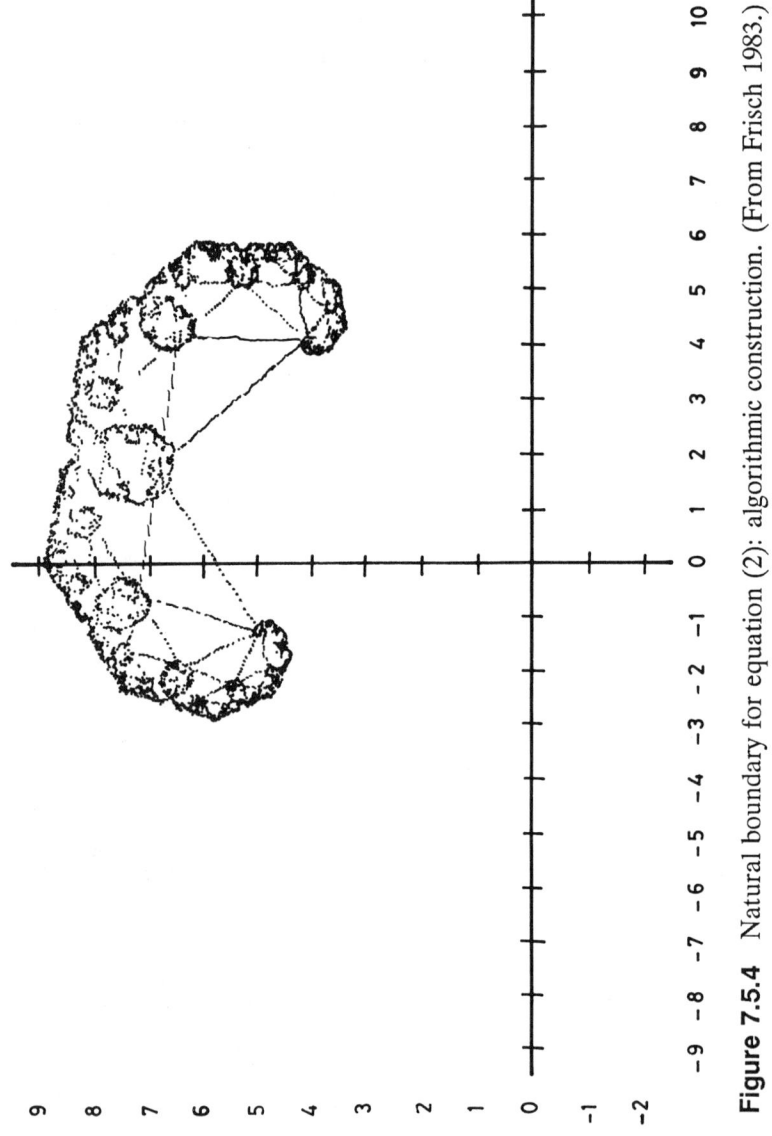

Figure 7.5.4 Natural boundary for equation (2): algorithmic construction. (From Frisch 1983.)

7.6 PAINLEVÉ PROPERTY FOR SOME OTHER SYSTEMS

Now we quote some other work on the Painlevé property of systems of DE. Lukashevich and Matatov (1973) considered the system

$$\frac{dw_1}{dt} = \tilde{a}_0 + \tilde{a}_1 w_1 + \tilde{a}_2 w_2 + \tilde{a}_3 w_1^2 + \tilde{a}_4 w_1 w_2 + \tilde{a}_5 w_2^2,$$
$$\frac{dw_2}{dt} = \tilde{b}_0 + \tilde{b}_1 w_1 + \tilde{b}_2 w_2 + \tilde{b}_3 w_1^2 + \tilde{b}_4 w_1 w_2 + \tilde{b}_5 w_2^2, \tag{1}$$

where $\tilde{a}_j(t)$ and $\tilde{b}_j(t)$ ($j = 0, 1, \ldots, 5$) are analytic functions of t in some region G. System (1) was first transformed by writing

$$w_1 = x + \beta(t)y,$$
$$w_2 = y, \tag{2}$$

where $\beta(t)$ is a root of the equation

$$\tilde{b}_3 \beta^3 + (\tilde{b}_4 - \tilde{a}_3)\beta^2 + (\tilde{b}_5 - \tilde{a}_4)\beta - \tilde{a}_5 = 0. \tag{3}$$

This transformation is used to arrange that $\tilde{a}_5(t) = 0$ so that (1) maybe written in the form

$$\frac{dx}{dt} = a_0 + a_1 x + a_2 y + a_3 x^2 + a_4 xy,$$
$$\frac{dy}{dt} = b_0 + b_1 x + b_2 y + b_3 x^2 + b_4 xy + b_5 y^2, \tag{4}$$

where $a_0(t), \ldots, b_5(t)$ are analytic functions that depend on the coefficients of (1).

Assuming $a_4(t) \neq 0$, we can further simplify (4) by the ansatz

$$x = \xi - \frac{a_2}{a_4}, \quad y = \frac{1}{a_4}\eta - \frac{a_3}{a_4}\xi - \frac{a_1}{a_4} + \frac{2a_2 a_3}{a_4^2}. \tag{5}$$

We arrive at the system

$$\frac{d\xi}{dt} = p_0(t) + \xi\eta,$$
$$\frac{d\eta}{dt} = q_0(t) + q_1(t)\xi + q_2(t)\eta + q_3(t)\xi^2 + q_4(t)\xi\eta + q_5(t)\eta^2, \tag{6}$$

where $p_0(t), \ldots, q_5(t)$ are analytic functions of t that depend on the coefficients of system (4). Eliminating η from (6), we obtain the second-order equation

$$\xi\xi'' = (1 + q_5)\xi'^2 + (q_4\xi^2 + q_2\xi - 2p_0 q_5 - p_0)\xi' + q_3\xi^4 + q_1\xi^3$$
$$+ (q_0 - p_0 q_4)\xi^2 + (p_0' - p_0 q_2)\xi + p_0^2 q_5. \tag{7}$$

It was shown by Lukashevich and Matatov that (7) has poles as its only movable singularities provided one of the following conditions holds: (i) $1 + q_5 = 0$, (ii) $q_5 = 0$, (iii) $mq_5 + 1 = 0$ $(m = 2, 3, \ldots)$. Equation (7) belongs to the class studied by Ince (1956).

A very special case of (1), namely,

$$\frac{dw_1}{dt} = -w_1^2 + aw_1w_2 + \alpha w_2 + \beta w_2 + \lambda,$$

$$\frac{dw_2}{dt} = -w_2^2 + bw_1w_2 + \gamma w_1 + \delta w_2 + \mu,$$

where a, α, β, λ, b, γ, δ, and μ are constants, was studied in great detail by Ramani et al. (1984). Their main conclusion is that, while Painlevé analysis is a most useful guide, integrable cases also exist which do not possess the Painlevé property.

Finally, we refer to the work of Kolesnikova and Lukashevich (1972), who found conditions for the third-order equation

$$a_0(z)ww''' + a_1(z)w'w'' + a_2(z)w''w''' + a_3(z)w''^2 + a_4(z)w'w''$$
$$+ a_5(z)ww'' + a_6(z)w'^2 + a_7(z)ww' + a_8(z)w^2 = 0 \qquad (10)$$

to be of P-type. This equation was first reduced to one of second order by the transformation $w' = vw$, and then studied following the work of Ince (1956).

8
Painlevé Transcendents

8.1 INTRODUCTION

We have discussed in Chapter 3 the diversity of singularities that the solutions of nonlinear differential equations can manifest. Linear and nonlinear differential equations differ in one important respect (apart from the fact of the principle of linear superposition not holding for the latter). Linear DE have fixed singular points, namely those of the coefficients and, as a rule, the point at infinity. For nonlinear DE, on the other hand, the position (and the type) of the singularity cannot in general be determined from the differential equation. Indeed, the singularity may be placed at any desired point; the nature of the singularity is, however, not arbitrary. For example, the DE

$$y' = y^2 \tag{1}$$

with initial condition $y = y_0$ for $x = x_0$ has the solution

$$y = \frac{y_0}{y_0 x_0 - x y_0 + 1}. \tag{2}$$

This solution has a simple pole at the point

$$x = \frac{x_0 y_0 + 1}{y_0}, \tag{3}$$

which cannot be visualized from (1). Such singular points are called *movable*. The number of movable singular points may be finite or infinite, and may be of extremely complex character. The movability of singularity points can itself become an important criterion for distinguishing various types of nonlinear DE and their solutions. The subsets of movable singular points can belong to one of the following:

1. Single-valued ones, for which the solution does not change its value in going round a given initial point z_0.
2. Non-single-valued or branch points.
3. The points for which the solution has a finite or infinite limit as $z \to z_0$.
4. The points for which the solution $y = y(z)$ has no limit as $z \to z_0$—the essential singular points.
5. Isolated and nonisolated singular points.

A singular point may have more than one of the above attributes. For example, it may be a simple essential singularity, possessing properties 1 and 4, or it may be a simple pole belonging to subsets 1 and 3.

We have already met with an example of a movable simple pole for (1). The DE

$$w' = -w(\ln w)^2$$

has the solution

$$w = \exp \frac{1}{z-a}.$$

This has a movable singularity, and the solution does not tend to any limit as z tends to a along a vertical line: $z = a + i\epsilon$, $\epsilon \to 0$. The DE

$$\frac{d^2 w}{dz^2} = \frac{2w-1}{w^2+1} \left(\frac{dw}{dz}\right)^2 \tag{4}$$

has the general solution

$$w = \tan\{\ln(Az - B)\},$$

where A and B are arbitrary constants. As z tends to B/A (in any arbitrary manner), w does not tend to any limit, finite or infinite. The point B/A is a *moving singularity*—a logarithmic branch point as well as a cluster point of poles.

The nature of singularities of first-order nonlinear DE has been discussed fairly completely by Ince (1956) and Hille (1969). (See also Chapter 3). For example, the Painlevé theorem regarding

$$\frac{dw}{dz} = f(z,w) = \frac{P(z,w)}{Q(z,w)}, \tag{5}$$

Painlevé Transcendents

where $P(z,w)$ and $Q(z,w)$ are polynomials in w with analytic coefficients, ensures that the solutions of (5) have no movable essential singularities. That simplifies the study of (5) considerably. Indeed it has been proven (see Hille 1969) that movable singular points of (5) may be poles and/or algebraic branch points—that is, points in the neighborhood of which the solution can be written as a series of powers of $(z-z_0)^{1/n}$, where $n \geq 1$ is a positive integer *and* where the number of negative powers of this quantity may only be finite.

Even so, the determination of the number of movable singular points and their nature is a formidable task. In contrast to the case of first-order DE, there is no theorem analogous to Painlevé's for a second-order nonlinear DE. One can merely state that for such an equation a singular point z_0 is such that the solution $w(z)$ tends as $z \to z_0$ to the boundary of the domain $D(z,w)$ of the continuity of the (vector) function $f(z,w)$, where $dw/dz = f(z,w)$. The point z_0 may be such that $w(z)$ has no limit as $z \to z_0$ but simply tends to the boundary of the domain of continuity of $f(z,w)$. For scalar equations, however, $w \to w_0$, finite or infinite, as $z \to z_0$.

Thus the task of classifying nonlinear second-order DE presents severe difficulties. Painlevé and Gambier considered the second-order equations

$$\frac{d^2w}{dz^2} = R(z,w,w'), \tag{6}$$

where R is a rational function of w and w' and is analytic in z. Equation (6) covers a very important class as we shall presently show, but the restriction of R to the class of functions rational in w and w' should be noted. We shall see in Section 8.4 that this restriction avoids further complications in analysis. The problem tackled by Painlevé and Gambier was to characterize equations (6) which have fixed critical points. The latter refer to branch points and essential singularities. The only movable singularities that the solutions of (6) could possess were poles.

A similar query with reference to the first-order equations was posed and responded to in Section 3.4, and we were led to the Riccati equation. But, as we remarked earlier, the situation for the second-order equation is much more involved. The Painlevé theorem does not hold, and the movable singular points can include essential singularities as well as transcendental critical points. For example, the equation

$$[ww'' - (w')^2]^2 + 4w(w')^3 = 0 \tag{7}$$

has the general solution

$$w = C_1 \exp(z - C_2)^{-1}, \qquad C_1, C_2 \text{ arbitrary},$$

with a movable essential singularity at $z = C_2$. The equation

$$ww'' = \alpha(w')^2, \qquad \alpha \neq 1, \tag{8}$$

has the general solution

$$w = C_2(z - C_1)^{1/(1-\alpha)}$$

with a movable transcendental critical point unless α is rational. We have already encountered another complicated movable singularity in the solution of (4).

The method to identify the conditions under which (6) has fixed critical points has been given in a lucid and masterly fashion by Ince (1956) (see references there to the original works of the French mathematicians, in particular, Painlevé and Gambier). Here, we content ourselves with a summary of the so-called α-method. The method comprises two stages. First, a set of (two) necessary conditions for the absence of movable critical points is obtained. Then, a comprehensive set of equations satisfying these necessary conditions is derived. These turn out to be 50 in number. It is then shown, by direct integration or by reduction to simpler equations, that the general solutions of these equations are free from movable critical points. This proves the sufficiency of conditions. Six of these 50 equations required rather elaborate proofs to show the absence of movable critical points. These are referred to as *Painlevé transcendents* and shall occupy most of this chapter.

For the purpose of deriving the necessary conditions, Painlevé introduced an auxiliary parameter α in (6) in such a way that for $\alpha = 0$, a simpler equation results, which is integrable or from which the absence or presence of movable critical points can be inferred by inspection. If the transformed equation has single-valued solutions for small $|\alpha|$ except possibly for $\alpha = 0$, then the following lemma helps to conclude that $\alpha = 0$ can be no exception: if the general solution of (6) is uniform in z for all values of α in D except (possibly) $\alpha = 0$, then it will be uniform also for $\alpha = 0$. Here D is a domain over which R as a function of α is analytic. The point $\alpha = 0$ lies in the interior of this domain.

Using this artifice, Painlevé showed that the function R (which was assumed to be rational in w' and w) should be a polynomial in w' of degree ≤ 2. Thus, (6) reduces to

$$w'' = L(z,w)w'^2 + M(z,w)w' + N(z,w), \tag{9}$$

where L, M, and N are rational in w.

Next it was shown that the functions L, M, and N must satisfy two further conditions:

Painlevé Transcendents

I. $L(z,w)$ must either be identically zero or have one of the following forms:

$$\frac{m+1}{m(w-a_1)} + \frac{m-1}{m(w-a_2)}, \quad m \geq 1, \tag{10A}$$

$$\frac{1}{2}\sum_{n=1}^{4}\frac{1}{w-a_n}, \tag{10B}$$

$$\frac{2}{3}\sum_{n=1}^{3}\frac{1}{w-a_n}, \tag{10C}$$

$$\frac{3}{4}\left(\frac{1}{w-a_1} + \frac{1}{w-a_2}\right) + \frac{1}{2}\left(\frac{1}{w-a_3}\right), \tag{10D}$$

$$\frac{1}{6}\sum_{n=1}^{3}\frac{n+2}{w-a_n}. \tag{10E}$$

The quantities a_n are arbitrary functions of z. They are not necessarily unequal and any of them may be infinite.

II. The coefficients $L(z,w)$, $M(z,w)$ and $N(z,w)$ must satisfy certain conditions. If $L = 0$, then M must be linear in w while N is a polynomial in w of degree ≤ 3. If $L \neq 0$, let $D(z,w)$ be the least common denominator of the partial fractions in $L(z,w)$ and of degree δ, $2 \leq \delta \leq 4$, in w. Then

$$M(z,w) = \frac{\mu(z,w)}{D(z,w)}, \quad N(z,w) = \frac{\nu(z,w)}{D(z,w)},$$

where μ and ν are polynomials in w of degree $\leq \delta + 1$ and $\leq \delta + 3$, respectively.

This exhausts the sets of necessary conditions, but it was a long haul to separate equations with the desired property from the total class of equations which satisfied conditions I and II. Even the special simpler case $L = 0$, which has M linear in w and N a cubic in w, led to 10 equations (see appendix 1 of Davis 1962). Eight of these could be solved explicitly, while the remaining two constitute the first two of the six Painlevé transcendents to be listed below. For example, the sixth of these 10 equations,

$$\frac{d^2w}{dz^2} = [3w + q(z)]\frac{dw}{dz} - q(z)w^2 - w^3, \tag{11}$$

can be linearized by the (logarithmic) nonlinear transformation

$$w = -\frac{u'}{u}$$

to
$$u''' + q(z)u'' = 0. \tag{12}$$

The case $L \neq 0$ leads to 40 equations of which all but 4 can be solved in terms of classical transcendents or reduced to linear equations. For example, the equation

$$(w - w^2)\frac{d^2w}{dz^2} = \frac{3}{4}(1 - 2w)\left(\frac{dw}{dz}\right)^2 \tag{13}$$

changes, by the transformation $z = z(x)$, to

$$\frac{d^2w}{dx^2} - \left(\frac{z''}{z'}\right)\frac{dw}{dx} = \frac{3}{4}\frac{1-2w}{w(1-w)}\left(\frac{dw}{dx}\right)^2. \tag{14}$$

Now, we set $z''/z' = q(x)$ in this equation and have

$$4(w - w^2)\frac{d^2w}{dx^2} = 3(1 - 2w)\left(\frac{dw}{dx}\right)^2 + 4q(x)(w - w^2)\frac{dw}{dx}. \tag{15}$$

Thus, while equation (15) is not included in the list of 50 equations of Painlevé, it can be related to (13) through the transformation $z = z(x)$, where $z'' - q(x)z' = 0$. Equation (13) is 43rd in Painlevé's list (see appendix 1 of Davis 1962) and has the solution

$$w = \frac{1}{1 - \mathbf{P}^2(z)}, \tag{16}$$

where $\mathbf{P}(u) = \mathbf{P}(u, 4, 0)$ is the elliptic function of Weierstrass, with $g_2 = 4$, $g_3 = 0$ (see Davis 1962, p. 184).

This example illustrates that a much larger number of equations, with solutions satisfying the fundamental criterion of absence of movable critical points, can be generated from the 50 equations of Painlevé by applying to each member the following transformations:

$$w = \frac{a + b\tilde{w}}{c + d\tilde{w}}, \quad \Delta = ad - bc \neq 0, \tag{17}$$

$$z = z(t) \tag{18}$$

where a, b, c, d are analytic functions of z and $z(t)$ is an analytic function of t.

We list below the six equations whose solutions are the Painlevé transcendents:

$$P_I \quad w'' = 6w^2 + z, \tag{19a}$$
$$P_{II} \quad w'' = 2w^3 + zw + \alpha, \tag{19b}$$

Painlevé Transcendents

$$P_{III} \quad w'' = \frac{1}{w}(w')^2 - \frac{1}{z}w' + \frac{1}{z}(\alpha w^2 + \beta) + \gamma w^3 + \frac{\delta}{w}, \tag{19c}$$

$$P_{IV} \quad w'' = \frac{1}{2w}(w')^2 + \frac{3}{2}w^3 + 4zw^2 + 2(z^2 - \alpha)w + \frac{\beta}{w}, \tag{19d}$$

$$P_V \quad w'' = \left\{\frac{1}{2w} + \frac{1}{w-1}\right\}(w')^2 - \frac{1}{z}w' + \frac{(w-1)^2}{z^2}\left\{\alpha w + \frac{\beta}{w}\right\}$$
$$+ \frac{\gamma w}{z} + \frac{\delta w(w+1)}{w-1}, \tag{19e}$$

$$P_{VI} \quad w'' = \frac{1}{2}\left\{\frac{1}{w} + \frac{1}{w-1} + \frac{1}{w-z}\right\}(w')^2$$
$$- \left\{\frac{1}{z} + \frac{1}{z-1} + \frac{1}{w-z}\right\}w'$$
$$+ \frac{w(w-1)(w-z)}{z^2(z-1)^2}\left\{\alpha + \frac{\beta z}{w^2} + \frac{\gamma(z-1)}{(w-1)^2} + \frac{\delta z(z-1)}{(w-z)^2}\right\}. \tag{19f}$$

Equations P_I–P_{II} have no critical points at all, while P_{III} has logarithmic branch points (see Section 8.2). If we set $z = e^t$ in P_{IV} and P_V, the solutions become single-valued functions of t. Equation P_{VI} has three critical points: 0, 1, and ∞. Actually, only equations P_I–P_{III} were discovered by Painlevé. Gambier added P_{IV} and P_V. Equation P_{VI} is generic of the other five: they arise from it either by passage to a limit or by coalescence.

It is well known that second-order DE (in contrast to those of other orders) play a predominant role in the description of physical phenomena. It was not realized until a few years ago that nonlinear ODE which do not possess movable branch points or essential singularities have any special physical significance. Ablowitz, Ramani, and Segur (1978) referred to these nonlinear ODE as Painlevé type or P type and made the following conjecture: every nonlinear ODE obtained by an exact reduction of a nonlinear PDE solvable by some inverse scattering transform has the Painlevé property. It is well established that the PDE which are exactly linearized via IST display the soliton property (see Whitham 1974); that is, two solitary waves interacting under one of these PDE come out nearly clean (and unchanged) apart from a phase shift.

We give a few examples of PDE which are solvable by IST and permit exact reduction to ODE of P type.

EXAMPLE 1 The Boussinesq equation

$$u_{tt} = u_{xx} + 6(u^2)_{xx} + u_{xxxx} \tag{20}$$

is a nonlinear PDE solvable by IST. A traveling wave solution of (20),
$$u(x,t) = f(x - ct), \tag{21}$$
say, is governed by the equation
$$f^{(iv)} - (c^2 - 1)f'' + 12ff'' + 12f'^2 = 0. \tag{22}$$
This equation can be integrated twice and, depending upon the choice of the constants of integration, reduces in terms of $w = f + (1 - c^2)/12$ to
$$w'' + 6w^2 + C_1 z + C_2 = 0. \tag{23}$$
Its special case $C_1 = 0$ gives
$$w'' = -6w^2 - C_2. \tag{24}$$
Equation (23) defines the *first Painlevé transcendent* (after simple scalings), while (24) has solutions expressible in terms of elliptic functions. In either case, the only movable singularities are poles.

EXAMPLE 2 The modified Korteweg–deVries (KdV) equation
$$u_t - 6u^2 u_x + u_{xxx} = 0 \tag{25}$$
can be solved by IST (Wadati 1972). If we look for similarity solution of (25) in the form
$$u(x,t) = (3t)^{-1/3} w(z), \quad z = \frac{x}{(3t)^{1/3}}, \tag{26}$$
then the function w is governed by
$$w''' - 6w^2 w' - (zw)' = 0. \tag{27}$$
This equation integrates to give
$$w'' = 2w^3 + zw + \alpha. \tag{28}$$
This is just P_{II}.

EXAMPLE 3 The Sine–Gordon equation
$$u_{xt} = \sin u \tag{29}$$
can be solved by IST (Ablowitz et al. 1973). It has a self-similar solution
$$u(x,t) = f(z), \quad z = xt. \tag{30}$$
If we set $w(z) = \exp(if)$, it easily follows that
$$w'' = \frac{1}{w} w'^2 - \frac{1}{z} w' + \frac{1}{2z}(w^2 - 1). \tag{31}$$

Painlevé Transcendents

This is a special case of P_{III}.

EXAMPLE 4 The KdV equation

$$u_t + 6uu_x + u_{xxx} = 0 \qquad (32)$$

through the transformation

$$u = \alpha t + U(z), \qquad z = x - 3\alpha t^2 \qquad (33)$$

becomes

$$U''' + 6UU' + \alpha = 0. \qquad (34)$$

This equation, on integration, takes the form

$$U'' + 3U^2 + \alpha z = 0 \qquad (35)$$

if the constant of integration is put equal to zero. Equation (35) is essentially P_I.

The KdV equation (32) also admits another similarity transformation

$$u = (3t)^{-2/3} U(z), \qquad z = x(3t)^{-1/3}, \qquad (36)$$

so it now becomes

$$K_1(U) \equiv U''' + 6UU' - (2U + zU') = 0. \qquad (37)$$

This equation can be integrated once using the identity $[(2U - z)K_2(U)]' = (2U - z)K_1(U)$, where

$$K_2(U) = U'' + 2U^2 - zU + \frac{\nu + U' - U'^2}{2U - z}. \qquad (38)$$

Here ν is an arbitrary constant. Equation (38) is essentially the same as that called P_{XXIV} by Ince (1956) and is related by a one-to-one map to P_{II}.

Ablowitz and his co-workers checked numerous examples. In every case checked, PDE that can be solved by IST reduce to ODE of P type, and PDE that are not solvable by IST (e.g., this may be determined by observing numerically that two solitary waves do not interact like solitons) reduce to ODE that are not P type. We note that the reduced ODE does not have to be one of the six Painlevé equations, but should merely enjoy the Painlevé property. Here then is an instance of a mathematical classification of DE corresponding so neatly to the physics of a certain kind. We discussed in Chapter 7 the "weak Painlevé property" and other related matters in the context of chaotic and near chaotic behavior of solutions of nonlinear DE.

There is another interesting way of viewing these equations—that is, to pick up one of the 50 Painlevé equations and show how its special cases cover exact traveling or similarity solutions of several nonlinear model PDE. This was done by Kawamoto (1985). We illustrate this with the equation (twelfth in the list of Davis 1962)

$$y \frac{d^2y}{dx^2} = \left(\frac{dy}{dx}\right)^2 + a_0 + a_1 y + a_3 y^3 + a_4 y^4, \tag{39}$$

where a_0, a_1, a_3, and a_4 are arbitrary constants. By writing

$$y(x) = \beta \exp(\alpha f(x)), \tag{40}$$

we change equation (39) to

$$\frac{d^2 f}{dx^2} = \frac{a_0}{\alpha \beta^2} \exp(-2\alpha f) + \frac{a_1}{\alpha \beta} \exp(-\alpha f)$$
$$+ \frac{\beta a_3}{\alpha} \exp(\alpha f) + \frac{\beta^2 a_4}{\alpha} \exp(2\alpha f). \tag{41}$$

Equation (41) can be integrated once by multiplying it by df/dx:

$$\left(\frac{df}{dx}\right)^2 = -\frac{a_0}{\alpha^2 \beta^2} \exp(-2\alpha f) - \frac{2a_1}{\alpha^2 \beta} \exp(-\alpha f)$$
$$+ \frac{2\beta a_3}{\alpha^2} \exp(\alpha f) + \frac{\beta^2 a_4}{\alpha^2} \exp(2\alpha f) + k, \tag{42}$$

where k is the constant of integration. Again, we introduce the transformation

$$g(x) = \gamma \exp(\alpha f(x)) \tag{43}$$

so that (42) becomes

$$\left(\frac{dg}{dx}\right)^2 = A_0 + A_1 g + A_2 g^2 + A_3 g^3 + A_4 g^4 \equiv -F(g) \geq 0, \tag{44}$$

say, where

$$A_0 = -\frac{a_0 \gamma^2}{\beta^2}, \quad A_1 = -\frac{2a_1 \gamma}{\beta}, \quad A_2 = \alpha^2 k,$$
$$A_3 = \frac{2\beta a_3}{\gamma}, \quad A_4 = \frac{\beta^2 a_4}{\gamma^2}. \tag{45}$$

With arbitrary constants, α, β, and γ, equation (44) may be interpreted as an equation of motion of a particle with potential function $F(g)$ or as an equation of an anharmonic oscillator. We now show how special cases of (44) arise in the one-soliton solutions of well-studied soliton equations.

Painlevé Transcendents

I

$A_0 = A_1 = A_4 = 0$.

Equation (44) becomes

$$\left(\frac{dg}{dx}\right)^2 = g^2(A_2 + A_3 g) \equiv -F(g) \geq 0. \tag{46}$$

The potential function $F(g)$ and the solitary wave solutions in this case under the condition $A_2 > 0$ and $A_3 < 0$ are illustrated in Figure 8.1.1. Equation (46) can be integrated, and we obtain two solutions

$$g_1(x) = -\frac{A_2}{A_3} \operatorname{sech}^2\left\{\frac{\sqrt{A_2}(x + x_0)}{2}\right\} \quad \text{for } 0 \leq g(x) \leq -\frac{A_2}{A_3}, \tag{47}$$

$$g_2(x) = \frac{A_2}{A_3} \operatorname{csch}^2\left\{\frac{\sqrt{A_2}(x + x_0)}{2}\right\} \quad \text{for } g(x) \leq 0, \tag{48}$$

where x_0 is the constant of integration. Solution (48) is of exploding type.

Now, the stationary solutions of the KdV equation $u_t + \alpha u u_x + u_{xxx} = 0$, the nonlinear Schrödinger equation (NLS) $i u_t + u_{xx} + \gamma |u|^2 u = 0$, the Boussinesq equation $u_{xx} - u_{tt} + \alpha |u^2|_{xx} + u_{xxxx} = 0$, etc., are governed, under vanishing boundary conditions $g = 0$ at $|x| = \infty$, by (46), and have the form (47) or (48). For example, the KdV equation

$$u_t + \alpha u u_x + u_{xxx} = 0 \tag{49}$$

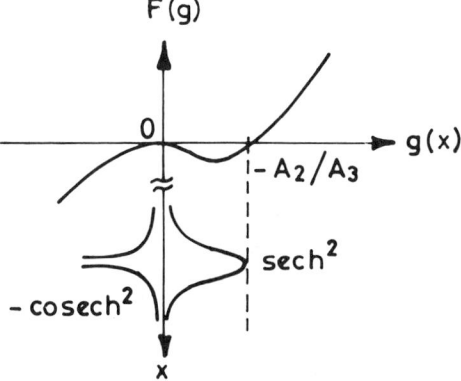

Figure 8.1.1 The potential function $F(g) = -g^2(A_2 + A_3 g)$ and the solitary solutions $A_2 > 0$ and $A_3 < 0$ (see equation (46)). (From Kawamoto 1985.)

has the solution $u = U(x - ct) \equiv U(X)$ if

$$(-c + \alpha U)U_X + U_{XXX} = 0. \tag{50}$$

Integrating (50), we obtain

$$-cU + \frac{\alpha U^2}{2} + U_{XX} = 0 \tag{51}$$

if we use the vanishing conditions on U and its derivatives at $|X| = \infty$. Multiplying (51) by U_X and integrating it once again with respect to X immediately reduces this equation to the form (46), provided the constant of integration is again set equal to zero.

II

$$A_0 = A_1 = A_3 = 0.$$

Equation (44) now reduces to

$$\left(\frac{dg}{dx}\right)^2 = g^2(A_2 + A_4 g^2) \equiv -F(g). \tag{52}$$

This situation is depicted in Figure 8.1.2. The solutions do not explode in the present case and, for $A_2 > 0$ and $A_4 < 0$, have the form

$$g(x) = \pm \sqrt{\left|\frac{A_2}{A_4}\right|} \operatorname{sech}\left\{\sqrt{A_2}(x + x_0)\right\}. \tag{53}$$

One may check as for the KdV equation in I that the stationary solutions of the modified KdV equation

$$u_t + \alpha u^2 u_x + u_{xxx} = 0 \tag{54}$$

are governed by (52) and have the form (53).

III

$$A_0 = A_1 = 0.$$

Equation (44) takes the form

$$\left(\frac{dg}{dx}\right)^2 = g^2(A_2 + A_3 g + A_4 g^2) \equiv -F(g). \tag{55}$$

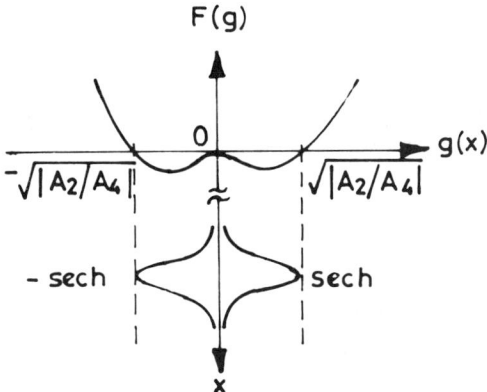

Figure 8.1.2 $F(g) = -g^2(A_2 + A_4 g^2)$ and the solitary wave solutions $A_2 > 0$ and $A_4 < 0$ (see equation (52)). (From Kawamoto 1985).

The function $F(g)$ is shown in Figure 8.1.3. When $A_2 < 0$ and $A_3 > 0$, $A_4 > 0$, the solitary wave solutions of (55) are

$$g(x) = \pm \frac{2A_2}{\sqrt{A_3^2 - 4A_2 A_4} \, \exp(X) \mp 2A_3/\sqrt{A_3^2 - 4A_2 A_4} + \exp(-X)}, \quad (56)$$

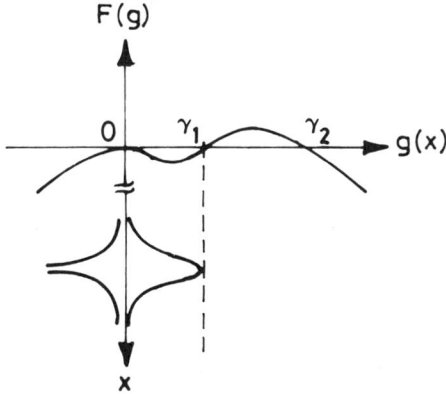

Figure 8.1.3 $F(g) = -g^2(A_2 + A_3 g + A_4 g^2)$ and the solitary wave solutions. γ_1 and γ_2 are roots of the equation $A_2 + A_3 g + A_4 g^2 = 0$: $A_2 < 0, A_3 > 0$, $A_4 > 0$, and $\gamma_1 < \gamma_2$ (see equation (55)). (From Kawamoto 1985.)

where $X \equiv \sqrt{|A_2|}(x + x_0)$. The upper signs in equation (56) correspond to expoloding solutions while the lower signs give "combined" one-soliton solutions. These characterize the stationary solutions of the "combined" KdV equation $u_t + (\alpha u + \beta u^2)u_x + u_{xxx} = 0$.

We indicate how the classical Boussinesq equations

$$u_t + ((1 + \alpha u)v)_x + \beta u_{xxx} = 0, \tag{57}$$
$$v_t + u_x + \alpha v v_x = 0 \tag{58}$$

have stationary solutions governed by (55). We may choose $\alpha = \beta = 1$ for convenience. Substituting $u = f(\zeta)$, $v = g(\zeta)$, $\zeta = x - ct$ into (57) and (58), eliminating $f(\zeta)$, and integrating the equation for $g(\zeta)$ twice, we obtain

$$(g')^2 = Ag^4 + Bg^3 + Cg^2, \tag{59}$$

where

$$A = \frac{1}{4}, \quad B = -c, \quad C = (c^2 - 1), \tag{60}$$

and the constants of integration have been ignored. Equation (59) is of the form (55).

IV

Either $A_0 = A_4 = 0$ (i.e., $a_0 = a_4 = 0$) or $A_1 = A_3 = 0$ (i.e., $a_1 = a_3 = 0$). We consider the case $a_0 = a_4 = 0$. Equation (39) becomes

$$y\frac{d^2y}{dx^2} = \left(\frac{dy}{dx}\right)^2 + a_1 y + a_3 y^3. \tag{61}$$

Introducing the transformation

$$y(x) = \sqrt{\left|\frac{a_1}{a_3}\right|} \exp(i\alpha f(x)) \tag{62}$$

into (61), we obtain

$$\frac{d^2 f}{dx^2} = -\left(2\frac{\sqrt{|a_1 a_3|}}{\alpha}\right) \sin \alpha f. \tag{63}$$

The one-kink solution of (63) is

$$f(x) = \pm \frac{4}{\alpha} \tan^{-1} \left\{ \exp\left(\sqrt{2\sqrt{|a_1 a_2|}}(x + x_0)\right) \right\}. \tag{64}$$

Painlevé Transcendents

As an example of this kind, we consider the Sine–Gordon (SG) equation

$$\phi_{tt} - \phi_{xx} + \sin \phi = 0. \tag{65}$$

The stationary solutions of this equation $\phi = \phi(X), X = x - Ut$, are governed by

$$(1 - U^2)\phi_{XX} - \sin \phi = 0. \tag{66}$$

This is of the form (63) and, under suitable conditions, has solution of type (64).

V

$A_1 = A_4 = 0$ (i.e., $a_1 = a_4 = 0$) or $A_0 = A_3 = 0$ (i.e., $a_0 = a_3 = 0$). For $a_1 = a_4 = 0$ and $\alpha = \beta = \gamma = a_0 = -a_3 = 1$, equation (41) becomes

$$\frac{d^2 f}{dx^2} = \exp(-2f) - \exp(f), \tag{67}$$

and, correspondingly, equation (44) has the form

$$\left(\frac{dg}{dx}\right)^2 = -1 + kg^2 - 2g^3$$
$$= -2(g - m_1)(g - m_2)(g - m_3), \quad m_1 > m_2 > m_3. \tag{68}$$

The solution of (68) and hence of (67) can be expressed in terms of Jacobi's sn-function:

$$f(x) = \ln\left\{m_1 - (m_1 - m_3)\operatorname{sn}^2\left(\frac{\sqrt{m_1 - m_3}}{2}x, \sqrt{\frac{m_1 - m_2}{m_1 - m_3}}\right)\right\}. \tag{69}$$

The equation of motion of the Toda cyclic chain with three particles is identical to (67) and has the solution (69).

Kawamoto (1985) has also considered generalizations of (39) when the terms $a_5 y^5$ and $a_5 y^5 + a_6 y^6$, respectively, are added to the right-hand side, and has identified solutions to more general soliton equations.

Originally it was thought that Painlevé equations represent entirely new transcendental functions in the sense that they cannot be expressed in terms of first-order differential equations of the Fuchsian class

$$A_0(w, z)w' + A_1(w, z) = 0 \tag{70}$$

not having movable branch points, or reduced to linear equations of arbitrary order, or solved in terms of (known) solutions of some other equations. This surmise was shown to be incorrect, first by Soviet investigators. Indeed, many *special* cases of equations P_I–P_{VI} can be explicitly solved. There are important transformations connecting these equations and their solutions. We shall deal with these matters in the next section and thus bring the Painlevé transcendents to a more familiar level before we take up their general analytic study, with or without reference to some initial or boundary conditions arising from physical problems.

8.2 SPECIAL SOLUTIONS OF PAINLEVÉ EQUATIONS

As we remarked in Section 8.1, the notion that Painlevé equations P_I–P_{VI} represented entirely new functions underwent a change around 1960. Actually, Ince (1956) did remark that

> P_{VI} contains, in reality, the first five equations, which, may be derived from it by a process of coalescence. As it can be proved that the solutions of P_I are indeed new transcendents, it follows that the solutions of the remaining five equations cannot (*except* possibly for special values of α, β, γ, and δ) be expressible in terms of the classical transcendental functions alone.

Before we take up a systematic study of the transformations of Painlevé transcendents due to Fokas and Ablowitz (1982) in section 8.3, we show how P_{III} and P_{VI} can be solved for some special values of the parameters in a simple manner (Lukashevich 1965). If we write P_{III} as

$$z^2 \frac{d^2 w}{dz^2} - \frac{1}{w}\left(z \frac{dw}{dz}\right)^2 + z \frac{dw}{dz} - z(\alpha w^2 + \beta) - \gamma z^2 w^3 - \frac{\delta z^2}{w} = 0, \quad (1)$$

the homogeneity with respect to z in the differentiated terms immediately suggests the change of variable

$$z = e^t, \qquad w = v e^{kt}, \tag{2}$$

where k is a constant. This constant may be suitably chosen to bring about necessary simplification. Equation (1) now becomes

$$v'' + (2k-1)v' + k(k-1)v = \frac{1}{v}(v' + kv)^2 - (v' + kv) + \alpha v^2 e^{(k+1)t}$$
$$+ \beta e^{(1-k)t} + \gamma v^3 e^{(2k+2)t} + \frac{\delta}{v} e^{(2-2k)t}. \tag{3}$$

The two choices of the parameters α, β, γ, δ and k render this equation autonomous:

i. $\alpha = \gamma = 0$; $k = 1$. Equation (3) becomes

$$vv'' = v'^2 + \beta v + \delta. \tag{4}$$

Painlevé Transcendents

Let $v' = u$; then

$$vu \frac{du}{dv} = u^2 + \beta v + \delta. \tag{5}$$

This is a linear equation in u^2 with solution

$$v' = u = \pm\sqrt{C_1 v^2 - 2\beta v - \delta} \tag{6}$$

or

$$\int \frac{dv}{\sqrt{C_1 v^2 - 2\beta v - \delta}} = \pm t + C_2. \tag{7}$$

This integral can be expressed in terms of elementary functions, the form depending on the constants C_1, β, and δ.

ii. $\beta = \delta = 0$; $k = -1$. Then, (3) becomes

$$vv'' = v'^2 + \alpha v^3 + \gamma v^4. \tag{8}$$

Again setting $v' = u$, we have

$$vu \frac{du}{dv} = u^2 + \alpha v^3 + \gamma v^4. \tag{9}$$

This equation is linear in u^2 and has solution

$$u = \pm v\sqrt{\gamma v^2 + 2\alpha v + C_1},$$

that is,

$$\int \frac{dv}{v\sqrt{\gamma v^2 + 2\alpha v + C_1}} = \pm t + C_2. \tag{10}$$

This is an elementary integral (see Gradshteyn and Ryzhik 1965, p. 80). In particular for $C_1 = \alpha = 0$, (10) admits the solution

$$w = \frac{1}{\gamma^{1/2} z(\pm \ln z + C_2)}, \tag{11}$$

displaying both algebraic and logarithmic branch points. This proves false the statement in Hille (1969) and Ince (1956) that the solutions to P_I–P_{III} have no branch points. There are other simple solutions of P_{III} with branch points: (i) if $\beta = \gamma = 0$, $\alpha\delta \neq 0$, then there exist solutions $w = hz^{1/3}$, where h is any root of the equation $\alpha h^3 + \delta = 0$; (ii) if $\alpha = \delta = 0$, $\beta\gamma \neq 0$, then $w = hz^{-1/3}$ are solutions where $\gamma h^3 + \beta = 0$.

We proved in Section 3.4 that if the solution of the equation $dy/dx = P(x,y)/Q(x,y)$ is free from movable branch points, it must be a Riccati equation. Painlevé equations were derived on the basis of their freedom from movable branch points and essential singularities. It is therefore natural to expect that the solutions of Painlevé equations, in some special cases, may be governed by some Riccati equation. This matter was studied by Lukashevich and Yablonskii (1967) with reference to P_{VI}, the most general of the six Painlevé equations. They sought a Riccati equation

$$w' = a(z)w^2 + b(z)w + c(z), \tag{12}$$

all of whose solutions are also solutions of P_{VI}:

$$w'' = \frac{1}{2}\left(\frac{1}{w} + \frac{1}{w-1} + \frac{1}{w-z}\right)w'^2 - \left(\frac{1}{z} + \frac{1}{z-1} + \frac{1}{w-z}\right)w'$$
$$+ \frac{w(w-1)(w-z)}{z^2(z-1)^2}\left[\alpha + \beta\frac{z}{w^2} + \gamma\frac{(z-1)}{(w-1)^2} + \frac{\delta z(z-1)}{(w-z)^2}\right].$$

The coefficients of $a(z)$, $b(z)$, and $c(z)$ in (12) are to be chosen so that the above assertion is true. Differentiating (12), substituting for w'' and w' in P_{VI}, and equating to zero the coefficients of various powers of w, we get

$$z^2(z-1)^2 a^2(z) = 2\alpha, \tag{13a}$$

$$z^2(z-1)^2[a'(z) - (z+1)a^2(z)]$$
$$+ z(z-1)(2z-1)a(z) + 2\alpha(z+1) = 0, \tag{13b}$$

$$z^2(z-1)^2[b^2(z) + 2a(z)c(z) - 2b'(z) - 3za^2(z) + 2(z+1)a'(z)$$
$$+ 2(z+1)a(z)b(z)] - 2z(z-1)[(2z-1)b(z) + (1-2z-z^2)a(z)]$$
$$+ 2\alpha(z^2 + 4z + 1) + 2\beta z + 2\gamma(z-1) + 2\delta z(z-1) = 0, \tag{13c}$$

$$z(z-1)^2[(z+1)b'(z) + 2b(z)c(z) - c'(z) - za'(z) - 2za(z)b(z)]$$
$$- (z-1)[(2z-1)c(z) + (1-2z-z^2)b(z) + z^2 a(z)]$$
$$- 2\alpha(z+1) - 2\beta(z+1) - 2\gamma(z-1) - 2\delta(z-1) = 0, \tag{13d}$$

$$z(z-1)^2[3c^2(z) - 2(z+1)b(z)c(z) + 2(z+1)c'(z) - zb^2(z)$$
$$- 2za(z)c(z) - 2zb'(z)] - 2(z-1)[(1-2z-z^2)c(z) + z^2 b(z)]$$
$$+ 2\alpha z + 2\beta(z^2 + 4z + 1) + 2\gamma z(z-1) + 2\delta(z-1) = 0, \tag{13e}$$

$$(z-1)^2[zc'(z) + (z+1)c^2(z)] + z(z-1)c(z) + 2\beta(z+1) = 0 \tag{13f}$$

$$(z-1)^2 c^2(z) + 2\beta = 0. \tag{13g}$$

Equations (13a) and (13g) immediately determine the functions $a(z)$ and $c(z)$:

$$a(z) = \frac{\sqrt{2\alpha}}{z(z-1)}, \quad c(z) = \frac{\sqrt{-2\beta}}{z-1}, \tag{14}$$

Painlevé Transcendents

where we have chosen the plus sign for the square roots. Equations (14) satisfy (13b) and (13f) identically. Equations (13c) and (13e) for $b(z)$ are of Riccati type:

$$-2b' + b^2 + 2b\left[(z+1)a - \frac{2z-1}{z(z-1)}\right] + 2ac - 3a^2z + 2(z+1)a'$$

$$+ \frac{2(z^2 + 2z - 1)a}{z(z-1)}$$

$$+ \frac{2\alpha(z^2 + 4z + 1) + 2\beta z + 2\gamma(z-1) + 2\delta z(z-1)}{z^2(z-1)^2} = 0, \quad (15)$$

$$-2zb' - zb^2 - 2b\left[(z+1)c + \frac{z}{z-1}\right] - 2zac + 3c^2 + 2(z+1)c'$$

$$+ \frac{2(z^2 + 2z - 1)}{z(z-1)}c(z)$$

$$+ \frac{2\alpha z + 2\beta(z^2 + 4z + 1) + 2\gamma z(z-1) + 2\delta(z-1)}{z(z-1)^2} = 0. \quad (16)$$

Equation (13d) is linear in $b(z)$:

$$(z+1)b' + b\left[2c - 2za + \frac{z^2 + 2z - 1}{z(z-1)}\right] - c' - za' - \frac{(2z-1)c}{z(z-1)} - \frac{az^2}{z(z-1)}$$

$$- \frac{2\alpha(z+1) + 2\beta(z+1) + 2\gamma(z-1) + 2\delta(z-1)}{z(z-1)^2} = 0. \quad (17)$$

We thus have three DE for $b(z)$. We eliminate b^2 from (15) and (16) to get another linear DE for $b(z)$:

$$-4zb'(z) + \frac{2b(z)}{z-1}\left[(z+1)\left(\sqrt{2\alpha} - \sqrt{-2\beta}\right) - 3z + 1\right] + \frac{2(\sqrt{-2\beta} - z\sqrt{2\alpha})}{z(z-1)}$$

$$+ \frac{2(\alpha + \beta)(z+1)^2 + 2(\gamma + \delta)(z^2 - 1)}{z(z-1)^2} = 0, \quad (18)$$

where $a(z)$ and $c(z)$ have been substituted from (14). It may be checked that any root $b(z)$ of the quadratic

$$zb^2 + \frac{b}{z-1}\left[(z+1)\left(\sqrt{2\alpha} + \sqrt{-2\beta}\right) - z + 1\right]$$

$$+ \frac{4\sqrt{-\alpha\beta}}{(z-1)^2} - \frac{1}{z(z-1)}\left(z\sqrt{2\alpha} + \sqrt{-2\beta}\right)$$

$$- \frac{(\beta - \alpha)(z^2 + 1) + (\gamma - \delta)(z-1)^2}{z(z-1)^2} = 0 \quad (19)$$

satisfies (18) identically.

It is also easy to verify that the expression

$$b(z) = \frac{\lambda z + \mu}{z(z-1)} \qquad (20)$$

satisfies (16), provided

$$\lambda = \frac{\sqrt{2\alpha} - (\alpha + \beta + \gamma + \delta)}{\sqrt{2\alpha} - \sqrt{-2\beta} - 1},$$

$$\mu = \frac{\sqrt{-2\beta} - (\alpha + \beta - \gamma - \delta)}{\sqrt{2\alpha} - \sqrt{-2\beta} - 1}, \qquad (21)$$

$$\sqrt{2\alpha} - \sqrt{-2\beta} - 1 \neq 0.$$

Equation (19) is a consequence of (17) and (18). Equation (20) satisfies (19) provided the following relations between the coefficients are satisfied:

$$\lambda^2 + \lambda(\sqrt{2\alpha} + \sqrt{-2\beta} - 1) - \sqrt{2\alpha} + \alpha - \beta - \gamma + \delta = 0, \qquad (22a)$$

$$2\lambda\mu + \lambda(\sqrt{2\alpha} + \sqrt{-2\beta} + 1) + \mu(\sqrt{2\alpha} + \sqrt{-2\beta} - 1)$$
$$+ 4\sqrt{-\alpha\beta} - \sqrt{-2\beta} + \sqrt{2\alpha} + 2(\gamma - \delta) = 0, \qquad (22b)$$

$$\mu^2 + \mu(\sqrt{2\alpha} + \sqrt{-2\beta} + 1) + \sqrt{-2\beta} + \alpha - \beta - \gamma + \delta = 0. \qquad (22c)$$

If follows from (21) that

$$\lambda + \mu + \sqrt{2\alpha} + \sqrt{-2\beta} = 0. \qquad (23)$$

Adding (22a)–(22c), we obtain

$$(\lambda + \mu + \sqrt{2\alpha} + \sqrt{-2\beta})^2 = 0;$$

that is, equations (22a)–(22c) are dependent because of (23). The difference of (22c) and (22a), $(\lambda - \mu - 1)(\lambda + \mu + \sqrt{2\alpha} + \sqrt{-2\beta})$, also vanishes on account of (23). After substituting for λ from (21) into (22a) and simplifying, we get

$$2\sqrt{2\alpha}(-\alpha + 3\beta + \gamma - \delta) + 2\sqrt{-2\beta}(3\alpha - \beta - \gamma + \delta)$$
$$+ 4\sqrt{-\alpha\beta}(-\alpha + \beta + \gamma - \delta - 1) + 2(\alpha - \beta - \gamma) + \alpha^2 - 6\alpha\beta$$
$$- 2\alpha\gamma + 2\alpha\delta + \beta^2 + 2\beta\gamma - 2\beta\delta + \gamma^2 + 2\gamma\delta + \delta^2 = 0, \qquad (24)$$

where we take $\sqrt{2\alpha}\sqrt{-2\beta}$ to mean $2\sqrt{-\alpha\beta}$.

If condition (24) is satisfied by the coefficients occurring in the sixth Painlevé equation, we are assured of a solution $b(z)$ of (15)–(17). Thus, we choose $\sqrt{2\alpha}$ and $\sqrt{-2\beta}$, with any sign, satisfying (24) and determine λ and μ according to (21); then P_{VI} has a family of solutions satisfying the

Painlevé Transcendents

Riccati equation

$$w' = \frac{\sqrt{2\alpha}}{z(z-1)}w^2 + \frac{\lambda z + \mu}{z(z-1)}w + \frac{\sqrt{-2\beta}}{z-1}. \tag{25}$$

This equation can be linearized (see Davis 1962) by the transformation

$$w = -\frac{z(z-1)}{\sqrt{2\alpha}}\frac{v'}{v}, \qquad \alpha \neq 0, \tag{26}$$

to

$$v'' + \frac{(2-\lambda)z - \mu - 1}{z(z-1)}v' + \frac{\sqrt{-\alpha\beta}}{z(z-1)^2}v = 0. \tag{27}$$

This equation is Fuchsian with singular points $z = 0$, $z = 1$, and $z = \infty$. In terms of the Riemann P-function, we have

$$v = P \left\{ \begin{array}{ccc} 0 & 1 & \infty \\ 0 & -\sqrt{-2\beta} & 0 \\ -\mu & -\sqrt{+2\alpha} & 1-\lambda \end{array} \; z \right\}. \tag{28}$$

Equation (27) can be transformed into a Gaussian hypergeometric equation

$$\tau(\tau-1)v'' + [(1+\alpha_1+\beta_1)\tau - \gamma_1]v' + \alpha_1\beta_1 v = 0, \qquad ' = \frac{d}{d\tau}, \tag{29}$$

by the substitution

$$\tau = \frac{1}{1-z}, \quad \alpha_1 = +\sqrt{2\alpha}, \quad \beta_1 = +\sqrt{-2\beta}, \quad \gamma_1 = \lambda. \tag{30}$$

We summarize the above results: if $v(\tau)$ is the solution of the hypergeometric equation (29) and the parameters α, β, γ, and δ are chosen according to (30), (21), and (24), then

$$w(z) = -\frac{z(z-1)}{\sqrt{2\alpha}}\frac{v'_2(1/(1-z))}{v(1/(1-z))} \tag{31}$$

is a solution of the sixth Painlevé equation. Solution (31) holds for $\alpha \neq 0$. If $\alpha = 0$, (25) becomes a linear DE

$$w' = \frac{\lambda z + \mu}{z(z-1)}w + \frac{\sqrt{-2\beta}}{z-1} \tag{32}$$

with

$$\lambda = \frac{\beta + \gamma + \delta}{\sqrt{-2\beta}+1}, \qquad \mu = \frac{-\sqrt{-2\beta}+\beta-\gamma-\delta}{\sqrt{-2\beta}+1}.$$

Another constraint on the parameters, as is apparent from (21), is that $\sqrt{2\alpha} - \sqrt{-2\beta} - 1 \neq 0$. If

$$\sqrt{2\alpha} - \sqrt{-2\beta} + 1 = 0, \tag{33}$$

then for $b(z)$ to exist the numerators in the expressions for λ and μ in (21) must also vanish. This leads to the condition

$$\gamma + \delta = \frac{1}{2} \tag{34}$$

In this case the two (linear) DE (17) and (18) coincide and have the solution

$$b(z) = \frac{\lambda z + \mu}{z(z - 1)}, \tag{35}$$

where λ now is found by substitution of (35) into (19) with $\mu = -\lambda - 2\sqrt{-2\beta} - 1$.

Conditions (22a)–(22c) still hold, while (24) is replaced by (33) and (34). The present singular case also covers the special singular situation $\alpha = 0$, $\sqrt{-2\beta} + 1 = 0$ encountered in (32) in view of the presence of the factor $\sqrt{-2\beta} + 1$ in the denominators of the expression for λ and μ there.

8.3 TRANSFORMATIONS AND SPECIAL SOLUTIONS OF PAINLEVÉ EQUATIONS

The examples in Section 8.2 point to the direct and simple approach to Painlevé equations in the Soviet literature. Although sufficient motivation was not provided, many interesting results were obtained, as summarized by Erugin (1976a, b). The main points were as follows:

1. P_I is strictly irreducible, but for certain choices of the parameters, P_{II}–P_V admit one-parameter families of solutions expressible in terms of classical transcendental functions: Airy, Bessel, Weber–Hermite, and Whittaker (Lukashevich 1967, 1968).
2. P_{II}–P_V admit transformations, mapping solutions of a given Painlevé equation to the solution of the same equation with different values of the parameters (Lukashevich 1971, Gromak 1975, 1976).
3. Property (2) helps construction of various elementary solutions of P_{II}–P_V for certain choices of the parameters. These special solutions are either rational or functions which are related (through repeated differentiation and multiplication) to the above-mentioned classical transcendental functions.
4. P_{VI}, from which the other five Painlevé equations follow through some limiting processes, admits solutions in terms of hypergeomet-

ric functions, which are also solutions of a Riccati equation (see Section 8.2).

Now we give a more systematic treatment due to Fokas and Ablowitz (1982), who recovered most results of Soviet investigators and found some new ones. Specifically, they developed an algorithm to find explicit transformations (i) between a given Painlevé equation and the same equation with different parameters, (ii) between two different Painlevé equations, and (iii) between an equation of the type investigated by Painlevé (i.e., linear in the second derivative) and an equation belonging to the same class, which, however, is quadratic in the second derivative. They also found rational solutions of P_{VI}, and related P_{III} and P_{VI} to certain new equations which are quadratic in the second derivative and are of Painlevé type. In the following, we closely follow Fokas and Ablowitz (1982).

Suppose we are given one of the 50 equations of Painlevé and his group in the form

$$v'' = P_1 v'^2 + P_2 v' + P_3, \tag{1}$$

where P_1, P_2, P_3 depend on v, z, and a set of parameters which we denote by the vector α. We wish to find transformations

$$\hat{v}(z;\hat{\alpha}) = F(v(z,\alpha),z), \tag{2}$$

where the function F is such that if $v(z;\alpha)$ solves (1) with parameters α, then $\hat{v}(z;\hat{\alpha})$ solves (1) with parameters $\hat{\alpha}$. It is well known (see Ince 1956) that the only transformation of type (2) which preserves the Painlevé property is the Möbius transformation

$$\hat{v}(z;\hat{\alpha}) = \frac{a_1 v + a_2}{a_3 v + a_4}, \tag{3}$$

where $a_1, \ldots, a_4$ are functions of z only. A very detailed discussion of (3) with its applications to transformation theory of Riccati equations and second-order nonlinear differential equations with polynomial coefficients, namely,

$$A(y)\frac{d^2 y}{dx^2} + B(y)\frac{dy}{dx} + C(y)\left(\frac{dy}{dx}\right)^2 + D(y) = 0, \tag{4}$$

has been given by Davis (1962). Here,

$$A(y) = A_0 + A_1 y, \quad B(y) = B_0 + B_1 y,$$
$$C(y) = C_0, \quad D(y) = D_0 + D_1 y + D_2 y^2 + D_3 y^3$$

where A_i, B_i, C_0, and D_i are functions of x. One may generalize (2) by including the first derivative in the dependence of F so that

$$\hat{v}(z;\hat{\alpha}) = F(v'(z;\alpha), v(z;\alpha), z). \tag{5}$$

Moreover, if we are interested in relating a solution of a Painlevé equation not only to its other solution but also to those of some other Painlevé equations, we may replace (5) by

$$u(z;\hat{\alpha}) = F(v'(z;\alpha), v(z;\alpha), z), \tag{6}$$

where F is such that u satisfies some (other) second-order equation of Painlevé type. The only transformation of type (6) which is linear in v' and preserves the Painlevé property is the one equivalent to a Riccati equation; i.e.,

$$u(z;\hat{\alpha}) = \frac{v' + av^2 + bv + c}{dv^2 + ev + f}, \tag{7}$$

where $a, b, \ldots, f$ depend on z only. (There are more general first-order equations of Fuchsian type, nonlinear in v' and possessing no movable critical points which might generalize (7); see Hille 1969). The transformation (7) extends (12) in Section 8.2 for P_{VI}.

We now find the functions $a, b, c, \ldots, f$ such that (7) defines a one-to-one invertible map between the solutions v of (1) and solution u of some second-order equation of Painlevé type. In this process, the latter equation is completely determined. We write

$$\mathcal{J} = dv^2 + ev + f, \quad Y = av^2 + bv + c \tag{8}$$

so that (7) becomes

$$\mathcal{J}u = v' + Y. \tag{9}$$

Differentiating (9), replacing v'' by (1), and v' by $\mathcal{J}u - Y$, we obtain

$$\mathcal{J}u' = [P_1\mathcal{J}^2 - 2d\mathcal{J}v - e\mathcal{J}]u^2$$
$$+ [-2P_1\mathcal{J}Y + \mathcal{J}P_2 + 2av\mathcal{J} + b\mathcal{J} + 2dvY + eY - (d'v^2 + e'v + f')]u$$
$$+ [P_1Y^2 - P_2Y + P_3 - 2avY - bY + a'v^2 + b'v + c']. \tag{10}$$

Here $\mathcal{J}$ and Y are to be replaced by (8). Two possibilities arise from (10):

A. Find $a, \ldots, f$ such that (10) is linear in v,

$$A(u', u, z)v + B(u', u, z) = 0, \text{ say.} \tag{11}$$

Now, upon substituting $v = -B/A$ in (7), we get a second-order equation of Painlevé type (in fact, it is one of the 50 equations referred to earlier), linear in d^2u/dz^2.

B. Find $a, \ldots, f$ such that (10) reduces to a quadratic in v,

$$A(u',u,z)v^2 + B(u',u,z)v + C(u',u,z) = 0, \text{ say.} \tag{12}$$

Again solving for v and substituting in (7) yields a Painlevé equation for u which is quadratic in d^2u/dz^2.

Fokas and Ablowitz (1982) point out that P_{II}–P_V admit transformations of types A and B, but P_{VI} does not admit a transformation of type A. In the process of applying the above transformations to (1), one obtains equations of the same type with different parameters α. When (11) and (12) break down (i.e., where $A = B = C = 0$), we obtain one-parameter families of solutions of (1). These points are best illustrated by specific examples. We now treat P_{II} in detail and quote the results for P_{III}, P_{IV}, P_V, and P_{VI}.

Painlevé II

Gambier (1910) and Lukashevich (1971) proved the following results. Let $v(z;\alpha)$ be a solution of P_{II},

$$v'' = 2v^3 + zv + \alpha. \tag{13}$$

Then $\bar{v}(z;\bar{\alpha})$ is also a solution of P_{II}, where either

$$\bar{v}(z;\bar{\alpha}) = -v(z,\alpha), \qquad \bar{\alpha} = -\alpha, \tag{14}$$

or

$$\bar{v}(z;\bar{\alpha}) = -v(z;\alpha) - \frac{1+2\alpha}{2v^2 + 2v' + z}, \qquad \bar{\alpha} = \alpha + 1, \alpha \neq -\frac{1}{2}. \tag{15}$$

For the exceptional case, it was proved that P_{II} admits a one-parameter family of solutions characterized by

$$v' + v^2 + \frac{z}{2} = 0, \tag{16}$$

iff $\alpha = -\frac{1}{2}$.

A second theorem connects P_{II} with equation PXXXIV of Ince (1956). Let $v(z;\alpha)$ be a solution of P_{II}; and let $u(z;\gamma)$ be a solution of

$$u'' + 2u^2 - zu + \frac{\gamma + u' - (u')^2}{2u - z} = 0, \qquad \gamma = \alpha(\alpha + 1). \tag{17}$$

Then there exists the following one-to-one correspondence between the solutions of (13) and (17):

$$u = -v' - v^2, \qquad v = \frac{u' + \alpha}{2u - z}. \tag{18}$$

Equation (17) under the further transformation $w = (u - z/2)/(4\alpha + 1)$ reduces to the form PXXXIV of Ince (1956).

Equation (13) is a special case of (1) with $P_1 = P_2 = 0$ and $P_3 = 2v^3 + zv + \alpha$. The transformation (7) is too general. For our purposes it suffices to consider only the linear terms in the denominator of (7) so that $d = 0$. Therefore, equation (10) becomes

$$u' = -eu^2 + \left[\frac{3aev^2 + (2af + 2eb - e')v + (bf + ec - f')}{ev + f}\right]u \qquad (19)$$

$$+ \frac{2(1-a^2)v^3 + (a' - 3ab)v^2 + (z + b' - b^2 - 2ac)v + (\alpha - bc + c')}{ev + f}.$$

We have to choose a, b, c, e, and f such that (19) becomes linear in v. To that end, the numerators should exactly divide out $ev + f$; moreover, to remove the cubic term we need $a^2 = 1$. Again, two choices are possible: either $e \neq 0$ or $e = 0$. The former connects P_{II} with PXXXV of Ince (1956). For the present case, $e = 0$ proves adequate. In addition, we choose $f = -1$ and $c = 0$. Equation (19), with $c = d = e = 0$, $f = -1$, reduces to

$$u' = (2av + b)u + 3abv^2 - (z + b' - b^2)v - \alpha. \qquad (20)$$

To remove the nonlinear term $3abv^2$, we put $b = 0$ so that (20) becomes

$$u' + \alpha = v(2au - z), \qquad (21)$$

while (7) reduces to

$$(v' + av^2) = -u, \qquad a^2 = 1. \qquad (22)$$

Choosing $a = 1$, and substituting $v = (u' + \alpha)/(2u - z)$ into (22), we get (17). The transformation $v = (u' + \alpha)/(2u - z)$ breaks down iff $u = z/2$. For finiteness, the numerator must also vanish, so $u' + \alpha = 0$, implying that $\alpha = -1/2$. Actually $u = z/2$, $\alpha = -1/2$ is a particular solution of (17).

Now we derive the connection (15) between solutions of P_{II} for different values of the parameters. This connection occurs because $\nu = \alpha(\alpha + 1)$ in (17) so that there exist two values of α, namely α and $-(\alpha + 1)$, for which ν has the same value; hence $u(z; \alpha) = u(z; -(\alpha + 1))$. Now, from (18),

$$\bar{v}(z; -(\alpha + 1)) = \frac{u'(z; -(\alpha + 1)) - (\alpha + 1)}{2u(z; -(\alpha + 1)) - z} = \frac{u'(z; \alpha) - (\alpha + 1)}{2u(z; \alpha) - z}$$

$$= \frac{u' + \alpha}{2u(z; \alpha) - z} - \frac{2\alpha + 1}{2u(z; \alpha) - z}$$

$$= v(z; (\alpha)) - \frac{2\alpha + 1}{2u(z; \alpha) - z}. \qquad (23)$$

Painlevé Transcendents

Now, we replace $\bar{v}(z; -(\alpha + 1))$ by $-v(z; \alpha + 1)$ and u by $-(v^2 + v')$ (see (22)) to obtain (15).

We may now use (14) and (15) to obtain some known elementary solutions of P_{II}. We can also use (18) to construct elementary solutions of (17). For the latter, we note that

$$u(z; \alpha) = -v'(z; \alpha) - v^2(z; \alpha)$$

so that

$$\bar{u}(z; -\alpha) = -\bar{v}'(z; -\alpha) - \bar{v}^2(z; -\alpha) = v'(z; \alpha) - v^2(z; \alpha). \tag{24}$$

Similarly,

$$u(z; \alpha - 1) = -v'(z; \alpha - 1) = v^2(z; \alpha - 1). \tag{25}$$

From (24) and (25),

$$\bar{u}(z; -\alpha) = u(z; \alpha - 1), \tag{26}$$

since

$$v(z; \alpha) = -v(z; \alpha - 1) \quad \text{and} \quad v'^2(z; \alpha) = v'^2(z; \alpha - 1).$$

Also,

$$\bar{u}(z, \alpha + 1) = -\bar{v}'(z, \alpha + 1) - \bar{v}^2(z, \alpha + 1) = v'(z, \alpha) - v^2(z, \alpha)$$
$$= -u(z, \alpha) - 2v^2(z, \alpha) = -u(z, \alpha) - 2v^2(z, -(\alpha + 1))$$
$$= -u(z, \alpha) - 2\left[\frac{u'(z, \alpha) - (\alpha + 1)}{2u(z; \alpha) - z}\right]^2. \tag{27}$$

Equations (26) and (27) help generate new solutions.

To construct rational solutions of (13), we note that $v = 0$ is a solution if $\alpha = 0$. Using (15), we can obtain a rational solution of (13) for every positive integral value of α:

$$v(z; 0) = 0, \quad v(z; 1) = \frac{-1}{z},$$

$$v(z; 2) = \frac{1}{z} - \frac{3}{2/z^2 + 2/z^2 + z} = \frac{1}{z} - \frac{3z^2}{z^3 + 4}, \ldots \tag{28}$$

Equation (17) again has zero as a solution if $\alpha = 0$. Thus, using (26) and (27), we can also find rational solutions of (17):

$$u(z; 0) = 0, \quad u(z; 1) = -\frac{2}{z^2},$$

$$u(z; 2) = -\frac{6z(z^3 - 8)}{(z^3 + 4)^2}, \ldots \tag{29}$$

It is clear that the hierarchies of solutions (28) and (29) are connected by (18).

Finally for $\alpha = -1/2$, $u = z/2$ is a solution of (17), as we noted earlier. Substituting this solution in (18), we have

$$v' + v^2 + \frac{z}{2} = 0. \tag{30}$$

This is a Riccati equation which transforms by $v(z; -1/2) = y'/y$ into the Airy equation

$$y'' + (z/2)y = 0. \tag{31}$$

The solutions of (31) cannot be used directly in (15) to generate new solutions, since in this case the latter does not hold. What is done, instead, is to use (14) first and then (15). The following hierarchy of solutions results:

$$v\left(z; -\frac{1}{2}\right) = \frac{y'}{y}, \quad v\left(z; \frac{1}{2}\right) = -\frac{y'}{y},$$

$$v\left(z; \frac{3}{2}\right) = \frac{y'}{y} - \frac{y^2}{2y'^2 + zy^2}, \dots \tag{32}$$

We can also generate Airy-type solutions for (17):

$$u\left(z; -\frac{1}{2}\right) = \frac{z}{2}, \quad u\left(z; \frac{1}{2}\right) = -\frac{2y'^2}{y^2} - \frac{z}{2}, \dots \tag{33}$$

Since the derivations of all the results—namely equations (17) and (18), the quadratic dependence of γ on α leading to (15), characterizing one-parameter family of solutions of P_{II} when the transformations (18) break down, and the generation of elementary solutions—are very similar for Painlevé equations II–VI, we merely state the results for P_{III}–P_{VI} and refer the reader to Fokas and Ablowitz (1982).

Painlevé III

Let $v(z; \alpha, \beta, \gamma, \delta)$ be a solution of P_{III}

$$v'' = \frac{v'^2}{v} - \frac{1}{z}v' + \frac{1}{z}(\alpha v^2 + \beta) + \gamma v^3 + \frac{\delta}{v}. \tag{34}$$

Then $\bar{v}(z; \bar{\alpha}, \bar{\beta}, \bar{\gamma}, \bar{\delta})$ are also solutions of P_{III}, where

$$v(z; \bar{\alpha}, \bar{\beta}, \bar{\gamma}, \bar{\delta}) = -v(z; \alpha, \beta, \gamma, \delta),$$
$$\bar{\alpha} = -\alpha, \quad \bar{\beta} = -\beta, \quad \bar{\gamma} = \gamma, \quad \bar{\delta} = \delta, \tag{35}$$
$$\bar{v}(z; \bar{\alpha}, \bar{\beta}, \bar{\gamma}, \bar{\delta}) = [v(z; \alpha, \beta, \gamma, \delta)]^{-1},$$

Painlevé Transcendents

$$\bar{\alpha} = -\beta, \quad \bar{\beta} = -\alpha, \quad \bar{\gamma} = -\delta, \quad \bar{\delta} = -\gamma, \tag{36}$$

$$\bar{v}(z; \bar{\alpha}, \bar{\beta}, \bar{\gamma}, \bar{\delta}) = \frac{\gamma^{1/2}}{\bar{\gamma}^{1/2}} v$$

$$\times \left[1 + \frac{2 + \beta(-\delta)^{1/2} + \alpha(\gamma)^{-1/2}}{z(v'/v + \gamma^{1/2}v + (-\delta)^{1/2}/v) - 1 - \beta(-\delta)^{-1/2}} \right], \tag{37a}$$

$$\bar{\alpha} = -[2 + \beta(-\delta)^{-1/2}]\gamma^{-1/2}, \quad \bar{\beta} = -[2 + \alpha\gamma^{-1/2}](-\delta)^{1/2}\left(\frac{\gamma^{1/2}}{\bar{\gamma}^{1/2}}\right),$$

$$(-\bar{\delta})^{1/2} = (-\delta)^{1/2}\frac{\gamma^{1/2}}{\bar{\gamma}^{1/2}}. \tag{37b}$$

In (37) it has been assumed that

$$\gamma \neq 0 \quad \text{and} \quad 2 + \alpha(\gamma)^{-1/2} + \beta(-\delta)^{1/2} \neq 0. \tag{37c}$$

If $\gamma = 0$, then (37) is replaced by

$$\bar{v}(z; \bar{\alpha}, \bar{\beta}, \bar{0}, \bar{\delta}) = z\frac{v'}{v^2} - (1 + \beta(-\delta)^{1/2})v + \frac{z(-\delta)^{1/2}}{v^2},$$

$$\bar{\alpha} = (-\delta)^{1/2}, \quad \bar{\beta} = \alpha(2 - \beta(-\delta)^{1/2}), \quad \bar{\delta} = -\alpha^2. \tag{38}$$

The case $2 + \beta(-\delta)^{-1/2} + \alpha(\gamma)^{-1/2} = 0$ gives rise to a one-parameter family of solutions characterized by

$$\frac{1 - \alpha\gamma^{-1/2}}{z} = \frac{v'}{v} + \gamma^{1/2}v + \frac{(-\delta)^{1/2}}{v}. \tag{39}$$

Another special case occurs if $\beta = \delta = 0$, and the solution of P_{III} is given by

$$v = \frac{\phi'}{\gamma^{1/2}\phi + \alpha + \gamma^{1/2}}, \quad \int \frac{d\phi}{\phi^2/2 + \phi + c_1} = \ln z + c_2, \tag{40}$$

where c_1, c_2 are arbitrary constants.

Using (36), similar results can be obtained for the case $\alpha = \gamma = 0$. Elementary solutions of P_{III} can be derived in the same manner as for P_{II}.

Painlevé IV

Let $v(z; \alpha, \beta)$ be a solution of P_{IV}:

$$v'' = \frac{v'^2}{2v} + \frac{3}{2}v^3 + 4zv^2 + 2(z^2 - \alpha)v + \beta v^{-1}. \tag{41}$$

Then $\bar{v}(z; \bar{\alpha}, \bar{\beta})$ is also a solution of P_{IV}, where

$$\bar{v} = \frac{v' - v^2 - 2zv - (-2\beta)^{1/2}}{2v},$$

$$\bar{\alpha} = \frac{1}{4}[2 - 2\alpha + 3(-2\beta)^{1/2}],$$

$$\bar{\beta} = -\frac{1}{2}\left[1 + \alpha + \frac{(-2\beta)^{1/2}}{2}\right]^2, \tag{42}$$

provided that

$$1 + \alpha + (-2\beta)^{1/2}/2 \neq 0. \tag{43}$$

When the equality sign holds in (43), we get a single-parameter family of solutions.

Painlevé V

Let $v(z; \alpha, \beta, \gamma, \delta)$ be a solution of P$_\mathrm{V}$:

$$v'' = \frac{3v-1}{2v(v-1)}v'^2 - \frac{1}{z}v' + \frac{\alpha}{z^2}v(v-1)^2$$

$$+ \frac{\beta}{z^2}\frac{(v-1)^2}{v} + \frac{\gamma}{z}v + \frac{\delta v(v+1)}{v-1}. \tag{44}$$

Then $\bar{v}(z; \bar{\alpha}, \bar{\beta}, \bar{\gamma}, \bar{\delta})$ is also a solution of P$_\mathrm{V}$, where

$$\bar{v} = 1 - \frac{2(-2\delta)^{1/2}zv}{zv' - (2\alpha)^{1/2}v^2 + [(2\alpha)^{1/2} - (-2\beta)^{1/2} + (-2\delta)^{1/2}z]v + (-2\beta)^{1/2}},$$

$$\bar{\alpha} = -\frac{1}{16\delta}[\gamma + (-2\delta)^{1/2}(1 - (-2\beta)^{1/2} - (2\alpha)^{1/2})]^2,$$

$$\bar{\beta} = \frac{1}{16\delta}[\gamma - (-2\delta)^{1/2}(1 - (-2\beta)^{1/2} + (-2\delta)^{1/2})]^2,$$

$$\bar{\gamma} = (-2\delta)^{1/2}[(-2\beta)^{1/2} - (2\alpha)^{1/2}], \qquad \bar{\delta} = \delta, \tag{45}$$

provided that $\delta \neq 0$ and

$$(-2\delta)^{1/2}[1 - (-2\beta)^{1/2} - (2\alpha)^{1/2}] \neq \gamma. \tag{46}$$

The equality sign in (46) would again lead to a one-parameter family of solutions.

Painlevé VI

Let $v(z; \alpha, \beta, \gamma, \delta)$ be a solution of P$_\mathrm{VI}$:

$$v'' = \frac{1}{2}\left(\frac{1}{v} + \frac{1}{v-1} + \frac{1}{v-z}\right)v'^2 - \left(\frac{1}{z} + \frac{1}{z-1} + \frac{1}{v-z}\right)v'$$

Painlevé Transcendents

$$+ \frac{v(v-1)(v-z)}{z^2(z-1)^2}\left[\alpha + \frac{\beta z}{v^2} + \frac{\gamma(z-1)}{(v-1)^2} + \frac{\delta z(z-1)}{(v-z)^2}\right]. \quad (47)$$

Then $\bar{v}(z; \bar{\alpha}, \bar{\beta}, \bar{\gamma}, \bar{\delta})$ are also solutions of P$_{VI}$, where

$$\bar{v}(z; \bar{\alpha}, \bar{\beta}, \bar{\gamma}, \bar{\delta}) = zv\left(\frac{1}{z}; \alpha, \beta, \gamma, \delta\right); \quad (48)$$

$$\bar{\alpha} = \alpha, \quad \bar{\beta} = \beta, \quad \bar{\gamma} = -\delta + \frac{1}{2}, \quad \bar{\delta} = -\gamma + \frac{1}{2},$$

$$\bar{v}(z; \bar{\alpha}, \bar{\beta}, \bar{\gamma}, \bar{\delta}) = 1 - v(1-z; \alpha, \beta, \gamma, \delta); \quad (49)$$

$$\bar{\alpha} = \alpha, \quad \bar{\beta} = -\gamma, \quad \bar{\gamma} = -\beta, \quad \bar{\delta} = \delta,$$

$$\bar{v}(z; \bar{\alpha}, \bar{\beta}, \bar{\gamma}, \bar{\delta}) = 1 - (1-z)v\left(\frac{1}{1-z}; \alpha, \beta, \gamma, \delta\right); \quad (50)$$

$$\bar{\alpha} = \alpha, \quad \bar{\beta} = \delta - \frac{1}{2}, \quad \bar{\gamma} = -\beta, \quad \bar{\delta} = -\gamma + \frac{1}{2},$$

$$\bar{v} = v + 2((z+1)v - 2z)$$

$$\times \left[-\frac{2z(z-1)}{\kappa}\frac{\Phi'}{\Phi} + \frac{(z-1)I}{\kappa\Phi} - (z+1)\right]^{-1}; \quad (51a)$$

$$\bar{\alpha} = \frac{1}{2}[(-2\beta)^{1/2} - 1]^2, \quad \bar{\beta} = -\frac{1}{2}[(2\alpha)^{1/2} + 1]^2, \quad (51b)$$

$$\bar{\gamma} = \gamma + \frac{\kappa\mu}{4}, \quad \bar{\delta} = \delta + \frac{\kappa\mu}{4},$$

where, in (51), Φ, I, κ, μ are defined by

$$\Phi = z\frac{v'}{v} + \frac{\lambda - \kappa - 1}{2(z-1)}v + \frac{(\lambda + \kappa + 1)z}{2(z-1)}\frac{1}{v} - \frac{\lambda}{2}\frac{z+1}{z-1} - \left(\frac{1}{2} + \frac{\mu}{4}\right), \quad (52)$$

$$I \doteq \Phi^2 + \frac{\mu}{2}\Phi + \nu, \quad (53)$$

$$\kappa = (-2\beta)^{1/2} - (2\alpha)^{1/2} - 1, \quad \lambda = (-2\beta)^{1/2} + (2\alpha)^{1/2}, \quad (54)$$

$$\mu = \frac{4}{\kappa}\left(\frac{1}{2} - \gamma - \delta\right), \quad \nu = 2\delta - 1 + \left(\frac{\mu}{4} + \frac{\kappa}{2}\right)^2. \quad (55)$$

In (51), we have assumed that

$$\Phi \neq 0, \quad \kappa \neq 0, \quad \nu \neq 0. \quad (56)$$

Here the notation $\doteq$ stands for definition.

P_{VI} is related to another equation quadratic in the second derivative. Let $v(z; \alpha, \beta, \gamma, \delta)$ be a solution of P_{VI}, and let $\Phi(z; \kappa^2, \lambda, \mu, \nu)$ be a solution of

$$(z-1)^2 \Omega^2 = \frac{1}{z^2}\left(\Phi'^2 + \frac{I^2 - \kappa^2 \Phi^2}{z(z-1)^2}\right)\Psi^2, \tag{57}$$

where

$$\Omega \doteq \Phi'' + \frac{(3z-1)\Phi'}{2z(z-1)} + \frac{2\Phi I + \mu I/2 - \Phi \kappa^2}{z(z-1)^2},$$

$$\Psi \doteq (z+1)\Phi + \frac{\mu}{4}(z+1) + \frac{\lambda}{2}(z-1)^2, \tag{58}$$

and I, κ, λ, μ, and ν are defined by (53)–(55). Equations (52) and (59) establish a one-to-one correspondence between solutions of (47) and (57):

$$v = \left(-\frac{z+1}{z}\Phi' + \frac{2\kappa}{z(z-1)}\Phi - (z-1)^2 \frac{\Omega}{\Psi}\right)$$

$$\times \left(-\frac{2\Phi'}{z} + \frac{I}{z^2} + \frac{\kappa(z+1)\Phi}{z^2(z-1)}\right)^{-1}, \tag{59}$$

where we assume that (56) holds.

Finally, Fokas and Ablowitz (1982) state the result on the solutions of P_{VI} expressible in terms of hypergeometric functions, which we have already proved in Section 8.2. These special solutions are employed to derive an infinite hierarchy of rational solutions. It may be remarked that Fokas and Ablowitz (1982) obtain a new one-parameter family of solutions of P_{VI} not found earlier by the Soviet investigators.

We shall prove some of the results quoted in this section, as we analyze P_I–P_V individually in Sections 8.5–8.9.

8.4 SOLUTIONS OF SECOND-ORDER NONLINEAR DE WITH IRRATIONAL RIGHT SIDES

An important assumption in the Painlevé study of the second-order equations

$$w'' = f(w', w, z) \tag{1}$$

is that f is rational in w and w'. If this assumption is waived, fundamentally different behavior of the solution is observed. This is natural because if $f(w', w, z)$ is an irrational function, (1) can be changed into a system of equations of order greater than 2 with rational right-hand sides. It is well known that the investigation of a system of two equations differs essentially

Painlevé Transcendents

from that of a system of three or more equations. Moreover, the equivalent system may not possess all the solutions which the original system does, and, in particular, a singular solution may disappear in the transition to a system of n (> 2) equations. We shall discuss some of these intricate points with the help of several examples, following Erugin (1980). We shall deal with DE in the complex plane. But first we make a few remarks regarding singular solutions.

If $D = D(w', w, z)$ is the region over which $f(w', w, z)$ is a holomorphic function of w', w and z, and $(w'_0, w_0, z_0) \in D$, then (1) has a unique solution passing through this point and holomorphic in the neighborhood of z_0:

$$w = w_0 + w'_0(z - z_0) + \alpha_2(z - z_0)^2 + \cdots \qquad (2)$$

The boundary of D is defined to be the set of points of $\bar{D}(w', w, z)$ satisfying the relation $\Phi(w', w, z) = 0$ in whose neighborhood $f(w', w, z)$ is not holomorphic. This boundary may depend only on w, w': $\Phi = \Phi(w, w')$, all points w, w' satisfying this relation form the boundary of D, on which $f(w', w, z)$ ceases to be holomorphic, for arbitrary value of z_0. In exceptional circumstances, this boundary may consist of isolated planes $w = w_k$, $k = 1, 2, \ldots$, or $w' = w'_k$, $k = 1, 2, \ldots$

A solution

$$w = \phi(z, C_1, C_2) \qquad (3)$$

of (1) involving two arbitrary constants C_1 and C_2 is called a *general solution* in D if, for each $(w'_0, w_0, z_0) \in D$, there exists an appropriate choice of C_1, C_2 such that (3) represents the solution of this initial value problem in the neighborhood of z_0. That is, C_1 and C_2 can be uniquely determined from

$$w_0 = \Phi(z_0, C_1, C_2), \qquad w'_0 = \Phi'_z(z_0, C_1, C_2), \qquad (4)$$

and the corresponding (unique) solution is realized in the form (2). If, in addition to a general solution there exists a solution

$$w = \phi(z, C), \qquad (5)$$

involving another arbitrary constant (or possibly no constant) which cannot be obtained from the general solution as a special case, that solution is called a *singular solution*.

If the initial conditions $(\bar{w}'_0, \bar{w}_0, z_0)$ satisfy the relations $\bar{w}_0 = \phi(z_0, C)$ and $\bar{w}'_0 = \phi'_z(z_0, C)$ according to (5), then $(\bar{w}'_0, \bar{w}_0, z)$ is not contained in $D(w', w, z)$ and $f(w', w, z)$ is not holomorphic in the neighborhood of $(\bar{w}'_0, \bar{w}_0, z_0)$, If, on the contrary, $(\bar{w}'_0, \bar{w}_0, z)$ belonged to D we would obtain this solution from $w = \phi(z, C_1, C_2)$ by a suitable choice of C_1 and C_2 and this solution would not be unique. However, it is possible that the solution $w = \phi(z, C)$ is also holomorphic in the neighborhood of $z = z_0$, and

$\bar{w}_0 = \Phi(z_0, C_1, C_2)$ and $\bar{w}'_0 = \Phi'_z(z_0, C_1, C_2)$ (see equation (4)) but the domain of holomorphicity of this solution is different. In other words, the same initial conditions can give rise to two different holomorphic solutions so that $[w = \phi(z, C), w']$ is not contained in $D(w', w, z)$ and $\Phi(z, C_1, C_2) \neq \phi(z, C)$ except for certain points.

The nature of DE with irrational right sides and the physical significance of singular solutions are not fully understood. However, we may agree that the latter exist on the boundary of the domain D of existence and uniqueness of the solutions, and their presence or absence characterizes the behavior of integral curves in the neighborhood of the boundary of D. Erugin asserts that "in mathematics and, in particular, in differential equations encountered in physics, all results are of physical importance!"

EXAMPLE 1 Consider the DE

$$w'' = (w' - 1)^{1/2}, \tag{6}$$

with initial conditions $z = z_0$, $w = w_0$, $w'_0 = 1$. At $w' = 1$, the right-hand side of (6) is not holomorphic. We thus have singular initial conditions. The general solution of this equation is

$$w = z + \frac{1}{12}(z + C_1)^3 + C_2 \tag{7}$$

(obtained by writing $w''/(w' - 1)^{1/2} = 1$ and integrating twice, etc.) Here, C_1 and C_2 are arbitrary constants. Equation (6) also has a singular solution

$$w = z + C, \tag{8}$$

where C is another constant; this solution cannot be derived from (7) for any choice of C_1 and C_2. Through every point (z_0, w_0, w'_0) of the singular solution, $w_0 = z_0 + C$, $w'_0 = 1$, we can draw an integral curve belonging to the general solution (7) by choosing C_1 and C_2 such that

$$z_0 + C = z_0 + \frac{1}{12}(z_0 + C_1)^3 + C_2$$

and

$$w'_0 = 1 = 1 + \frac{1}{4}(z_0 + C_1)^2;$$

that is,

$$C_1 = -z_0, \qquad C_2 = C.$$

We can recast equation (6) in the form

$$w' = u, \qquad u' = v, \qquad v' = \frac{1}{2}, \tag{9}$$

Painlevé Transcendents

where

$$v = (w' - 1)^{1/2} = (u - 1)^{1/2}, \tag{10}$$

so that the right sides of system (9) are now rational in w, u, v. This system is linear and has no singularity at a finite point.

Equation (9) can be integrated (from the last equation backward) to yield

$$v = \frac{1}{2}z + C_1,$$

$$u = \frac{1}{4}z^2 + C_1 z + C_2,$$

$$w = \frac{1}{12}z^3 + C_1 \frac{z^2}{2} + C_2 z + C_3. \tag{11}$$

These functions must satisfy the (consistency) algebraic relation (10):

$$\frac{1}{2}z + C_1 = \left(\frac{1}{4}z^2 + C_1 z + C_2 - 1\right)^{1/2},$$

or

$$C_2 = C_1^2 + 1.$$

Equations (11) now become

$$v = \frac{1}{2}z + C_1,$$

$$u = \frac{1}{4}z^2 + C_1 z + C_1^2 + 1,$$

$$w = \frac{1}{12}z^3 + C_1 \frac{z^2}{2} + (C_1^2 + 1)z + C_3. \tag{12}$$

We must also examine the special system

$$u = 1, \quad v = 0, \quad w = z + C \tag{13}$$

when the equation $v' = 1/2$ does not arise. This, as we have seen, is a singular solution not contained in (12). If we consider only system (9), we would skip this singular solution.

Suppose we obtain an intermediate integral of (9) by writing

$$2vv' - u' = 2\left(\frac{1}{2}v\right) - v = 0$$

and integrating, so that

$$v^2 - (u - 1) = C_4. \tag{14}$$

Choosing $C_4 = 0$, we have $v = (u - 1)^{1/2}$, and system (9) changes to

$$w' = u,$$
$$u' = (u - 1)^{1/2}. \tag{15}$$

The general solution of system (15) is

$$u = 1 + \frac{(z + C_5)^2}{4},$$
$$w = z + \frac{(z + C_5)^3}{12} + C_6, \tag{16}$$

involving arbitrary constants C_5 and C_6.

System (15) has a singular solution $u = 1$, $w = z + C_7$, which again is not contained in the general solution of (9). This is because $v = (u - 1)^{1/2}$ and $v' \neq 1/2$, contradicting (9).

EXAMPLE 2 Consider the DE

$$w'' = (w' - 1)^{1/2} + (w' - 2)^{1/2}. \tag{17}$$

Here, the boundary of the domain of existence and uniqueness of solutions consists of the "planes" $w' = 1$ and $w' = 2$ (where the right side is not holomorphic) but $w = z + C_1$ and $w = 2z + C_2$ are not solutions.

Since (17) is autonomous, we write $w' = p$, $w'' = p(dp/dw)$ so that it becomes

$$dw = \frac{p\,dp}{(p - 1)^{1/2} + (p - 2)^{1/2}}, \tag{18}$$

or

$$dz = \frac{dp}{(p - 1)^{1/2} + (p - 2)^{1/2}}.$$

The solution, therefore, is given parametrically by

$$w = \int_{p_0}^{p} \frac{p\,dp}{(p - 1)^{1/2} + (p - 2)^{1/2}} + w_0,$$
$$z = \int_{p_0}^{p} \frac{dp}{(p - 1)^{1/2} + (p - 2)^{1/2}} + z_0. \tag{19}$$

Clearly, w and $z \to \infty$ as $p \to \infty$. Indeed, for large $|p|$,

$$w = \int^{p} \frac{p\,dp}{p^{1/2}[(1 - 1/p)^{1/2} + (1 - 2/p)^{1/2}]} \approx \frac{1}{3}p^{3/2}, \qquad z \approx p^{1/2}.$$

Therefore, the solution is $w \approx z^3/3$.

Painlevé Transcendents

We easily see by differentiating (17) that $w''' \to \infty$ as $w' \to 1$ or $w' \to 2$, indicating the possibility of an algebraic singularity in the neighborhood of these points for arbitrary z_0.

To find the solution in the neighborhood of $w' = 2$, we put $u = (p-2)^{1/2}$, $u^2 + 2 = p$, $2u\,du = dp$, and $(p-1)^{1/2} = (u^2+1)^{1/2}$. Equations (19) become

$$w = \int_0^u \frac{(u^2+2)2u\,du}{(1+u^2)^{1/2}+u} + w_0, \qquad z = \int_0^u \frac{2u\,du}{(1+u^2)^{1/2}+u} + z_0. \tag{20}$$

When $u \sim 0$, the neighborhood of the singularity, we write

$$[(1+u^2)^{1/2}+u]^{-1} = \left[1 + u + \frac{1}{2}u^2 - \frac{1}{8}u^4 + \cdots\right]^{-1}$$

$$= 1 - u + \frac{1}{2}u^2 - \frac{1}{8}u^4 + \cdots,$$

$$2u[(1+u^2)^{1/2}+u]^{-1} = 2u - 2u^2 + u^3 + 0u^4 + \cdots,$$

$$(4u + 2u^3)[(1+u^2)^{1/2}+u]^{-1} = 4u - 4u^2 + 4u^3 - 2u^4 + \cdots.$$

The solution (20) now becomes

$$z = u^2 - \frac{2}{3}u^3 + \frac{u^4}{4} + \cdots + z_0, \tag{21}$$

$$w = 2u^2 - \frac{4}{3}u^3 + u^4 + \cdots + w_0. \tag{22}$$

If we write $v = (z - z_0)^{1/2}$, (21) becomes

$$v = u\left[1 + \left(\frac{u^2}{4} - \frac{2}{3}u\right) + \cdots\right]^{1/2} \tag{23a}$$

or

$$v = u - \frac{1}{3}u^2 + \frac{5}{(8)(9)}u^3 + \frac{5}{(8)(27)}u^4 + \cdots. \tag{23b}$$

Inverting this series (see Abramowitz and Stegun 1964), we get

$$u = v + \frac{1}{3}v^2 + \frac{11}{(8)(9)}v^3 + \alpha_4 v^4 + \cdots. \tag{24}$$

Substituting this series into (22), we get

$$w - w_0 = 2v^2 + \frac{1}{2}v^4 + \alpha v^5 + \cdots$$

$$= 2(z - z_0) + \frac{1}{2}(z - z_0)^2 + \alpha(z - z_0)^{5/2} + \cdots. \tag{25}$$

We conclude that z_0 is a nonstationary double-valued algebraic singularity. Here $w'|_{z=z_0} = 2$, $w''|_{z=z_0} = 1$ (this must be so by (17)), but $w'''|_{z=z_0} = \infty$.

The singular point $w' = 1$ can be discussed in the same manner, but now $w''(z_0) = (-1)^{1/2}$ in view of (17).

To obtain the general solution of (17), we set

$$\zeta = (w' - 1)^{1/2} + (w' - 2)^{1/2}.$$

Squaring it, etc., we have

$$2w' - 3 + 2[(w' - 1)(w' - 2)]^{1/2} = \zeta^2,$$
$$4(w' - 1)(w' - 2) = (\zeta^2 + 3 - 2w')^2,$$

or

$$\zeta^4 + 2\zeta^2(3 - 2w') + 1 = 0; \tag{26}$$

obviously, $\zeta \neq 0$. Now solving for w' from (26), we have

$$p = w' = \frac{\zeta^4 + 6\zeta^2 + 1}{4\zeta^2} = \frac{\zeta^2}{4} + \frac{3}{2} + \frac{1}{4}\zeta^{-2}.$$

Therefore,

$$\frac{dp}{\zeta} = \frac{dw'}{\zeta} = \frac{1}{2\zeta}(\zeta - \zeta^{-3})d\zeta = dz.$$

Integrating this equation, we get the parametric representation of the solution

$$z = \frac{1}{2}\int(1 - \zeta^{-4})d\zeta = \frac{1}{2}\left(\zeta + \frac{1}{3}\zeta^{-3}\right) + C_3,$$

$$w = \frac{1}{2}\int\left(\frac{3}{2} + \frac{1}{4}\zeta^{-2} + \frac{1}{4}\zeta^2\right)(1 - \zeta^{-4})d\zeta$$

$$= \frac{1}{2}\left(\frac{3}{2}\zeta + \frac{1}{12}\zeta^3 + \frac{1}{2}\zeta^{-3} + \frac{1}{20}\zeta^{-5}\right) + C_4. \tag{27}$$

We may eliminate ζ to get an explicit solution

$$w = w(z, C_3, C_4).$$

EXAMPLE 3 We discuss briefly the equation

$$w'' = (2w' + 1)^{1/2} - (w' + 2)^{1/2}. \tag{28}$$

Again, writing

$$w' = p, \quad \frac{dw'}{dz} = p\frac{dp}{dw} = \sqrt{2p + 1} - \sqrt{p + 2},$$

Painlevé Transcendents

the solution can be found in parametric form

$$w = \int \frac{p((2p+1)^{1/2} + (p+2)^{1/2})}{p-1} dp + C_1, \tag{29a}$$

$$z = \int \frac{(2p+1)^{1/2} + (p+2)^{1/2}}{p-1} dp + C_2 \tag{29b}$$

so z and w tend to ∞ as $p \to 1$. Evidently

$$(2w' + 1)^{1/2} - (w' + 2)^{1/2} = 0$$

when $w' = 1$, so $w = z + C$ is a solution of (28).

It easily follows from (29) that

$$w = \int \frac{p\, dp}{(2p+1)^{1/2} - (p+2)^{1/2}}$$

$$= \int \frac{dp}{(2p+1)^{1/2} - (p+2)^{1/2}} + \int \frac{(p-1)\, dp}{(2p+1)^{1/2} - (p+2)^{1/2}}$$

$$= z + 2(3)^{1/2}(p-1) + \frac{3^{1/2}}{4}(p-1)^2 + \cdots + C \tag{30}$$

(see (29b)). Thus, $w \to z + C$ as $p \to 1$, or

$$w \approx z + C \tag{31}$$

for $p \approx 1$.

All solutions of (28) approximate the solution (31) when $p \to 1$ or $w' \to 1$. Indeed, since the right-hand side of (28) is holomorphic about $w' = 1$, we can easily find a unique solution of (28) satisfying $w|_{z=0} = C$, $w'|_{z=0} = 1$. The solution is simply $w = z + C$. Thus, the solution is not singular. It is a special case of the general solution (29) and is obtained from it in the limit $p \to 1$. The boundary of the domain of existence and uniqueness of the solution of (28) is formed by $2w' + 1 = 0$ and $w' + 2 = 0$, since the right side of (28) is not holomorphic on these planes. It is easily checked that $w''' \to \infty$ as $w' \to -1/2$ or $w' \to -2$. Again writing $u = (2p+1)^{1/2}$ and taking u to be small, we can obtain from (29) the solution

$$w = \left(\frac{2}{3}\right)^{1/2} \left[\frac{u^2}{4} + \frac{1}{6}\left(\frac{2}{3}\right)^{1/2} u^3 + \cdots\right], \tag{32}$$

where

$$u = [6^{1/2}(z_0 - z)]^{1/2} + \sum_{k=2}^{\infty} \alpha_k \left\{[6^{1/2}(z_0 - z)]^{1/2}\right\}^k, \tag{33}$$

showing that $w = w(z)$ is a double-valued solution in the neighborhood of each point $z = z_0$.

Erugin has also considered the equation

$$w'' = -w^3 w' + ww'(4w' + w^4)^{1/2} \tag{34}$$

to show that if w_0 and w'_0 are complex and satisfy $4w' + w^4 = 0$, so that (34) is not holomorphic in the neighborhood of (w_0, w'_0), there exist two solutions

$$w_1 = \left(\frac{3}{4}(z - C)\right)^{-1/3} \tag{35}$$

and

$$w_2 = A \tan(A^3 z + B), \tag{36}$$

where C, A, and B are arbitrary constants. Both solutions possess movable singularities. While (35) has three-valued branch points, (36) has an infinite number of poles.

Here we can move to each of the integral curves from a given point (z_0, w_0, w'_0). Equation (34) illustrates nonuniqueness in the neighborhood of the boundary of the domain of existence and uniqueness, namely $4w' + w^4 = 0$. The solutions considered here are in the complex domain.

These examples suffice to demonstrate the diversity and complexity of solutions when the right-hand side of (1) is an irrational function of w and w'.

8.5 THE FIRST PAINLEVÉ EQUATION

The first Painlevé equation

$$w'' = 6w^2 + z \tag{1}$$

is the simplest and yet has no special solutions expressible in terms of known classical transcendents. First, we notice that if the initial conditions $w(z_0) = w_0$, $w'(z_0) = w_1$ are finite, then, since the right side of (1) is a holomorphic function, we can find a unique solution $w(z; z_0, w_0, w_1)$, which is holomorphic in some neighborhood of z_0 and assumes these initial values there. Indeed, Davis (1962) has found this Taylor series solution to 15 terms in the form

$$w = w_0 + w'_0(z - z_0) + w''_0 \frac{(z - z_0)^2}{2!} + \cdots \tag{2}$$

for a slightly more general equation

$$w'' = 6w^2 + \lambda z \tag{1a}$$

which, however, can easily be transformed into (1) by writing $z = \lambda^{-1/5} t$, $w = \lambda^{2/5} y$.

If, on the other hand, $z = z_0$ is to be an algebraic singularity, then it is easy to see that at least one of $w(z)$ and $w'(z)$ tends to infinity as $z \to z_0$. If $w'(z_0)$ is infinite, then so is $w''(z_0)$. It follows from (1) that $w(z)$ is also infinite. If $w(z_0)$ is infinite, it follows that $w'(z_0)$ is also infinite. What needs to be proved, however, is that the singularity at $z = z_0$ is a second-order pole, with z_0 arbitrary. We first construct such a solution and show that it converges in a certain neighborhood of $z = z_0$: $0 < |z - z_0| < R$.

We assume an expansion of the form

$$w(z) = \sum_{n=0}^{\infty} a_n (z - z_0)^{\alpha_n} \tag{3}$$

for the solution of (1). Here, $\alpha_0 < \alpha_1 < \alpha_2 < \cdots$ and $\alpha_0 < 0$. The α_i are possibly rational numbers having a common denominator if the solution has an algebraic singularity. It will turn out, however, that α_i's are integers. Substituting (3) into (1), we have

$$\sum_{n=0}^{\infty} \alpha_n(\alpha_n - 1) a_n (z - z_0)^{\alpha_n - 2} = 6 \left[\sum_{n=0}^{\infty} a_n (z - z_0)^{\alpha_n} \right]^2 + z_0 + z - z_0. \tag{4}$$

The lowest terms balance if

$$\alpha_0(\alpha_0 - 1) a_0 (z - z_0)^{\alpha_0 - 2} = 6 a_0^2 (z - z_0)^{2\alpha_0}; \tag{5}$$

that is, if

$$\alpha_0 = -2, \quad a_0 = 1. \tag{6}$$

This is consistent with the assumption that $\alpha_0 < 0$. The next lowest term on the left is

$$\alpha_1(\alpha_1 - 1) a_1 (z - z_0)^{\alpha_1 - 2}, \tag{7}$$

where $\alpha_1 > -2$ by assumption, while the next lowest terms on the right are

$$12 a_1 (z - z_0)^{\alpha_1 - 2} + z_0. \tag{8}$$

For arbitrary z_0, (7) and (8) balance iff

$$\alpha_1 = 2, \quad a_1 = -\frac{1}{10} z_0. \tag{9}$$

At the next stage, terms of order $(z - z_0)^{\alpha_2-2}$ and $z - z_0$ balance:
$$\alpha_2(\alpha_2 - 1)a_2(z - z_0)^{\alpha_2-2} = 12a_2(z - z_0)^{\alpha_2-2} + (z - z_0), \tag{10}$$
where $\alpha_2 > \alpha_1 = 2$. Equation (10) requires
$$\alpha_2 = 3, a_2 = -\frac{1}{6}. \tag{11}$$
When $n = 3$, we have
$$\alpha_3(\alpha_3 - 1)a_3(z - z_0)^{\alpha_3-2} = 12a_3(z - z_0)^{\alpha_3-2}. \tag{12}$$
This gives $a_3 = h$, an arbitrary constant, and $\alpha_3 = 4$. Thus, the exponents can be checked to be integers, and expansion (3) has the specific form
$$w(z) = (z - z_0)^{-2} - \frac{1}{10}z_0(z - z_0)^2 - \frac{1}{6}(z - z_0)^3 + h(z - z_0)^4$$
$$+ \frac{1}{300}z_0^2(z - z_0)^6 + \frac{1}{150}z_0(z - z_0)^7 + \cdots. \tag{13}$$
This formal series solution has to be shown to possess a positive radius of convergence. For this purpose, we choose a constant M such that the first seven coefficients in (13) satisfy the inequality $|a_k| < M^k$. This is so if
$$1 < M, \quad \frac{1}{10}|z_0| < M, \quad |h| < M^3. \tag{14}$$

The recurrence relation for the general coefficient a_n is found from (1) by comparing the coefficients of powers of $(z - z_0)^{\alpha_n-2}$:
$$[\alpha_n(\alpha_n - 1) - 12]a_n = 6\sum_{j=1}^{n-1} a_j a_{n-j}. \tag{15}$$
(The coefficient $-12a_n$ on the left comes from the nonlinear term on the right of (1).) The sum on the right contains at most $n-1$ terms. If $|a_k| < M^k$ for $k = 1, 2, \ldots, n - 1$, (15) implies that
$$|a_n| \leq \frac{6(n-1)}{(n-3)(n+4)}M^n \leq M^n, \quad n \geq 6. \tag{16}$$
Here we have used the fact that $\alpha_n \geq n + 1$. Since $|a_k| < M^k$ for $0 \leq k < 6$, it is valid for all n. We conclude that the formal series (13) has a finite radius of convergence
$$0 < |z - z_0| < M^{-1}. \tag{17}$$
Hence it actually solves (1). This series involves arbitrary constants h and z_0, the latter being the arbitrary position of the pole of second order. The relation of h to initial conditions will be presently discussed.

Painlevé Transcendents

We shall now give a second proof of the existence of a solution $w(z)$ of (1), with a double pole at $z = z_0$ and an arbitrary parameter h, by introducing some auxiliary functions due to Painlevé. Making the substitution $z_0 = (z_0 - z) + z$ in the coefficients in series (13), we obtain

$$w = \frac{1}{(z-z_0)^2} - \frac{z}{10}(z-z_0)^2 - \frac{(z-z_0)^3}{15}$$
$$+ h(z-z_0)^4 + \frac{z^2}{300}(z-z_0)^6 + \cdots; \quad (18)$$

hence we have

$$w' = -\frac{2}{(z-z_0)^3} - z\frac{z-z_0}{5} - \frac{3}{10}(z-z_0)^2$$
$$+ 4h(z-z_0)^3 + \frac{z^2}{50}(z-z_0)^5 + \cdots. \quad (19)$$

We write $w = y^{-2}$, so (18) implies that

$$\frac{1}{y} = \frac{1}{z-z_0}\left[1 - \frac{z}{10}(z-z_0)^4 - \frac{(z-z_0)^5}{15} + h(z-z_0)^6\right.$$
$$\left. + \frac{z^2}{300}(z-z_0)^8 + \cdots\right]^{1/2}$$
$$= \frac{1}{z-z_0}\left[1 - \frac{z}{20}(z-z_0)^4 - \frac{1}{30}(z-z_0)^5 + \frac{h}{2}(z-z_0)^6\right.$$
$$\left. + \frac{z^2}{2400}(z-z_0)^8 + \cdots\right].$$

Therefore,

$$y = (z-z_0)\left[1 - \frac{z}{20}(z-z_0)^4 - \frac{1}{30}(z-z_0)^5 + \frac{h}{2}(z-z_0)^6\right.$$
$$\left. + \frac{z^2}{2400}(z-z_0)^8 + \cdots\right]^{-1}$$
$$= (z-z_0)\left[1 + \frac{z}{20}(z-z_0)^4 + \frac{1}{30}(z-z_0)^5 - \frac{h}{2}(z-z_0)^6\right.$$
$$\left. + \frac{z^2}{480}(z-z_0)^8 + \cdots\right]. \quad (20)$$

This series may be inverted (see Abramowitz and Stegun 1964, p. 16) so that we have

$$z - z_0 = y - \frac{z}{20}y^5 + \cdots. \quad (21)$$

Substituting (21) into (19) gives

$$w' = -\frac{2}{\epsilon y^3} - z\epsilon\frac{y}{2} - \frac{y^2}{2} + 7\epsilon y^3 + \cdots, \qquad \epsilon = \pm 1. \tag{22}$$

Thus, we change (1) into the system

$$w = y^{-2}, \qquad w' = -\frac{2}{y^3} - \frac{zy}{2} - \frac{y^2}{2} + uy^3, \tag{23}$$

where the second equation of (23) is (22) with uy^3 representing the remaining terms in the series. We choose ϵ to be $+1$. Differentiating (23) and using (1), we get

$$w' = -\frac{2}{y^3}y', \tag{24a}$$

$$w'' = \frac{6}{y^4} + z = \frac{6}{y^4}y' - \frac{y}{2} - \frac{zy'}{2} - yy' + 3y^2y'u + y^3u'. \tag{24b}$$

We eliminate w' from (22) and (24a). Combining the resulting equation with (24b), we arrive at a system of two equations in y and u:

$$\begin{aligned}\frac{dy}{dz} &= 1 + \frac{zy^4}{4} + \frac{y^5}{4} - \frac{uy^6}{2}, \\ \frac{du}{dz} &= \frac{z^2y}{8} + \frac{3zy^2}{8} + y^3\left(\frac{1}{4} - uz\right) - \frac{5}{4}uy^4 + \frac{3}{2}u^2y^5.\end{aligned} \tag{25}$$

This system has a solution $(y(z), u(z))$ satisfying the initial conditions

$$y(z_0) = 0, \qquad u(z_0) = u_0 \quad \text{(an arbitrary constant)},$$

which is holomorphic in the neighborhood of $z = z_0$:

$$y(z_0) = z - z_0 + \sum_{k=5}^{\infty} \alpha_k(z - z_0)^k, \tag{26a}$$

$$u(z_0) = u_0 + \frac{z_0^2(z - z_0)^2}{16} + \sum_{k=3}^{\infty} \beta_k(z - z_0)^k. \tag{26b}$$

These series converge in $|z - z_0| < r$, the radius r depending on the choice of u_0. Hence series (13) converges for any value of h; this value, however, determines the region of convergence of series (13): $0 < |z - z_0| < R$. Hence, we conclude that nonstationary singularities of the solutions of (1) are second-order poles, which may be placed at any arbitrary point $z = z_0$. We note that (26a) may also be found for nonvanishing initial conditions: $y(z_0) = y_0$, $u(z_0) = u_0$.

Painlevé Transcendents

We now introduce yet another function. First, we write (22) in the two alternative forms for $\epsilon = \pm 1$, following from (23):

$$u = w^{3/2}\left[w' + \frac{1}{2w}\right] + 2w^3 + \frac{zw}{2}, \tag{27a}$$

$$u = -w^{3/2}\left[w' + \frac{1}{2w}\right] + 2w^3 + \frac{zw}{2}; \tag{27b}$$

that is,

$$w' + \frac{1}{2w} + w^{3/2}\left[2 + \frac{z}{2w^2} - \frac{u}{w^3}\right] = 0, \tag{28a}$$

$$w' + \frac{1}{2w} - w^{3/2}\left[2 + \frac{z}{2w^2} - \frac{u}{w^3}\right] = 0. \tag{28b}$$

Multiplying (28a) and (28b), we get

$$w'^2 - 4w^3 - 2zw + \frac{w'}{w} + 4u + \frac{1}{w}\left[\frac{1}{4w} - \frac{z^2}{4} - \frac{u^2}{w^2} + \frac{zu}{w}\right] = 0. \tag{29}$$

We set

$$v = w'^2 - 4w^3 - 2zw + \frac{w'}{w} + z, \tag{30}$$

so (29) becomes

$$v - z + 4u + \frac{1}{w}\left[\frac{1}{4w} - \frac{z^2}{4} - \frac{u^2}{w^2} + \frac{zu}{w}\right] = 0. \tag{31}$$

It is easy to check by substituting (13) into (30) that v is finite at the point z_0 and has the form

$$v(z) = -28h + z_0 + (z - z_0) + (z - z_0)^2[\cdot] \tag{32}$$

where $[\cdot]$ denotes a finite quantity at $z = z_0$. It follows from (32) that

$$v(z_0) = -28h_0 + z_0, \quad \frac{dv}{dz}(z_0) = 1. \tag{33}$$

Hence $u(z)$ and $v(z)$ are finite at the pole $z = z_0$ of the solution $w(z)$. Using the definition (30) of v and one of the values of u in (28), we conclude that, in general, $v(z)$ and $u(z)$ are either both bounded or both unbounded in the neighborhood of $z = z_0$ for every $w(z)$ satisfying $0 < \rho \leq w(z)$. Indeed, (31) gives

$$u = \frac{w^{1/2}[z^2w - 1 - w^2(v - z)]}{2[4w^{5/2} + zw^{1/2} \mp \{16w^5 + 8w^3z + 1 + 4w^2(v - z)^{1/2}\}]}, \tag{34}$$

showing that $|u(z)|$ is bounded if $|v(z)|$ is, and $u \to -(v-z)/8$ when $|w| \to \infty$.

Yet another function used by Painlevé was

$$W(z) = \frac{v'(z)}{v(z)} = \frac{2w'w'' - 12w^2w' - 2zw' - 2w + w''/w - w'^2/w^2 + 1}{w'^2 - 4w^3 - 2zw + w'/w + z}$$

$$= \frac{4w^3 - w'^2 + w^2 + zw}{w(ww'^2 - 4w^4 - 2zw^2 + w' + zw)}. \tag{35}$$

This follows easily from equations (30) and (1).

Making use of the DE for the auxiliary functions u, v, and w, Painlevé proved that the solutions of (1) have no singularities other than poles in the finite plane (see Ince 1956, Hille 1969, and Erugin 1976a for rigorous proofs of this theorem).

There is another set of variables due to Erugin (1976b) which gives an asymptotic expression for the solutions of (1). The latter is equivalent to the system

$$\frac{dw}{dz} = y, \qquad \frac{dy}{dz} = 6w^2 + z. \tag{36}$$

Writing

$$u = yw^{-3/2}, \qquad v = w^{-1/2} \tag{37}$$

(the motivation for this transformation is rather circuitous; Erugin 1976b), we arrive at the system

$$v\frac{du}{dv} = \frac{3u^2 - 12}{u} - \frac{2zv^4}{u}, \tag{38a}$$

$$\frac{dz}{dv} = -\frac{2}{u}. \tag{38b}$$

Here we have used (36), (37), and the differential relation $y\,dy = 6w^2\,dw + z\,dw = 6w^2\,dw + zy\,dz$. Further, changing the variables in (38) to

$$u = 2 + \tau, \qquad \theta = z - z_0, \tag{39}$$

we get

$$v\frac{d\tau}{dv} = \frac{3(\tau^2 + 4\tau) - 2(z_0 + \theta)v^4}{2 + \tau}, \tag{40a}$$

$$\frac{d\theta}{dv} = \frac{-2}{2 + \tau}. \tag{40b}$$

We easily see that $\tau \to 0$, $\theta \to 0$ as $v \to 0$. Now, if we write $\tau = v^3\tilde{w}$, equations (40) become

$$v\frac{d\tilde{w}}{dv} = \frac{6\tilde{w} - 2(z_0 + \theta)v}{2 + v^3\tilde{w}}. \tag{41a}$$

Painlevé Transcendents

$$\frac{d\theta}{dv} = -\frac{2}{2+v^3\tilde{w}}. \tag{41b}$$

Finally, the transformation

$$\zeta = \theta + v, \qquad \eta = \tilde{w} - \frac{1}{2}z_0 v \tag{42}$$

changes system (41) to

$$v\frac{d\eta}{dv} = \frac{6\eta + 2v^2 - 2v\zeta - (z_0 v^4/2)(z_0 v/2 + \eta)}{2 + v^3(z_0 v/2 + \eta)}, \tag{43a}$$

$$v\frac{d\zeta}{dv} = \frac{v^4(z_0 v/2 + \eta)}{2 + v^3(z_0 v/2 + \eta)}. \tag{43b}$$

Transformation (42) has eliminated the first powers of v in (41a) so that (43) is in a canonical form:

$$v\frac{d\eta}{dv} = 3\eta + v^2 - \zeta v + \omega_5(v, \eta, \zeta), \tag{44a}$$

$$v\frac{d\zeta}{dv} = \psi_5(v, \eta, \zeta), \tag{44b}$$

where ω_5 and ψ_5 are power series, convergent in some neighborhood of the point $v = 0$, $\eta = 0$, $\zeta = 0$ and starting with a term of fifth degree. A solution of (44) vanishing in the limit as $v \to 0$ is found to be

$$\eta = -v^2 + Cv^3 + \sum_{k \geq 4} \eta_k v^k,$$
$$\zeta = \sum_{k \geq 5} \zeta_k v^k. \tag{45}$$

These series, in view of the form (44), converge in the neighborhood of $v = 0$.

Returning now to the variables w, y, and z (see (37), (39), and (42)), we have

$$y = 2v^{-3} + \frac{1}{2}z_0 v - v^2 + \frac{1}{4}Cv^3 + \cdots, \tag{46a}$$

$$z = z_0 - v + \frac{1}{20}z_0 v^5 - \frac{1}{12}v^6 + \frac{1}{56}Cv^7 + \cdots, \tag{46b}$$

$$v = w^{-1/2}. \tag{46c}$$

Here we have replaced the coefficient C of v^3 by $C/4$. Thus, we have obtained a series solutions of (1) satisfying the conditions

$$w(z) \longrightarrow \infty, \quad y(z) \longrightarrow \infty \quad \text{as } z \longrightarrow z_0. \tag{47}$$

Besides, these power series in $v = w^{-1/2}$ are convergent. The series (46b) can be inverted, and we finally arrive at the solution

$$w = \frac{1}{(z-z_0)^2}\left(1 + \sum_{k=1}^{\infty} w_k(z-z_0)^k\right),$$

$$\frac{dw}{dz} = y = \frac{1}{(z-z_0)^3}\left(-2 + \sum_{k=1}^{\infty} y_k(z-z_0)^k\right),$$

(48)

where w_k and y_k are constants, and the series converge for $|z - z_0| < r$, $r > 0$. The solution for w has a pole of order 2 and involves an arbitrary parameter C.

Given an initial value problem for (1), $w = w_i$, $w' = w_i'$ at $z = z_i$, say, how does one find the location of a pole z_0 and the arbitrary constant h in series (13)? This is accomplished as follows. We substitute $w = w_i$ and $w' = w_i'$ in (13) and its derivative, respectively:

$$w_i = \frac{1}{v_0^2} - \frac{z_0}{10}v_0^2 - \frac{1}{6}v_0^3 + hv_0^4 + \frac{z_0^2}{300}v_0^6 + \frac{z_0}{150}v_0^7 + \cdots,$$

$$w_i' = -\frac{2}{v_0^3} - \frac{z_0}{5}v_0 - \frac{1}{2}v_0^2 + 4hv_0^3 + \frac{z_0^2}{50}v_0^5 + \frac{7z_0^2}{150}v_0^6 + \cdots,$$

(49)

where $v_0 = z_i - z_0$. Equations (49) determine the unknown parameters h and v_0 in terms of the given initial values z_i, w_i, and w_i'.

From v_0 one can find the location of the pole, namely $z_0 = z_i - v_0$. This value delimits the region of convergence of the Taylor series solution of (1) about $z = z_i$,

$$w = w_i + w_i'(z - z_i) + \frac{w_i''}{2!}(z-z_i)^2 + \frac{w_i'''}{3!}(z-z_i)^3 + \cdots,$$

(50)

namely $|z - z_i| < |z_0 - z_i|$. Davis (1962) has computed h and z_0 (by truncating the series (49)) for a particular choice of the initial conditions $z_i = 0$, $w_i = 1$, $w_i' = 0$. For this purpose, he resorted to analytic continuation of (50) whose convergence slows down as z approaches z_0, the location of the pole. See Davis (1962) for computational details.

It is possible to show that solutions to (1) possess an infinite number of poles and to predict their spacing for large z from their asymptotic behavior. This is accomplished by introducing the transformation due to Boutroux (1914)

$$w = z^{1/2}u(s), \qquad s = \frac{4}{5}z^{5/4},$$

(51)

Painlevé Transcendents

into (1). We obtain

$$\frac{d^2u}{ds^2} = 6u^2 + 1 - \frac{1}{s}\frac{du}{ds} + \frac{4}{25s^2}u. \tag{52}$$

For large s, it was shown by Boutroux that u is asymptotic to the solution of the equation

$$\frac{d^2u}{ds^2} = 6u^2 + 1, \tag{53}$$

the last two terms on the right side of (52) being small.

Equation (53) is an elliptic equation whose solution may be expressed in terms of the Weierstrass **P** function

$$W = \mathbf{P}(Cs, k), \tag{54}$$

where the constant C and k are expressed in terms of the roots of the equation

$$4s^3 + 2s + g = 0, \tag{55}$$

where the constant g is arbitrary (see Davis 1962). The elliptic functions are periodic and their poles are separated by a constant period P. If the distance between consecutive poles in the variable z is denoted by Δ, then according to (51),

$$(z + \Delta)^{5/4} - z^{5/4} \sim \frac{5}{4}P$$

or

$$\Delta \sim Pz^{-1/4}. \tag{56}$$

We have thus shown that there are an infinite number of second-order poles with a cluster point about $z = \infty$. They are separated for large z by a distance Δ given by (56).

Bender and Orszag (1978) have carried out a numerical study of (1). Their numerical solution shows that the quantity $\bar{z}_n^{1/4}\Delta_n$ approaches P with a discrepancy $O(10^{-7})$ as the sixth pole is approached. Here, z_n is the nth pole, $\Delta_n = z_{n+1} - z_n$, and $\bar{z}_n = (z_n + z_{n+1})/2$. The solution of the first-order equation

$$\frac{dw}{dz} = w^2 + z,$$

a close kin of the first Painlevé equation (1), shows a remarkably similar structure with an infinite number of poles (see again Bender and Orszag 1978 for a numerical study of this equation).

Actually, it was proved by Yablonskii (1964) that (1) has a unique solution passing through each point of the region $-\infty < z < 0$, $0 < w < \infty$. We have already shown that an infinite number of poles exist for large positive values of z. (Here we take z to be real.) Now we prove some other simple qualitative results to demonstrate that no solution can be continued to the whole real positive axis. We consider the initial value problem for (1) with

$$w(z_0) = w_0, \quad w'(z_0) = k, \quad z_0 \geq 0, \tag{57}$$

where k is an arbitrary real number (Bartashevich 1973).

1. First, we show that if $k < 0$, the solution $w(z; z_0, w_0, k)$ of (1) and (57) takes its minimum value for a finite $z \in (z_0, \infty)$. Since $w'(z_0) = k < 0$, $w(z)$ decreases in the small interval $(z_0, z_0 + \epsilon)$. Let us assume that $w(z)$ decreases in the entire interval (z_0, ∞). Since $w'' = 6w^2 + z > 0$, we rule out the possibilities that either $w' \to -\infty$ or $w' \to c \leq 0$. This follows immediately if we write (1) as

$$w' = \int_{z_0}^{z} (6w^2 + z) dz + k. \tag{58}$$

Therefore, the assumption that w decreases throughout (z_0, ∞) is not valid, and w has a minimum somewhere in this interval.

2. Every integral curve $w(z; z_0, w_0, k)$ of (1) has a vertical asymptote at a finite distance to the right of z_0. To prove this, we consider the cases $w_0 > 0$ and $w_0 \leq 0$ separately. For the former, we consider a comparison equation

$$w'' = 6w^2 - 3\frac{w_0^2}{2} \tag{59}$$

with initial conditions at $z = z_0$, $w(z_0) = w_0$, $w'(z_0) = 0$. Its solution is

$$w = w_0 + \frac{3}{2} w_0 \tan^2 \left[\sqrt{\frac{3w_0}{2}} (z - z_0) \right]. \tag{60}$$

Comparing the right side of (1) and (59) and the initial conditions at z_0 for both these equations, we check that the integral curve $w(z; z_0, w_0, k)$ for $z_0 < z < \sqrt{\pi/6z_0}$ lies above the integral curve (60). The first vertical asymptote of (60) is situated at

$$z = z_0 + \frac{\pi}{(6w_0)^{1/2}}. \tag{61}$$

Therefore, the vertical asymptote of (1) is located at $z = z_0 + m$, where $0 < m \leq \pi/(6w_0)^{1/2}$. In the limit $w_0 \to 0$, the asymptote (61) of (60) moves off to infinity. To consider this case, take a point (z_1, w_1) on the curve $w(z; z_0, w_0, k)$ with a sufficiently large ordinate w_1 and compare this curve

Painlevé Transcendents

with the integral

$$w = w_1 + \frac{3}{2}w_1 \tan^2\left[\left(\frac{3w_1}{2}\right)^{1/2}(z-z_1)\right] \tag{62}$$

of the equation $w'' = 6w^2 - (3/2)w_1^2$ with initial conditions $z = z_1$, $w(z_1) = w_1$, $w'(z_1) = 0$. Now, following the same argument as above, we conclude that the integral curve for (1) goes to infinity at a finite point.

The second case concerns either $z_0 \geq 0$, $w_0 \leq 0$, and k any real number or $z_0 \geq 0$, $w_0 > 0$, and $k \leq 0$. Now since $w'' > 0$, the form $w' = \int_{z_0}^z (6w^2 + z)dz + k$ of (1), and result 1 above imply that there is a point (z_1, w_1) on the curve $w = w(z; z_0, w_0, k)$ with a positive ordinate at which this curve is increasing. The rest of the reasoning is the same as for result 1.

3. We show that there exists a point $\tilde{z}$, $\tilde{z} > z_0 \geq 0$, beyond which no solution of (1) satisfying $w(z_0) = w_0$ can be continued. Changing the role of dependent and independent variables, we can write (1) as

$$\frac{d^2z}{dw^2} = -\left(\frac{dz}{dw}\right)^3 (6w^2 + z). \tag{63}$$

The integral curves for (1) and (63) passing through the point (z_0, w_0) are identical with the exception of the curve $z = z_0$, which satisfies only (63). For $w > w_0$, consider the part of the integral curve of (63) that lies above $z = z_0$; on integration of (63) we have

$$z(w) = z_0 + \int_{w_0}^w \frac{dw}{\{k^2 + 4(w^3 - w_0^3) + 2\int_{w_0}^w z(w)dw\}^{1/2}}. \tag{64}$$

Hence, for arbitrary k and $b > w_0$, we have

$$z(b) < z_0 + \int_{w_0}^b \frac{dw}{2(w^3 - w_0^3)^{1/2}} < \infty; \tag{65}$$

since $\int_{w_0}^\infty \frac{dw}{2(w^3 - w_0^3)^{1/2}} < \infty$, we conclude, by allowing b to tend to infinity, that

$$z(\infty) < z_0 + \int_{w_0}^\infty \frac{dw}{2(w^3 - w_0^3)} < \infty. \tag{66}$$

This proves the existence of the point $\tilde{z}$. If $z = z_1$ is an asymptote of the curve $w = w(z; z_0, w_0, k)$, then $\tilde{z}$ is the maximum of the boundary function $z_1(k)$ as k varies from $-\infty$ to $+\infty$.

4. Now we show that the solution $w(z)$ of (1) can be expressed as the quotient of two integral functions. Multiply (1) by dw/dz and write it in

the form

$$\frac{d}{dz}\left[\frac{1}{2}\left(\frac{dw}{dz}\right)^2 - 2w^3 - zw\right] = -w. \tag{67}$$

Let

$$\frac{d}{dz}\eta = -w; \tag{68}$$

then $\eta(z)$ satisfies the equation

$$\frac{d^3\eta}{dz^3} + 6\left(\frac{d\eta}{dz}\right)^2 + z = 0. \tag{69}$$

(Equation (69) follows after substituting (68) into (1).) Using the expansion (13) in (68), we find that

$$\eta(z) = (z - z_0)^{-1} + O\left\{(z - z_0)^3\right\}. \tag{70}$$

Thus, $\eta(z)$ has simple poles at $z = z_0$. Writing

$$\zeta(z) = e^{\int \eta(z)\, dz}, \tag{71}$$

we observe that $\zeta(z)$ is regular; for, although the function $\int \eta\, dz$ has an infinite number of many-valued (logarithmic) branch points, its values differ by additive multiples of $2\pi i$. The function $\zeta(z)$, according to (71), has no poles (or branch points). It is therefore an integral function of z. By differentiating (71) twice, we have

$$-\frac{d\eta}{dz} = w = \frac{\zeta'^2 - \zeta\zeta''}{\zeta^2}. \tag{72}$$

Both numerator and denominator in (72) are integral functions of z; hence so is w.

Bender and Orszag (1978) have considered the variant

$$w'' = w^2 + e^z$$

of (1) categorized as "beyond Painlevé transcendants." It displays some features different from (1). In particular, its asymptotic expansion involves powers of $\ln(z - z_0)$ in addition to a second-order pole.

We refer to another variant of (1), namely,

$$w'' = w^2 - z. \tag{73}$$

A boundary value problem for (73), which occurs in the studies of natural convective flows with viscous dissipation, was analyzed by Holmes and Spence (1984), using shooting techniques (see Chapter 5).

8.6 THE SECOND PAINLEVÉ EQUATION

We discussed in Section 8.3 the transformation properties of the second Painlevé equation

$$\frac{d^2w}{dz^2} = 2w^3 + zw + \mu. \tag{1}$$

We also discussed, in Section 4.5, its special case $\mu = 0$,

$$\frac{d^2w}{dz^2} = 2w^3 + zw, \tag{2}$$

with reference to a boundary value problem over the real line via an integral equation formulation. Before studying further the boundary value problem for (2) as it arises in applications, we summarize the solution of (1) in the neighborhood of an ordinary nonsingular point and a pole (see Davis 1962 for details). As in Section 8.5, it can be easily seen that the expansion in the neighborhood of a pole $z = z_0$ is

$$w = \frac{a_{-1}}{v} + a_0 + a_1 v + a_2 v^2 + a_4 v^4 + \cdots, \quad v = z - z_0. \tag{3}$$

The poles $z = z_0$ are simple, and the coefficients a_i are determined by substitution of (3) into (1):

$$a_{-1} = 1, \quad a_0 = 0, \quad a_1 = -\frac{z_0}{6}, \quad a_2 = -\frac{1}{4}(1+\mu), \quad a_3 = h,$$

$$a_4 = \frac{1}{72}z_0(1+3\mu), \tag{4}$$

$$a_5 = \frac{1}{3024}(27 + 108\mu - 216hz_0 + 81\mu^2 - 2z_0^3), \ldots,$$

where z_0 and h are arbitrary constants.

Davis (1962) has calculated a_n up to $n = 15$ and given a recursion formula to generate higher a_n. He has also given the Taylor series

$$w = w_1 + w_1'(z - z_1) + \frac{w_1''}{2!}(z - z_1)^2 + \frac{w_1'''}{3!}(z - z_1)^3 + \cdots \tag{5}$$

about an initial point (z_1, w_1). The coefficients in (5) were calculated through $i = 10$. Again, the values of h and z_0 in (3)–(4) were computed from the initial conditions z_1 and w_1 for different values of the parameter μ by truncating (3) and (5) and solving them simultaneously. The method of analytic continuation was used to reduce inaccuracy in the Taylor series (5) when the initial point after continuation approached the pole.

Again, to prove the existence of infinite number of poles in the solution of (1) and find their spacing, at least for large real z, we resort to asymptotic analysis, using the Boutroux transformation

$$w = z^{1/2}y, \qquad t = \frac{2}{3}z^{3/2}. \tag{6}$$

Equation (1) assumes the form

$$\frac{d^2y}{dt^2} = 2y^3 + y - \frac{1}{t}\frac{dy}{dt} + \frac{1}{9}\frac{y}{t^2} + \frac{2}{3}\frac{\mu}{t}. \tag{7}$$

When $|z|$ and hence $|t|$ is large, (7) is asymptotic to the equation

$$\frac{d^2W}{dt^2} = 2W^3 + W. \tag{8}$$

Its general solution may be written in the form

$$W = C\,\mathrm{sn}(\lambda u, k), \qquad k^2 = -\left(1 + \frac{1}{\lambda^2}\right), \; C^2 = -(\lambda^2 + 1), \; u = t - t_0 \tag{9}$$

(see Davis 1962, chapter 6), where the constants C and λ depend on the initial conditions on W. The function sn is periodic: $\mathrm{sn}(z + \Omega) = \mathrm{sn}(z)$. Therefore, the period of W is $\Omega_1 = \Omega/C$. Since $t = t_0$ is a pole of W, there exist infinite values, $t = t_0 + m\Omega_1$, with m an integer, for which W has a pole.

As in Section 8.5, we note that if the difference between consecutive poles in the variables z is denoted by Δ, then, for large z,

$$\Omega = \frac{2}{3}[(z + \Delta)^{3/2} - (z)^{3/2}]$$

$$= \frac{2}{3}\left[z^{3/2}\left(1 + \frac{3}{2}\frac{\Delta}{z} + \cdots\right) - z^{3/2}\right] \sim \Delta z^{1/2}, \tag{10}$$

where we have used the transformation (6). Hence the distance between two consecutive poles is

$$\Delta \sim \Omega_1 z^{-1/2}, \tag{11}$$

which approaches zero asymptotically as the square root of the distance of the poles from the origin. The parameter μ in (1) has no effect on the asymptotic formula (11).

Now we detail some results, both analytic and numerical, for the second Painlevé transcendent in direct relation to the Korteweg–de Vries equation, due to Rosales (1978) (see also Section 4.5). We recall that the normalized KdV equation

$$u_t + 6uu_x + u_{xxx} = 0 \tag{12}$$

has a self-similar solution

$$u(x,t) = (3t)^{-2/3}f(z), \qquad z = \frac{x}{(3t)^{1/3}}, \tag{13}$$

so (12) becomes

$$f''' + 6ff' - 2f - zf' = 0. \tag{14}$$

The solutions of primary interest should decay exponentially as $z \to \infty$. In this limit, the linearized form of (14) is valid so that

$$f''' - 2f - zf' = 0. \tag{15}$$

The relevant solution of this equation in terms of Airy function is

$$f \sim a Ai'(z), \tag{16}$$

where a is an amplitude parameter (other solutions of (15), $Bi'(z)$ and $Gi'(z)$, either do not decay or decay algebraically).

The earliest numerical study of (14) and (16) is due to Berezin and Karpman (1964), who showed that when a is small enough f becomes oscillatory as $z \to -\infty$, but otherwise f may develop singularities at finite z. Rosales (1978) showed that there is a critical value a_1 of a which separates the oscillatory from the singular solutions. In fact, when $|a| = a_1$, $f \sim (1/2)z$ as $z \to -\infty$. For $|a| < a_1$, $f(z)$ is oscillatory as $z \to -\infty$, while, for $a > a_1$, $f(z)$ develops a singularity at a finite z. Numerical studies of Rosales and others indicated that $a_1 = 1 + O(10^{-13})$. Hastings and McLeod (1980) later proved analytically that $a_1 = 1$.

Now we transform (14) via the Ansatz

$$f = F' - F^2, \tag{17}$$

which is credited to G. B. Whitham. This is suggested by the relation between the Korteweg–de Vries and the modified Korteweg–de Vries equations. Now (14) becomes

$$(F'' - zF - 2F^3)'' - 2F(F'' - zF - 2F^3)' = 0. \tag{18}$$

This equation can be integrated once to give

$$(F'' - zF - 2F^3)' = \alpha \exp\left\{-2\int_z^\infty F(y)\,dy\right\}, \tag{19}$$

where α is an arbitrary constant. Since we require f (and hence F) to decay exponentially as $z \to \infty$, α must be zero. We, therefore, get

$$F'' - zF - 2F^3 = 0, \tag{20}$$

a special case of (1) with $\mu = 0$. One of its solutions in the neighborhood of a polar singularity is given by (3) with $w \equiv F$. The second solution with a negative sign,

$$F = -\left\{\frac{1}{z - z_0} - \frac{z_0}{6}(z - z_0) - \frac{1}{4}(z - z_0)^2 + h(z - z_0)^3 + \cdots\right\}, \tag{21}$$

leads to a function f which is regular at z_0. Equation (21) with a plus sign before the brace (which is just (3)) produces a double pole in f at $z = z_0$. This is the pole which develops when the amplitude parameter a exceeds a_1, as we shall presently discuss. According to transformation (17), the asymptotic behavior of F is

$$F(z,a) \sim aA_i(z) \quad \text{as } z \longrightarrow \infty. \tag{22}$$

As we noted in Section 4.5, it suffices to consider $a > 0$ since $F(z,-a) = -F(z,a)$.

We shall now discuss some qualitative features of the solution of (20), as a varies from 0 to ∞, with the aid of the equation itself and its numerical solution subject to (22). If we write equation (20) in the form

$$F'' = (z + 2F^2)F, \tag{23}$$

then, depending on the sign of F'' (which determines the curvature of F), we divide the (z,F) plane into four regions (see Figure 8.6.1): (I) $F > 0$, $z + 2F^2 > 0$; (II) $F > 0$, $z + 2F^2 < 0$; (III) $F < 0$, $z + 2F^2 < 0$; and (IV) $F < 0, z + 2F^2 > 0$. The solution is strictly concave, $F'' > 0$, in regions I and III and strictly convex in the other regions. Equation (22) shows that as z decreases from ∞, $F(z,a) > 0$ increases, while $F'(z,a) < 0$ ($F'' > 0$ in I). The following cases arise as a varies (see Figure 8.6.1); some of the inferences are drawn from the numerical solution of the problem:

CASE 1 If a is "large enough," the integral $F = F(z,a)$ rises above the parabola $z + 2F^2 = 0$, always remaining in region I and at some finite distance $z = s(a)$ develops a singularity of polar type with residue equal to 1. The dependence on a in the region is monotonic, and the solutions are nested: $F(z,a') > F(z,a'')$ and $F'(z,a') < F'(z,a'')$ if $a' > a''$. The location of the singularity $s(a)$ moves to infinity as $a \to \infty$ in a strict monotonic manner.

CASE 2 If a_2 is the infimum of a's for which case 1 is true, then the solution $F = F(z,a_2)$ has the form described in case 1, except that now there exists a point of tangency of the integral curve with the parabola $z + 2F^2 = 0$, $z = z_T$, say. As a decreases to a_2^+ (from above), the location of the singularity $s(a)$ recedes monotonically to assume a finite limit $s(a_2)$.

CASE 3 Now consider the range $0 < a < a_2$. In this case, as z decreases from ∞, the integral curve crosses the parabola $z + 2F^2 = 0$ at a point $z = z_c(a)$, going from region I to region II. Then the solution becomes convex (F'' changes sign). In the special circumstance when a is close to a_2, the integral curve remains very close to the parabola $z + 2F^2 = 0$ as z

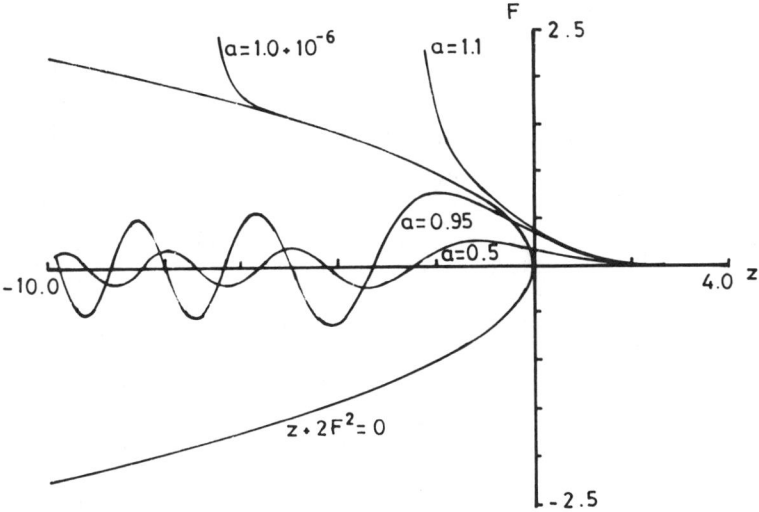

Figure 8.6.1 Solutions of (22)–(23) for $a = 1.1$, $a = 1.0 + 10^{-6}$, $a = 0.95$, and $a = 0.5$. (From Rosales 1978.)

decreases. The curvature of the integral curve is very small—smaller than that of $z + 2F^2 = 0$ itself. The integral curve crosses the parabola $z + 2F^2 = 0$ at $z = z_e(a)$, exiting back to region I. Thereafter, $F = F(z,a)$ remains in I, behaves as in cases 1 and 3, and develops a singularity at some finite z. Denoting the infimum of a's which show this behavior by a_1, we have $a_1 < a < a_2$ and $z_e(a) < z_T < z_c(a)$; moreover z_e, s, and z_c are monotonic in a. As $a \to a_2^-$, $z_e \to z_T^-$, $z_c \to z_T^+$, and $s \to s(a_2)^-$; when $a \to a_1^+$, $z_e \to -\infty$ and $s \to -\infty$ monotonically.

CASE 4 At the lower end of the range, $a = a_1$, the integral curve $F = F(z,a)$ crosses the parabola at only one point, $z = z_c(a_1)$. It remains in region II for all $z < z_c(a_1)$ and asymptotes to $z + 2F^2 = 0$ from below as $z \to -\infty$.

CASE 5 For $0 < a < a_1$, the convexity being proportional to F'' becomes large enough to make $F'(z,a) = 0$ at a point $z = z_1(a) < z_c(a)$. Thereafter, $F = F(z,a)$ turns downward, crosses the line $F = 0$, enters region III, and then assumes a minimum. Then it turns back, crosses $F = 0$ again, has a maximum, and so on. (These conclusions are drawn from the numerical solution.) The solution thus becomes oscillatory. As $z \to -\infty$, the amplitude and the wavelength of the oscillations tend to zero since $z + 2F^2 \to -\infty$.

This is because this factor in $F'' = F(z + 2F^2)$ makes the integral curve highly convex in region II, where F is positive and bounded, and highly concave in region III, where F is negative and bounded. The solutions in this range have close similarity with the Airy functions shifted to the left, the shift being larger the closer a is to a_1. (Compare these qualitative features with those deduced by Miles 1978 and described in Section 4.5.)

The solution F to (20) is to be found subject to the boundary condition (22). Rosales sought an expansion in the amplitude parameter a, as $z \to \infty$, in the form

$$F(z,a) \sim \sum_{n=0}^{\infty} a^{2n+1} \psi_n(z), \qquad (24)$$

where $\psi_0 = A_i(z)$. Substituting (24) into (20) and equating the coefficients of various powers of a to zero, we get

$$\psi_n'' - z\psi_n = 2 \sum_{i+j+k=n-1} \psi_i \psi_j \psi_k \qquad (n \geq 1), \qquad (25)$$

subject to

$$\psi_n = o(A_i(z)) \quad \text{as } z \to \infty. \qquad (26)$$

We note the expansions of $A_i(z)$ and $A_i'(z)$ as $z \to \infty$ (Abramowitz and Stegun 1965, p. 448):

$$A_i(z) \sim \frac{1}{2\sqrt{\pi}} z^{-1/4} \exp\left(-\frac{2}{3} z^{3/2}\right) \sum_{k=0}^{\infty} (-1)^k c_k \left(\frac{2}{3} z^{3/2}\right)^{-k}, \qquad (27)$$

$$A_i'(z) \sim -\frac{1}{2\sqrt{\pi}} z^{1/4} \exp\left(-\frac{2}{3} z^{3/2}\right) \sum_{k=0}^{\infty} (-1)^k d_k \left(\frac{2}{3} z^{3/2}\right)^{-k},$$

$$z \to \infty, \qquad (28)$$

where

$$c_k = \frac{\Gamma(3k + 1/2)}{54^k k! \Gamma(k + 1/2)},$$

$$d_k = \frac{-(6k + 1)}{(6k - 1)c_k} \qquad (29)$$

with

$$c_k \sim \frac{1}{(2\pi k)^{1/2}} \left(\frac{k}{2e}\right)^k \quad \text{as } k \to \infty. \qquad (30)$$

Since $\psi_n = O(A_i(z))$ and $A_i(z)$ has expansion (27) as $z \to \infty$, Rosales (1978) sought ψ_n in the form

$$\psi_n(z) \sim \left\{ \frac{1}{2\sqrt{\pi}} \exp\left(-\frac{2}{3}z^{3/2}\right) \right\}^{2n+1} z^{-1/4} \sum_{j=n}^{\infty} \alpha_{jn} \left(\frac{2}{3}z^{3/2}\right)^{-j} \qquad (31)$$

as $z \to \infty$ ($n = 0, 1, 2, \ldots$).

For $n = 0$, (31) coincides with (27) for $Ai(z)$. Substituting (31) into (25) leads to the determination of the coefficients α_{jn}, $j \geq n \geq 0$. Some of the coefficients are

$$\alpha_{j0} = \frac{(-1)^j \Gamma(3j + 1/2)}{54^j \Gamma(j + 1/2)\Gamma(j + 1)} \qquad (32)$$

$$= (-1)^j (2j + 1)(2j + 3) \cdots \frac{(6j - 1)}{(216)^j j!}, \qquad j \geq 0.$$

$$\alpha_{nn} = \left(\frac{1}{6}\right)^n, \qquad n \geq 0, \qquad (33)$$

$$\alpha_{n+1,n} = -\left(\frac{1}{6}\right)^n \left(\frac{35}{36}n - \frac{1}{72}\right) - \frac{1}{12}\delta_{n,0}, \qquad n \geq 0, \qquad (34)$$

where $\delta_{n,0}$ is Kronecker's delta. Expression (31) is nonconvergent and asymptotic, with terms alternating in sign. The first term itself with α_{nn} given by (33) gives a good approximation to the solution in the limit $z \to \infty$.

We now give some numerical results, in particular those relating to the accurate evaluation of the "critical" amplitude parameters a_1 and a_2 at which the nature of the solution changes. Equation (23) was solved numerically by using a fourth-order Runge–Kutta procedure with double precision and step size 0.001, which introduced a truncation error of only $O(10^{-15})$. The initial conditions were chosen to be $F(10) = aAi(10)$, $F'(10) = aAi'(10)$ for various values of a. Expansions (27), (28) were used for Ai and Ai' up to and including the fifteenth term. The error committed is less than the first neglected term. Now,

$$c_{16} \approx 3.16 \times 10^6 \qquad \text{and} \qquad d_{16} \approx -3.23 \times 10^6;$$

therefore,

$$c_{16}\left(\frac{2}{3}10^{3/2}\right)^{-16} \approx 2.08 \times 10^{-15},$$
$$-d_{16}\left(\frac{2}{3}10^{3/2}\right)^{-16} \approx 2.12 \times 10^{-15}, \qquad (35)$$

and there are at least 14 significant digits in the initial conditions. The values F and F' resulting from numerical integration at $z = 6$ were again compared with $aAi(6)$ and $aAi'(6)$, using 15 terms in (27) and (28). The agreement was found good to nine significant figures. Actually, Rosales also found a perturbation solution of (23) in the form (cf. Miles 1978)

$$F(z,a) = aAi(z)\{1 + \epsilon(z,a)\}, \tag{36}$$

$$\epsilon(z,a) \sim \frac{a^2}{16\pi z^{3/2}} \exp\left(-\frac{4}{3}z^{3/2}\right) \quad \text{as } z \longrightarrow \infty.$$

He computed some values of ϵ,

$$\begin{aligned}
\epsilon(4,a) &\approx 5.8 \times 10^{-8} a^2, & \epsilon(5,a) &\approx 6.0 \times 10^{-10} a^2, \\
\epsilon(6,a) &\approx 4.2 \times 10^{-12} a^2, & \epsilon(10,a) &\approx 3.1 \times 10^{-22} a^2,
\end{aligned} \tag{37}$$

and concluded that $F(z,a) = aAi(z)$ is a very good approximation, even for moderately sized z. The same was found to be the case for $F'(z,a) = aAi'(z)$.

The conclusions of the numerical study were that for $a \leq 1 - 10^{-9}$, the solutions became oscillatory as $z \to -\infty$, as shown in Figures (8.6.1)–(8.6.4). For $a \gtrsim 1 + 10^{-9}$, the solution had unbounded growth for $z \to -\infty$, as shown in Figure 8.6.1. For $a \geq 1.02$, the solution completely avoided the parabola

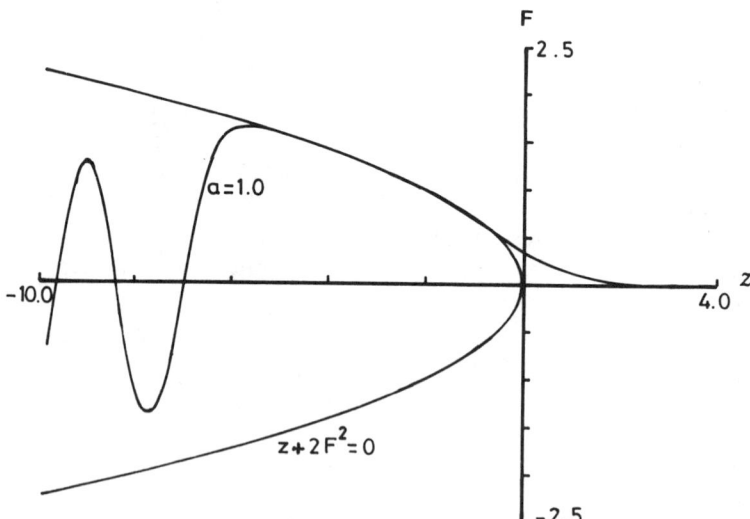

Figure 8.6.2 Solution of (22)–(23) for $a = 1.0$. (From Rosales 1978.)

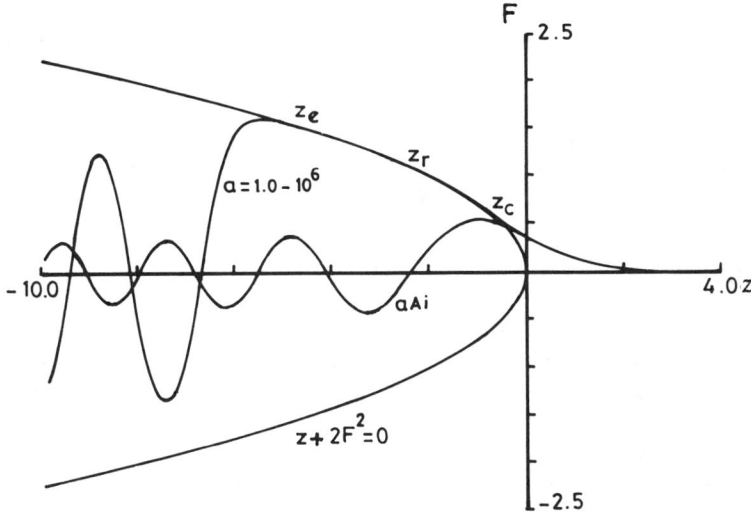

Figure 8.6.3 Comparison of the solution of (22)–(23) for $a = 1.0 - 10^{-6}$ with $aAi(z)$. (From Rosales 1978.)

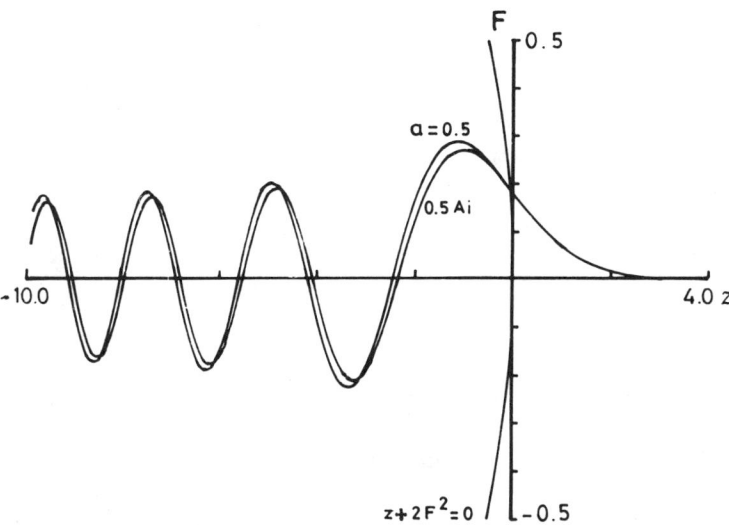

Figure 8.6.4 Comparison of the solution of (22)–(23) for $a = 0.5$ with $aAi(z)$. (From Rosales 1978.)

$z + 2F^2 = 0$, while for $a \leq 1.0175$ it did not. Therefore, it was concluded that

$$a_1 = 1 + O(10^{-9}), \qquad 1.0175 < a_2 < 1.02. \tag{38}$$

These numbers seem to have been confirmed by other investigators (see Hastings and McLeod 1980).

Finally, we note that the solution ψ_n of (25) was written by Rosales (1978) in the form

$$\psi_n = \frac{(-1)^n}{(2\pi)^{2n+1}}$$
$$\times \int_{\Gamma^{2n+1}} \frac{\exp\{i \sum_1^{2n+1}(k_j z + (1/3)k_j^3)\}}{\prod_1^{2n}(k_j + k_{j+1})} dk_1 dk_2 \cdots dk_{2n+1}, \tag{39}$$

where Γ is any path in the upper complex plane going from $k = -\infty$ to $k = \infty$. Since these solutions are motivated by the connection of F with the KdV equation and the form of the solutions of the latter obtained by using inverse scattering techniques, its derivation would fall much beyond the scope of this book (see Rosales 1978 for a verification of the solution).

8.7 THE FOURTH PAINLEVÉ EQUATION

We consider the fourth Painlevé equation

$$2ww'' = w'^2 + 3w^4 + 8zw^3 + 4(z^2 - \alpha)w^2 + 2\beta, \tag{1}$$

where α and β are (constant) parameters. It is obvious that $w(-z) = -w(z)$. As for the first and second Painlevé equations, it is not difficult to verify that (1) has a first-order movable pole with residue $+1$ or -1 with the corresponding Laurent series as

$$w(z) = \frac{1}{z - z_0} - z_0 + \frac{1}{3}(z_0^2 + 2\alpha - 4)(z - z_0) + h(z - z_0)^2 + \cdots \tag{2}$$

or

$$w(z) = -\frac{1}{z - z_0} - z_0 - \frac{1}{3}(z_0^2 + 2\alpha + 4)(z - z_0) + h(z - z_0)^2 + \cdots, \tag{3}$$

where h is an arbitrary constant and the rest of the coefficients in expansions (2) and (3) can be uniquely determined in terms of α, β, z_0, and h. Series (2) and (3) can be shown to be convergent as in the case of the first Painlevé transcendent.

In the following analysis, we follow Lukashevich (1967); the elementary methods used here are applicable to other Painlevé equations as well.

Painlevé Transcendents

We show that any solution of (1) is representable as a ratio of two entire functions,

$$w = \frac{u(z)}{v(z)}. \tag{4}$$

We identify the function v and derive a system of DE for u and v equivalent to (1). From (2) and (3) we have

$$z + w(z) = \frac{a_{-1}}{z - z_0} + (z - z_0)\phi(z, z_0), \tag{5}$$

where a_{-1} is 1 or -1, and $\phi(z, z_0)$ is a holomorphic function in the neighborhood of $z = z_0$. Squaring (5), we have

$$w^2 + 2zw = \frac{1}{(z - z_0)^2} + \Phi(z) - z^2, \tag{6}$$

where $\Phi(z)$ is another holomorphic function. If we change the sign on both sides of (6) and integrate with respect to z twice, we may identify the denominator in (4) as the entire function

$$v = \exp\left\{-\int_{z_0}^{z} d\tau \int_{\tau_0}^{\tau} [w^2(\tau) + 2\tau w(\tau)] d\tau\right\}, \tag{7}$$

provided the path of integration in (7) does not pass through the singular points of $w(z)$. Differentiating (7) twice with respect to z, we get

$$v^2 w^2 + 2v^2 zw = v'^2 - vv''. \tag{8}$$

Solving for $w(z)$, we have

$$w(z) = \frac{-zv + \{z^2 v^2 + v'^2 - vv''\}^{1/2}}{v} = \frac{u}{v}, \text{ say}. \tag{9}$$

This is the form (4) of w if v is chosen according to (7).

If we simply substitute (9) into (1), we get a fourth-order equation in v. Instead, we obtain a simultaneous system of DE for u and v. Actually, the first of this system is simply (8) if we replace w by u/v; the second is obtained from (1), (7), and (9) as

$$(u'v - uv')^2 - (vv'' - v'^2)^2 + 4\alpha u^2 v^2 + 2\beta v^4 - 4uv^2 v' = 0. \tag{10}$$

To show the equivalence of (8) and (10) with (1), we derive the latter from the former. In view of (1), (4) and (7), (10) can be written as

$$w'^2 + 4\alpha w^2 + 2\beta - 4w\left(\frac{v'}{v}\right) = (w^2 + 2zw)^2. \tag{11}$$

Differentiating (11) with respect to z and eliminating v'/v and $(v'/v)'$ with the help of (11) and (8) rewritten as

$$\left(\frac{v'}{v}\right)' = -(w^2 + 2zw), \tag{12}$$

we obtain (1). We obtain another form of (8) and (10) by eliminating u^2 and $2uu'$ from (10) with the help of (8):

$$vv'' - v'^2 + u^2 + 2zuv = 0, \tag{13}$$

$$v'v''' - v''^2 + 2zu'v' - 2uv' + u'^2 + 4\alpha u^2 + 2\beta v^2 = 0. \tag{14}$$

Given nonsingular initial conditions $w(z_0) = w_0$, $w'(z_0) = w'_0$, we have $w_0 = u(z_0)/v(z_0)$ and $w'_0 = (u/v)'|_{z=z_0}$. We can find the solution for u and v from (13) and (14) in the form

$$u(z) = \sum_{j=0}^{\infty} a_j (z - z_0)^j, \quad v(z) = \sum_{j=0}^{\infty} b_j (z - z_0)^j, \tag{15}$$

where the coefficients a_j and b_j can be determined in terms of α, β, z_0, w_0, and w'_0. However, if $v(z_0) = 0$, then by substituting (15) into (13) and (14), we have

$$v(z) = a_0(z - z_0) + (a_1 + z_0)(z - z_0)^2$$
$$+ \frac{1}{6}(3a_1^2 + 4z_0^2 + 4a_1 z_0 - 4\alpha - 2)(z - z_0)^3 + \cdots, \tag{16}$$

where a_0 is either 1 or -1 while the coefficient a_1 remains arbitrary (cf. (5)).

We derive another system from (1) which gives solutions subject to the condition $w(z_0) = 0$. It is obvious from (1) that the solutions with this condition exist only if $w'(z_0) = w'_0$ satisfies the relation

$$(w'_0)^2 + 2\beta = 0. \tag{17}$$

Writing (1) as

$$\frac{w''}{w} = \frac{1}{2}\left(\frac{w'}{w}\right)^2 + \frac{3}{2}w^2 + 4zw + 2(z^2 - \alpha) + \frac{\beta}{w^2}$$

and defining

$$s = -\frac{1}{w}\frac{dw}{dz} + 2 + \frac{1}{w}(-2\beta)^{1/2}, \tag{18}$$

it is possible to form a system in w and s equivalent to (1):

$$\frac{dw}{dz} = 2w - sw + (-2\beta)^{1/2}, \tag{19}$$

Painlevé Transcendents

$$\frac{ds}{dz} = \frac{1}{2}s^2 - \frac{3}{2}w^2 - 2s - 4zw - 2(z^2 - \alpha) + 2. \tag{20}$$

System (19)–(20) has holomorphic right-hand sides. The solution satisfying the conditions $w(z_0) = 0$, $s(z_0) = s_0$ is

$$w(z) = \sum_{j=1}^{\infty} p_j(z - z_0)^j, \qquad s(z) = \sum_{j=0}^{\infty} s_j(z - z_0)^j, \tag{21}$$

where s_0 is arbitrary and $(w_0')^2 + 2\beta = 0$. Subject to this latter condition the function v can be obtained from (7) as

$$v(z) = 1 - \frac{2z_0 w_0'}{3!}(z - z_0)^3 + \frac{4\beta - 2w_0'}{4!}(z - z_0)^4 + \cdots. \tag{22}$$

It is possible to obtain two other systems equivalent to (1) by writing it in a certain fashion. For example, one can write (1) as

$$\frac{d}{dz}(2ww') = 3(w' - w^2 - 2zw)^2 + (w^2 + 2zw)(6w' - 4zw) - 2(2\alpha w^2 - \beta).$$

This form is not particularly suggestive of the equivalent forms. On the other hand, the transformation (42a) of Section 8.3 suggests more directly that we may make the Ansatz

$$\frac{dw}{dz} - w^2 - 2zw = \xi$$

or

$$\frac{dw}{dz} = w^2 + 2zw + \xi, \tag{23}$$

where ξ is the new unknown function. Then (1) can be rewritten in terms of ξ and w as

$$2w\frac{d\xi}{dz} = \xi^2 - 2[2(1+\alpha) + \xi]w^2 + 2\beta. \tag{24}$$

Thus, (23) and (24) are equivalent to (1). A similar argument leads to the system

$$\frac{dw}{dz} = -w^2 - 2zw + \eta, \tag{25}$$

$$2w\frac{d\eta}{dz} = \eta^2 + 2[2(1-\alpha) + \eta]w^2 + 2\beta. \tag{26}$$

If we eliminate ξ from (23)–(24) or η from (25)–(26), we get (1). On the other hand, if we eliminate w from (23)–(24), we obtain the equation

$$D(\xi) \equiv [2z\xi' - \xi'' - 2\beta - 8(1+\alpha)\xi - 5\xi^2](\xi'' + 2z\xi' + \xi^2 + 2\beta)$$

$$+ [8z\xi\xi' + 8z^2\xi^2 + 16\beta z - 4\xi^3 - 8(1+\alpha)\xi^2][2(1+\alpha) + \xi]$$
$$= 0. \tag{27}$$

A similar equation for η may be obtained by eliminating w from (25) and (26). We denote the latter equation also by $D(\eta) = 0$. The relation between the solutions of system (23)–(24) and those of (27) is obvious:

$$\xi(z) = w'(z) - w^2(z) - 2zw(z) \tag{28}$$

and

$$w(z) = -\frac{\xi'' + 2z\xi' + \xi^2 + 2\beta}{4[2(1+\alpha) + \xi]z} - \frac{\xi}{2z}. \tag{29}$$

The latter follows from differentiating (24) and eliminating dw/dz with the help of (23). Equation (29) holds provided

$$\xi + 2(1+\alpha) \neq 0. \tag{30}$$

A similar relationship can be established between $w(z)$ and $\eta(z)$:

$$\eta(z) = w'(z) + w^2(z) + 2zw(z) \tag{31}$$

and

$$w(z) = \frac{-\eta'' + 2z\eta' + \eta^2 + 2\beta}{4[2(1-\alpha) + \eta]z} + \frac{\eta}{2z}, \tag{32}$$

provided

$$2(1-\alpha) + \eta \neq 0. \tag{33}$$

The initial conditions for $w(z)$, namely,

$$w(z_0) = w_0, \qquad w'(z_0) = w'_0, \tag{34}$$

are related to those of $\xi(z)$ by (24) and (28); that is,

$$\xi_0 = w'_0 - w_0^2 - 2z_0 w_0,$$
$$\xi'_0 = \frac{\xi_0^2 - 2[2(1+\alpha) + \xi_0]w_0^2 + 2\beta}{2w_0}, \tag{35}$$

provided

$$\xi_0 + 2(1+\alpha) \neq 0 \tag{36}$$

and $w_0 \neq 0$. If the latter condition is not satisfied, the numerator in (35) must also vanish and we have

$$\xi_0^2 + 2\beta = 0 \quad \text{if } w_0 = 0 \tag{37}$$

Painlevé Transcendents

(cf. (17)). In a similar manner, the initial conditions for $w(z)$ and $\eta(z)$ are related by

$$\eta_0 = w_0' + w_0^2 + 2z_0 w_0,$$

$$\eta_0' = \frac{\eta_0^2 + 2[2(1-\alpha) + \eta_0]w_0^2 + 2\beta}{2w_0}, \quad \text{if } w_0 \neq 0, \tag{38}$$

and

$$\eta_0^2 + 2\beta = 0, \quad \text{if } w_0 = 0. \tag{39}$$

The expansions for ξ about the pole $z = z_0$ are easily obtained from (2) or (3), and (23). Corresponding to (2), we have

$$\xi(z) = -\frac{2}{(z-z_0)^2} + \frac{2}{3}(z_0^2 - \alpha - 1) + 2z_0(z-z_0) + \cdots, \tag{40}$$

while corresponding to (3),

$$\xi(z) = -2(\alpha+1) + 2(z_0 + 2h)(z-z_0) + \cdots, \tag{41}$$

so the latter solution has no pole (cf. the discussion of the second Painlevé equation in Section 8.6).

Similarly, corresponding to (2), the Laurent series expansion for $\eta(z)$ is obtained from (31) as

$$\eta(z) = \frac{2}{(z-z_0)^2} + a_0 + a_1(z-z_0) + \cdots, \tag{42}$$

while the one corresponding to (3) is holomorphic, as may easily be checked.

Now, we consider the special degenerate cases when the equality sign holds in (30) and (33). For the former, $\xi = -2(1+\alpha)$. Substituting this in (24), we obtain the condition

$$\beta + 2(1+\alpha)^2 = 0 \tag{43}$$

on the coefficients α and β in (1). System (23)–(24) now reduces to just the Riccati equation

$$w' = w^2 + 2zw - 2(1+\alpha), \tag{44}$$

all solutions of which are solutions of (1). It is easily seen that the Laurent series solutions of (44) about $z = z_0$ has first term $-1/(z-z_0)$; thus the residue at the pole is -1. Hence, we have shown that, when (43) holds, any solution of (1) which is not rational and which has its initial conditions connected by the relation

$$w_0' = w_0^2 + 2z_0 w_0 - 2(1+\alpha) \tag{45}$$

has an infinite number of poles, the residue at each pole being -1.

For the function $\eta(z)$, analogous results hold. When $\eta = -2(1-\alpha)$, (26) gives

$$\beta + 2(1-\alpha)^2 = 0, \tag{46}$$

while (25) becomes

$$w' = -w^2 - 2zw - 2(1-\alpha). \tag{47}$$

Again, any solution of (1) which is not rational and whose initial conditions satisfy the relation

$$w'_0 = -w_0^2 - 2z_0 w_0 - 2(1-\alpha) \tag{48}$$

has an infinite number of poles, and the residue at each of these is 1.

Now we turn our attention to rational solutions of (1). A necessary condition for the solutions of (1) to be rational is that the point at $z = \infty$ be either a pole or holomorphic. Substituting $z = 1/t$ in (1), we get

$$2t^6 w \frac{d^2w}{dt^2} = t^6 \left(\frac{dw}{dt}\right)^2 - 4t^5 w \frac{dw}{dt} + 3t^2 w^4 + 8tw^3 \\ + 4(1 - \alpha t^2) w^2 + 2\beta t^2. \tag{49}$$

A simple balancing argument shows that the solution of (49) as a Laurent series is

$$w(t) = \frac{a_{-1}}{t} + a_1 t + a_3 t^3 + \cdots, \tag{50}$$

where the coefficients, by substitution into (49), are obtained from the system

$$\begin{aligned} 3a_{-1}^2 + 8a_{-1} + 4 &= 0, \\ (3a_{-1}^2 + 6a_{-1} + 2)a_1 - \alpha a_{-1} &= 0, \ldots. \end{aligned} \tag{51}$$

The first of (51) gives rise to two possibilities:

$$a_{-1} = -2, \quad a_1 = -\alpha, \ldots, \tag{52a}$$

$$a_{-1} = -\frac{2}{3}, \quad a_1 = \alpha, \ldots. \tag{52b}$$

If we assume that the solution of (49) is holomorphic at $t = 0$, we can check that the expansion

$$w(t) = b_0 + b_1 t + b_2 t^2 + \cdots \tag{53}$$

holds, where

$$b_0 = 0, \quad \beta + 2b_1^2 = 0, \quad b_2 = 0, \quad 2b_3 = \alpha b_1 - 2b_1^2, \text{ etc.} \tag{54}$$

Painlevé Transcendents

Expansions (50) and (53), when written in terms of $z = 1/t$, show that if rational solutions exist they must have the form

$$w(z) = \lambda z + \frac{p(z)}{q(z)}, \tag{55}$$

where $\lambda = 0, -2,$ or $-2/3$, and

$$p(z) = \sum_{j=0}^{n-1} c_j z^j, \quad q(z) = \sum_{j=0}^{n} d_j z^j, \quad d_n = 1, \tag{56}$$

$$c_{n-2} = c_{n-4} = c_{n-6} = \cdots = 0, \quad d_{n-1} = d_{n-3} = \cdots = 0. \tag{57}$$

For the simple case $p(z) \equiv 0$, we easily verify that (i) $w = 0$ when $\beta = 0$, (ii) $w = -2z$ when $\alpha = 0, \beta = -2$, (iii) $w = -(2/3)z$ when $\alpha = 0, \beta = -2/9$.

Now, we derive the result which was merely stated in Section 8.3, namely the case when the coefficients satisfy the relation

$$1 + \alpha + \frac{(-2\beta)^{1/2}}{2} \equiv b \neq 0. \tag{58}$$

Writing (42a) of Section 8.3 as

$$\frac{dw}{dz} = a + 2zw + w^2 + 2wv, \tag{59}$$

where (see (17))

$$a^2 + 2\beta = 0, \tag{60}$$

we can rewrite (1) as

$$\frac{dv}{dz} = -b - 2zv - v^2 - 2wv, \tag{61}$$

where $b = a/2 + \alpha + 1$. Another system similar to (59) and (61) can be written as

$$\frac{dw}{dz} = p - 2zw - w^2 - 2wv, \tag{62}$$

$$\frac{dv}{dz} = q + 2zv + v^2 + 2wv, \tag{63}$$

where

$$p^2 + 2\beta = 0, \quad q = \alpha - 1 - \frac{1}{2}p. \tag{64a,b}$$

Systems (59), (61), and (62)–(64) are derived from (1) under constraints (60) and (64a). Now, if we eliminate w from each of them, we obtain the following equations for v:

$$2vv'' = v'^2 + 3v^4 + 8zv^3 + 4(z^2 - \gamma)v^2 + 2\delta, \tag{65}$$

where

$$4\gamma = 2 - 2\alpha + 3a, \qquad 2\delta + b^2 = 0, \tag{66}$$

and

$$2vv'' = v'^2 + 3v^4 + 8zv^3 + 4(z^2 - \epsilon)v^2 + 2\mu, \tag{67}$$

where

$$4\epsilon = -(2 + 2\alpha + 3p), \qquad 2\mu + q^2 = 0. \tag{68}$$

Equations (65) and (67) are of the form (1) and are related to it through (66) and (68). This proves results (41)–(43) of Section 8.3 and more. In several alternative forms of systems we have derived, such as (59), (61), and (62)–(63), the right-hand sides have differed in sign, apart from the difference in parameters such as a, b and p, q. This is because the right-hand side of (1) is quadratic in w', w^2, zw, etc.

Now, we derive some rational solutions. First we note that if $w_0(z)$ is a rational solution of (1) with $\alpha = \alpha_0$, $\beta = \beta_0$, then, from (59),

$$w_1(x) = \frac{w_0' - a_0 - 2zw_0 - w_0^2}{2w_0}, \qquad a_0^2 + 2\beta_0 = 0, \tag{69}$$

is a rational solution of (1) with parametric values

$$\alpha = \alpha_1 = \frac{1}{4}(2 - 2\alpha_0 + 3a_0),$$
$$\beta = \beta_1 = -\frac{1}{2}\left(\frac{1}{2}a_0 + \alpha_0 + 1\right)^2, \tag{70}$$

(see (66)). Similarly, it follows from (62) that

$$w_0^* = -\frac{w_0' - p_0 + 2zw_0 + w_0^2}{2w_0}, \qquad p_0^2 + 2\beta_0 = 0, \tag{71}$$

is a rational solution of (1) with parametric values

$$\alpha = \alpha_1^* = -\frac{1}{4}(2 + 2\alpha_0 + 3p_0),$$
$$\beta = \beta_1^* = -\frac{1}{2}\left(\alpha_0 - 1 - \frac{1}{2}p_0\right)^2, \tag{72}$$

(see (68)). Here asterisks denote not complex conjugates but other distinct solutions with distinct parameters.

To illustrate the construction of rational solutions, we choose the initial solutions of (1) to be $w_0(z) = -2z$ when $\alpha = 0$, $\beta = -2$ so that $a_0^2 = 4$.

Painlevé Transcendents

We choose $a_0 = 2$. The choice $a_0 = -2$ gives $w_1 = 0$. Then we obtain successively the following solutions from (69) and (70):

$$w_1(z) = \frac{1}{z}, \qquad \alpha = 2, \quad \beta = -2;$$

$$w_2 = -2z - \frac{1}{z}, \qquad \alpha = 1, \quad \beta = -8; \tag{73}$$

$$w_3 = \frac{4z}{2z^2 + 1}, \qquad \alpha = 3, \quad \beta = -8;$$

$$w_3 = \frac{4z}{2z^2 + 1}, \qquad \alpha = 3, \quad \beta = -9;$$

$$w_4 = -2z - \frac{4}{2z^2 + 1}, \qquad \alpha = 2, \quad \beta = -18.$$

These solutions indicate that for every integral α, $0 \leq \alpha \leq 3$, and for $\beta = -2(\alpha \pm 1)^2$, a rational solution exists and has one of the following forms:

i. $w_n = S'_n(z)/S_n(z)$, $n = 0, 1, 2$; $\alpha_n = n + 1$, $\beta_n = -2n^2$; S_n is a polynomial of degree n.
ii. $w_{n+1} = -2z - S'_n/S_n$, $n = 0, 1, 2$; $\alpha_n = n$, $\beta_n = -2(n + 1)^2$; S_n is a polynomial of degree n.

We attempt to set up a general formula for generating these solutions. Writing them as

$$w_n = \frac{S'_n}{S_n}, \qquad w_{n+1} = -2z - \frac{S'_n}{S_n} \tag{74}$$

and substituting into (69), rewritten as

$$w_{n+1} = \frac{w'_n - a_0 - 2zw_n - w_n^2}{2w_n}, \tag{75}$$

we obtain an equation for $S_n(z)$:

$$S''_n + 2zS'_n - 2nS_n = 0. \tag{76}$$

Its solution is

$$S_n(z) = z^n + \frac{1!^n}{2}C_2 z^{n-2} + \frac{3!^n}{2^3}C_4 z^{n-4} + \frac{5!^n}{2^6}C_6 z^{n-6} + \cdots. \tag{77}$$

The functions w_n and w_{n+1} are governed respectively by the equations

$$w' = -w^2 - 2zw + 2n, \tag{78}$$

$$w' = w^2 + 2zw - 2(n + 1). \tag{79}$$

This is easily shown by eliminating S_n from (76) with the help of the expressions $w_n = S'_n/S_n$ and $w_{n+1} = -2z - S'_n/S_n$. Equations (78) and (79) are Riccati type.

Thus we have demonstrated that a rational solution exists under the condition $\beta + 2\alpha^2 = 0$, where α is any integer. Moreover, (77) shows that when n is even, $S_n(z)$ has no real root. For n odd, its only real root is $z = 0$.

Now, we summarize the results for the second case, equations (71)–(72). If $w_0 = -2z$ again, then, successively,

$$w_1 = -\frac{1}{z}, \qquad \alpha = -2, \qquad \beta = -2;$$

$$w_2 = -2z + \frac{1}{z}, \qquad \alpha = -1, \qquad \beta = -8;$$

$$w_3 = -\frac{4z}{2z^2 - 1}, \qquad \alpha = -3, \qquad \beta = -8; \qquad (80)$$

$$w_4 = -2z + \frac{4z}{2z^2 - 1}, \qquad \alpha = -2, \qquad \beta = -18.$$

It is evident from (80) that for every negative α-value, such that $-3 \leq \alpha \leq 0$, $\beta = -2(\alpha \pm 1)^2$, a rational solution exists having one of the following forms:

$$w_n = -\frac{S'_n}{S_n}, \quad n = 0, 1, 2, 3; \qquad \alpha_n = -(n+1); \quad \beta_n = -2n^2,$$
$$w_{n+1} = -2z - \frac{S'_n}{S_n}; \quad n = 0, 1, 2, 3; \qquad \alpha_n = -n; \quad \beta_n = -2(n+1)^2, \qquad (81)$$

where S_n is a polynomial of degree n. Thus, writing

$$w_n = -\frac{S'_n}{S_n}, \qquad w_{n+1} = -2z + \frac{S'_n}{S_n}, \qquad (82)$$

we get rational solutions of (1) for the choices $\alpha_n = -(n+1)$, $\beta_n = -2n^2$ and $\alpha_n = -n$, $\beta_n = -2(n+1)^2$. The function $S_n(z)$ is governed by

$$S'' - 2zS' + 2nS = 0. \qquad (83)$$

Equation (83) is just the Hermite equation with polynomial solutions

$$S_n(z) = z^n - \frac{1!}{2}\binom{n}{2} C_2 z^{n-2} + \frac{3!}{2^2}\binom{n}{4} C_4 z^{n-4} - \frac{5!}{2^3}\binom{n}{6} C_6 z^{n-6} + \cdots. \qquad (84)$$

All the roots of $S_n(z)$ are real.

Again by suitable elimination from (82) and (83), we can derive the equations

$$w' = -w^2 - 2zw - 2n,$$

Painlevé Transcendents

$$w' = -w^2 - 2zw - 2(n + 1), \tag{85}$$

satisfied by w_n and w_{n+1}, respectively.

Finally, we remark that because of the availability of the connection formulas (65)–(66) and (67)–(68), it suffices to construct the general solution of (1) for all values of α and β contained in the intervals $[\alpha_0, \alpha_1]$, $[\beta_0, \beta_1]$ and $[\alpha_0, \alpha_1^*]$, $[\beta_0, \beta_1^*]$ in order to get the solution for any admissible values of α and β. Here α_0 and β_0 are any preassigned values for which the general solution of (1) is $w_0(z)$, while α_1, α_1^*, β_1 and β_1^* are determined by relations (70) and (72).

8.8 THE FIFTH PAINLEVÉ EQUATION

The fifth Painlevé equation

$$2z^2 w(w-1)w'' = z^2(3w-1)w'^2 - 2zw(w-1)w' + 2\alpha w^2(w-1)^3 \\ + 2\beta(w-1)^3 + 2\gamma zw^2(w-1) + 2\delta z^2 w^2(w+1), \tag{1}$$

where α, β, γ, and δ are constants, is rather complicated but shows some nice analytic properties. We shall relate some of its solutions to those of P_{III}. First we consider the nature of the solution of (1) in the neighborhood of fixed singularities $z = 0$ and $z = \infty$ (Lukashevich 1968). We assume a Taylor series solution about $z = 0$:

$$w(z) = \sum_{j=0}^{\infty} a_j z^j. \tag{2}$$

Substituting (2) into (1) and equating coefficients of various powers of z, we have

$$(\alpha a_0^2 + \beta)(a_0 - 1) = 0, \tag{3}$$

$$[a_0 - B(a_0 - 1)](a_0 - 1)a_1 = \gamma a_0^2(a_0 - 1), \tag{4}$$

$$2[a_0 - B(a_0 - 1)](a_0 - 1)a_n = P_n(a_0, a_1, \ldots, a_{n-1}), \quad n = 2, 3, \ldots, \tag{5}$$

where

$$B = 5a_0^2 \alpha - 2a_0 \alpha - 3\beta \tag{6}$$

and P_n is a polynomial in a_j, $j \leq n - 1$. It is clear from (3) that two cases arise: (i) $a_0 \neq 1$, (ii) $a_0 = 1$.

Let $a_0 \neq 1$. Then, from (4), either $a_0 \neq 0$ or $a_0 = 0$. If $a_0 \neq 0$, $\alpha a_0^2 + \beta = 0$, and the coefficients a_n, $n > 0$, are determined successively from

$$2[(n^2 + 2\alpha - 2\beta)a_0 + 4\beta](a_0 - 1)a_n = P_n(a_0, a_1, \ldots, a_{n-1}) \\ \equiv \sigma_n(\alpha, \beta, \gamma, \delta). \tag{7}$$

For a_n to be uniquely determined, we require the magnitude of the term in brackets, $(n^2 + 2\alpha - 2\beta)^2 + 16\alpha\beta \neq 0$ (since $a_0^2 = -\beta/\alpha$). Thus, for every root a_0 of $\alpha a_0^2 + \beta = 0$, a unique solution of (1) in the form (2) exists. If, on the other hand, for a certain positive integer n, $(n^2 + 2\alpha - 2\beta)^2 + 16\alpha\beta = 0$, then for the solution to exist the right-hand side of (7), namely $\sigma_n(\alpha, \beta, \gamma, \delta)$, must also vanish. In this case a_n is arbitrary, and there exists a one-parameter family of solutions in the form (2) (see Section 3.7 for a similar discussion for the Briot–Bouquet equation).

Now we consider the subcase $a_0 = 0$. Since $a_0^2\alpha + \beta = 0$, we require $\beta = 0$. In the present case, the coefficients a_j are given by

$$(2\alpha - 1)a_1 = 0, \tag{8}$$

$$2[2\alpha - n^2 + n - 1]a_1 a_n = P_n(a_1, a_2, \ldots, a_{n-1}), \quad n = 2, 3, \ldots. \tag{9}$$

Equations (8) and (9) show that if $\alpha \neq 1/2$, then $a_1 = 0$ and the unique solution holomorphic in the neighborhood of $z = 0$ will be $w \equiv 0$. Suppose $a_0 = 1$. Then equation (5) for $n = 2$ gives $\gamma a_1 + 2\delta = 0$. The higher coefficients a_n, $n > 1$, are

$$[(n-1)^2 a_1 - \gamma]a_n = P_n(a_1, a_2, \ldots, a_{n-1}) \equiv \sigma_n^*(\alpha, \beta, \gamma, \delta), \text{ say.} \tag{10}$$

Therefore, by substituting $-2\delta/\gamma$ for a_1, we find that the coefficient of a_n is nonzero for any n if $2\delta(n-1)^2 + \gamma^2 \neq 0$. In this case, we have a unique solution in the neighborhood of $z = 0$ in the form (2). Again, if $2\delta(n-1)^2 + \gamma^2 = 0$ for some n, then σ_n^* must also vanish for a solution to exist. We again have a single-parameter family of solutions, depending on a_n.

To prove the convergence of series (2), we transform (1) into a Briot–Bouquet system and use the standard results pertaining to the latter. First we write (1) as

$$z\frac{d\xi}{dz} = -h - (a+c)\xi + \frac{1}{4}\eta - \xi^2\eta, \tag{11}$$

$$z\frac{d\eta}{dz} = \gamma z + 2\delta z^2 \xi + (a+c)\eta + \xi\eta^2, \tag{12}$$

where

$$w = \frac{2\xi + 1}{2\xi - 1}, \quad h = \frac{a-c}{2}, \quad a^2 = 2\alpha, \quad c^2 + 2\beta = 0, \tag{13}$$

and

$$\eta = \frac{1}{w}[zw' - c(w-1) - aw(w-1)]. \tag{14}$$

The motivation for (11)–(12) is not quite clear but elimination of η from (11) and (12) and use of (13) yields (1). Also, eliminating η from (11) and

Painlevé Transcendents

(14) and using (13) lead to (1). We assume $a + c \neq 0$ and make the change of variables in (11)–(12) according to

$$\xi = \epsilon_0 + u, \qquad \eta = v, \tag{15}$$

where

$$\epsilon_0(a + c) + h = 0,$$

and obtain

$$z\frac{du}{dz} = -(a+c)u + \left(\frac{1}{4} - \epsilon_0^2\right)v - v(2\epsilon_0 u + u^2), \tag{16}$$

$$z\frac{dv}{dz} = \gamma z + 2\delta\epsilon_0 z^2 + 2\delta z^2 u + (a+c)v + (\epsilon_0 + u)v^2. \tag{17}$$

System (16)–(17) is of Briot–Bouquet type (see Section 3.7) with eigenvalues $\lambda_{1,2} = \pm(a + c)$; in consequence, series (2) converges and $u \to 0$, $v \to 0$ as $z \to 0$.

The case $a + c = 0$ can be handled by introducing the transformation

$$z\xi = u + \epsilon_0, \qquad \eta = z(v + \epsilon_1)$$

and choosing ϵ_0 and ϵ_1 suitably so that we again obtain a Briot–Bouquet system.

To find the conditions for the existence of poles at $z = 0$, assume the form

$$w(z) = Az^{-k}$$

and substitute in (1). It is easy to check that the necessary condition for the existence of a pole at $z = 0$ is $\alpha = 0$; the actual balance of terms gives $2\beta + k^2 = 0$. The positive integer k must satisfy this condition. The constant A remains arbitrary, and the expansion

$$w(z) = \frac{a_{-k}}{z^k} + \frac{a_{(-k-1)}}{z^{k-1}} + \cdots \tag{18}$$

is uniquely determined in terms of the arbitrary constant $a_{-k} = A$. It is clear that in general, however, the point $z = 0$ is an algebraic branch point or a transcendental critical point.

To know the nature of the other fixed point $z = \infty$, we change the independent variable in (1) to $t = 1/z$. We arrive at the equation

$$2t^4 w(w-1)\frac{d^2w}{dt^2} = t^4(3w-1)\left(\frac{dw}{dt}\right)^2 - 2t^3 w(w-1)\frac{dw}{dt}$$
$$+ 2\alpha t^2 w^2(w-1)^3 + 2\beta t^2 (w-1)^3$$
$$+ 2\gamma t w^2(w-1) + 2\delta w^2(w+1). \tag{19}$$

Now, if we seek an expansion about $t = 0$ (corresponding to $z = \infty$), namely,
$$w = b_0 + b_1 t + b_2 t^2 + \cdots,$$
and substitute it into (19), we encounter three possibilities:

$$b_0 = 0, \quad \delta b_1^2 = \beta, \tag{20a}$$

$$b_0 = -1, \quad \delta \neq 0, \quad b_1 = \frac{2\gamma}{\delta}, \tag{20b}$$

$$b_0 = 1, \quad \delta = 0, \quad b_1 = 0. \tag{20c}$$

Thus, the solution is holomorphic at $z = \infty$ under any of these conditions.

To investigate the possibility of a pole at $z = \infty$, we substitute
$$w = B t^{-k}$$
in (19). It is easy to verify by balancing the most dominant terms in (19) that the point $z = \infty$ is a simple pole (i.e., $k = 1$) only if $\alpha\delta \neq 0$ and the residue a_{-1} at the pole is either $\sqrt{-\delta/\alpha}$ or $-\sqrt{-\delta/\alpha}$. The point $z = \infty$ may, in general, be any other critical point (an algebraic branch point or a transcendental critical point).

Now we turn our attention to movable singularities at $z = z_0$, which may be poles, zeros, or unities (i.e., where $w(z_0) = 1$); since the analysis is similar to that for other Painlevé equations, though rather involved, we just summarize the results.

The point $z = z_0 \neq 0$ is a simple pole with expansion

$$w(z) = \frac{a_{-1}}{z - z_0} + h_0 + a_1(z - z_0) + \cdots, \tag{21}$$

where h_0 is an arbitrary constant and $a_{-1} = \pm z_0/\sqrt{2\alpha}$, provided $\alpha \neq 0$. If $\alpha = 0$, the function $w(z)$ has a double pole at $z = z_0$, with the corresponding expansion

$$w(z) = \frac{z_0 a_{-1}}{(z - z_0)^2} + \frac{a_{-1}}{z - z_0} + a_0 + a_1(z - z_0) + \cdots, \tag{22}$$

where $z_0 \neq 0$ and a_{-1} is an arbitrary constant.

Similarly, we can write an expansion of $w(z)$ about $z = z_0$ such that $w(z_0) = 0$:

$$w = a_k(z - z_0)^k + a_{k+1}(z - z_0)^{k+1} + \cdots, \quad k \geq 1, \ a_k z_0 \neq 0. \tag{23}$$

Here two possibilities arise. If $\beta \neq 0$, then $k = 1$ and the coefficient a_1 is a root of the equation $z_0^2 a_1^2 = 2\beta$. If $\beta = 0$, then $k = 2$, and the coefficient a_2 remains arbitrary.

Finally, the solution of (1) with initial condition $w(z_0) = 1$ has the form

$$w = 1 + a_k(z - z_0)^k + a_{k+1}(z - z_0)^{k+1} + \cdots. \tag{24}$$

Painlevé Transcendents

Again, two cases arise. If $\delta \neq 0$, then $k = 1$ and a_1 is a root of $a_1^2 + 2\delta = 0$. If $\delta = 0$, $\gamma \neq 0$, then $k = 2$ and a_2 is determined from the equation $2z_0 a_2 + \gamma = 0$.

Referring to transformations of (1), we note that if $w = w_0(z; \alpha, \beta, \gamma, \delta)$ is a solution of (1), then $w = w_0^{-1}(z; -\beta, -\alpha, -\gamma, \delta)$ is also a solution (see also Section 8.3).

Now we write some systems equivalent to (1) (see also (16)–(17)). First, it is easily seen that (1) is equivalent to

$$w' = tw + \frac{(aw-c)(w-1)}{z}, \tag{25}$$

$$t' = \frac{w+1}{w-1}\left(\frac{t^2}{2} + \delta\right) + \frac{\gamma - t(1-a+c)}{z}, \tag{26}$$

where $\alpha = a^2/2$, $\beta = -c^2/2$. If we assume t to be a constant equal to ϵ, say, and $\delta = -\epsilon^2/2$, $\gamma = \epsilon(1-a+c)$, then it follows from (25) that solutions of

$$zw' = aw^2 + (\epsilon z - a - c)w + c \tag{27}$$

are solutions of (1). Equation (27) is of Riccati type and can be linearized to a standard form by writing

$$w = -\frac{z}{a}\frac{s'}{s}, \quad s = uz^{-p/2}\exp\left(\frac{\epsilon z}{2}\right), \tag{28}$$

namely,

$$z^2 u'' + \left(A^2 z^2 + Abz + \frac{1}{4} - h^2\right)u = 0, \tag{29}$$

where

$$A = \frac{\epsilon i}{2}, \quad b = -pi, \quad 2h = a-c, \quad i^2 + 1 = 0, \quad p = 1 + a + c. \tag{30}$$

Equation (29) is the radial wave equation, which can in certain cases be integrated in a closed form (see Murphy 1960, p. 343). In particular, if $c = n$, where n is zero or a positive integer, (27) has rational solutions. Also, if $a = c = 0$, $\delta\gamma \neq 0$, (27) has an exponential form of solution.

We transform (25)–(26) into another system by replacing w with $(y+1)/(y-1)$:

$$y' = -\frac{t}{2}y^2 + \frac{t}{2} - \frac{(a-c)y + a + c}{z}, \tag{31}$$

$$t' = y\left(\frac{t^2}{2} + \delta\right) + \frac{\gamma}{z} - \frac{t}{z}(1-a+c). \tag{32}$$

In particular, if $\delta = \gamma = 0$, system (31)–(32) can be changed to a special case of P_{III}. For this purpose, we put $\delta = \gamma = 0$, $a - c - 1 = b$, $t = 2Y$ in (31)–(32) and obtain

$$y' = -Yy^2 + Y - \frac{(a+c) + (b+1)y}{z}, \tag{33}$$

$$Y' = yY^2 + \frac{bY}{z}. \tag{34}$$

Eliminating y from (33) and (34), we get

$$Y'' = \frac{Y'^2}{Y} - \frac{Y'}{z} - \frac{a+c}{z}Y^2 + Y^3, \tag{35}$$

which is a special case of P_{III}:

$$w'' = \frac{w'^2}{w} - \frac{w'}{z} + \frac{1}{z}(\alpha w^2 + \beta) + \gamma w^3 + \frac{\delta}{w}. \tag{36}$$

Equation (35) coincides with (36), if, in the latter, $\alpha = -(a+c)$, $\beta = 0$, $\gamma = 1$, and $\delta = 0$.

It is possible to write (1) in the following form which involves a quadrature:

$$\frac{z^2 w'^2}{2w(w-1)^2} - \alpha w + \frac{\beta}{w} + \frac{\gamma z}{w-1} + \frac{\delta z^2 w}{(w-1)^2}$$

$$= \int \left[\frac{\gamma}{w-1} + \frac{2\delta zw}{(w-1)^2} \right] dz. \tag{37}$$

From the previous discussion on the expansion of w about the zeros and poles, of any meromorphic solution of (1), it is not difficult to see that right-hand side of (37) is holomorphic at these points. It is singular only at $w = 1$. By substituting (24), etc., in (37), it can be easily seen that it has a pole of first order with principal part $z_0/(z - z_0)$. Therefore, we may write, using the results of first-order nonlinear ODE (Hille 1969),

$$w = \frac{u}{v}, \tag{38}$$

where

$$v(z) = \exp\left\{ \int_{z_0}^{z} \frac{1}{\tau} \int_{\tau_0}^{\tau} \left[\frac{\gamma}{w-1} + \frac{2\delta tw}{(w-1)^2} \right] dt\, d\tau \right\} \tag{39}$$

is an entire function, provided the paths of integration in (39) do not pass through singular points of w. If we substitute (38) into (1), we get

$$2z^2 u(u-v)(u''v - uv'') = z^2(3u-v)(u'v - v'u)^2$$
$$- 2zu(u-v)(v - 2zv')(u'v - v'u)$$

Painlevé Transcendents

$$+ 2\alpha u^2(u-v)^3 + 2\beta v^2(u-v)^3$$
$$+ 2\gamma zu^2v^2(u-v) + 2\delta z^2u^2v^2(u+v), \qquad (40)$$

while (37) simply becomes

$$z^2(u'v - v'u)^2 = 2zu(u-v)^2 v' + 2\alpha u^2(u-v)^2 - 2\beta v^2(u-v)^2$$
$$- 2\gamma zuv^2(u-v) - 2\delta z^2 u^2 v^2, \qquad (41)$$

where we have written $z(v'/v)$ for the right side of (37). Another variant for u, v equations is obtained by first differentiating (39) twice with respect to z and replacing w by u/v. We obtain

$$z(u-v)^2(vv'' - v^2) + vv'(u-v)^2 = \gamma v^3(u-v) + 2\delta zuv^3. \qquad (42)$$

Now eliminating the terms involving γ from (41) and (42), we get

$$2z^2 u(u-v)^2(vv'' - v'^2) + z^2 v(u'v - v'u)^2$$
$$= 2\delta z^2 u^2 v^3 + 2(\alpha u^2 - \beta v^2) v(u-v)^2. \qquad (43)$$

Equations (42)–(43) constitute an alternative system for u and v.

To discuss the rational solutions, we let

$$w = \frac{u}{v} = \frac{P_n(z)}{P_m(z)} \qquad (44)$$

(where P_n and P_m are polynomials of degree n and m, respectively), and substitute P_n and P_m for u and v in (40)–(41). Then, equating coefficients of highest powers of z and considering the cases $m < n$, $m > n$ and $m = n$, we arrive at the following possibilities:

(i) $n = m + 1$, provided $\delta \neq 0$.
(ii) $m = n + 1$, provided $\beta\delta \neq 0$.
(iii) $n = m$.

Considering case (i) we may state than any rational solution of (1) for $|\gamma| + |\delta| > 0$ has the form

$$w(z) = az + b + \frac{P_{n-1}(z)}{Q_n(z)}, \qquad (45)$$

where a and b are constants; $P_{n-1}(z)$ and $Q_n(z)$ are polynomials of degree $n - 1$ and n, respectively.

The following special simple cases of solutions of (1) may be noted:

1. $w \equiv 0$ if $\beta = 0$.
2. $w \equiv -1$ if $\gamma = 0$, $\alpha + \beta = 0$, $\delta \neq 0$.
3. $w \equiv 1$ if $\delta = 0$.
4. $w = -(2\delta/\gamma)z + 1$ provided $\beta = -1/2$, $4\alpha\delta + \gamma^2 = 0$, $\gamma \neq 0$.

5. $w = az + b$, where $a^2 + 2\delta = 0$, $b^2 + 2\beta = 0$, provided $\alpha = 1/2$, $\gamma^4 + 64\delta^2 + 16\beta^2\delta^2 - 8\beta\gamma^2\delta + 64\gamma^2\beta + 16\gamma^2\delta = 0$.

Now, we show that if $\delta = 0$, $\gamma \neq 0$, then (1) does not have rational solutions. Since $\delta = 0$, $\gamma \neq 0$, we have case (iii), so the rational solution if it exists has the form

$$w_r(z) = \frac{z^n + a_{n-1}z^{n-1} + \cdots + a_0}{z^n + b_{n-1}z^{n-1} + \cdots + b_0}. \tag{46}$$

Now if we substitute $z = 1/t$ in (1), we have (19). We may then write (46) as

$$w_r(t) = 1 + \sum_{j=1}^{\infty} c_j t^j. \tag{47}$$

Substituting (47) into (19) and equating various powers of t leads, for $\delta = 0$, to $c_j = 0$, $j = 1, 2, \ldots$. We thus have $w_r(z) \equiv 1$. This in turn implies that $a_j = b_j$. The case $\beta = 0$ leads to the trivial solution $w \equiv 0$.

A more general form of the rational solution $w = u/v$, where

$$u(z) = P_n(z)\exp g(z), \quad v(z) = Q_m(z)\exp g(z) \tag{48}$$

($g(z)$ being an entire function), leads, on substitution into (42) and (43), to cases (i)–(iii) with some constraints on $g(z)$:

(i)' $n = m + 1$, $g'(z) = \lambda$, where λ is a root of the equation $\lambda^2 + 4\alpha\delta = 0$.
(ii)' $m = n + 1$, $g'(z) = \mu$, where $\mu \neq \gamma$ satisfies the relation $(\mu - \gamma)^2 = 4\beta\delta$.
(iii)' $n = m$, $g'(z) = \nu z + h$, where ν and h are constants and ν satisfies the relation $2\nu + \delta = 0$.

The functions $P_n(z)$ and $Q_m(z)$ must satisfy rather complicated equations obtained by substituting (48) into (42) and (43).

For $\delta \neq 0$, rational solutions of (1) different from those covered by (27) can be shown to exist. We give two examples:

$$w_1(z) = \frac{-\delta z + \gamma}{\delta z + \gamma} \quad \text{with } a = b = \frac{\gamma}{\delta} \text{ and } \alpha = -\beta = \frac{1}{8}, \tag{49}$$

$$w_2(z) = \frac{-\delta\gamma z + 4(\alpha + \beta) + \gamma^2}{\delta\gamma z - 4(\alpha + \beta) + \gamma^2}, \tag{50}$$

where the coefficients α, β, γ, and δ satisfy the relations

$$\begin{aligned}&\alpha a^2 + \beta b^2 = 0, \\ &\gamma(\alpha + \beta)(2\delta + \gamma^2) = 0, \\ &\gamma^2(4\alpha - 4\beta - 1) + 8\delta(\alpha + \beta)^2 = 0.\end{aligned} \tag{51}$$

It is easy to verify that (49) and (50) do not satisfy (27).

Finally, we note that, apart from the types of solutions of (1) discussed above, there may exist rational solutions in the variable z^s where s is not an integer. For instance,

$$w(z) = a\sqrt{z} + 1, \quad \alpha = -\beta = \frac{1}{8}, \quad \delta = 0, \quad \gamma = -\frac{1}{8}a^2, \quad a \neq 0, \quad (52)$$

and

$$w(z) = \frac{1}{a\sqrt{z} + 1}, \quad \beta = -\alpha = \frac{1}{8}, \quad \delta = 0, \quad \gamma = \frac{1}{8}a^2, \quad a \neq 0, \quad (53)$$

are both solutions of (1).

To end the discussion of special solutions of (1), we mention two degenerate cases (Lukashevich 1965).

Case 1. $\alpha = \beta = 0$, $\gamma^2 + 2\delta = 0$. The solution of (1) has the form

$$w = c \exp(\pm\sqrt{-2\delta}z). \quad (54)$$

Case 2. $\delta = \gamma = 0$, α, β arbitrary. Equation (1) has an implicit form of the solution

$$\int \frac{dw}{(w-1)[\alpha w^2 + C_1 w - \beta]^{1/2}} = \pm 2^{1/2}(C_2 + \ln z), \quad (55)$$

where C_1 and C_2 are arbitrary constants. Solution (55) can be easily obtained by noting that, for $\delta = \gamma = 0$, equation (1) can be rendered autonomous if we write $x = \ln z$, say. It can then be integrated by writing $dw/dx = p$, etc.

8.9 THE THIRD PAINLEVÉ EQUATION

We have delayed the discussion of the third Painlevé equation because of its closeness to P_V, which we have discussed in Section 8.8 (see equation (35)). The general theory of even special cases of P_{III} is rather elaborate, so we shall just state some of these results. However, it is possible to give a simple discussion of P_{III} for some special choices of the parameters.

Recall that the special cases $\alpha = \gamma = 0$ or $\delta = \beta = 0$ of the third Painlevé equation

$$zww'' = zw'^2 - ww' + \alpha w^3 + \beta w + \gamma z w^4 + \delta z \quad (1)$$

were discussed in Section 8.2. The solutions were obtained in terms of simple quadratures. Now we assume that either (i) $\gamma = 0$ and $\alpha\delta \neq 0$ or (ii) $\delta = 0$, $\beta\gamma \neq 0$. In either case, equation (1) can be reduced to one involving a single parameter. For case (i), we make the change of variables

$$z = \lambda\tau, \quad w = \mu u, \tag{2}$$

where

$$\lambda = \left(-\frac{1}{\alpha^2\delta}\right)^{1/4}, \quad \mu = \left(-\frac{\delta}{\alpha^2}\right)^{1/4}; \tag{3}$$

then equation (1) transforms to

$$\tau u u'' = \tau u'^2 - u u' + u^3 + \beta_1 u - \tau, \tag{4}$$

where $\beta_1 = \beta(1/\delta)^{1/2}$.

Similarly, for case (ii), we write

$$z = \lambda\tau, \quad w = \frac{\mu}{u} \tag{5}$$

and define

$$\lambda = \left(\frac{1}{\beta^2\gamma}\right)^{1/4}, \quad \mu = -\left(\frac{\gamma}{\beta^2}\right)^{-1/4}. \tag{6}$$

Then (1) becomes (4) with $\beta_1 = \alpha(1/\gamma)^{1/2}$. Thus, for cases (i) and (ii) we may consider the equation

$$zww'' = zw'^2 - ww' + w^3 + \beta w - z \tag{7}$$

in place of (1). This equation can be written as the system

$$\frac{dw}{dz} = a_0 + \frac{a_0 - \beta}{a_0 z} w - \frac{1}{a_0 z} w^2 v, \tag{8}$$

$$\frac{dv}{dz} = -a_0 - \frac{a_0 - \beta}{a_0 z} + \frac{1}{a_0 z} wv^2, \tag{9}$$

where

$$a_0^2 = 1. \tag{10}$$

Elimination of w from (8) and (9) leads to an equation in v:

$$zvv'' = zv'^2 - vv' + v^3 + pv - z, \tag{11}$$

where

$$p = \beta - 2a_0. \tag{12}$$

Equations (7) and (11) are the same except that the coefficient of v in the latter is $\beta - 2a_0$ instead of β. We thus arrive at the result that if $w_0 = w_0(z, \beta_0)$ is a solution of (7) for some fixed $\beta = \beta_0$, then

$$w_1(z, \beta_1) = \frac{(a_0 - \beta_0)w_0 + z - a_0 z w_0'}{w_0^2}, \quad a_0^2 = 1, \tag{13}$$

Painlevé Transcendents

is a solution of (7) for

$$\beta_1 = \beta_0 - 2a_0, \tag{14}$$

The above result implies that if $w_0(z, \beta_0)$ is a rational solution of (7), then so is $w_1(z)$ for β_1 given by (14). Also, if $w_0 = w_0(z, \beta_0)$ is the general solution of (7) for some fixed value $\beta = \beta_0$, then to find the general solution of (7) it suffices to construct it in the strip $[\operatorname{Re} \beta_0, \operatorname{Re} \beta_1]$, where β_0 is any given number and β_1 is given by (14). As an example we may construct rational solutions with the help of (13) and (14). We may verify that $w_0(z) = z^{1/3}$ is a solution of (7) for $\beta_0 = 0$. It easily follows from (13) and (14) that

$$w_1(z) = \frac{\mp 2 + 3z^{2/3}}{3z^{1/3}}, \tag{15}$$

$$w_2(z) = \frac{\mp 24z + 20z^{1/3} + 9z^{5/3}}{(2 \mp 3z^{2/3})^2}$$

are solutions of (7) for $\beta_1 = \pm 2$ and $\beta_1 = \pm 4$, respectively.

Now we consider the transformation which under the condition

$$\alpha(-\delta)^{1/2} + \beta \gamma^{1/2} = 0 \tag{16}$$

changes (1) into an equation involving only one parameter. Note that, unlike cases (i) and (ii), neither γ nor δ is assumed to be zero. If we write

$$w(z) = A\eta(\theta), \qquad z = B^{-1}\theta, \tag{17}$$

where A and B are some constants, and substitute into (1), we get

$$\frac{d^2\eta}{d\theta^2} = \frac{1}{\eta}\left(\frac{d\eta}{d\theta}\right)^2 - \frac{1}{\theta}\frac{d\eta}{d\theta} + \frac{\alpha A}{B}\frac{1}{\theta}\eta^2$$
$$+ \frac{\beta}{AB}\frac{1}{\theta} + \frac{\gamma A^2}{B^2}\eta^3 + \frac{\delta}{A^2 B^2}\frac{1}{\eta}. \tag{18}$$

This equation is of the form (1) with only one parameter ν, that is,

$$zww'' = zw'^2 - ww' + 2\nu(w^3 - w) + zw^4 - z, \tag{19}$$

provided

$$\frac{\alpha A}{B} = \frac{-\beta}{AB} = 2\nu \tag{20}$$

and

$$\frac{\gamma A^2}{B^2} = -\frac{\delta}{A^2 B^2} = 1. \tag{21}$$

Solving for A^2 and B^2 from (21) and substituting into (20), we find that

$$2\nu = \alpha/\gamma^{1/2} = \frac{-\beta}{(-\delta)^{1/2}}. \tag{22}$$

Equation (22) immediately leads to (16). We shall return to the discussion of equation (19) presently.

It is possible to write equation (1) as a system of Riccati equations in three ways:

$$\begin{aligned} w' &= -\frac{aw}{z} - cw^2 + g, \\ g' &= \frac{1}{z}\left[\frac{z}{w}(\delta + g^2) - (1+a)g + \beta\right], \end{aligned} \tag{23}$$

with $\gamma = c^2$ and $\alpha = (a-1)c$, or

$$\begin{aligned} w' &= \frac{aw}{z} + cw^2 + g, \\ c' &= -c^2 w + \gamma w + \frac{\alpha}{z} - \frac{a+1}{z}c, \end{aligned} \tag{24}$$

with $\beta = g(1-a)$ and $\delta = -g^2$, or

$$\begin{aligned} w' &= cw^2 + g, \\ g' &= \frac{g^2}{w} - \frac{g}{z} + \frac{\beta}{z} + \frac{\delta}{w}, \\ c' &= -c^2 w - \frac{c}{z} + \frac{\alpha}{z} + \gamma w. \end{aligned} \tag{25}$$

To prove the equivalence of systems (23)–(25) to (1), we may differentiate the first of these systems and substitute into (1). The Riccati systems (23)–(25) may help our analysis of equation (1).

Actually, it is possible to show by scaling (see the derivation of (7) and (19)) that we may take $\delta = -1$, $\gamma = 1$ in (1) without loss of generality. Then we may write (1) as the system

$$z\frac{dw}{dz} = (\alpha\epsilon - 1)w + zv + \epsilon zw^2, \tag{26a}$$

$$zw\frac{dv}{dz} = \beta w - z + (\alpha\epsilon - 2)wv + zv^2, \quad \epsilon^2 = \frac{1}{\gamma} = 1. \tag{26b}$$

System (26) is the same as (23) if $\delta = -1$ and $\gamma = 1$ and other parameters are suitably defined. Since v, by way of (26a), can be expressed in terms of w and w', it has no movable critical points. Eliminating w from (26),

Painlevé Transcendents

we get an equation for v:

$$v'' = \frac{v}{v^2-1}v'^2 + \frac{v'}{z} + \theta(z,v), \qquad (27)$$

where

$$\theta(z,v) = -\epsilon \frac{v^4}{v^2-1} - \left[\frac{\beta(\alpha\epsilon-2)}{z^2} - 2\epsilon\right]\frac{v^2}{v^2-1}$$
$$- \left[\frac{\beta(\alpha\epsilon-2)}{z^2} + \epsilon\right]\frac{1}{v^2-1} - \frac{\beta^2 + (\alpha\epsilon-2)^2}{z^2}\frac{v}{v^2-1}.$$

Now if we write

$$v(z) = -\frac{\xi(z)+1}{\xi(z)-1}, \qquad z = (2\tau)^{1/2}$$

equation (27) changes to

$$\xi'' = \frac{3\xi-1}{2\xi(\xi-1)}\xi'^2 - \frac{1}{\tau}\xi' + \frac{(\beta-\alpha\epsilon+2)^2}{32\tau^2}\xi(\xi-1)^2$$
$$- \frac{\beta+\alpha\epsilon-2}{32\tau^2}\frac{(\xi-1)^2}{\xi} - \frac{\epsilon}{\tau}\xi, \qquad \epsilon^2 = 1. \qquad (28)$$

Equation (28) is a special case of the Painlevé fifth equation

$$u'' = \frac{3u-1}{2u(u-1)}u'^2 - \frac{u'}{\tau} + \frac{a}{\tau^2}u(u-1)^2 + \frac{b}{\tau^2}\frac{(u-1)^2}{u}$$
$$+ \frac{c}{\tau}u + \frac{du(u+1)}{u-1}, \qquad (29)$$

with

$$a = \frac{1}{32}(\beta-\alpha\epsilon+2)^2, \quad b = -\frac{1}{32}(\beta+\alpha\epsilon-2), \quad c = -\epsilon, \quad d = 0. \qquad (30)$$

We thus prove a result a little different from that quoted in Section 8.3, namely, that if $w = w(z)$ is a solution of equation (1) with some values of α and β and $\gamma = 1$ and $\delta = -1$ such that

$$\frac{dw}{dz} - \epsilon w^2 - \frac{1}{z}(\alpha\epsilon-1)w + 1 \neq 0, \qquad (31)$$

then

$$u(\tau) = 1 - \frac{2}{dw/dz - \epsilon w^2 - (1/z)(\alpha\epsilon-1)w + 1}, \qquad (32)$$
$$2\tau = z^2,$$

is a solution of (29) with a, b, c, and d given by (30). Conversely, we may show that if $u = u(\tau)$ is a solution of (29) for some values of a and b, $c = \pm 1$ and $d = 0$, such that

$$\tau \frac{du}{d\tau} - \sqrt{2a}u^2 + (\sqrt{2a} + \sqrt{-2b})u - \sqrt{-2b} \neq 0, \tag{33}$$

then

$$w(z) = \frac{\sqrt{2\tau}u}{\tau(du/d\tau) - \sqrt{2a}u^2 + (\sqrt{2a} + \sqrt{-2b})u - \sqrt{-2b}}, \tag{34}$$

$$2\tau = z^2,$$

is a solution of (1) for

$$\alpha = 2c(\sqrt{2a} - \sqrt{-2b} - 1), \quad \beta = 2\sqrt{2a} + 2\sqrt{-2b}, \quad \gamma = 1, \quad \delta = -1. \tag{35}$$

Now writing

$$R = \frac{dw}{dz} - \epsilon w^2 - \frac{1}{z}(\alpha\epsilon - 1)w + 1 \neq 0, \tag{36}$$

and substituting $u(\tau)$ from (32) into (34), we arrive at the result that if $w(z)$ is a solution of (1) for some values of α and β, and $\gamma = 1$, $\delta = -1$, then the function

$$w_1(z) = \frac{2zR(R-2)}{2z(dR/dz) - \eta(\beta + \alpha\epsilon - 2)R - 2\sigma(\beta - \alpha\epsilon + 2)} \tag{37}$$

is a solution of (1) for

$$\alpha_1 = \frac{\epsilon}{2}[\eta(\beta + \alpha\epsilon - 2) - \sigma(\beta - \alpha\epsilon + 2) + 4],$$

$$\beta_1 = \frac{\eta}{2}(\beta + \alpha\epsilon - 2) + \frac{\sigma}{2}(\beta - \alpha\epsilon + 2), \quad \gamma_1 = 1, \quad \delta_1 = -1, \tag{38}$$

$$\epsilon^2 = \eta^2 = \sigma^2 = 1.$$

Now we return to a discussion of equation (19). A special case of this equation with $\nu = 0$ was encountered by Wu et al. (1976) in the study of spin-spin correlation functions for the two-dimensional Ising model. We give some of their analysis, relevant to (19). First, they found a local expansion of the solution of (19) with $\nu = 0$; that is, the solution of

$$\frac{d^2w}{dz^2} = \frac{1}{w}\left(\frac{dw}{dz}\right)^2 - \frac{1}{w} + w^3 - \frac{1}{z}\frac{dw}{dz} \tag{39}$$

near $z = 0$. Actually a dominant balance argument with $w = Az^m$ shows that all terms in (39) except w^3 dominate and balance if $m = 1$, but that then $A \to \infty$. This indicates the need for a psi series with both powers and logarithm of z appearing in the double sum. In any case, Wu et al. showed that the solution of (39) can be written as

$$w(z) = \sum_{n=0}^{\infty} z^{4n+1} \rho_n, \tag{40}$$

where

$$\rho_n = \sum_{k=0}^{2n+1} b_{n,k} \Omega^k, \tag{41}$$

and

$$\Omega = \ln\left(\frac{z}{4}\right) + \gamma_E.$$

Here γ_E is the Euler constant. The first few coefficients $b_{n,k}$ are given in Table 8.9.1.

To study the large real z behavior, we note that $w = 1$ is a solution of (39). A perturbation about this point, with $w = 1 + y$, say, where y is small shows that $y = K_0(2z)$, where $K_0(2z)$ is the Bessel function of order zero. Extending the perturbation series, one easily shows that for large real z,

$$w(z) = 1 - \frac{2}{\pi} K_0(2z) + \frac{2}{\pi^2} K_0^2(2z) + O(e^{-6z}). \tag{42}$$

Wu et al. studied a boundary problem for (39) over $(0, \infty)$ with asymptotic behavior (40) and (42) near zero and infinity, respectively. This is the appropriate behavior for the physical problem they discussed. They numerically integrated (39), starting both from small z ($= 0.013$) and large z ($= 10$) values, using expansions (40) and (42). The results for $w(z)$ and $w'(z)$ are shown in Figure 8.9.1. The two solutions agreed in the overlap region ($z \simeq 8$) to 13 decimal places.

The more general equation (19) with $\nu \neq 0$ was treated by McCoy, Tracy, and Wu (1977). First, they noted that $z = 0$ is the only point in the finite z-plane for which a branch point or an essential singularity of (19) can occur. They found that the solution near $z = 0$ has the form

$$w(z) = B(2z)^\sigma \{1 - \nu B^{-1}(1-\sigma)^{-2}(2z)^{1-\sigma} + B\nu(1+\sigma)^{-2}(2z)^{1+\sigma}$$
$$+ \left[\frac{1}{4}\nu^2 B^{-2}(1-\sigma)^{-4} - \frac{1}{16}B^{-2}(1-\sigma)^{-2}\right](2z)^{2-2\sigma}$$
$$+ O(z^2)\}, \tag{43}$$

where $-1 < \text{Re}\,\sigma < 1$, but otherwise σ and B are arbitrary.

In general, the solution behaving according to (43) near $z = 0$ will be unbounded as $z \to \infty$. McCoy, Tracy, and Wu (1977) proved some general

Table 8.9.1 Coefficients $b_{n,k}$ in the Small (Real) z Expansion of $w(z)$ (See Equations (40)–(41))

n	k	$b_{n,k}$
0	0	0
0	1	-1
1	0	-2^{-7}
1	1	2^{-5}
1	2	-2^{-4}
1	3	2^{-4}
2	0	-145×2^{-20}
2	1	145×2^{-17}
2	2	-129×2^{-15}
2	3	121×2^{-14}
2	4	-2^{-7}
2	5	2^{-8}
3	0	$-28195 \times 3^{-5} \times 2^{-25}$
3	1	$25603 \times 3^{-4} \times 2^{-23}$
3	2	$-82729 \times 3^{-3} \times 2^{-24}$
3	3	$9539 \times 3^{-2} \times 2^{-21}$
3	4	-477×2^{-19}
3	5	275×2^{-18}
3	6	-3×2^{-12}
3	7	2^{-12}

Source: Wu et al. (1976).

results for a one-parameter family of solutions of (19) that remain bounded as $z \to \infty$ along the positive real axis. The proofs of the theorems given here are essentially by substitution, but are lengthy and cumbersome. We therefore refer the reader to the original paper.

First it is convenient to change (19) by the transformation

$$\frac{1 - w(z; \nu, \lambda)}{1 + w(z; \nu, \lambda)} = G(t; \nu, \lambda), \qquad (44)$$

$$t = 2z,$$

Painlevé Transcendents

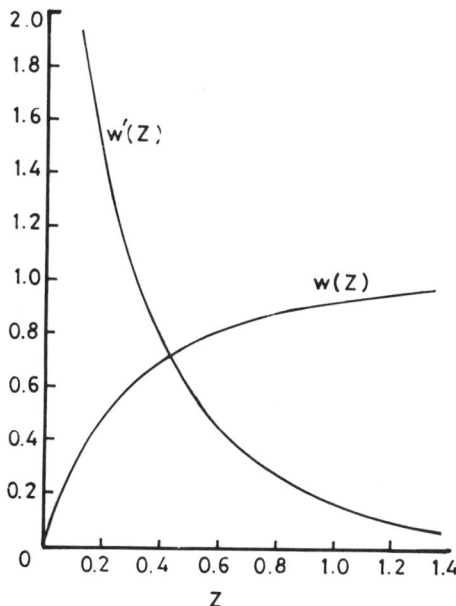

Figure 8.9.1 Third Painlevé function $w(z)$ and its derivative $w'(z)$. (From Wu et al. (1976).)

where $w(z;\nu,\lambda)$ denotes the one-parameter family of solutions with λ as the parameter; G satisfies the equation

$$G'' + \frac{1}{t}G' - \left(1 + \frac{2\nu}{t}\right)G = G''G^2 - 2(G'^2)G$$
$$+ \frac{1}{t}G'G^2 + G^3 - \frac{2\nu}{t}G^3, \qquad (45)$$

where the prime denotes differentiation with respect to t. The left-hand side contains linear terms, while nonlinear ones have been pushed to the right.

THEOREM 8.9.1 The function $w(z;\nu,\lambda)$ satisfies (19), and for sufficiently large positive z and $\mathrm{Re}\,\nu > -1/2$, $w(z;\nu,\lambda)$ has the representation (44) where

$$G(t;\nu,\lambda) = \sum_{n=0}^{\infty} \lambda^{2n+1} g_{2n+1}(t;\nu), \qquad (46)$$

$$g_1(t;\nu) = \int_1^{\infty} dy\, \frac{\exp(-ty)}{(y^2-1)^{1/2}} \left(\frac{y-1}{y+1}\right)^{\nu}, \qquad (47)$$

and, for $n \geq 1$,

$$g_{2n+1}(t;\nu) = (-1)^n \int_1^\infty dy_1 \cdots \int_1^\infty dy_{2n+1} \left[\prod_{j=1}^{2n+1} \frac{\exp(-ty_j)}{(y_j^2-1)^{1/2}} \left(\frac{y_j-1}{y_j+1}\right)^\nu \right]$$
$$\times \left[\prod_{j=1}^{2n} (y_j + y_{j+1})^{-1} \right] \left[\prod_{j=1}^n (y_{2j}^2 - 1) \right]. \tag{48}$$

The parameter λ is subject to the condition $|\lambda| < R(t)$, where $R(t)$ is the radius of convergence of (46) viewed in the complex λ plane. The restriction $\operatorname{Re}\nu > -\frac{1}{2}$ can be lifted by first changing the contour of integration in (47) and (48) to C which begins at infinity and loops around the branch point at $y = 1$.

For the next theorem, we define another function $\psi(t;\nu,\lambda)$ such that

$$w(z;\nu,\lambda) = \exp[-\psi(t;\nu,\lambda)], \qquad t = 2z,$$
and $\tag{49}$
$$\psi(t;\nu,\lambda) \longrightarrow 0 \qquad \text{as } t \longrightarrow \infty.$$

THEOREM 8.9.2 For t sufficiently large and $\operatorname{Re}\nu > -\frac{1}{2}$,

$$\psi(t;\nu,\lambda) = \sum_{n=0}^\infty \lambda^{2n+1} \psi_{2n+1}(t;\nu), \tag{50}$$

where

$$\psi_1(t;\nu) = 2g_1(t;\nu) \tag{51}$$

and

$$\psi_{2n+1}(t;\nu) = \frac{2}{2n+1} \int_1^\infty dy_1 \cdots \int_1^\infty dy_{2n+1} \left[\prod_{j=1}^{2n+1} \frac{\exp(-ty_j)}{y_j + y_{j+1}} \right]$$
$$\times \left[\prod_{j=1}^{2n+1} \left(\frac{y_j-1}{y_j+1}\right)^{\nu-1/2} + \prod_{j=1}^{2n+1} \left(\frac{y_j-1}{y_j+1}\right)^{\nu+1/2} \right] \tag{52}$$

for $n \geq 1$; $y_{2n+2} \equiv y_1$ in (52). The restriction $\operatorname{Re}\nu > -1/2$ can again be lifted by using the contour C noted above in the discussion of (48).

It is clear from the above discussion that, apart from the set of parameters α, β, γ, δ (see equation (30)), for which the third Painlevé equation is connected with the fifth Painlevé equation and for which the analysis of the

Painlevé Transcendents

latter report in Section 8.8 becomes available, there is still scope for carrying out a more general study of Painlevé third equation for arbitrary values of α, β, γ, and δ. Of course, as we pointed out earlier, one may here choose $\gamma = 1$ and $\delta = -1$, with α and β arbitrary, without any loss of generality.

In conclusion we note that our account of Painlevé equations has been rather simple; it relies on methods discussed in the text. More advanced and comprehensive treatment may be found in current literature on differential equations, specifically that appearing in Soviet journals.

9
Applications of the Theory of Nonlinear Ordinary Differential Equations to Solutions of Partial Differential Equations—Some Physical Problems

9.1 INTRODUCTION

We encountered similarity solutions in Chapter 6, which are governed by nonlinear ODE derived via similarity transformations. These are mathematically "degenerate" solutions in which both the independent variables tend either to zero or infinity. On the other hand, these solutions are significant since they form intermediate asymptotics to which a large class of other solutions arising from "appropriate" initial/boundary conditions tend under some limiting processes. The similarity solutions per se represent rather special physical solutions and hold in restricted spatial and temporal domains. This, however, does not detract from their usefulness. Two distinct classes of similarity solutions and their intermediate asymptotic character have been discussed in great detail by Zeldovich and Raizer (1966), Barenblatt (1979), and Sachdev (1987).

We take up the matter of constructing more general, sometimes global, solutions in the form of infinite series (in time) with coefficients which are functions of the similarity variable. The problem "reduces" to the solution of an infinite system of ordinary differential equations with appropriate initial/boundary conditions. The series solution can be constructed with any number of terms with the aid of a computer (Van Dyke 1984). There arise, of course, the questions of the convergence of these series and their

Applications of Nonlinear ODE to PDE

summation. These matters are best discussed with the help of physical problems.

The examples chosen derive mainly from the author's own interests in wave phenomena and gas dynamics. We treat in succession the following types of PDE: a single second-order nonlinear parabolic equation (the nonplanar Burgers equation), a system of two nonlinear hyperbolic equations describing one-dimensional compressible isentropic flows, and a system of three nonlinear hyperbolic equations for the nonisentropic compressible flows. Each problem we consider has an exact solution for plane symmetry. The solutions for nonplanar symmetries are obtained by an infinite series in time with coefficients as functions of the similarity variable. The planar solution provides the motivation and suggests transformations of the variables for subsequent use in the series solution.

We also treat the climb of a bore up a sloping beach, for which a usual self-similar solution does not exist; an intuitive argument, however, leads to the right solution, which is a generalized similarity solution (Barker and Whitham 1980). The analysis of this problem again involves solutions of an infinite systems of ODE with suitable initial conditions. The problem is different from the others discussed in that here we discover its nature as a generalized similarity soltution; we do not treat the intermediate asymptotic nature of this solution. The latter has been discussed numerically by Keller, Levine, and Whitham (1960).

9.2 N-WAVE SOLUTIONS OF NONPLANAR BURGERS EQUATION

The generalized Burgers equation

$$u_t + uu_x + \frac{ju}{2t} = \frac{\delta}{2}u_{xx} \tag{1}$$

describes the propagation of weak nonlinear disturbances which are attenuated, in addition, by a small viscosity and spherical ($j = 2$) or cylindrical ($j = 1$) spreading. The case $j = 0$ in (1) corresponds to the standard Burgers equation with plane symmetry. The variable u denotes excess wave velocity, and x and t are the spatial and time variables, the former in a frame of reference which moves with the undisturbed speed of sound; δ is the coefficient of diffusivity of sound. The derivation of (1) is found in Sachdev (1987). Many initial/boundary value problems may be related to (1), but we restrict ourselves to sawtooth "balanced" initial conditions

$$u(x,t_i) = \begin{cases} x & \text{for } |x| < 1, \\ 0 & \text{for } |x| > 1, \end{cases} \tag{2}$$

which themselves would arise from a certain source, emitting waves which strengthen to shock.

First we discuss the evolution from initial conditions (2) under the Burgers equation

$$u_t + u u_x = \frac{\delta}{2} u_{xx}. \tag{3}$$

If the initial conditions (2) were followed under (3), they would evolve so that the front and tail of the N-wave smoothen in the so-called embryonic shock regime until Taylor shocks form from a balance of viscous and nonlinear terms in (3). The structure of these shocks is found by substituting $X = x - Ut$ in (3) and looking for traveling waves which satisfy constant end conditions at $X = \pm\infty$. In any case, from the Taylor shock regime onward, it becomes possible to find an exact solution of (3) by the Hopf–Cole transformation

$$u = -\delta (\ln \phi)_x. \tag{4}$$

This changes (3), by substitution and integration with respect to x, to the heat equation

$$\phi_t = \frac{\delta}{2} \phi_{xx}. \tag{5}$$

Now, the initial value problem for the heat equation can be solved quite simply. In particular, we seek an even (with respect to x about zero) solution of (5) which, by way of (4), gives an odd solution of (3) vanishing at $x = \pm\infty$, the N-wave we seek. The appropriate solution of (5) is

$$\phi = 1 + \left(\frac{t_0}{t}\right)^{1/2} \exp\left(-\frac{x^2}{2\delta t}\right), \tag{6}$$

where t_0 is a constant. The corresponding N-wave solution of (3) is

$$u = -\delta \frac{\phi_x}{\phi} = \frac{x/t}{1 + (t/t_0)^{1/2} e^{x^2/2\delta t}}. \tag{7}$$

We define the Reynolds number as the area under one of the two (equal) lobes of the N-wave divided by δ,

$$R = \frac{A}{\delta} = \frac{1}{\delta} \int_0^\infty u\, dx = \ln \phi(0, t) = \ln\left(1 + \left(\frac{t_0}{t}\right)^{1/2}\right). \tag{8}$$

This expression clearly describes how the wave decays with time, vanishing at $t = \infty$. The solution (7) may be rewritten in terms of $R = R(t)$ as

$$u = \frac{x/t}{1 + e^{x^2/2\delta t}/(e^R - 1)}. \tag{9}$$

Applications of Nonlinear ODE to PDE

This form shows more clearly how the solution appears in the different Reynolds number regimes. It describes uniformly (for all x from $-\infty$ to $+\infty$) the structure of the N-wave for each Reynolds number. For example, when the wave has propagated far away and decayed sufficiently so that $R \ll 1$, solution (9) may well be approximated by

$$u = (e^R - 1)\left(\frac{x}{t}\right) e^{-x^2/2\delta t} = \frac{t_0^{1/2}}{t^{3/2}} x e^{-x^2/2\delta t}. \tag{10}$$

On the other hand, when $R \gg 1$ so that the N-wave is relatively strong during its initial evolution, we may approximate (9) by

$$u \simeq \frac{x}{t}\left\{1 + e^{x^2/2\delta t - R}\right\}^{-1}, \tag{11}$$

which in turn may be approximated by

$$u \sim \begin{cases} \dfrac{x}{t}, & |x| < (2\delta R t)^{1/2}, \\ 0, & |x| > (2\delta R t)^{1/2}. \end{cases} \tag{12}$$

This is the inviscid solution of (3) with $\delta = 0$.

The solution (7) combines the behaviors (10) and (12) in the appropriate time regimes, and is a uniformly valid solution on the whole real line and for all time from the moment of Taylor shock formation onward. The shocks (at the front and tail) are thin initially; they broaden as the wave evolves until they are too thick to be called shocks. Finally, the N-wave is so weak that nonlinear effects are unimportant; it then decays under pure diffusion according to (10).

Now we turn to the more general case with $j = 1, 2$ in (1). The presence of the geometrical term $ju/2t$ makes any Hopf–Cole transformation for exact linearization unlikely. However, there is some information for this equation which we may use to construct a solution analogous to (7) for $j = 0$. Equation (1) with $\delta = 0$ has the exact (inviscid) solution

$$u = \begin{cases} \dfrac{x}{2t} & \text{for } j = 1, \\ \dfrac{x}{t \ln t} & \text{for } j = 2 \end{cases} \tag{13}$$

(cf. (12)), as may be easily verified by substitution. Moreover, there also exists an asymptotic diffusive solution of (1) in the limit $t \to \infty$, namely

$$u = \begin{cases} C_c x t^{-2} \exp\left(-\dfrac{x^2}{2\delta t}\right) & \text{for } j = 1, \tag{14a} \\ C_s x t^{-5/2} \exp\left(-\dfrac{x^2}{2\delta t}\right) & \text{for } j = 2, \tag{14b} \end{cases}$$

when the nonlinear term is unimportant, and the geometric and diffusive terms dominate; here C_c and C_s are constants which may be obtained by numerically solving (1), (2) and matching that solution with (14) when t is large.

Now we wish to use (13) and (14) to write a solution similar to (7) for $j = 1, 2$. First, we write (7) in the form

$$u = (2\delta)^{1/2} \frac{x/(2\delta t)^{1/2}}{t^{1/2}\{1 + t^{1/2}[1/t_0^{1/2} + (x^2/2\delta t)(1/t_0^{1/2}) + (1/2!t_0^{1/2})(x^2/2\delta t)^2 + \cdots]\}} \tag{15}$$

to motivate the solution for the more general case. It is thus convenient to introduce first the variables

$$v = (2\delta)^{-1/2}u, \qquad \xi = \frac{x}{\sqrt{2\delta t}}, \qquad T = t^{1/2}, \tag{16}$$

so that (1) becomes

$$Tv_T + (2Tv - \xi)v_\xi + jv = \frac{1}{2}v_{\xi\xi}. \tag{17}$$

Now we write

$$v = \frac{\xi}{V} \tag{18}$$

to "peel off" the factor $x/\sqrt{t}$ from u and get an equation for the "inverse function" V (cf. (15)). It is even better to choose $\eta = x^2/2\delta t = \xi^2$ and T as independent variables. Thus, (17) changes to

$$V(jV - TV_T) + (2T - V)(V - 2\eta V_\eta) + 3VV_\eta + 2\eta VV_{\eta\eta} - 4\eta V_\eta^2 = 0. \tag{19}$$

Before we consider the cases $j = 1, 2$, we note that, for the Burgers equation with $j = 0$, the solution (7) can be written as

$$u = (2\delta)^{1/2}v = (2\delta)^{1/2} \frac{\xi}{T(1 + aTe^{\xi^2})} \equiv (2\delta)^{1/2} \frac{\xi}{V(\xi, T)}, \tag{20}$$

where $a = t_0^{-1/2}$ and

$$V = T(1 + aTe^{\xi^2}) = T(1 + aTe^\eta)$$
$$= T + aT^2 + aT^2\eta + aT^2\frac{\eta^2}{2!} + aT^2\frac{\eta^3}{3!} + \cdots. \tag{21}$$

Now, we seek solutions of (19) analogous to (21) for $j = 1, 2$:

$$V = f_0(T) + f_1(T)\eta + f_2(T)\frac{\eta^2}{2!} + f_3(T)\frac{\eta^3}{3!} + \cdots \tag{22}$$

# Applications of Nonlinear ODE to PDE

$$= \sum_{i=0}^{\infty} f_i(T) \frac{\eta^i}{i!}.$$

Here the functions $f_i(T)$ should be such that they combine the behaviors (13) and (14) for V after all the transformations and reductions have been made. For the plane case $j = 0$, $f_0(T)$ is simply the sum of the remnant of the inviscid solution, namely T, and of the asymptotic time behavior aT^2; all higher-order $f_i(T)$ just correspond to the asymptotic time factor aT^2.

For $j = 1, 2$, $f_i(T)$ are more complicated. Substituting (22) into (19) and equating the coefficients of various powers of η to zero, we have

$$3f_1 + 2T - Tf_0' + (j-1)f_0 = 0 \qquad (23)$$

$$5f_0 f_2 + 2j f_0 f_1 - f_1^2 - T(f_0 f_1' + f_1 f_0') - 2Tf_1 = 0, \qquad (24)$$

$$7f_0 f_3 + [2(j+1)f_0 - Tf_0' - 3f_1 - 6T]f_2 - Tf_0 f_2' + 2f_1[(j+1)f_1 - Tf_1'] = 0, \ldots \qquad (25)$$

It is easy to check that for $j = 0$ the functions $f_0 = T + aT^2$, $f_1 = f_2 = f_3 = aT^2$ satisfy (23)–(25). Now we consider the cylindrical and spherical cases separately.

Cylindrical Case

Just as for the plane case, what remain for the functions $f_i(T)$, after peeling off, etc., from the inviscid and asymptotic behaviors (13) and (14), are the terms $2T$ and T^3, respectively. Quite plausibly, we may look for solutions of (23)–(25) with $j = 1$ in the form

$$f_0 = 2T + b_2 T^2 + b_3 T^3,$$
$$f_1 = c_1 T + c_2 T^2 + c_3 T^3, \qquad (26)$$
$$f_2 = d_1 T + d_2 T^2 + d_3 T^3, \ldots.$$

Substituting (26) into (23)–(25), we find that $c_1 = d_1 = 0$, $c_2 = 2b_2/3$, $c_3 = b_3$, $b_2 = \pm 3\sqrt{b_3}$, $d_3 = b_3$, $d_2 = (4/15)b_2$, so the functions f_0, f_1, f_2 become known in terms of the constant b_3, which itself may be found by matching the present analytic solution to the numerical solution of (1)–(2). That would also complete the asymptotic behavior (14a). Choosing the positive sign for b_2 to avoid any singularity at a finite time, we have

$$f_0 = 2T + 3\sqrt{b_3} T^2 + b_3 T^3. \qquad (27)$$

All f_i, $i > 1$, can now be found by differentiation and algebraic operations from equations (23)–(25), etc., and hence, in principle, the solution is found analytically. Thus, f_1 and f_2 are given by (26) with c_i and d_i ($i = 1, 2, 3$)

known in terms of b_3. Similarly, f_3 can be found by substituting f_1 and f_2 in (25):

$$f_3 = \frac{(16/105)b_2T^2 + (54/35)b_3T^3 + (27/35)b_2b_3T^4 + b_3^2T^5}{2 + b_2T + b_3T^2} \qquad (28)$$

$$\sim b_3T^3 \quad \text{as } T \longrightarrow \infty.$$

Now the solution for $j = 1$ can be written as

$$u = \frac{x}{t^{1/2}} \frac{1}{f_0(t^{1/2}) + f_1(t^{1/2})\eta + f_2(t^{1/2})(\eta^2/2!) + \cdots} \qquad (29)$$

so that

$$\left. \frac{\partial u}{\partial x} \right|_{x=0} = \frac{1}{t^{1/2}f_0(t^{1/2})}, \qquad (30)$$

where $f_0(t^{1/2})$ is given by (27). To get the rate of decay of the wave, we obtain an ODE for the lobe Reynolds number as defined by (8). This is achieved by integrating (1) for $j = 1$ with respect to x from 0 to ∞ and noting that $u_x(\infty, t) = 0$ and $u_x(0, t) = 1/t^{1/2}f_0(t^{1/2})$. We thus have

$$\frac{dR}{dt} + \frac{R}{2t} = -\frac{1}{2}u_x(0, t)$$

$$= -\frac{1}{2t^{1/2}} \frac{1}{2t^{1/2} + 3b_3^{1/2}t + b_3t^{3/2}}. \qquad (31)$$

Equation (31) can be integrated to yield

$$R = \left(\frac{t_0}{t}\right)^{1/2} + \left(\frac{1}{b_3t}\right)^{1/2} \ln \frac{t^{1/2} + 2/b_3^{1/2}}{t^{1/2} + 1/b_3^{1/2}}, \qquad (32)$$

where t_0 is another constant. Sachdev, Tikekar, and Nair (1986) solved (1)–(2) numerically and found the values of the constants $t_0^{1/2}$ and $b_3^{-1/2}$ by matching to be -4.716 and 26.44, respectively. A comparison of (32) with the numerical solution is shown in Table 9.2.1. The agreement is good from $f \simeq 300$ onward and improves as t increases. Formula (32) holds up to $t = \infty$.

Spherical Symmetry

In the present case, $j = 2$, the analysis was not quite successful. We peel off the factor T from f_i by writing

$$f_0 = TF_0, \quad f_1 = TF_1, \quad f_2 = TF_2, \ldots \qquad (33)$$

Table 9.2.1 N-Wave Reynolds Number for Cylindrical Burgers Equation at Different Times: numerical and analytic according to (32) with $t_0^{1/2} = -4.716$, $b_3^{-1/2} = 26.44$.

t	R (num)	R (Eq. (32))
5.0	6.52471	5.61662
10.0	4.46658	3.84512
20.0	3.01153	2.59951
50.0	1.72225	1.50794
100.0	1.08011	0.97085
150.0	0.80369	0.73931
200.0	0.64448	0.60448
250.0	0.53965	0.51440
350.0	0.40882	0.39940
450.0	0.32989	0.32781
650.0	0.23854	0.24178
850.0	0.18711	0.19106
1050.0	0.15374	0.15726
1350.0	0.12160	0.12320
1650.0	0.10055	0.10024
2000.01	0.08364	0.08133

Source: From Sachdev, Tikekar and Nair (1986).

and introduce the variable $\tau = \ln T$. The system (23)–(24), etc., becomes autonomous. We have

$$3F_1 - \frac{dF_0}{d\tau} + 2 = 0,$$

$$5F_0 F_2 - F_1(2 + F_1) - F_0 \frac{dF_1}{d\tau} - F_1 \frac{dF_0}{d\tau} + 2F_0 F_1 = 0, \ldots \quad (34)$$

We may now find F_i as a power series in τ, that is, in $\ln T$; the functions F_1, F_2, etc., can be expressed in terms of F_0, but the latter itself cannot be found explicitly. It appears that the solution for $j = 2$ is much more intricate, possibly involving a double series in t and $\ln \tau$. This case needs further investigation.

9.3 COLLAPSE OF A SPHERICAL CAVITY IN A PERFECT GAS—THE GLOBAL SOLUTION

This problem has been studied in planar symmetry by Stanyukovich (1960). We studied the asymptotic self-similar solution of this problem in Section 6.11. Here we consider the global solution.

The physical problem is as follows. A uniform stationary gas occupies the region exterior to a sphere of radius R_0 and is allowed to expand freely into complementary vacuous space at $t = 0$. There are two moving boundaries: a sound wave propagates into the quiescent gas with undisturbed characteristic speed c_0, while the gas-vacuum interface proceeds into vacuum. The speed with which this interface moves is not known in advance, but we shall show that it is constant and is equal to $2c_0/(\gamma - 1)$. This is the speed with which the planar gas-vacuum interface moves.

The Eulerian equations of motion for an isentropic gas flow with spherical symmetry are

$$u_t + uu_r + \beta cc_r = 0, \tag{1}$$

$$c_t + uc_r + \beta^{-1}\left(cu_r + \frac{2uc}{r}\right) = 0, \tag{2}$$

where $\beta = 2/(\gamma - 1)$ and $\gamma = c_p/c_v$, the adiabatic heat exponent. System (1)–(2) is in nondimensional form. The spatial variable r and time t have been normalized by the initial cavity radius R_0 and R_0/c_0, respectively. The radial flow velocity u and sound speed c have been rendered nondimensional by c_0.

The boundary condition on the leading characteristic moving into the quiescent gas is

$$u = 0 \quad \text{on the curve } c = 1. \tag{3}$$

The second boundary condition is that

$$c = 0 \tag{4}$$

on the gas-vacuum interface. This condition locates the trajectory of the interface.

In the absence of effects of spherical contraction—i.e., when the term $2\beta^{-1}(uc/r)$ in (2) is absent—we have a simple wave solution of problem (1)–(4):

$$u = \frac{\beta}{1+\beta}(1+\eta), \tag{5}$$

$$c = \frac{1}{1+\beta}(\beta - \eta), \tag{6}$$

Applications of Nonlinear ODE to PDE

where $\eta = r/t$ and $-1 \leq \eta \leq \beta$. The region $\eta \geq \beta$ corresponds to a vacuum, while $\eta \leq -1$ describes the space occupied by the undisturbed gas where $u = 0$ and $c = 1$. It is evident from (5)–(6) that in this planar case the interface $c = 0$ moves with a constant speed $u = \beta$.

We shall describe the solution for the spherical case for which no simple wave solution exists (a more comprehensive treatment of this problem for both spherical and cylindrical geometries may be found in Sachdev, Gupta, and Ahluwalia 1990). The gas-vacuum interface, on initiation of the collapse, accelerates instantaneously to move with a constant speed β. On the interface, $c = 0$, and $dr/dt = u$, so the interface is a particle path. Moreover, the slope of the C_+ and C_- characteristics on the interface is $dr/dt = u$, so the latter is also a genuine member of both families of characteristics. Finally, it can be shown (see Sachdev et al. 1990) that all derivatives of u and c remain finite at all nonzero times before collapse, in particular immediately behind the front so that $\lim cc_r = 0$ as $c \to 0$.

It may appear that the spherical cavity would begin to accelerate as it converges. This is, however, not the case for $\gamma < 5/3$. To see this we write the equation of motion in the mixed Euler–Lagrangian form

$$u_t = -\beta cc_r \tag{7}$$

and note that the acceleration at the interface will be finite near the cavity surface $c \sim 0$ only if c_r tends to infinity there in such a way that $\lim_{c \to 0} cc_r \neq 0$. The wavefront analysis in Sachdev et al. (1990) shows the infinite gradients behind the vacuum front may develop before the collapse time only if $\gamma > 5/3$. We therefore restrict ourselves to the range $1 < \gamma \leq 5/3$ for which the flow behind the interface remains free from infinite gradients.

In the initial stages of collapse, a sound wave propagates outward into the quiescent gas as the gas-vacuum interface moves inward with a constant speed. The initial radius of the cavity is unity. This fact and the similarity form of the planar solution (5)–(6) suggest the introduction of the similarity variable

$$\eta = \frac{r-1}{t} \tag{8}$$

and t as the independent variables. We also introduce the Riemann invariants as they appear in the planar problem, namely,

$$\phi = u + \beta c, \\ \psi = u - \beta c, \tag{9}$$

as the new dependent variables in (1)–(2). We thus have

$$(1 + \eta t)\left[t\phi_t - \eta\phi_\eta + \frac{1}{2\beta}\left\{(\beta + 1)\phi + (\beta - 1)\psi\right\}\phi_\eta\right]$$

$$+ \frac{t}{2\beta}(\phi^2 - \psi^2) = 0, \tag{10}$$

$$(1 + \eta t)\left[t\psi_t - \eta\psi_\eta + \frac{1}{2\beta}\{(\beta-1)\phi + (\beta+1)\psi\}\psi_\eta\right]$$
$$- \frac{t}{2\beta}(\phi^2 - \psi^2) = 0. \tag{11}$$

Boundary condition (3) on the leading characteristic of the rarefaction wave $\eta = 1$ (see (8)) becomes

$$\phi(1,t) = \beta, \qquad \psi(1,t) = -\beta. \tag{12}$$

Following Greenspan and Butler (1960), we introduce t and

$$z = \frac{\eta + \beta}{1 + \beta} \tag{13}$$

as new independent variables in (10)–(11). This change of variable reduces the interval of interest to $0 \le z \le 1$: $z = 0$ corresponds to the gas-vacuum interface, while $z = 1$ represents the leading characteristic of the rarefaction wave. Equations (10)–(11) now become

$$\{1 + ((1 + \beta)z - \beta)t\}\left[t\phi_t + \left\{\frac{1}{2\beta}((\beta+1)\phi + (\beta-1)\psi) + \beta\right.\right.$$
$$\left.\left. - (1+\beta)z\right\}\phi_z\right] + \frac{t}{2\beta}(\phi^2 - \psi^2) = 0, \tag{14}$$

$$\{1 + ((1 + \beta)z - \beta)t\}\left[t\psi_t + \left\{\frac{1}{2\beta}((1+\beta)\psi + (\beta-1)\phi) + \beta\right.\right.$$
$$\left.\left. - (1+\beta)z\right\}\psi_z\right] - \frac{t}{2\beta}(\phi^2 - \psi^2) = 0. \tag{15}$$

The boundary conditions (12) on the leading characteristic $z = 1$ become

$$\phi(1,t) = \beta, \qquad \psi(1,t) = -\beta. \tag{16}$$

We seek the solution of system (14)–(15) subject to boundary conditions (16) in the form

$$\phi(z,t) = \sum_{i=0}^{\infty} f_i(z)t^i, \qquad \psi(z,t) = \sum_{i=0}^{\infty} g_i(z)t^i. \tag{17}$$

We restrict ourselves to the interval $0 < t < 1$. It is easily checked from (13) that it covers the collapse time $t = (\gamma - 1)/2$ for $1 < \gamma < 3$. Substituting (17) into (14)–(15) and equating the coefficients of t^i, $i = 0, 1, 2, \ldots$, to

Applications of Nonlinear ODE to PDE

zero, we have

$$\left[\beta - (1+\beta)z + \frac{1}{2\beta}\{(1+\beta)f_0 + (\beta-1)g_0\}\right]\frac{df_0}{dz} = 0, \quad (18)$$

$$\left[\beta - (1+\beta)z + \frac{1}{2\beta}\{(1+\beta)g_0 + (\beta-1)f_0\}\right]\frac{dg_0}{dz} = 0, \quad (19)$$

and

$$S_i(\{f_n\},\{g_n\}) + ((\beta+1)z - \beta)S_{i-1}(\{f_n\},\{g_n\}) = -T_{i-1}(\{f_n\},\{g_n\}), \quad (20)$$
$$S_i(\{g_n\},\{f_n\}) + ((\beta+1)z - \beta)S_{i-1}(\{g_n\},\{f_n\}) = T_{i-1}(\{f_n\},\{g_n\}), \quad (21)$$

for $i = 0$ and $i \geq 1$, respectively, where

$$S_i(\{f_n\},\{g_n\}) = if_i - \frac{1}{1+\beta}(\beta - (\beta+1)z)\frac{df_i}{dz}$$

$$+ \frac{1}{2\beta(\beta+1)}\sum_{k=0}^{i}\{(\beta+1)f_{i-k} + (\beta-1)g_{i-k}\}\frac{df_k}{dz}, \quad (22)$$

$$T_i(\{f_n\},\{g_n\}) = \frac{1}{2\beta}\sum_{k=0}^{i}(f_{i-k}f_k - g_{i-k}g_k). \quad (23)$$

The relevant boundary conditions for f_i and g_i are

$$f_0(1) = \beta, \quad g_0(1) = -\beta \quad (24)$$

and

$$f_i(1) = 0, \quad g_i(1) = 0, \quad i \geq 1. \quad (25)$$

Equations (20)–(21) can be rewritten as

$$z\frac{dg_i}{dz} - \frac{i(\beta+1)}{2}g_i - \frac{1}{4\beta}\sum_{k=1}^{i-1}\{(\beta+1)g_{i-k} + (\beta-1)f_{i-k}\}\frac{dg_k}{dz}$$

$$= -\frac{1+\beta}{2}\sum_{k=1}^{i}\{\beta - (1+\beta)z\}^{k-1}T_{i-k}, \quad (26)$$

$$(i+1)f_i + \frac{\beta-1}{\beta+1}g_i + \frac{1}{2\beta(\beta+1)}\sum_{k=1}^{i-1}\{(\beta+1)f_{i-k} + (\beta-1)g_{i-k}\}\frac{df_k}{dz}$$

$$= -\sum_{k=1}^{i}\{\beta - (1+\beta)z\}^{k-1}T_{i-k}. \quad (27)$$

The solution of the nonlinear system (18)–(19) subject to (24) is

$$f_0(z) = \beta(2z - 1), \quad g_0 = -\beta. \tag{28}$$

This is the complete solution of the planar problem, namely the escape of a slab of gas into vacuum. The functions $g_i(z)$, $i \geq 1$ are governed by inhomogeneous linear first-order equations (26), subject to the (second) homogeneous boundary conditions in (25). The functions $f_i(z)$, $i \geq 1$, are found from (27) by simple algebraic operations. These higher-order coefficients yield the effect of spherical geometry.

Solving (26)–(27) for $i = 1$ with homogeneous boundary conditions, we have

$$f_1(z) = \begin{cases} 2\beta\left\{z + \dfrac{\beta-2}{3-\beta}z^2 - \dfrac{z^\alpha}{3-\beta}\right\}, & \gamma \neq 3, \dfrac{5}{3}, \\ 3\{2z(1-z) + z^2 \ln z\}, & \gamma = \dfrac{5}{3}; \end{cases} \tag{29}$$

$$g_1(z) = \begin{cases} 2\beta(1+\beta)\left\{\dfrac{z}{1-\beta} - \dfrac{z^2}{3-\beta} - \dfrac{2z^\alpha}{(1-\beta)(3-\beta)}\right\}, & \gamma \neq 3, \dfrac{5}{3}, \\ 12(-z + z^2 - z^2 \ln z), & \gamma = \dfrac{5}{3} \end{cases} \tag{30}$$

where $\alpha = (\gamma + 1)/2(\gamma - 1)$. Note that $f_1(z)$ automatically satisfies the boundary condition $f_1(1) = 0$. We shall see that $f_i(1) = 0$ for all $i \geq 2$.

For $i > 1$, we integrate (26) and use the second boundary condition in (25) to obtain

$$g_i(z) = z^{i\alpha}\left[\int_1^z \left(\sum_{k=1}^{i-1} M_{i-k} - \sum_{k=1}^{i} N_{i-k}\right) z^{-i\alpha-1} dz\right], \tag{31}$$

where

$$M_{i-k} = \frac{1}{4\beta}\{(\beta + 1)g_{i-k} + (\beta - 1)f_{i-k}\}\frac{dg_k}{dz}$$

and

$$N_{i-k} = \frac{\beta + 1}{2}\{\beta - (1 + \beta)z\}^{k-1} T_{i-k}.$$

Once the structure of the functions $g_i(z)$ in (31) and hence of $f_i(z)$ from (27) is clearly established, the convergence of the series solution (17) may be demonstrated. As the solution (29)–(30) for $f_1(z)$ and $g_1(z)$ shows, the general form of $g_i(z)$ and $f_i(z)$ would be quite complicated and would depend crucially on the value of γ. The higher coefficient functions $g_i(z)$

Applications of Nonlinear ODE to PDE

and $f_i(z)$ possess either algebraic branch point singularities or logarithmic branch point singularities or both. The case $\gamma = 1.4$ is exceptional as we shall see, for which $f_i(z)$ and $g_i(z)$ are polynomials.

We now show that, for $1 < \gamma < 3$, $g_i(z)$ (and hence $f_i(z)$) has the general form

$$g_i(z) = P_i(z, z^\alpha, z^k \ln z), \tag{32}$$

where $\alpha = (\gamma + 1)/2(\gamma - 1)$, $k \geq 2$ and P_i is a finite series in its arguments. We first show by an induction argument that, for $1 < \gamma < 3$, the leading power of z in (31) is $\max[i + 1, i\alpha]$. This is evident for $i = 1$ from (29). For $i > 1$, we first prove that z is always a factor of $g_i(z)$ and that there is no term with a negative power of z in the representation (32). This is shown by the following argument. When the integral in (31) is evaluated and the resultant is multiplied by $z^{i\alpha}$, the structure in (31) is preserved. The power $i\alpha$ of z in (31) is always greater than 1, since $\alpha > 1$ for $\gamma < 3$ and $i \geq 1$. Moreover, if there is a term of the form $z^{c_1}(\ln z)^{c_2}$ ($c_1 \geq 2$, $c_2 \geq 1$) in the summands under the integral sign in (31), the integration of $z^{-i\alpha + c_1 - 1}(\ln z)^{c_2}$ and then multiplication by $z^{i\alpha}$ leaves the power c_1 of z intact. It is obvious from the expression $\alpha = (\gamma + 1)/2(\gamma - 1)$ that there is a countable set of values of γ for which either α or $i\alpha$ is an integer. Three distinct forms of P_i may be identified:

1. $\alpha \neq (i + 1)/i$ is fractional; P_i is a multinomial in z and $z^{\alpha-1}$ and no logarithmic terms appear.
2. $\alpha = (i + 1)/i$; i.e., $\gamma = 2 + (i - 1)/(i + 2)$; P_i contains logarithmic terms. It is therefore evident that logarithmic terms appear only for the value $\gamma = 5/3$ in the range $1 < \gamma < 2$. We consider the integral in (31) for $i < i_0$, $i = i_0$, and $i > i_0$ separately, where $i_0 = 2(\gamma - 1)/(3 - \gamma)$. When $i < i_0$, the highest integral power of z in $g_i(z)$ is $i + 1$, while the highest fractional power is $(i-1)\alpha$. When $i = i_0$ so that $i_0\alpha = i_0 + 1$, $g_i(z)$, after evaluation of the integral on the right-hand side of (31), has $z^{i_0+1} \ln z$, $i_0 \geq 1$, as the term with the highest degree of z. It may be verified that even for the third case ($i > i_0$) the minimum power of z which multiplies $\ln z$ is $i_0 + 1$ and is greater than or equal to 2. If the coefficients of different powers of z and $z^{c_1}(\ln z)$, $c_1 > 1$ appearing in the functions $g_i(z)$ are known, the latter can be computed for $0 < z < 1$ and the convergence of series (17) demonstrated. When $i \neq 2(\gamma - 1)/(3 - \gamma)$, $1 < \gamma < 3$, $f_i(z)$ and $g_i(z)$, $i \geq 2$, may be found to have the form of a finite psi series in z and $z^{\alpha-1}$:

$$f_i(z) = \sum_{k=1}^{i+1} a_{k,i} z^k + \sum_{k=1}^{i} \sum_{p=2}^{i+1} a_{ki+p,i} z^{p-1} (z^{\alpha-1})^k \tag{33}$$

Table 9.3.1 Partial Sums u_i for the Series Solutions for u as Given by (9), (17), (37), and (38) for $\gamma = 1.4$ and $\nu = 2$ at $t = 0.099$, 0.149, and 0.199.

			$t = 0.099$			
z	0.0	0.01	0.21	0.51	0.81	1.0
u_0	−5.00000	−4.95000	−3.95000	−1.95000	−0.95000	0.00000
u_1	−5.00000	−4.95240	−3.97381	−1.93705	−0.92638	0.00000
u_2	−5.00000	−4.95313	−3.97746	−1.93378	−0.92535	0.00000
u_3	−5.00000	−4.95339	−3.97805	−1.93331	−0.92534	0.00000
u_4	−5.00000	−4.95348	−3.97813	−1.93325	−0.92534	0.00000
u_5	−5.00000	−4.95352	−3.97813	−1.93324	−0.92534	0.00000
u_6	−5.00000	−4.95353	−3.97812	−1.93324	−0.92534	0.00000
u_7	−5.00000	−4.95354	−3.97812	−1.93324	−0.92534	0.00000
u_8	−5.00000	−4.95354	−3.97812	−1.93324	−0.92534	0.00000
u_9	−5.00000	−4.95354	−3.97812	−1.93324	−0.92534	0.00000
u_{10}	−5.00000	−4.95354	−3.97812	−1.93324	−0.92534	0.00000
u_{11}	−5.00000	−4.95354	−3.97812	−1.93324	−0.92534	0.00000
u_{12}	−5.00000	−4.95354	−3.97812	−1.93324	−0.92534	0.00000
u_{13}	−5.00000	−4.95354	−3.97812	−1.93324	−0.92534	0.00000
u_{14}	−5.00000	−4.95354	−3.97812	−1.93324	−0.92534	0.00000
u_{15}	−5.00000	−4.95354	−3.97812	−1.93324	−0.92534	0.00000
			$t = 0.149$			
z	0.0	0.01	0.21	0.51	0.81	1.0
u_0	−5.00000	−4.95000	−3.95000	−1.95000	−0.95000	0.00000
u_1	−5.00000	−4.95361	−3.98584	−1.93050	−0.91446	0.00000
u_2	−5.00000	−4.95528	−3.99410	−1.92311	−0.91212	0.00000
u_3	−5.00000	−4.95613	−3.99612	−1.92151	−0.91208	0.00000
u_4	−5.00000	−4.95661	−3.99653	−1.92117	−0.91207	0.00000
u_5	−5.00000	−4.95689	−3.99652	−1.92110	−0.91207	0.00000
u_6	−5.00000	−4.95705	−3.99644	−1.92108	−0.91207	0.00000
u_7	−5.00000	−4.95716	−3.99637	−1.92108	−0.91207	0.00000
u_8	−5.00000	−4.95722	−3.99631	−1.92108	−0.91207	0.00000
u_9	−5.00000	−4.95726	−3.99628	−1.92108	−0.91207	0.00000
u_{10}	−5.00000	−4.95729	−3.99626	−1.92108	−0.91207	0.00000
u_{11}	−5.00000	−4.95731	−3.99624	−1.92108	−0.91207	0.00000
u_{12}	−5.00000	−4.95732	−3.99624	−1.92108	−0.91207	0.00000
u_{13}	−5.00000	−4.95733	−3.99623	−1.92108	−0.91207	0.00000
u_{14}	−5.00000	−4.95733	−3.99623	−1.92108	−0.91207	0.00000
u_{15}	−5.00000	−4.95734	−3.99623	−1.92108	−0.91207	0.00000

Table 9.3.1 (continued)

			$t = 0.199$			
z	0.0	0.01	0.21	0.51	0.81	1.0
u_0	−5.00000	−4.95000	−3.95000	−1.95000	−0.95000	0.00000
u_1	−5.00000	−4.95483	−3.99787	−1.92396	−0.90253	0.00000
u_2	−5.00000	−4.95779	−4.01260	−1.91076	−0.89837	0.00000
u_3	−5.00000	−4.95984	−4.01741	−1.90696	−0.89827	0.00000
u_4	−5.00000	−4.96135	−4.01870	−1.90589	−0.98924	0.00000
u_5	−5.00000	−4.96253	−4.01870	−1.90558	−0.98924	0.00000
u_6	−5.00000	−4.96348	−4.01823	−1.90550	−0.98924	0.00000
u_7	−5.00000	−4.96426	−4.01766	−1.90548	−0.98924	0.00000
u_8	−5.00000	−4.96492	−4.01711	−1.90547	−0.98924	0.00000
u_9	−5.00000	−4.96548	−4.01664	−1.90547	−0.98924	0.00000
u_{10}	−5.00000	−4.96596	−4.01625	−1.90547	−0.98924	0.00000
u_{11}	−5.00000	−4.96639	−4.01594	−1.90547	−0.98924	0.00000
u_{12}	−5.00000	−4.96676	−4.01570	−1.90547	−0.98924	0.00000
u_{13}	−5.00000	−4.96710	−4.01551	−1.90547	−0.98924	0.00000
u_{14}	−5.00000	−4.96740	−4.01537	−1.90547	−0.98924	0.00000
u_{15}	−5.00000	−4.96766	−4.01526	−1.90547	−0.98924	0.00000

Table 9.3.2 Partial Sums c_i for the Series Solutions for c as Given by (9), (17), (37), and (38) for $\gamma = 1.4$ and $\nu = 2$ at $t = 0.099$, 0.149, and 0.199.

			$t = 0.099$			
z	0.0	0.01	0.21	0.51	0.81	1.0
c_0	0.00000	0.01000	0.21000	0.61000	0.81000	1.00000
c_1	0.00000	0.01243	0.24416	0.64015	0.82341	1.00000
c_2	0.00000	0.01328	0.25283	0.64283	0.82339	1.00000
c_3	0.00000	0.01361	0.25536	0.64312	0.82340	1.00000
c_4	0.00000	0.01374	0.25615	0.64315	0.82340	1.00000
c_5	0.00000	0.01380	0.25641	0.64316	0.82340	1.00000
c_6	0.00000	0.01383	0.25649	0.64316	0.82340	1.00000
c_7	0.00000	0.01384	0.25652	0.64316	0.82340	1.00000
c_8	0.00000	0.01384	0.25653	0.64316	0.82340	1.00000
c_9	0.00000	0.01385	0.25653	0.64316	0.82340	1.00000
c_{10}	0.00000	0.01385	0.25653	0.64316	0.82340	1.00000
c_{11}	0.00000	0.01385	0.25654	0.64316	0.82340	1.00000
c_{12}	0.00000	0.01385	0.25654	0.64316	0.82340	1.00000
c_{13}	0.00000	0.01385	0.25654	0.64316	0.82340	1.00000
c_{14}	0.00000	0.01385	0.25654	0.64316	0.82340	1.00000
c_{15}	0.00000	0.01385	0.25654	0.64316	0.82340	1.00000

Table 9.3.2 (continued)

			$t = 0.149$			
z	0.0	0.01	0.21	0.51	0.81	1.0
c_0	0.00000	0.01000	0.21000	0.61000	0.81000	1.00000
c_1	0.00000	0.01366	0.26142	0.65537	0.83018	1.00000
c_2	0.00000	0.01557	0.28105	0.66144	0.83014	1.00000
c_3	0.00000	0.01670	0.28968	0.66243	0.83016	1.00000
c_4	0.00000	0.01740	0.29373	0.66261	0.83016	1.00000
c_5	0.00000	0.01785	0.29570	0.66264	0.83016	1.00000
c_6	0.00000	0.01815	0.29669	0.66264	0.83016	1.00000
c_7	0.00000	0.01835	0.29719	0.66265	0.83016	1.00000
c_8	0.00000	0.01848	0.29745	0.66265	0.83016	1.00000
c_9	0.00000	0.01857	0.29758	0.66265	0.83016	1.00000
c_{10}	0.00000	0.01864	0.29765	0.66265	0.83016	1.00000
c_{11}	0.00000	0.01868	0.29769	0.66265	0.83016	1.00000
c_{12}	0.00000	0.01871	0.29771	0.66265	0.83016	1.00000
c_{13}	0.00000	0.01873	0.29772	0.66265	0.83016	1.00000
c_{14}	0.00000	0.01875	0.29773	0.66265	0.83016	1.00000
c_{15}	0.00000	0.01876	0.29773	0.66265	0.83016	1.00000

			$t = 0.199$			
z	0.0	0.01	0.21	0.51	0.81	1.0
c_0	0.00000	0.01000	0.21000	0.61000	0.81000	1.00000
c_1	0.00000	0.01489	0.27867	0.67060	0.83695	1.00000
c_2	0.00000	0.01830	0.31370	0.68143	0.83688	1.00000
c_3	0.00000	0.02098	0.33425	0.68379	0.83692	1.00000
c_4	0.00000	0.02321	0.34713	0.68434	0.83693	1.00000
c_5	0.00000	0.02512	0.35552	0.68447	0.83693	1.00000
c_6	0.00000	0.02681	0.36112	0.68450	0.83693	1.00000
c_7	0.00000	0.02832	0.36492	0.68451	0.83693	1.00000
c_8	0.00000	0.02969	0.36753	0.68451	0.83693	1.00000
c_9	0.00000	0.03094	0.36934	0.68452	0.83693	1.00000
c_{10}	0.00000	0.03210	0.37061	0.68452	0.83693	1.00000
c_{11}	0.00000	0.03317	0.37150	0.68452	0.83693	1.00000
c_{12}	0.00000	0.03416	0.37213	0.68452	0.83693	1.00000
c_{13}	0.00000	0.03510	0.37258	0.68452	0.83693	1.00000
c_{14}	0.00000	0.03598	0.37289	0.68452	0.83693	1.00000
c_{15}	0.00000	0.03680	0.37312	0.68452	0.83693	1.00000

$$g_i(z) = \sum_{k=1}^{i+1} b_{k,i} z^k + \sum_{k=1}^{i} \sum_{p=2}^{i+1} b_{ki+p,i} z^{p-1} (z^{\alpha-1})^k. \tag{34}$$

For the special value $\gamma = 5/3$, $f_i(z)$, and $g_i(z)$ for $i \geq 2$ may be shown to have the form

$$f_i(z) = \sum_{k=1}^{i+1} \sum_{j=0}^{k-1} a_{kj,i} z^k (\ln z)^j + \sum_{k=i+2}^{2i} \sum_{j=0}^{2i-k} a_{kj,i} z^k (\ln z)^j, \tag{35}$$

$$g_i(z) = \sum_{k=1}^{i+1} \sum_{j=0}^{k-1} b_{kj,i} z^k (\ln z)^j + \sum_{k=i+2}^{2i} \sum_{j=0}^{2i-k} b_{kj,i} z^k (\ln z)^j. \tag{36}$$

3. For the important case $\gamma = 7/5$ (belonging to case $\alpha \neq \frac{i+1}{i}$ an integer), $f_i(z)$ and $g_i(z)$ simplify and have the polynomial form

$$f_i(z) = \sum_{k=1}^{2i} a_{k,i} z^k, \tag{37}$$

$$g_i(z) = \sum_{k=1}^{2i} b_{k,i} z^k. \tag{38}$$

The rapid numerical convergence of the series solution for u and c as given by (9), (17), (37), and (38) is demonstrated in Tables 9.3.1–9.3.2. Similar results for $\gamma = 1.45, 5/3$, for which (33)–(34) and (35)–(36), respectively, have to be used, may be found in Sachdev et al. (1990).

9.4 THE CONVERGING SHOCK WAVE FROM A SPHERICAL OR CYLINDRICAL PISTON

The asymptotic (close to center or axis) collapse of a converging shock wave is a classical problem in gas dynamics (see Zeldovich and Raizer 1966), often referred to as Guderley's problem and illustrates the self-similar solutions of the second kind for which the dimensional analysis or group properties of the PDE do not fully determine the self-similar form of the problem. The problem requires a global solution of an eigenvalue problem (Barenblatt 1979). The exponent in the definition of the similarity variable turns out, in general, to be an irrational number. Guderley found that, for $\gamma = c_p/c_v = 1.4$, this exponent α in the similarity variable $\xi = rt^{-\alpha}$ was 0.717 for the spherical shock. This value was subsequently refined by several investigators.

Our concern here is not with the local behavior of the solution described by the similarity solution. Instead, we describe a global problem

(Van Dyke and Guttmann 1982) arising out of the motion of a spherical or cylindrical piston that collapses with constant inward speed. This motion is such that the base solution for small time is just that describing the flow produced by impulsive motion of a plane piston. The piston motion is assumed to be so great that the shock generated is of infinite strength. This motion is rather contrived, but it can be shown that several other "similar" piston motions asymptotically lead to Guderley's self-similar solution near the axis (center). The present piston motion lends itself to a fairly precise and convenient global analysis. More importantly, it illustrates the method of computer-extended series (Van Dyke 1984), which delegates the mounting arithmetic to a computer, which in a few minutes calculates 40 terms of the expansion in powers of time. The radius of convergence considerations show that the 40-term expansion describes the whole field accurately up to the instant the shock wave collapses onto the axis (center).

We envisage that a spherical or cylindrical container of initial radius R_0 encloses a perfect gas at rest with uniform density ρ_0 and adiabatic constant γ. At time $t = 0$ the container contracts with a very large velocity V, emitting ahead of it a shock wave with radius $R(t)$. The problem is to find its propagation law $R = R(t)$. The equations of continuity, momentum, and energy for cylindrical and spherical symmetry, with $j = 1, 2$, respectively, are

$$\rho_t + (\rho v)_r + \frac{j\rho v}{r} = 0, \tag{1}$$

$$v_t + v v_r + \frac{1}{\rho} p_r = 0, \tag{2}$$

$$(p\rho^{-\gamma})_t + v(p\rho^{-\gamma})_r = 0. \tag{3}$$

Here ρ and p are density and pressure, and v is the outward (radial) particle velocity. The Rankine–Hugoniot relations across an infinitely strong shock connect the conditions immediately behind the shock with those ahead:

$$v = \frac{2}{\gamma + 1}\dot{R}, \tag{4a}$$

$$\rho = \frac{\gamma + 1}{\gamma - 1}\rho_0, \tag{4b}$$

$$p = \frac{2}{\gamma + 1}\rho_0 \dot{R}^2. \tag{4c}$$

The boundary condition at the piston requires that the particle speed there be equal to the piston speed so that

$$v = -V \quad \text{at } r = R_0 - Vt. \tag{5}$$

Applications of Nonlinear ODE to PDE

A shift in the origin of r is convenient; the coordinate $x = R_0 - r$ measures the distance inward from the initial position of the piston. The velocity v is thus changed into $-u$. Then the flow in the (x, t) diagram is shown in Figure 9.4.1. The strong shock hypothesis leads, for small time, to a constant flow between a uniformly moving shock with speed $(\gamma + 1)V/2$ and the piston; this flow has the particle speed $u = V$, density $[(\gamma + 1)/(\gamma - 1)]\rho_0$, and pressure $2\rho_0 V^2/(\gamma + 1)$.

A "similarity variable" (cf. Section 9.3)

$$\xi = \frac{2}{\gamma - 1}\left(\frac{x}{Vt} - 1\right) \tag{6}$$

is introduced so that the piston position is given by $\xi = 0$, while the basic position of the shock corresponds to $\xi = 1$; i.e., $x = [(\gamma + 1)/2]Vt$ (cf. (4a)). All the variables are rendered dimensionless by referring length to R_0, speed to V, density to ρ_0, pressure to $\rho_0 V^2$, and time to R_0/V. The governing PDE (1)–(3), in terms of the independent variables t and ξ, become

$$\left[1 - \left(1 + \frac{\gamma - 1}{2}\xi\right)t\right]\left[\rho u_\xi + \left(u - 1 - \frac{1}{2}(\gamma - 1)\xi\right)\rho_\xi + \frac{1}{2}(\gamma - 1)t\rho_t\right]$$

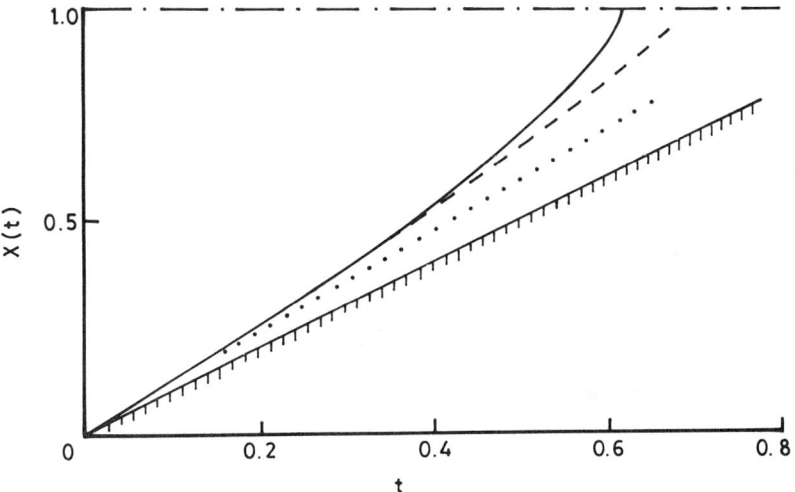

Figure 9.4.1 History of converging shock wave in (x, t) plane for spherical piston with $\gamma = 7/5$: path of piston, ////////, one-term (planar) approximation to shock wave, ---, three-term approximation (22); ———, full solution. (From Van Dyke and Guttmann 1982.)

$$= \frac{1}{2}(\gamma - 1)jt\rho u, \tag{7}$$

$$\rho\left(u - 1 - \frac{1}{2}(\gamma - 1)\xi\right)u_\xi + \frac{1}{2}(\gamma - 1)t\rho u_t + p_\xi = 0, \tag{8}$$

$$\left(u - 1 - \frac{1}{2}(\gamma - 1)\xi\right)(\rho p_\xi - \gamma p \rho_\xi) + \frac{1}{2}(\gamma - 1)t(\rho p_t - \gamma p \rho_t) = 0. \tag{9}$$

The boundary conditions (4) and (5) at the shock and piston, respectively, become

$$u = \frac{2}{\gamma + 1}\dot{X}, \quad \rho = \frac{\gamma + 1}{\gamma - 1}, \quad p = \frac{2}{\gamma + 1}\dot{X}^2$$

$$\text{at } \xi = \frac{2}{\gamma - 1}\left[\frac{X(t)}{t} - 1\right] \tag{10}$$

and

$$u = 1 \quad \text{at } \xi = 0. \tag{11}$$

The solution between the shock and the piston may be assumed to be analytic in time. We may assume that the (unknown) shock trajectory is described by

$$X(t) = \sum_{n=1}^{\infty} X_n t^n. \tag{12}$$

The other flow variables may be expanded as

$$u = \sum_{n=1}^{\infty} U_n(\xi)t^{n-1}, \quad \rho = \sum_{n=1}^{\infty} R_n(\xi)t^{n-1}, \quad p = \sum_{n=1}^{\infty} P_n(\xi)t^{n-1}, \tag{13}$$

where the base flow or solution is given by the uniform flow conditions behind the shock, namely,

$$U_1 = 1, \quad R_1 = \frac{\gamma + 1}{\gamma - 1}, \quad P_1 = \frac{1}{2}(\gamma + 1), \quad X_1 = \frac{1}{2}(\gamma + 1). \tag{14}$$

The functions U_i, R_i, P_i, $i > 1$, can be obtained from the solution of the linear ODE obtained by substituting (13) into (7)–(9) and equating to zero the coefficients of different powers of t. For the second approximation, the ODE are

$$\frac{\gamma + 1}{\gamma - 1} U_2' - \frac{1}{2}(\gamma - 1)\xi R_2' + \frac{1}{2}(\gamma - 1)R_2 = \frac{1}{2}(\gamma + 1)j, \tag{15}$$

$$-\xi U_2' + U_2 + \frac{2}{\gamma + 1}P_2' = 0, \tag{16}$$

Applications of Nonlinear ODE to PDE

$$\xi\left(P_2' - \frac{1}{2}\gamma(\gamma-1)R_2'\right) - \left(P_2 - \frac{1}{2}\gamma(\gamma-1)R_2\right) = 0. \tag{17}$$

The boundary condition (11) at the piston simply requires that $U_n(0) = 0$ for all $n > 1$. The shock conditions, on the other hand, need some care. At the shock we write

$$\xi = 1 - \left[\frac{\gamma+1}{\gamma-1} - \frac{2}{\gamma-1}\frac{X(t)}{t}\right], \tag{18}$$

where $X(t)$ is given by (12). We expand the left sides of (10) about $\xi = 1$ and write (12) for $X(t)$ on the right sides. Equating like powers of t on both sides, we obtain $U_n(1)$, $R_n(1)$, $P_n(1)$, etc. Thus, for $n = 2$,

$$U_2(1) = \frac{4}{\gamma+1}X_2, \quad R_2(1) = 0, \quad P_2(1) = 4X_2. \tag{19}$$

The solution of (15)–(17) with boundary conditions (19) and $U_2(0) = 0$ is

$$U_2 = \frac{\gamma(\gamma-1)}{2(2\gamma-1)}j\xi, \quad R_2 = \frac{\gamma+1}{2\gamma-1}j(1-\xi),$$

$$P_2 = \frac{\gamma(\gamma+1)(\gamma-1)}{2(2\gamma-1)}j, \quad X_2 = \frac{\gamma(\gamma+1)(\gamma-1)}{8(2\gamma-1)}j. \tag{20}$$

This solution suggests that we may seek solutions for U_n, R_n, and P_n in the form of polynomials in ξ of degree $n - 1$:

$$U_n(\xi) = \sum_{k=2}^{n} U_{nk}\xi^{k-1}, \quad R_n(\xi) = \sum_{k=1}^{n} R_{nk}\xi^{k-1},$$

$$P_n(\xi) = \sum_{k=1}^{n} P_{nk}\xi^{k-1}. \tag{21}$$

We replace $U_n(\xi)$, $R_n(\xi)$, and $P_n(\xi)$ in (13) by (21).

The shock conditions (10) are also expanded about $\xi = 1$. We may then obtain, by equating like powers of ξ and t, etc., the coefficients U_{nk}, R_{nk}, P_{nk}, etc. This requires, for each approximation of order n, the solution of a system of $3n$ linear algebraic equations whose right-hand sides (inhomogeneous terms) depend on the previously determined approximations. The coefficients X_i in the expansion for the shock trajectory are also obtained in the process. Thus, Van Dyke and Guttmann (1982) find that to the third approximation the shock trajectory is given by

$$X(t) = \frac{1}{2}(\gamma+1)t + \frac{\gamma(\gamma+1)(\gamma-1)}{8(2\gamma-1)}jt^2 + \frac{(\gamma+1)(\gamma-1)}{48(7\gamma-5)}$$

$$\times \left[(\gamma + 1)(3\gamma + 1)j + \frac{\gamma(13\gamma^3 - 21\gamma^2 + 13\gamma - 1)}{(2\gamma - 1)^2} j^2 \right] t^3 + \cdots. \quad (22)$$

The trajectory is drawn in Figure 9.4.1 for a spherical shock ($j = 2$) propagating into a gas with $\gamma = 7/5$.

Van Dyke and Guttmann extended the above series solution by writing a computer program to calculate the general term. The details of their computations may be found in their original paper. We give in Table 9.4.1 the end results—the 40 coefficients in the series for the shock trajectory (12) for different γ-values for both spherical and cylindrical geometries. Since all the coefficients in the series are positive, the singularity (if any) of the shock trajectory lies on the positive t-axis. Moreover, the coefficients increase steadily in magnitude, implying that the radius of convergence must be less than unity. This must be the case since the piston itself would reach the axis (center) with unit velocity at $t = 1$.

The coefficients grow faster for the spherical than for the cylindrical case, indicating that the focusing is more intense for the former. They also increase faster for larger γ, in conformity with the Newtonian theory of hypersonic flow according to which the shock hugs the piston in the limit γ tending to 1.

In order to estimate the radius of convergence of the series (12), one uses an approach due to Domb and Sykes (1957). If the series in the neighborhood of the nearest singularity has the form

$$X(t) = \sum_{n=0}^{\infty} X_n t^n \sim A_1 \left(1 - \frac{t}{t_c}\right)^{\alpha_1} \quad \text{as } t \longrightarrow t_c, \quad (23)$$

then

$$\frac{X_n}{X_{n-1}} \sim \frac{1}{t_c} \left(1 - \frac{1 + \alpha_1}{n}\right) \quad \text{as } n \longrightarrow \infty. \quad (24)$$

Figure 9.4.2 shows $1/n$ versus X_n/X_{n-1} for the spherical piston problem with $\gamma = 7/5$. A linear fit with $\alpha_1 = 0.717$, as given by Guderley, gives $1/t_c = 1.61$ (i.e., $t_c = 0.62$) to graphical accuracy. A more accurate fit by a polynomial in $1/n$ gave a value of $1/t_c$ as 1.609021, which agrees with Guderley's result to three significant figures.

To verify that the nearest singularity corresponds to collapse of the spherical shock wave to the center, equation (12) was solved for t_0 such that $X(t_0) = 1$. This value, for $\gamma = 1.4$, was found to be 0.62149604. Similar figures were computed for $\gamma = 5/3, 3$ for the spherical piston and for $\gamma = 1.4$ for the cylindrical piston.

Table 9.4.1 Coefficients X_n in Series (12) for Shock Wave

n	Spherical $\gamma = 7/5$	Spherical $\gamma = 5/3$	Spherical $\gamma = 3$	Cylindrical $\gamma = 7/5$
1	1.200000000000	1.333333333333	2.000000000000	1.200000000000
2	0.186666666667	0.317460317460	1.200000000000	0.0933333333333
3	0.188345679012	0.330964978584	1.833333333333	0.0730864197531
4	0.172851981806	0.351087328915	3.40035087719	0.0577257959714
5	0.172147226896	0.428702976041	7.24262900585	0.0497185254748
6	0.195748089820	0.581262688522	16.7325356185	0.0473867537972
7	0.239592510180	0.833416073327	40.8212062145	0.0487020337051
8	0.303219524757	1.24182040572	103.538798073	0.0525457596193
9	0.394337922617	1.90667020627	270.351164204	0.0586078973893
10	0.525663995528	2.99573095341	721.973134446	0.0670385267585
11	0.714271423746	4.79335492559	1962.93555769	0.0782473038694
12	0.985060389731	7.78505460535	5415.71134591	0.0928536362648
13	1.37561449412	12.8028036868	15125.3041521	0.111712634840
14	1.94193338406	21.2785506061	42681.0787588	0.135973603154
15	2.76700088699	35.6880234991	121509.247882	0.167161791445
16	3.97437632751	60.3290517468	348589.799633	0.207289223128
17	5.74887230925	102.690500277	1006783.95686	0.259004706586
18	8.36757126135	175.866803349	2925043.77126	0.325795437783
19	12.2467590407	302.827404305	8543150.61409	0.412256433109
20	18.0133655273	523.983733067	25069946.9513	0.524449933038
21	26.6137638895	910.630204719	73881275.4824	0.670384889712
22	39.4795522908	1588.86850668	218567708.399	0.860657162621
23	58.7805913213	2782.27435391	648869068.945	1.10930506416
24	87.8118838110	4888.12883923	1932484742.18	1.43495376450
25	131.585889835	8613.85327622	5772286224.84	1.86234753012
26	197.740487538	15221.5900368	17288250591.8	2.42440315613
27	297.932522944	26967.3254176	51908194965.1	3.16496440419
28	449.979223858	47890.4406560	156214990411	4.14250004943
29	681.152558004	85235.2928220	471129305758	5.43507307638
30	1033.25274612	152014.101220	1.42371888519 12	7.14702352524
31	1570.42985951	271633.152889	4.31039960943 12	9.41796318002
32	2391.25395142	486253.291668	1.30727810033 13	12.4348912563
33	3647.35908070	871915.946437	3.97126124722 13	16.4485262673
34	5572.26775610	1565938.77695	1.20824798683 14	21.7953372274
35	8525.99348990	2816584.06448	3.68138936143 14	28.9272839088
36	13064.1157676	5073195.16260	1.12320678078 15	38.4519908447
37	20044.8405790	9149924.83352	3.43136709434 15	51.1870510605
38	30795.0631275	16523403.6091	1.04955212172 16	68.2334756229
39	47368.1399675	29874379.4238	3.21397456855 16	91.0751000913
40	72944.3025390	54074091.6579	9.85273540521 16	121.713200768

(From Van Dyke and Guttmann (1982).)

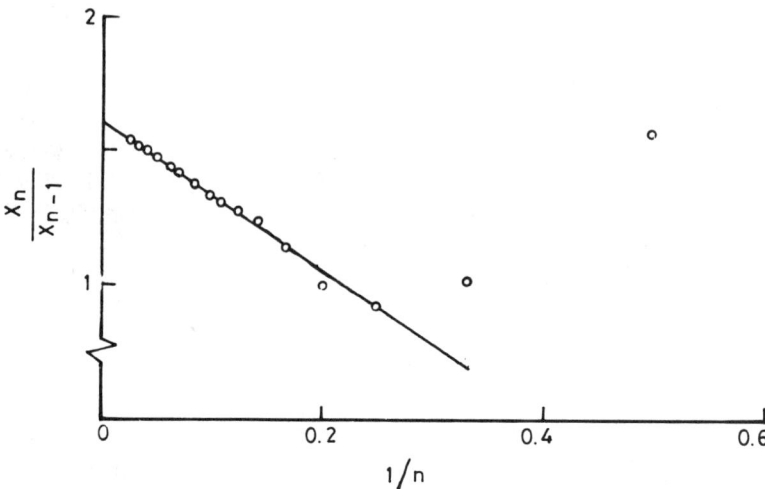

Figure 9.4.2 Graphical ratio test of Domb and Sykes for series (12) for position of shock wave: ———, $1.61(1 - 1.717/n)$ (From Van Dyke and Guttmann 1982.)

A series equivalent to (12) in the form

$$R(\tau) = \sum_{i=1}^{\infty} \frac{A_i}{1 + \alpha_i \tau} \qquad (25)$$

with $\tau = \ln(1 - t/t_c)^{-1}$ was constructed using a Padé approximation (see Bender and Orszag 1978). This was done with a view to extract the exponents α_i and amplitude coefficients A_i in the asymptotic form of Guderley's solution from the present global solution. For $\gamma = 7/5$, for the spherical piston, $\alpha_1 = 0.71717450$, $A_1 = 0.981706$. Some other values are shown in Table 9.4.2. The value of α_1 thus calculated is in excellent agreement with that obtained from the precise numerical solution of the basic PDE and the boundary conditions by Lazarus and Richtmyer (1977). Actually it was found that the three-term expansion (22) itself gives a very accurate (with error less than 1/2%) description of the converging shock wave over its entire course.

There are some questions regarding the intermediate asymptotic nature of Guderley's solution. If we alter the piston motion to one with constant acceleration instead of constant speed, the time t_c for the shock to reach the axis (center) and the amplitudes A_1 in Guderley's local expansion (23)

Applications of Nonlinear ODE to PDE

Table 9.4.2 Exponents and Amplitudes in Guderley's Local Expansion (see equation (25)).

Geometry	γ	α_1	α_2	α_3	A_1	A_2	A_3
Spherical	$\frac{7}{5}$	0.7171745	2.045	3.4	0.981706	0.0140	0.007
Spherical	$\frac{5}{3}$	0.6883768	1.885	3.1	0.989732	0.0055	0.006
Spherical	3	0.636411	1.638	2.5	1.016952	−0.0244	0.01
Cylindrical	$\frac{7}{5}$	0.835324	2.033	3	0.983865	0.0133	0.01

Source: From Van Dyke and Guttmann (1982).

would change, but the similarity exponent α_1 would not change. There is some uncertainty regarding the uniqueness of this exponent for different γ for real flows. The situation is not entirely clear; there are conflicting conjectures and claims in this regard both in Soviet and Western literature (see Van Dyke and Guttmann 1982 for details).

The reflection of the shock from the axis is not covered by this analysis.

Now we summarize some related work. The scheme of the infinite series solution for this problem was introduced by Lee (1968) for a cylindrical piston that moves with a speed proportional to a power of time. Later, Bach and Lee (1969) carried out the same scheme to four terms for spherical and cylindrical waves, arising from the instantaneous deposition of energy at a finite radius. The zeroth approximation here is simply the self-similar solution for a strong planar blast. Nakamura (1983) adopted essentially the same scheme as that of Van Dyke and Guttmann (1982) for cylindrical and spherical piston motions with the difference that the piston velocity was assumed to be quadratic in time. The first three terms of the series were used to determine the starting conditions for the numerical solution. The numerical method of characteristics was employed, and the transition of the non-self-similar motion of the shock wave to its self-similar asymptotic regime was analyzed.

A similar approach to obtain the initial conditions for a general piston motion with the trajectory

$$x(t) = \xi_1 t + \xi_2 t^2 + \cdots + \xi_n t^n, \qquad \xi_1 > 1, \tag{26}$$

was adopted by Kozmanov (1977). The shock wave was assumed to have the trajectory

$$x = c_1 t + c_2 t^2 + \cdots + c_n t^n. \tag{27}$$

The problem was reduced to the determination of $c_1, \ldots, c_n$ in conformity with the piston motion (26), and the flow between the shock wave and the piston was thus found. The series form for the required flow was taken to be

$$u = \sum_{k=0}^{\infty} u_k(t)\phi^k(x,t), \quad \rho = \sum_{k=0}^{\infty} \rho_k(t)\phi^k(x,t),$$

$$S = \sum_{k=0}^{\infty} S_k(t)\phi^k(x,t), \tag{28}$$

where

$$\phi(x,t) = x - c_1 t - c_2 t^2 - \cdots - c_n t^n. \tag{29}$$

This variable is similar to ξ introduced earlier; here $\phi(x,t) = 0$ is the shock trajectory and not a characteristic. Kozmanov considered plane, cylindrical, and spherical piston motions. In particular, he considered the constant acceleration case, $x = \xi_1 t + \xi_2 t^2$, for the planar piston (see also Nakamura 1983). A special choice $\xi_1 = 10$, $\xi_2 = 5$ led to the shock trajectory $X = 15.132t + 4.241t^2$. A comparison with the numerical solution showed that the numerical shock trajectory differed from its analytic description by less than 0.1% for $t < 0.3$.

9.5 SOLUTIONS OF THE SHALLOW-WATER EQUATIONS REPRESENTING GRAVITY-CURRENT RELEASES

Gravity currents consist of fluid of one density flowing under the influence of gravity into fluid of another density. They occur in the atmosphere and in the ocean. They also find use in pollution studies, say, when an industrial storage tank containing a heavier-than-air gas suddenly bursts and releases its contents into the atmosphere. We discuss the theory of the spreading rate of gravity currents owing to the release of a fixed volume of fluid, due to Grundy and Rottman (1985) (see references in this paper for other experimental and theoretical work on the subject).

To fix the ideas, one may visualize gravity currents produced in plane and axisymmetric geometries by the release of fixed volumes of salt water in channels filled with fresh water. The experiments by Rottman and Simpson (1984) showed that initially the rate of advance of the gravity current is a strong function of the release conditions, but, asymptotically, if the current does not become so thin that viscous effects become comparable with the inertia of the current, the rate of advance of the current approaches the self-similar solutions of the depth-averaged shallow-water equations. These experimental and theoretical results suggest the intermediate asymptotic nature of the similarity solutions with respect to a class of initial conditions, and this is what was demonstrated both analytically and numerically by Grundy and Rottman (1985).

By way of analysis, they constructed a large-time asymptotic expansion with the similarity solution as the leading term. This expansion also serves to test the stability of the similarity solution with respect to symmetric disturbances. The higher-order terms are obtained from the solution of a linear eigenvalue problem, possessing an infinite discrete spectrum of eigenvalues $\{\lambda_i\}$. These eigenvalues are shown to have $\text{Re}(\lambda_i) < 0$ for all λ_i, thus confirming stability and also determining the rate at which the similarity solution is approached. The numerical study for the shallow-water equations with "dam-break" initial conditions both for plane and axisymmetric problems leads to the conclusion that a large "class" of initial conditions finally merge into the similarity solution.

Referring to Figure 9.5.1, we imagine a finite volume of fluid with density ρ resting on the horizontal bottom boundary of another fluid with slightly lower density ρ_a. The two fluids are incompressible and miscible so that the surface tension effects are insignificant. Any mixing between them, however, is considered negligible. At $t = 0$, the heavier fluid is released from rest. We wish to study its motion under the influence of gravity.

If it is assumed that the thickness of the heavy fluid is small in comparison with its length and with the depth of the surrounding fluid and if the effects of viscosity are considered unimportant, then the motion of the heavy fluid is described approximately by the shallow-water equations:

$$h_t + uh_x + hu_x + n\frac{uh}{x} = 0, \tag{1}$$

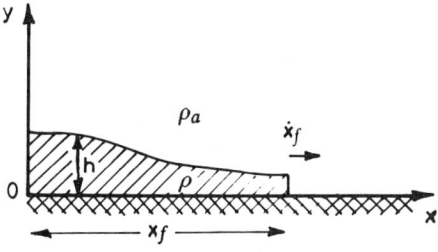

Figure 9.5.1 Schematic illustration of a heavy fluid with density ρ spreading at the base of a lighter fluid of density ρ_a. The position of the front of the spreading current is denoted by $x_f(t)$. (From Grundy and Rottman 1985.)

$$u_t + uu_x + g'h_x = 0. \tag{2}$$

The various quantities are shown in Figure 9.5.1; $h(x,t)$ is the thickness of the heavy fluid, $u(x,t)$ is the depth-averaged horizontal fluid speed, and $g' = g(\rho - \rho_a)/\rho_a$ is the reduced acceleration due to gravity. The densities ρ and ρ_a are both constant, and a Boussinesq approximation is implicit in the definition of g' so that ρ_a replaces ρ is the denominator of the ratio. The spatial variable x denotes the horizontal distance in the plane flow and the radial distance in symmetric flow. t represents the time, taken to be zero at the moment of release of the heavy fluid. The index n is zero for the plane symmetric flow and unity for the axisymmetric flow.

The boundary conditions on the flow are

$$u(0,t) = 0, \tag{3}$$

$$u(x_f,t) = \dot{x}_f, \tag{4}$$

$$\beta^2 g'h(x_f,t) = \dot{x}_f^2, \tag{5}$$

where $x_f(t)$ represents the position of the front of the heavy fluid and $\dot{x}_f$ is its speed. Conditions (3) and (4) imply that there is no inflow at either $x = 0$, the point of symmetry, or at the front $x = x_f$. Boundary condition (5) represents the balance between the buoyancy force $g'(\rho - \rho_a)h_f^2$ driving the current and the drag on the front due to the surrounding fluid, proportional to $\rho_a h_f \dot{x}_f^2$. These forces nearly balance if the acceleration of the front is small. β is a function of the constant of proportionality and has been found to be about unity when $(\rho - \rho_a)/\rho_a$ is small. Conditions (4) and (5) are analogous to shock conditions in gas dynamics. Condition (5) is necessary for current flows since the vertical acceleration at the front is not small and the shallow-water equations do not hold there.

Another important condition here arises from the release of a fixed volume of the heavy fluid. This is derived from (1) by multiplying throughout by x^n, writing it as

$$x^n h_t + nuhx^{n-1} + x^n \frac{\partial}{\partial x}(uh) = 0, \tag{6}$$

and integrating with respect to x from 0 to $x_f(t)$. Using the front conditions (3) and (4), we obtain

$$\int_0^{x_f(t)} h(x,t)(2\pi x)^n \, dx = Q, \tag{7}$$

where Q is the volume of the heavy fluid (volume per unit width for plane flow). Thus, we find that the volume of the heavy fluid is conserved.

In addition to boundary conditions (3)–(5), we need to specify initial conditions. For the physical problem under discussion, initial conditions, as in a "dam-break" problem, are

$$h(x,0) = \begin{cases} h_0(x), & 0 \le x \le x_0, \\ 0, & x_0 < x, \end{cases} \tag{8}$$

$$u(x,0) = 0, \tag{9}$$

where

$$x_0 = x_f(0). \tag{10}$$

The function $h_0(x)$ may, in particular, be taken to be constant, representing the "top-hat" condition.

First we observe that g' can be absorbed in h in equations (1)–(2), so we may deal with the variable $g'h$. We also note that the shallow-water equations (1)–(2) are equivalent to isentropic gas-dynamic equations with polytropic relation $p \propto \rho^2$ (see Courant and Friedrichs 1948, p. 34). The self-similar solutions referred to as quasi-simple waves for (1)–(2) with $n = 2$, the spherically symmetric case, have been discussed in great detail by Courant and Friedrichs (1948), pp. 416–429.

In the present problem, there are five dimensional parameters and variables: t, $g'Q$, x, x_0, and β. There are just two independent dimensions, length and time, so that the problem can be expressed in terms of three dimensionless parameters. These may be chosen to be

$$\xi = x(g'Q)^{-1/(3+n)} t^{-2/(3+n)}, \tag{11}$$

$$\tau = t(g'Q)^{1/2} x_0^{-(3+n)/2}, \tag{12}$$

and β. Here ξ may be interpreted as a dimensionless space variable and τ as a dimensionless time variable scaled by a time characteristic of the initial conditions. Therefore, in general, we may express the dependent variables as

$$\begin{aligned} g'h(x,t) &= (g'Q)^{2/(3+n)} t^{-2(1+n)/(3+n)} H(\xi, \tau, \beta); \\ u(x,t) &= (g'Q)^{1/(3+n)} t^{-(1+n)/(3+n)} U(\xi, \tau, \beta); \\ x_f(t) &= (g'Q)^{1/(3+n)} t^{2/(3+n)} A(\tau, \beta); \end{aligned} \tag{13}$$

where H, U and A are dimensionless functions. Now, the self-similar solutions are defined as degenerate solutions of the form (13) in the limit $\tau \to 0$ or $\tau \to \infty$ (see Barenblatt 1979, Sachdev 1987). It is not a priori clear that the solutions of the form (13) in the limit $\tau \to \infty$ and satisfying conditions (3)–(5) exist, but this is the limiting process we are interested in; if these solutions exist, we wish to show, in addition, that they form intermediate asymptotics to which the solutions of the initial value problems for a wide class of initial conditions converge in the limit $\tau \to \infty$. As we mentioned, the case of the self-similar form of (13) where H, U and A are independent of τ has been discussed by Courant and Friedrichs (1948) and Sedov (1982). It turns out that the powers of t in (13) correspond exactly to those relating to the case of strong explosion for various geometries (see Sedov 1982, p. 233). However, the similarity solutions for the two problems are quite different. In any case, the relevant solution for the present problem is one in which $A(\tau)$ is constant and the "velocity" U is a linear function of $\eta = \xi/A_0$:

$$A(\tau) = A_0, \tag{14}$$

$$H \equiv H_0 = \frac{(n+1)A_0^2}{(n+3)^2}(\eta^2 + B), \tag{15}$$

$$U \equiv U_0 = \frac{2A_0\eta}{n+3}, \tag{16}$$

with

$$B = \frac{4}{(n+1)\beta^2} - 1. \tag{17}$$

The form (13) with A, H, and U given by (14)–(17) constitutes for $0 < \beta < 2/(n+1)^{1/2}$ the unique similarity solution to (1)–(2) and (3)–(5) with A_0 determined by the mass-invariance condition (7):

$$A_0 = \left\{ \frac{(n+1)(n+3)^3\beta^2}{(2\pi)^n[4(n+3) - 2(n+1)\beta^2]} \right\}^{1/(n+3)}. \tag{18}$$

The above solution exists only when $0 < \beta < 2/(n+1)^{1/2}$. For $\beta \geq 2/(n+1)^{1/2}$, a similarity solution satisfying the boundary and mass invariance conditions does not exist.

Now we enquire how these similarity solutions are approached as $\tau \to \infty$. For this purpose we assume the expansions

$$H(\eta, \tau) = H_0(\eta) + \sum_{j=1}^{\infty} \tau^{\mu_j} H_j(\eta), \tag{19}$$

$$U(\eta, \tau) = U_0(\eta) + \sum_{j=1}^{\infty} \tau^{\mu_j} U_j(\eta), \tag{20}$$

$$A(\tau) = A_0 \left\{ 1 + \sum_{j=1}^{\infty} \tau^{\mu_j} A_j \right\}, \tag{21}$$

where the leading terms H_0, U_0 and A_0 describe the similarity solution (see (14)–(18)). The power law perturbations in (19)–(21) are one possible form: we investigate whether this is the appropriate form for the present

Applications of Nonlinear ODE to PDE

problem. A necessary condition for expansions (19)–(21) to converge is that $\text{Re}(\mu_j) < 0$. The actual substitution in PDE (1)–(2) and analysis of the resulting ODE may indicate the need for including logarithmic terms when the power of τ in an interaction term equals one of the μ_j. However, we find the expansions (19)–(21) adequate and the powers μ_j, which we rename γ_j, are found by solving eigenvalue problems for linear ODE.

Now we formally substitute (13), and (19)–(20) into (1) and (2). Equating the powers of τ^{γ_j}, etc., leads to a linear second-order equation for $H_j(\eta)$. This results from the elimination of U_j in favor of H_j'. The DE for H_j is

$$[(\eta^2 + B)\eta^n H_j']' + \frac{3+n}{1+n}[\gamma_j(n-1) - \gamma_j^2(n+3)]\eta^n H_j = 0, \tag{22}$$

where the prime denotes differentiation with respect to η. The functions $U_j(\eta)$ and H_j are related by

$$U_j(\eta) = -\frac{(n+3)H_j'(\eta)}{A_0[\gamma_j(n+3) - (n-1)]}. \tag{23}$$

Now we substitute (13) and (19)–(21) into the boundary conditions. To order τ^{γ_j}, (3) and (23) give

$$H_j'(0) = 0, \tag{24}$$

while (4) leads to

$$A_j = -\frac{(n+3)^2 H_j'(1)}{A_0^2[(n+3)\gamma_j - (n-1)](n+3)\gamma_j}. \tag{25}$$

Finally, (5) yields

$$\left[2(n+3)\gamma_j + 4 - (n+1)\beta^2\right]H_j'(1)$$
$$- \frac{1}{2}(n+3)\beta^2\gamma_j[(n-1) - \gamma_j(n+3)]\right]H_j(1) = 0. \tag{26}$$

The system of DE and boundary conditions (22)–(26) is linear and homogeneous; therefore the solution will determine the eigenvalues γ_j together with H_j, U_j, and A_j to within an arbitrary constant factor.

The following argument shows that $\text{Re}(\gamma_j) < 0$. H_j and H_j' are complex functions of γ_j. To obtain the result $\text{Re}(\gamma_j) < 0$, we first take the complex conjugate of equation (22) and multiply it by

$$\frac{1}{2}\gamma_j(n+3)\beta^2[(n-1) - \gamma_j(n+3)]\overline{H_j}.$$

The resulting equation is then integrated from $\eta = 0$ to $\eta = 1$. Integrating by parts the term involving $[(\eta^2 + B)\eta^n \overline{H_j'}]'H_j$ and using (24) and (26), we get a quadratic in γ_j:

$$\gamma_j^2 + D(\gamma_j)\gamma_j + E(\gamma_j) = 0, \tag{27}$$

where the coefficients

$$D(\gamma_j) = \frac{16|H_j'(1)|^2 + \beta^2(1-n^2)\int_0^1(\eta^2+B)\eta^n|H_j'|^2\,d\eta}{(1+n)(3+n)\beta^n\int_0^1(\eta^2+B)\eta^n|H_j'|^2\,d\eta}, \tag{28}$$

$$E(\gamma_j)$$
$$= \frac{8[4-\beta^2(n+1)] + 2(n+1)\beta^2|\gamma_j(n-1) - \gamma_j^2(n+3)|^2 \int_0^1 \eta^n|H_j|^2\,d\eta}{(1+n)(3+n)^2\beta^4\int_0^1(\eta^2+B)\eta^n|H_j'|^2\,d\eta}$$

are real and positive for any complex γ_j (for $n = 0$ or 1). Hence we conclude that $\mathrm{Re}(\gamma_j) < 0$ and the similarity solutions are stable to linear perturbations.

The linear second-order DE (22) can be easily solved in terms of the hypergeometric functions, and the relevant solution satisfying (24) is

$$H_j = K_j F\left(a, b; \frac{1}{2}(n+1); -\frac{\eta^2}{B}\right), \tag{29}$$

where

$$a + b = \frac{1}{2}(n+1), \qquad ab = \frac{\gamma_j(n+3)[(n-1) - \gamma_j(n+3)]}{4(n+1)}. \tag{30}$$

Here K_j are arbitrary constants, and $F(a, b; c; z)$ is the hypergeometric function. The equation for the eigenvalues is obtained by substituting the solution (29) into the boundary condition (26) at $\eta = 1$. The argument of the resulting hypergeometric functions $-1/B = \{1 - 4/(n+1)\beta^2\}^{-1}$ is transformed to $(1/4)\beta^2(n+1)$ for convenience in computing the series form of the hypergeometric function with its radius of convergence $0 < \beta^2 < 4/(n+1)$. Thus, the equation for the eigenvalues (wherein $F'(a, b; c; z)$ is changed to an undifferentiated form by suitable recurrence relation) is

$$F\left(a, a; \frac{1}{2}(n+1); \beta^2\frac{n+1}{4}\right) + \left[\frac{2\gamma_j(n+3) + 4 - \beta^2(n+1)}{2(n+1)}\right]$$
$$\times F\left(a+1, a; \frac{1}{2}(n+3); \frac{\beta^2(n+1)}{4}\right) = 0. \tag{31}$$

The equations for eigenvalues, which occur in complex conjugate pairs, must be solved numerically. The associated eigenfunctions are given by (29), where the constants K_j remain arbitrary and are to be determined in some way from the initial conditions. Tables 9.5.1 and 9.5.2 give the first 11 eigenvalues (when conjugates are included) for $n = 0$ and $\beta = 0.6(0.2)1.4$,

Applications of Nonlinear ODE to PDE 545

and $n = 1$, $\beta = 0.4(0.2)1.2$. Since $\text{Re}(\gamma_m)$ is independent of m for $n = 0$ and weakly dependent on m for $n = 1$, it follows that the rate of approach to similarity form is fairly insensitive to initial conditions. The eigenvalue $\gamma = -1$ follows from the invariance of PDE to translation in τ, so when $\tau + \tau_0$ is expanded for large τ the original similarity solutions are obtained as functions of τ along with a correction term $O(\tau^{-1})$.

Grundy and Rottman give the solutions of two initial value problems for (1) and (2), subject to (3), (4), and (5), taken from Rottman and Simpson (1983, 1984). The initial conditions were taken to be (8) and (9) with $h_0(x) = h_0$, a constant. This is typically the top-hat or dam-break form. Condition (7), however, distinguishes the present problem from the usual dam-break problem described, for example, by Stoker (1957). The solutions for the height $h(x,t)$ of the heavy fluid for early times after release are shown in Figure 9.5.2 for both geometries, $n = 0$ and $n = 1$. The two geometries show rather distinct behavior: the plane solution shows the mound of heavy fluid collapsing almost as a rectangular box, while the axisymmetric solution shows that most of the heavy fluid is, by time $t \sim 4x_0/(g'h_0)^{1/2}$, concentrated at the leading edge of the spreading current. Moreover, in the axisymmetric case, a backward-facing hydraulic jump forms just behind the front at $t \approx 2x_0/(g'h_0)^{1/2}$ and propagates back toward the axis of symmetry.

The computed height profile $H(\eta, T)$, for several values of dimensionless times T after release, of the dam-break problem is shown in Figure 9.5.3 for $n = 0$ and $n = 1$. The figures also display the similarity solution $H_0(\eta)$ as dashed curves. It is quite clear from the figures that the numerical results oscillate in time as they approach the similarity solution and the amplitude of the oscillations decreases rapidly with T. They become undistinguishable from the similarity solutions at $T \approx 6$ for $n = 0$ and $T \approx 10$ for $n = 1$. Thus, the similarity solutions are clearly seen to be large-time limits of the solutions of the initial value problems. The limits are approached as $t^{-1/2}$ in the plane case and $t^{-\gamma}$ ($\gamma \approx 0.2$) in the axisymmetric case, and weak dependence of the powers $-1/2$ and -0.2 on β and mode numbers indicates that the asymptotic rate of approach to the similarity form is quite insensitive to initial conditions.

9.6 GENERALIZED SIMILARITY SOLUTION FOR CLIMB OF A BORE OVER A SLOPING BEACH

We consider the asymptotic behavior of a bore (which is like a shock) as it approaches the shoreline on a sloping beach, situated at $x = x_0$. We take the bore to be propagating in the region $x \leq x_0$ towards the shoreline. We assume the slope of the beach to be uniform and the undisturbed depth to

Table 9.5.1 The First 11 Eigenvalues (including complex conjugates; see equation (31)) for $n = 0$, $\beta = 0.6(0.2)1.4$ Listed in Order of Increasing Imaginary Part.

$\beta = 0.6$		$\beta = 0.8$		$\beta = 1.0$		$\beta = 1.2$		$\beta = 1.4$	
Re(Y)	Im(Y)	Re(Y)	Im(Y)	Re(Y)	Im(Y)	Re(Y)	Im(Y)	Re(Y)	Im(Y)
−1.0	0.0	−1.0	0.0	−1.0	0.0	−1.0	0.0	−1.0	0.0
−0.5	3.323	−0.5	2.795	−0.5	1.803	−0.5	1.384	−0.5	1.056
−0.5	6.737	−0.5	5.696	−0.5	3.763	−0.5	2.961	−0.5	2.344
−0.5	10.130	−0.5	8.573	−0.5	5.686	−0.5	4.492	−0.5	3.575
−0.5	13.518	−0.5	11.445	−0.5	7.601	−0.5	6.013	−0.5	4.795
−0.5	16.905	−0.5	14.314	−0.5	9.512	−0.5	7.530	−0.5	6.009

Source: From Grundy and Rottman (1985).

Table 9.5.2 The First 11 Eigenvalues (including complex conjugates; see equation (31)) for $n = 1$, $\beta = 0.4(0.2)1.2$ Listed in Order of Increasing Imaginary Part.

$\beta = 0.4$		$\beta = 0.6$		$\beta = 0.8$		$\beta = 1.0$		$\beta = 1.2$	
Re(Y)	Im(Y)	Re(Y)	Im(Y)	Re(Y)	Im(Y)	Re(Y)	Im(Y)	Re(Y)	Im(Y)
−1.0	0.0	−1.0	0.0	−1.0	0.0	−1.0	0.0	−1.0	0.0
−0.244	4.627	−0.235	2.945	−0.220	2.054	−0.197	1.472	−0.155	1.025
−0.246	8.513	−0.240	5.451	−0.230	3.836	−0.213	2.777	−0.183	1.946
−0.246	12.358	−0.241	7.925	−0.232	5.587	−0.217	4.055	−0.190	2.849
−0.246	16.191	−0.241	10.388	−0.233	7.330	−0.218	5.325	−0.192	3.746
−0.246	20.020	−0.241	12.847	−0.233	9.068	−0.219	6.591	−0.194	4.639

Source: From Grundy and Rottman (1985).

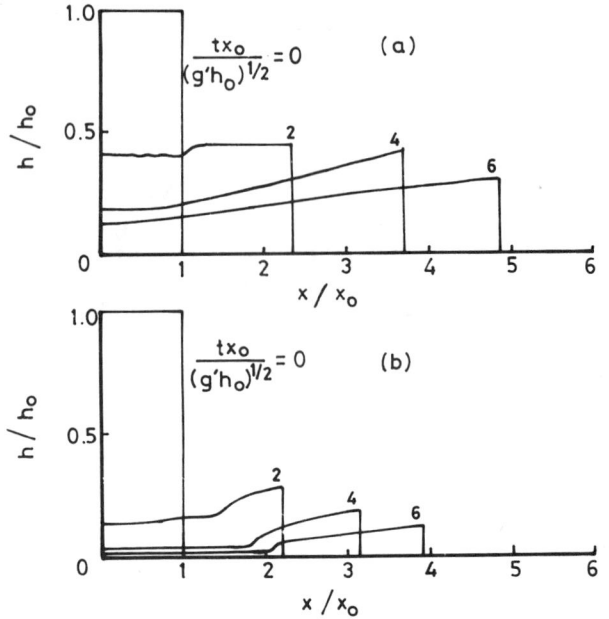

Figure 9.5.2 The computed height profile $h(x,t)$ of the heavy fluid for several times just after release for the "dam-break" problem: (a) plane ($n = 0$), (b) axisymmetric ($n = 1$). (From Grundy and Rottman 1985.)

be given by $h_0 = \beta(x_0 - x)$. The shallow-water equations are

$$u_t + uu_x + \hat{g}(h - h_0)_x = 0, \tag{1}$$

$$h_t + uh_x + hu_x = 0, \tag{2}$$

where u is the flow velocity in the x-direction and h is the water depth. The bore is assumed to reach the shoreline at time $t = t_0$. The term $\hat{g}$ denotes the acceleration due to gravity.

If we introduce the "sound" speed $c = (\hat{g}h)^{1/2}$ in (1) and (2), we obtain

$$u_t + uu_x + 2cc_x + \hat{g}\beta = 0, \tag{3}$$

$$c_t + uc_x + \frac{1}{2}cu_x = 0. \tag{4}$$

The bore conditions may be taken in the form

$$\frac{U}{(\hat{g}h_0)^{1/2}} = M(2M^2 - 1)^{1/2}, \tag{5}$$

Applications of Nonlinear ODE to PDE

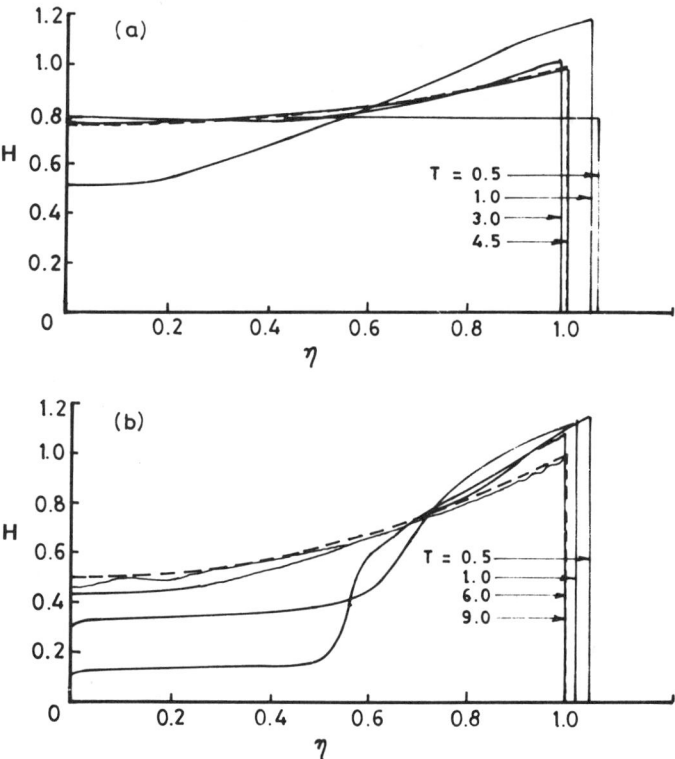

Figure 9.5.3 The computed height profiles $H(\eta, T)$ for several values of the dimensionless time T after release for the "dam-break" problem: (a) plane ($n = 0$) (b) axisymmetric ($n = 1$). The dashed curve in each plot is the similarity solution $H_0(\eta)$. (From Grundy and Rottman 1985).)

$$\frac{u}{(\hat{g}h_0)^{1/2}} = \frac{2M(M^2 - 1)}{(2M^2 - 1)^{1/2}}, \tag{6}$$

$$\frac{c}{(\hat{g}h_0)^{1/2}} = (2M^2 - 1)^{1/2}, \tag{7}$$

where U is the bore velocity and M is a parameter equal to $U/(\hat{g}h)^{1/2}$.

Numerical results of Keller, Levine, and Whitham (1960) for this problem indicated that the behavior of the bore near the shoreline is largely independent of the details of the initial motion producing the bore; indeed, the influence of initial motion reflects itself only in the fixing of an overall constant. This is typical of similarity solutions of the second kind. So,

Sachdev and Seshadri (1978) attempted a direct similarity solution. The notation used in this work is slightly different. The shoreline was assumed to be located at $x = 0$, and the sloping beach extended in the region $x > 0$. This would require changing β to $-\beta$ in (3). We shall take this change into account as we describe the work of Sachdev and Seshadri.

Introducing the nondimensional variables

$$\bar{x} = \frac{x\beta}{L}, \quad \bar{t} = t\beta \left(\frac{\hat{g}}{L}\right)^{1/2},$$
$$\bar{u} = \frac{u}{(\hat{g}L)^{1/2}}, \quad \bar{c} = \frac{c}{(\hat{g}L)^{1/2}}, \tag{8}$$

where L is some characteristic length, the system (3)–(4) was rewritten (with the bars dropped) in characteristic form;

$$(u + 2c)_t + (u + c)(u + 2c)_x - 1 = 0, \tag{9}$$
$$(u - 2c)_t + (u - c)(u - 2c)_x - 1 = 0. \tag{10}$$

Equations (9)–(10) admit a similarity form

$$u = tf(\eta), \quad c = tg(\eta), \quad \eta = \frac{x}{At^2}, \tag{11}$$

where A is an arbitrary (dimensional) constant. Thus, equations (9)–(10) become

$$\left(\frac{f+g}{A} - 2\eta\right)(f + 2g - 1)_\eta + (f + 2g - 1) = 0, \tag{12}$$

$$\left(\frac{f-g}{A} - 2\eta\right)(f - 2g - 1)_\eta + (f - 2g - 1) = 0. \tag{13}$$

The nondimensional form of the bore conditions is

$$uc^2 = U(c^2 - x), \tag{14}$$

$$u^2c^2 + \frac{1}{2}c^4 - \frac{1}{2}x^2 = Uuc^2, \tag{15}$$

where the uniform slope of the beach ahead of the bore has been explicitly introduced. If the bore path is assumed to be given by $\eta = 1$, the bore conditions (14)–(15) in terms of variables f and g become

$$f(1)g^2(1) = 2A(g^2(1) - A), \tag{16}$$

$$f^2(1)g^2(1) + \frac{1}{2}g^4(1) - \frac{1}{2}A^2 = 2Af(1)g^2(1). \tag{17}$$

Introducing the new variables

$$N = \left(\frac{1}{A} - 2\eta\right), \quad F(N) = \frac{f-1}{AN}, \quad \text{and} \quad G(N) = \frac{g}{AN}, \tag{18}$$

Applications of Nonlinear ODE to PDE

system (12)–(13) becomes

$$2N\frac{dF}{dN} = \frac{2G^2(F-1) - F(1+F)(1+2F)}{(1+F)^2 - G^2}, \tag{19}$$

$$4N\frac{dG}{dN} = \frac{4G^3 - 4G(1+F)^2 - FG + 2G(1+F)}{(1+F)^2 - G^2}. \tag{20}$$

The bore front now becomes

$$N = \frac{1}{A} - 2. \tag{21}$$

Dividing (19) by (20), we have a single equation

$$\frac{dF}{dG} = 2\frac{2G^2(F-1) - F(1+F)(1+2F)}{2G(1+F) - FG - 4G(1+F)^2 + 4G^3} \tag{22}$$

in the F-G (phase) plane. This equation has the following singular points, conveniently grouped with the coordinates (F, G) and with the type accredited to each:

$(-2, -1), (-2, 1)$, nodes.

$\left(-\frac{1}{3}, -\frac{1}{6}\right), \left(-\frac{1}{3}, \frac{1}{6}\right)$, saddle points.

$(0, 0)$, a nodal star. (23)

$(-1, 0)$, saddle point.

$\left(-\frac{1}{2}, 0\right)$, node.

It follows from (19) and (20) that dF/dN and dG/dN become infinite along the lines $F + G = -1$ and $F - G = -1$. Thus, for the solution to remain single-valued, the integral curves, starting from the bore point, must pass through a singular point of (22), where the numerators of (19) and (20) also vanish. It was found numerically, however, that the integral curves for all admissible A cross the singular lines $F + G = -1$ and $F - G = -1$ at points which are not singular. Hence it was concluded that the class of similarity solutions of the form (11) for shallow-water equations does not admit bores.

Proceeding from this point, Barker and Whitham (1980) argued that similarity solutions are often found after some appropriate (limiting) approximation has been made. This is certainly true of explosion and implosion problems for which strong shock approximation is made so that the ambient pressure ahead of the shock is assumed to be zero. Following this argument, they sought similarity solutions of a truncated form of the shallow-water equations (1)–(2) to have access to a larger class.

The solutions could subsequently be improved upon to take into account the neglected term. Apart from the strong bore approximation $M \gg 1$, they made the crucial assumption that $\hat{g} h_x = 2cc_x \gg \hat{g}\beta$ as the shoreline is approached. This implied that the term $\hat{g}\beta$ in (3), being small in comparison with other terms, might be dropped. Although h near the shoreline tends to zero, the surface slope h_x tends to infinity there. The beach slope β, however, continues to exert a controlling influence through the bore conditions (5)–(7). (In addition to the nonexistence result of Sachdev and Seshadri 1978, to the analytic results of Ho and Meyer 1962, who derived the bore path for the present problem to a very high degree of approximation using a different approach provided invaluable information. They carried out their analysis through a transformed system with t as the dependent variable and the characteristics as the independent variables.) Barker and Whitham rederived these results via a generalized similarity analysis, using a rather ingenious argument, as we now explain.

The approximated form of governing equations (3)–(4) is

$$u_t + uu_x + 2cc_x = 0, \tag{24}$$

$$c_t + uc_x + \frac{1}{2} cu_x = 0, \tag{25}$$

and, in the absence of the term $g\beta$, permits a much broader class of similarity solutions. Now introduce the variables

$$\tau = t_0 - t, \qquad \xi = x_0 - x - (t_0 - t)U_0, \tag{26}$$

where U_0 is the final velocity of the bore when it reaches the shoreline $x = x_0$ at time $t = t_0$. The analysis of Ho and Meyer (1962) suggests that the similarity solution may be taken in the form

$$u = U_0 + nb\tau^{n-1} zF(z), \qquad c = nb\tau^{n-1} zG(z), \tag{27}$$

where

$$z = \frac{\xi}{b\tau^n}; \tag{28}$$

n is an exponent to be determined. The arbitrary parameter b may be chosen such that the bore trajectory is given by $z = -1$. It is found in terms of U_0 from the bore conditions. The position and velocity of the bore $z = -1$ may be written as

$$x_0 - x = U_0(t_0 - t) - b(t_0 - t)^n, \tag{29}$$

$$U = U_0 - bn(t_0 - t)^{n-1}. \tag{30}$$

Applications of Nonlinear ODE to PDE

Substituting (27)–(28) into (24)–(25), we get the ODE for F and G:

$$[(1-F)^2 - G^2](zF' + F) = \frac{n-1}{n}\{(1-F)F + 2G^2\}, \tag{31}$$

$$\{(1-F)^2 - G^2\}(zG' + G) = \frac{n-1}{n}\left\{(1-F)G + \frac{1}{2}FG\right\}. \tag{32}$$

We find that system (31)–(32) is singular on the lines

$$(1-F)^2 - G^2 = 0. \tag{33}$$

We show that one of these is the limiting characteristic C_+ which reaches the bore as the bore reaches the shoreline. This characteristic plays an important role in the imploding shock solution of Guderley (1942) and determines the similarity exponent n. Now, if the similarity curve $z = z_c$, say, is also a characteristic, then

$$\frac{dx}{dt} = u + c \quad \text{or} \quad \frac{dx}{dt} = u - c \tag{34}$$

along it. By using the similarity variables (26) and (28), we change (34) to

$$U_0 + nbz_c\tau^{n-1} = U_0 + nbz_c\tau^{n-1}(F \pm G), \tag{35}$$

so

$$F \pm G = 1; \tag{36}$$

this is equivalent to (33). As we argued for equation (22), a solution starting at the bore $z = -1$ with given values of F and G will in general fold back on itself on passing through the point $z = z_c$ where (33) holds. It will be single-valued only if the right-hand sides of (31) and (32) vanish simultaneously with (33). This gives the appropriate values

$$F = 2, \quad G = -1, \quad \text{at } z = z_c. \tag{37}$$

Now approximating bore conditions (5)–(7) near the shoreline where u and U are finite and $h \to 0$ such that the various strength parameters $u/(\hat{g}h_0)^{1/2}$, $U/(\hat{g}h_0)^{1/2}$ and $h/h_0 \to \infty$, we have

$$\frac{U}{(\hat{g}h_0)^{1/2}} = 2^{1/2}M^2 + O(1), \tag{38}$$

$$\frac{u}{(\hat{g}h_0)^{1/2}} = 2^{1/2}M^2 + O(1), \tag{39}$$

$$\frac{c}{(\hat{g}h_0)^{1/2}} = 2^{1/2}M + O\left(\frac{1}{M}\right). \tag{40}$$

Now (30) and (38) yield

$$U_0 - nb\tau^{n-1} = (2\hat{g}h_0)^{1/2}M^2 + O(\tau^{1/2}). \tag{41}$$

Equation (39), in terms of the similarity function F, leads to
$$U_0 - nb\tau^{n-1}F(-1) = (2\hat{g}h_0)^{1/2}M^2 + O(\tau^{1/2}) \tag{42}$$
$$= U_0 - nb\tau^{n-1} + O(\tau^{1/2}). \tag{43}$$
Thus,
$$F(-1) = 1. \tag{44}$$
Similarly, for the function G,
$$-nb\tau^{n-1}G(-1) = (2\hat{g}h_0)^{1/2}M + O(\tau^{3/4})$$
$$= U_0^{1/2}(2\hat{g}h_0)^{1/4} + O(\tau^{3/4})$$
$$= U_0^{3/4}(2\beta\hat{g})^{1/4}\tau^{1/4} + O(\tau^{3/4}), \tag{45}$$
where we have used the relation $h_0 = \beta(x_0 - x) \simeq \beta U_0(t_0 - t)$ at the bore. Hence we arrive at the crucial determination that $n = 5/4$ and
$$G(-1) = -\frac{4}{5}\frac{(2\beta\hat{g})^{1/4}U_0^{3/4}}{b}. \tag{46}$$

It is interesting to compare the determination of the exponent n here merely from the bore conditions and the similarity form of the solution with that in Guderley's problem, for which an eigenvalue problem has to be solved. Equation (46) defines the relation between U_0 and b such that the solution, starting from the values $F(-1)$ and $G(-1)$ at the bore given by (44) and (46), passes through $z = z_c$ and satisfies conditions (37) there.

To get the explicit value of $z = z_c$ at the limiting characteristic and hence the flow behind the shock, Barker and Whitham (1980) used the solution
$$F + 2G = 0. \tag{47}$$
of (31)–(32), which automatically satisfies the characteristic condition (37). It satisfies bore condition (44) provided
$$G(-1) = -\frac{1}{2} \tag{48}$$
so that
$$b = \frac{8}{5}U_0^{3/4}(2\beta\hat{g})^{1/4}. \tag{49}$$
Using (47), we eliminate F from (31) or (32) and obtain
$$(1 + 3G)(zG' + G) = \frac{1}{5}G. \tag{50}$$
The solution of (50) satisfying (48) is
$$z = \frac{1}{2G}\left\{\frac{15G + 4}{7G}\right\}^{1/4}. \tag{51}$$

Applications of Nonlinear ODE to PDE

The value of z at the limiting characteristic where $G = -1$ is, therefore,

$$z_c = -\frac{1}{2}\left[\frac{11}{7}\right]^{1/4}. \tag{52}$$

The first approximation for the bore propagation has thus been obtained. We put $n = 5/4$ and use (49) in the bore description (29)–(30), and have

$$x_0 - x = U_0(t_0 - t) - \frac{8}{5}U_0^{3/4}(2\beta\hat{g})^{1/4}(t_0 - t)^{5/4}. \tag{53}$$

$$U = U_0 - 2U_0^{3/4}(2\beta\hat{g})^{1/4}(t_0 - t)^{1/4}$$
$$\simeq U_0 - 2^{5/4}U_0^{1/2}\{\hat{g}h_0(x)\}^{1/4}, \tag{54}$$

where we have used the relation $h_0 = \beta(x_0 - x) \simeq \beta U_0(t_0 - t)$.

To this approximation, the depth behind the bore is

$$h = U\left\{\frac{2h_0(x)}{\hat{g}}\right\}^{1/2} \tag{55}$$

(see (38)).

A more direct proof for the existence of the above solution follows from the phase plane analysis of (31)–(32) with $n = 5/4$ (see the discussion above (46)). The corresponding DE in the (F, G) plane is

$$\frac{dG}{dF} = \frac{(1 - F/2)G - 5\{(1 - F)^2 - G^2\}G}{(1 - F)F + 2G^2 - 5\{(1 - F)^2 - G^2\}F} \tag{56}$$

(cf. (22)). Figure 9.6.1 shows the limiting characteristic $F + G = 1$, the trajectory $F + 2G = 0$, and the bore line $z = -1$, which, according to (44) and (46), corresponds to a point B on $F = 1$ with $G < 0$. The integral curve, starting at such a point, must pass through the point C: $F = 2$, $G = -1$, which is the acceptable crossing point on the limiting characteristic $G = 1 - F$. The singular point $(2, -1)$ is a node. The trajectories drawn in Figure 9.6.1 clearly show that all the curves through the point C except the straight line $G = -F/2$ are unacceptable since they sweep around to cross the limiting characteristic $G = 1 - F$ at points where the right-hand sides of equations (31) and (32) do not vanish; thus they contain unacceptable folds. The only choice, therefore, is the curve $F + 2G = 0$ joining the points $B(1, -1/2)$ and $C(2, -1)$, as we have pointed out earlier.

Thus far, we have carried out the analysis with the exclusion of the term $g\beta$ from the basic equations (3)–(4). To include the effect of this term, we attempt a generalized similarity solution

$$u = U_0 + \sum_{m=1}^{\infty}\tau^{m/4}V_m(z), \tag{57}$$

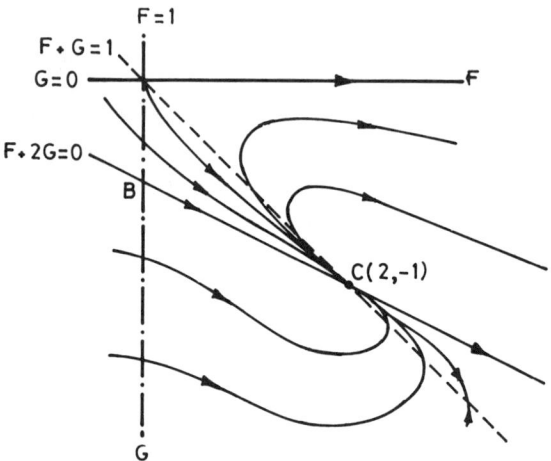

Figure 9.6.1 Phase plane (F, G) (see equation (56)): (---) limiting characteristic $F + G = 1$; (- . -) bore line; $z = -1$, (———) trajectory, arrow indicates increasing z. (From Barker and Whitham 1980).

$$c = \sum_{m=1}^{\infty} \tau^{m/4} C_m(z), \tag{58}$$

where

$$z = \frac{\xi}{B}, \quad B = \sum_{m=1}^{\infty} b_m \tau^{m+(1/4)}. \tag{59}$$

The basic terms are those obtained in the first approximation, namely,

$$V_1(z) = \frac{5}{4} bz F(z), \quad C_1(z) = \frac{5}{4} bz G(z), \quad b_1 = b; \tag{60}$$

expansions (57)–(59) extend this approximation to include the effect of $\hat{g}\beta$ in (3).

The integral powers of $\tau^{1/4}$ are suggested by the basic shock solution (53)–(54) and (28) with $n = 5/4$. The analysis may reveal other terms at a later stage. The work of Ho and Meyer (1962) suggests that the exact Riemann invariant

$$u + 2c + \hat{g}\beta(t - t_0)$$

is constant to a very high degree of accuracy, the deviation appearing at $m = 11$ in the expansions (57)–(59). Therefore, it is natural to write system

Applications of Nonlinear ODE to PDE

(3)–(4) along the positive characteristic direction

$$(u + 2c - \hat{g}\beta\tau)_t + (u + c)(u + 2c - \hat{g}\beta\tau)_x = 0 \qquad (61)$$

and substitute (57)–(59) into it. The quantities that appear in a combined way are those in the positive Riemann invariant, namely,

$$W_m = V_m + 2C_m, \qquad m \neq 4, \qquad (62)$$
$$W_4 = V_4 + 2C_4 - \hat{g}\beta\tau. \qquad (63)$$

The special case $m = 4$ arises naturally from the $\hat{g}\beta$ term in the basic equations. The W_m's are governed by

$$\left(V_1 + C_1 - \frac{5}{4}b_1 z\right) W'_m + \frac{1}{4} m b_1 W_m$$
$$= -\sum_{k=2}^{m} \left(V_k + C_k - \frac{4+k}{4} b_k z\right) W'_{m+1-k}$$
$$- \sum_{k=2}^{m} \frac{m+1-k}{4} b_k W_{m+1-k}, \qquad m \geq 2. \qquad (64)$$

For $m = 1$, the right-hand side is absent. System (64) is singular, where

$$V_1 + C_1 - \frac{5}{4} b_1 z = 0, \qquad (65)$$

which, expressed in terms of F and G, corresponds to the positive limiting characteristic $F + G = 1$. The equation for W_1 is

$$\left(V_1 + C_1 - \frac{5}{4} b_1 z\right) W'_1 + \frac{1}{4} b_1 W_1 = 0. \qquad (66)$$

To avoid a singularity on the limiting characteristic, we must choose

$$W_1 \equiv 0. \qquad (67)$$

Substituting (67) into (64) with $m = 2$ shows that W_2 satisfies the same equation as W_1; we must therefore argue as for W_1. Indeed, as long as the previous W's are identically zero, the equation for a new W_m is homogeneous:

$$(V_1 + C_1 - \frac{5}{4} b_1 z) W'_m + \frac{1}{4} m b_1 W_m = 0. \qquad (68)$$

In view of the relation $F = -2G$ between F and G and the expressions (60) for V_1 and C_1 in terms of F and G, we can write (68) as

$$\frac{z W'_m}{W_m} + \frac{2m}{5} \frac{1}{2-F} = 0. \qquad (69)$$

The singularity $F = 2$ corresponds to the limiting characteristic. Relations (51) and $F + 2G = 0$ give

$$z = -\frac{1}{F}\left(\frac{15F - 8}{7F}\right)^{1/4}. \tag{70}$$

Using (70), we can write (69) as

$$\frac{1}{W_m}\frac{dW_m}{dF} = -\frac{m}{2}\frac{1}{2-F}\left\{\frac{1}{F} - \frac{3}{15F - 8}\right\}, \tag{71}$$

which, on integration, gives

$$W_m = A_m F^{-m/4}(15F - 8)^{3m/44}(2 - F)^{2m/11}, \tag{72}$$

where A_m is an arbitrary constant. The singularity at $F = 8/15$ is not relevant, since it does not lie in the range of F that covers the flow field. F has the value 1 at the bore, increases to assume the value 2 at the limiting characteristic and tends to infinite as $z \to 0$. For $z > 0$, F assumes negative values.

It is seen from (72) that the derivatives of W_m are singular at $F = 2$ for $1 \leq m \leq 10$. Therefore, we must choose $A_m = 0$, $W_m \equiv 0$ for these values of m. The first nonzero W_m is allowed in this series at the remarkably high order $m = 11$. We may include W_{11} with an arbitrary coefficient A_{11}. This is a second arbitrary constant since the earlier coefficients $b_1, \ldots, b_{10}$ are all determined in terms of the single constant U_0, the final bore velocity at the shoreline. Using the bore conditions, the first few order terms in the bore law come out to be

$$b_1 = \frac{8}{5}(2\hat{g}\beta)^{1/4}U_0^{3/4}, \quad b_2 = -\frac{11}{5}(2\hat{g}\beta U_0)^{1/2},$$

$$b_3 = \frac{387}{175}(2\hat{g}\beta)^{3/4}U_0^{1/4}, \quad b_4 = -\frac{23{,}681}{7000}\hat{g}\beta. \tag{73}$$

If we include a nonzero W_m at $m = 11$, it will, in turn, introduce nonzero values in the subsequent W's. Additional sequences with arbitrary constants can be included at $m = 22, 33$, etc.

The numerical solution of Keller et al. (1960) and the above analysis agree very closely, particularly in their description of the bore trajectory (see also the work of Sachdev and Seshadri 1976 for a different approximate analytic treatment of this problem).

References

Abdelkader, M. A. (1969). Sequences of nonlinear differential equations with related solutions. Ann. Mat. Pura Appl. 81, 249–259.

Ablowitz, M. J., Kaup, D. J., Newell, A. C., and Segur, M. (1973). Method for solving the Sine-Gordon equation, Phys. Rev. Lett., *30*, 1262–1267.

Ablowitz, M. J., Ramani, A., and Segur, H. (1978). Nonlinear evolution equations and ordinary differential equations of Painlevé type, Lett. Nuovo Cimento *23*, 333–338.

Ablowitz, M. J., Ramani, A., and Segur, H. (1980). A connection between nonlinear evolution equations and ordinary differential equations of P-type. I, J. Math. Phys. *21*, 715–721.

Ablowitz, M. J., and Segur, H. (1981). Solitons and the inverse scattering transform. SIAM, Philadelphia.

Ablowitz, M. J. and Zeppetella, A. (1979). Explicit solutions of Fisher's equation for a special wave speed, Bull. Math. Biol. *41*, 835–840.

Abramowitz, M. and Stegun, I. A. (1964). *Handbook of Mathematical Functions*, Dover, New York.

Aizawa, Y. and Saito, N. (1972). On the stability of isolating integrals. I. Effect of the perturbation in the potential function, J. Phys. Soc. Japan *32*, 1636–1640.

Ames, W. F. (1968). *Nonlinear Ordinary Differential Equations in Transport Processes*. Academic Press, New York and London.

Andronov, A. A., Leontovich, E. A., Gordon, I. I., and Maier, A. G. (1973). *Qualitative Theory of Second Order Dynamic Systems*. Wiley, New York.

Bach, G. G. and Lee, J. H. (1969). Initial propagation of impulsively generated converging cylindrical and spherical shock waves, J. Fluid. Mech. *37*, 513–528.

Bailey, P. B., Shampine, L. F., and Waltman, P. E. (1968). *Nonlinear Two Point Boundary Value Problems*, Academic Press, New York.

Barenblatt, G. I. (1952). On some unsteady motions of fluids and gases in a porous medium, J. Appl. Math. Mech. *16*, 67–78.

Barenblatt, G. I. (1979). Similarity, self-similarity and intermediate asymptotics, Consultant Bureau.

Barker, J. W. and Whitham, G. B. (1980). The similarity solution for a bore on a beach. Comm. Pure Appl. Math. *33*, 447–460.

Bartashevich, D. A. (1973). The qualitative nature of real solutions of Painlevé's first equation, Diff. Eqns. 9, 714–716.

Bellman, R. (1953). *Stability Theory of Differential Equations*, McGraw-Hill, New York.

Bender, C. M. and Orszag, S. A. (1978). *Advanced Mathematical Methods for Scientists and Engineers*. McGraw-Hill, New York.

Berezin, Y. A. and Karpman, V. I. (1964). Theory of nonstationary finite amplitude waves in a low-density plasma, Soviet Physics, JETP *19*, 1265–1271.

Bergström, A. (1973). Electromagnetic theory of strong interaction, Phys. Rev. D *8*, 4394–4402.

Berkovich, L. M. (1979). Method of exact linearization of nonlinear autonomous differential equations of second order, J. Appl. Math. Mech. (PMM) *43*, 673–683.

References

Berkovich, L. M. and Rozov, N. Kh. (1972). Some remarks on differential equations of the form $y'' + a_0(x)y = \psi(x)y^\alpha$, Diff. Eqns. *8*, 1609–1612.

Birkhoff, G. and Rota, G. C. (1978). *Ordinary Differential Equations*, (3rd ed.) Wiley, New York.

Blum, E. K. (1972). *Numerical Analysis and Computation Theory and Practice*, Addison-Wesley, Reading, Mass.

Bountis, T. (1982). *On the Analytical Structure of Chaos in Dynamical Systems*, Lecture Notes in Physics 179, 179–227.

Bountis, T. and Segur, H. (1982). Logarithmic singularities and chaotic behaviour in Hamiltonian systems, in *Proceedings of the Workshop on Mathematical Systems*, Vol. 88, (ed. M. Tabor and Y. Treve), AIP, New York, p. 279.

Bountis, T., Segur, H., and Vivaldi, F. (1982). Integrable Hamiltonian systems and the Painlevé property, Phys. Rev. A *25*, 1257–1264.

Boutroux, M. P. (1914), Recherches sur les transcendants de M. Painlevé et l'étude asymptotique des équations differentielles du second ordre, Ann. Ecole Norm. Sup. *31*, 99–159.

Braude, Ya. (1967). Integrability conditions for certain nonlinear second-order differential equations, Diff. Eqns. *3*, 535–536.

Briot, A. A. and Bouquet, J. C. (1856). Recherches sur les propriétés des fonctions definies par des equations differentielles. J. Ecole Imp. Polytech. *21*, 133–197.

Briot, A. A. and Bouquet, J. C. (1875). Théorie des functions elliptiques, 2nd ed., Paris.

Byrd, P. F. and Friedman, M. D. (1954). *Handbook of Elliptic Integrals for Physicists and Engineers*, Springer, Berlin.

Callegari, A. J. and Reiss, L. E. (1968). Non-linear boundary value problems for the circular membrane, Arch. Rat. Mech. Anal. *31*, 390–400.

Canosa, J. (1969). Diffusion in nonlinear multiplicative media. J. Math. Physics *10*, 1862–1868.

Canosa, J. (1973). On a nonlinear diffusion equation describing population growth. IBM J. Res. Develop. 17, 307–313.

Chang, Y. F. and Corliss, G. (1980). Ratio-like and recurrence relation tests for convergence of series, J. Inst. Math. Appl. *25*, 349–359.

Chang, Y. F., Greene, J. M., Tabor, M., and Weiss, J. (1983). The analytic structure of dynamical systems and self-similar natural boundaries, Physica *8D*, 183–207.

Chang, Y. F., Tabor, M., and Weiss, J. (1982). Analytic structure of the Henon–Heiles Hamiltonian in integrable and nonintegrable regimes, J. Math. Phys. *23*, 531–538.

Chicone, C. and Jinjhuang, T. (1982). On general properties of quadratic systems, Am. Math. Monthly *89*, 167–178.

Cohen, A. (1931), *An Introduction to the Lie Theory of One Parameter Groups with Applications to the Solutions of Differential Equations*, Stechert, New York, pp. 86–89.

Coppel, W. A. (1966). A survey of quadratic systems, J. Diff. Eqns. *2*, 293–304.

Copson, E. T. (1967). *Asymptotic Expansions*, Cambridge University Press, Cambridge.

Courant, R. and Friedrichs, K. O. (1948). *Supersonic Flow and Shock Waves*, Interscience, New York.

Dasarathy, B. V. and Srinivasan, P. (1969). On the synthesis of a class of second order nonlinear differential equations, SIAM J. Appl. Math. *17*, 511–515.

Davies, T. V. and James, E. M. (1966). *Nonlinear Differential Equations*, Addison-Wesley, Reading, Mass.

Davis, H. T. (1962). *Introduction to Nonlinear Differential and Integral Equations*, Dover, New York.

De Boer, P. C. T. and Ludford, G. S. S. (1975). Spherical electric probe in a continuum gas, Plasma Phys. *17*, 29–43.

Dickson, R. J. and Perko, L. M. (1970). Bounded quadratic systems in the plane, J. Diff. Eqns. *7*, 251–273.

Domb, C. and Sykes, M. F. (1957). On the susceptibility of a ferromagnetic above the Curie point, Proc. Roy. Soc. Lond. A *240*, 214–228.

Dorizzi, B., Grammaticos, B., and Ramani, A. (1983). A new class of integrable systems, J. Math. Phys. *24*, 2282–2288.

Dresner, L. (1971). Phase-plane analysis of nonlinear second-order ordinary differential equations, J. Math. Phys. *12*, 1339–1348.

Eckmann, J. P. (1981). Roads to turbulence in dissipative dynamical systems, Rev. Mod. Phys. *53*, 643–654.

Ermakov, V. P. (1880). Universitetskie Izvestiya, Kiev, No. 9, 1–25.

Erugin, N. P. (1967). The analytic theory and problems of the real theory of differential equations connected with the first method and with the methods of the analytic theory, Diff. Eqns. *3*, 942–966.

Erugin, N. P. (1976a). Theory of nonstationary singularities of second-order equations. I, Diff. Eqns. *12*, 267–289.

Erugin, N. P. (1976b). Theory of singularities of second-order equations. II, Diff. Eqns. *12*, 579–598.

Erugin, N. P. (1980). The equation $w'' = f(w',w,z)$ with an irrational right side, Diff. Eqns. *16*, 236–250.

Fokas, A. S. and Ablowitz, M. J. (1982). On a unified approach to transformations and elementary solutions of Painlevé equations, J. Math. Phys. *23*, 2033–2042.

Fisher, R. A. (1936). The wave of advance of advantageous genes, Ann. Eugen. *7*, 355–369.

Frisch, U. (1983). The analytic structure of turbulent flows, 6th Kyoto Summer Institute on Chaos and Statistical Mechanics (preprint).

Fuchs, V. (1975). The influence of linear damping on nonlinearly coupled positive and negative energy waves, J. Math. Physics *16*, 1388–1392.

Gambier, B. (1910). Sur lés equations différentielles du second ordre et du premier degré dont l'integrale genéralé est á points critiques fixes, Acta Math. *33*, 1–55.

Gelfand, I. M. and Fomin, S. V. (1963). *Calculus of Variations*, Prentice-Hall, Englewood Cliffs, NJ.

Gilding, B. H. and Peletier, L. A. (1976). On a class of similarity solutions of the porous media equation, J. Math. Anal. Appl. *55*, 351–364.

Gleick, J. (1988). *Chaos: Making of a New Science*, Penguin, New York.

Godunov, S. K. and Kireeva, I. L. (1968). Some self-modelling motions of an ideal gas, U.S.S.R. J. Comp. Math. Math. Phys. *8*, 374–392.

Golomb, M. (1943). Zeros and poles of functions defined by Taylor series, Bull. Amer. Math. Soc., *49*, 581–592.

Gradshteyn, I. S. and Ryzhik, I. W. (1965). *Tables of Integrals, Series, and Products*, 4th ed., Academic Press, New York.

Greenspan, H. P. and Butler, D. S. (1962). On the expansion of a gas into vacuum, J. Fluid. Mech. *13*, 101–119.

Gromak, V. I. (1975). Theory of Painlevé's equation, Diff. Eqns. *11*, 285–287.

Gromak, V. I. (1976). Solutions of Painlevé's fifth problem, Diff. Eqns. *12*, 519–521.

Gromak, V. I. and Lukashevich, N. A. (1982). Special classes of solutions of Painlevé's equations, Diff. Eqns. *18*, 317–326.

Grundy, R. E. (1979). Similarity solutions of the nonlinear diffusion equation, Q. Appl. Math. *37*, 259–280.

Grundy, R. E. and Rottman, J. W. (1985). The approach to self-similarity of the solutions of the shallow-water equations representing gravity-current releases, J. Fluid. Mech. *156*, 39–53.

Guderly, G. (1942). Starke Kugelige und zylindrische Verdichtungsstösse in der Nahe des Kugelmittelpunktes bzw. der Zylinderachse, Luftfahrtforschung *19*, 302–312.

Hartman, P. (1964). *Ordinary Differential Equations.* Wiley, New York.

Hastings, S. P. and McLeod, J. B. (1980). A boundary value problem associated with the second Painlevé transcendent and the Korteweg-de Vries equation. Arch. Rat. Mech. Anal. *73*, 31–51.

Hille, E. (1969). *Lectures on Ordinary Differential Equations*, Addison-Wesley, Reading, Mass.

References

Hille, E. (1970). Some aspects of the Thomas-Fermi equation, J. Analyse Math. *23*, 147–170.

Hille, E. (1970a). Aspects of Emden's equation, J. Fac. Sci. Tokyo, Sect. 1. *17*, 11–30.

Hille, E. (1972/73). A note on quadratic systems, Proc. R.S.E. (A). *72*, 17–37.

Hille, E. (1976). *Ordinary Differential Equations in the Complex Domain*, Wiley, New York.

Ho, D. V. and Meyer, R. E. (1962). Climb of a bore on a beach, I. Uniform beach slope, J. Fluid. Mech. *14*, 305–318.

Hochstadt, H. (1963). *Differential Equations–A Modern Approach*, Dover, New York.

Holmes, P. and Spence, D. (1984). On a Painlevé-type boundary-value problem, Q. J. Mech. Appl. Math. *37*, 526–538.

Humi, M. (1986). Factorization of systems of differential equations, J. Math. Phys. *27*, 76–81.

Hunter, C. (1960). On the collapse of an empty cavity in water, J. Fluid Mech. *8*, 241–263.

Hunter, C. (1963). Similarity solutions for the flow into a cavity, J. Fluid Mech. *15*, 289–305.

Ince, E. L. (1956). *Ordinary Differential Equations*, Dover, New York.

Inselberg, A. (1969). Phase-plane solutions of Langmuir's equation, J. Math. Anal. Appl. *26*, 438–446.

Jindia, R. K. and Sachdeva, B. K. (1984). Simple pendulum revisited, Math. Edu. *18*, 38–41.

Jones, C. W. (1953). On reducible non-linear differential equations occurring in mechanics, Proc. Roy. Soc. Lond. *217A*, 327–343.

Jones, D. S. (1977). The scattering of sound by a simple shear layer, Phil. Trans. Roy. Soc. London. *A284*, 1323–1328.

Kamke, E. (1943). *Differential Gleichungen: Lösungsmethoden und Lösungen*. Akademische Verlaggesellschaft, Leipzig.

Kawamoto, S. (1985). Construction of stationary solitary wave solutions, J. Phys. Soc. Japan. *54*, 1701–1709.

Keller, H. (1968), *Numerical Methods for Two-Point Boundary Value Problems*, Blaisdell, Waltham, Mass.

Keller, H. B., Levine, D. A., and Whitham, G. B. (1960). Motion of a bore over a sloping beach, J. Fluid Mech. *7*, 302–316.

Klamkin, M. S. (1962). On the transformation of a class of boundary value problems into initial value problems for ordinary differential equations, SIAM Rev. *4*, 43–47.

Kochina, N. N. and Mel'nikova, N. S. (1958a). Strong point-blasts in a compressible medium, J. Appl. Math. Mech. *22*, 3–15.

Kochina, N. N. and Mel'nikova, N. S. (1958b). On the unsteady motion of gas driven outward by a piston, neglecting the counterpressure, J. Appl. Math. Mech. *22*, 444–451.

Kolesnikova, N. S. and Lukashevich. N. A. (1972). A class of third order differential equations with fixed critical singularities, Diff. Eqns. *8*, 1615–1619.

Kolmogoroff et al. (1937). Etude de l'équation de la diffusion avec croissance de la quantité de matiére et son application á un problém biologique. Bull Univ. Etat, Moscow, *A1*, 1–25.

Kozmanov, M. Iu. (1977). On the motion of piston in a polytropic gas, J. Appl. Math. Mech. *41*, 1152–1156.

Kravchenko, L. and Yablonskii, A. I. (1965). Solution of an infinite boundary-value problem in a third-order equation, Diff. Eqns. *1*, 248–249.

Kuramoto, Y. (1978). Diffusion-induced chaos in reaction systems, Suppl. Prog. Theor. Phys. *64*, 346–367.

Kurtz, J. C. (1981). A singular nonlinear boundary value problem, Rocky J. Math. *11*, 227–241.

Lazarus, R. B. and Richtmyer, R. D. (1977). Similarity solutions for converging shocks, Los Alamos Scientific Lab., Rep. LA-6823-MS.

Lee, E. S. (1968). *Quasilinearisation and Invariant Imbedding*, Academic Press, New York.

References

Lefschetz, S. (1963). *Differential Equations: Geometric Theory*. Dover, New York.

Levinson, N. (1970). Asymptotic behaviour of solutions of non-linear differential equations, Stud. Appl. Math. *49*, 285–297.

Lorenz, E. N. (1963). Deterministic nonperiodic flow, J. Atmos. Sci. *20*, 130–141.

Lukashevich, N. A. (1965). Elementary solutions of certain Painlevé equations, Diff. Eqns. *1*, 561–564.

Lukashevich, N. A. (1967). Theory of the fourth Painlevé equation, Diff. Eqns. *3*, 395–399.

Lukashevich, N. A. (1968). Solutions of the fifth equation of Painlevé, Diff. Eqns. *4*, 732–735.

Lukashevich, N. A. (1971). The second Painlevé equation, Diff. Eqns. *7*, 853–854.

Lukashevich, N. A. and Matatov, V. I. (1973). Second order systems without moving critical singularities, Diff. Eqns. *9*, 344–348.

Lukashevich, N. A. and Yablonskii, A. I. (1967). On a set of solutions of the sixth Painlevé equation. Diff. Eqns. *3*, 264–266.

Luke, Y. L. (1962). *Integrals of Bessel Functions*. McGraw-Hill, New York.

McCoy, B. M., Tracy, C. A., and Wu, T. T. (1977). Painlevé functions of the third kind, J. Math. Phys. 18, 1058–1092.

McLeod, J. B. (1977). The existence of a non-negative solution of an ordinary differential equation arising in electromagnetic theory. Nonlinear Anal. Theor. Meth. Appl. *1*, 679–689.

McLeod, J. B. and Serrin, J. (1968). The existence of similar solutions for some laminar boundary layer problems, Arch. Rat. Mech. Anal. *31*, 288–303.

Mambriani, A. (1929). Sur un teorema relativo alle equazioni differenziali ordinarie del 2° ordine, Rend. Accad. Naz. Lincei, Cl. Sci. Fis. Mat. Nat. (6) 9, 620–625. 142–144.

Merkin, J. H. (1984). A note on the solution of a differential equation arising in boundary-layer theory, J. Engg. Math. 18, 31—36.

Miles, J. W. (1978). On the second Painlevé transcendent. Proc. R. Soc. Lond. *361*, 277–291.

Miles, J. W. (1978a). An axisymmetric Boussinesq wave, J. Fluid Mech., 84, 181–191.

Miles, J. W. (1982). On a nonlinear Bessel equation, SIAM J. Appl. Math. *42*, 109–112.

Morioka, N. and Shimizu, T. (1978). Transition between turbulent and periodic states in the Lorenz model, Phys. Lett. *66A*, 447–449.

Murphy, G. M. (1960). *Ordinary Differential Equations and Their Solutions*, Van Nostrand Reinhold, Princeton.

Murray, J. D. (1976). On traveling wave solutions in a model for the Belousov–Zhabotinskii reaction, J. Theor. Biol. *56*, 329–353.

Murray, J. D., (1977). *Lectures on Nonlinear Differential Equation Models in Biology*, Clarendon Press, Oxford.

Nakamura, Y. (1983). Analysis of self-similar problems of imploding shock waves by the method of characteristics, Phys. Fluids. *26*, 1234–1239.

Newton, T. A. (1978). Two dimensional homogeneous quadratic differential systems, SIAM Rev. *20*, 120–138.

Pinney, E. (1950). The nonlinear differential equation $y'' + p(x)y + cy^{-3} = 0$, Proc. Amer. Math. Soc. *1*, 681.

Ramani, A., Dorizzi, B., Grammaticos, B., and Bountis, T. (1984). Integrability and the Painlevé property for low-dimensional systems, J. Math. Phys. *25*, 878–883.

Ramani, A., Grammaticos, B., and Bountis, T. (1988). The Painlevé property and singularity analysis of integrable and non-integrable systems (preprint).

Reid, J. L. (1973). Homogeneous solution of a nonlinear differential equation, Proc. Amer. Math. Soc. *38*, 532–536.

Reyn, J. W. (1964). Classification and description of the singular points of a system of three linear differential equations, J. Appl. Math. Phys. (ZAMP) *15*, 540–557.

Robbins, K. A. (1979). Periodic solutions and bifurcation structure at high R in the Lorenz model, SIAM J. Appl. Math. *36*, 457–472.

Rosales, R. (1978). The similarity solution for the Korteweg–deVries equation and the related Painlevé transcendent, Proc. Roy. Soc. London. A*361*, 265–275.

Rottman, J. W. and Simpson, J. E. (1983). Gravity current produced by instantaneous releases of a heavy fluid in a rectangular channel, J. Fluid. Mech. *135*, 95–110.

Rottman, J. W. and Simpson, J. E. (1984). The initial development of gravity currents from fixed-volume releases of heavy fluids, in *Proc. IUTAM Symposium on Atmospheric Dispersion of Heavy Gases and Small Particles*, Delft, The Netherlands.

Ruelle, D. (1980). Measures describing turbulent flow, Ann. N.Y. Acad. Sci. *357*, 1–9.

Sachdev. P. L. (1984). Nonlinear travelling waves in fluids, in *Wave Phenomenon: Modern Theory and Applications*, C. Rogers and T. B. Moodie (eds.) Elsevier, New York.

Sachdev. P. L. (1987). *Nonlinear Diffusive Waves*, Cambridge University Press, Cambridge.

Sachdev. P. L., Tikekar, V. G., and Nair, K. R. C., (1986). Evolution and decay of spherical and cylindrical N waves, J. Fluid. Mech. *172*, 347–371.

Sachdev, P. L., Gupta, Neelam, and Ahluwalia, D. S. (1990). Global and asymptotic solution describing the collapse of a spherical or cylindrical cavity. (to be published).

Sachdev, P. L. and Seshadri, V. S. (1976). Motion of a bore over a sloping beach: an approximate analytical approach, J. Fluid Mech. *78*, 481–487.

Sachdev, P. L. and Seshadri, V. S. (1978). On the non-existence of similarity solution for climb of a bore over a sloping beach (unpublished).

Sachdev, P. L., Nair, K. R. C., and Tikekar, V. G. (1986) Generalized Burgers equations and Euler-Painlevé transcendents. I., J. Math. Phys. 27, 1506–1522.

Sachdev. P. L. and Nair, K R. C. (1987). Generalized Burgers equation and Euler–Painlevé transcendents. II, J. Math. Phys. *28*, 997–1004.

Sachdev, P. L., Nair, K. R. C. and Tikekar, V. G. (1988) Generalized Burgers equations and Euler-Painlevé transcendents. III, J. Math. Phys. 29, 2397–2404.

Sansone, G. and Conti, R. (1952). *Nonlinear Differential Equations*, Pergamon, New York.

Schlichting, H. (1960). *Boundary Layer Theory*. McGraw-Hill, New York.

Sedov, L. I. (1982). *Similarity and Dimensional Methods in Mechanics*, "Mir" Moscow.

Serebriakova, N. N. (1963). Qualitative investigation of a system of differential equations of the theory of oscillations, J. Appl. Math. Mech. *27*, 227–237.

Segur, H. (1980). Soliton and inverse scattering transform, Lectures given at the International School of Physics, "Enrico Fermi," Varenna, Italy (preprint).

Segur, H. (1982). Solitons and the inverse scattering transform. *Topics in Ocean Physics*, LXXX Corso, Soc. Italiana di Fisica, Bologna, Italy.

Seshadri, V. S. (1977). Some analytic approaches to nonlinear wave propagation in non-uniform media, Ph.D. Thesis, IISc. Bangalore.

Seshadri, V. S. and Sachdev. P. L. (1977). Quasi-simple wave solutions for acoustic gravity waves, Phys. Fluids. *20*, 888–894.

Sevruk, I. G. (1958). Laminar convection over a linear heat source. J. Appl. Math. Mech. *22*, 807–812.

Shidfar, A. A. and Sadeghi, A. A. (1986). Some series solutions of the anharmonic motion equation, J. Math. Anal. Appl. *120*, 488–493.

Shilov, G. E. (1977). *Linear Algebra*, Dover, New York.

Shimizu, T. and Morioka, N. (1978). Chaos and limit cycles in the Lorenz model, Phys. Lett. *66A*, 182–184.

Simmons, G. F. (1972). *Differential Equations with Applications and Historical Notes*. McGraw-Hill, New York.

Smith, R. A. (1973/74). Singularities of solutions of certain plane autonomous systems, Proc. R.S.E. (A)., *72*, 307–315.

References

Sommerfield, A. (1932). Asymptotische Integration der Differential-gleichung des Thomas–Fermischen Atoms, Z. Physik, 78, 283–308.

Sparrow, C. (1982). *Chaos and Strange Attractors. Applied Mathematical Sciences*, vol. 41, Springer-Verlag, New York.

Stakgold, I. (1967). *Green's Function and Boundary Value Problems*, Wiley, New York.

Standard Mathematical Tables, 14th ed., (1969). The Chemical Rubber Company, Cleveland, Ohio.

Stanyukovich, K. P. (1960). *Unsteady Motion of Continuous Media*, Pergamon, New York.

Steeb, W. H., Kunick, A., and Strampp, W. (1983). The Rikitake two-disc dynamo system and the Painlevé property, J. Phys. Soc. Japan 52, 2649–2653.

Stoker, J. J. (1957). *Water Waves*, Wiley, New York.

Tabor, M. and Weiss, J. (1981). Analytic structure of the Lorenz system, Phys. Rev. A 24, 2157–2167.

Takayama, K. (1986). A class of solvable second-order ordinary differential equations with variable coefficients, J. Math Phys. 27, 1747–1749.

Tam, K. K. (1975). On the Lagerstrom model for flow at low Reynolds numbers, J. Math. Anal. Appl. 49, 286–294.

Thomas, L. P., Pais, V., Gratton, R., Diez, J. (1986). A numerical study on the transition to self-similar flow in collapsing cavities, Phys. Fluids 29, 676–679.

Thual, O. and Frisch, U. (1984). Natural boundaries in the Kuramoto model (preprint).

Treve, Y. M. (1967). On a class of solutions of the Blasius equation with applications, SIAM J. Appl. Math. 15, 1209–1227.

Van Dyke, M. and Guttmann, A. J. (1982). The converging shock wave from a spherical or cylindrical piston, J. Fluid. Mech. 120, 451–462.

Van Dyke, M. (1984). Computer extended series, Ann. Rev. Fluid Mech. 287–309.

Vein, P. R. (1967). Functions which satisfy Abel's differential equation, SIAM J. Appl. Math. *15*, 618–623.

Wadati, M. (1972). The modified Korteweg–deVries equation, J. Phys. Soc. Japan *32*, 1681.

Whitham, G. B. (1974). *Linear and Nonlinear Waves*. Wiley-Interscience, New York.

Wu, T. T., McCoy, B. M., Tracy, C. A., and Barouch, E. (1976). Spin-spin correlation functions for the two-dimensional Ising model. Exact theory in the scaling region, Phys. Rev. B *13*, 316–374.

Yablonskii, A. I. (1964). Dokl. Akad. Nauk BSSR *8*, 964.

Yablonskii, A. I. (1967). Movable singularities of systems of differential equations, Diff. Eqns. *3*, 383–389.

Yablonskii, A. I. (1972). Asymptotic properties of regular solutions of Painlevé first and second problems, Diff. Eqns. *8*, 870–872.

Zakharov, V. E. and Synakh, V. S. (1976). The nature of the self-focussing singularity, Sov. Phys. JETP 41, 465–468.

Zeldovich, Ya. B. and Raizer, Yu. P. (1966). *Physics of Shock Waves and High Temperature Hydro-dynamic Phenomena*, vol. II, Academic Press, New York.

Zhidkov, E. P. and Shirikov, V. P. (1964). Boundary value problem for ordinary second order differential equations, Zh. Vych. Mat. *4*, 804–816.

Index

Abel's equation, 30, 65
  relation to linear DE, 30
Abel's formula, 11
Asymptotic behaviour of nonlinear DE, 151
Asymptotic series
  operations with, 122
  solutions of a boundary layer problem - in terms of, 122-127
Asymptotic solutions of cylindrical and spherical Burgers equation, 516-519
Attractors
  definition of, 354
  strange, 356
Autonomous DE, 5

Bellman's equation, 54
Benedixon's reduction, 92
Binomial DE of first order, 86
Blasius equation, 56
  BVP for, 217-218

[Blasius equation]
  series solution with IC at infinity, 99-100
  solution for BVP relating to, 217
Bore over a sloping beach
  approximate treatment of, 552
  generalized similarity solution for, 552-558
  nonexistence of a self-similar solution, 549-551
  numerical solution for, 549
Boundary layer equations
  asymptotic series solution of BVP for, 122-127
Boundary layer theory
  BVP for, 118
  BVP over an infinite domain, 117-121
  equations of, 117
Boundary value problems (BVP), 8
  invariance of DE and, 217

[Boundary value problems]
  methods for existence of solutions of, 213
  number of possible solutions for, 214
  scaling and, 217
  shooting methods for, 213, 215-216
Boussinesq equation, 427, 428, 431, 434
Briot-Bouquet equation, 88
  equations reducible to, 90-92
  psi series solutions of, 96
  series solutions of, 93-96
  simple cases of, 88-90
  singularities of, 88
Briot-Bouquet system and Pv, 494-495
Burgers equation, plane N wave solutions of, 514

Cauchy's majorant, 111
Cavity collapse (spherical)
  of an accelerating cavity, 318
  phase plane analysis of, 320-324
  self-similar solutions describing, 317
  of a uniformly moving cavity, 318
Chaotic solutions, 352
Classification of singular points of linear DE, 69
Combined k-dV equation, 434
Contraction principle, 227
Converging shock wave
  global solutions for, 529
  Guderley's solution, 534
  trajectory of, 532
Critical point, 246

Dam break initial conditions, 539-540
Deformation of a circular membrane
  BVP for, 243-244
Degree of DE, 3
Dirichlet series for BVP in boundary layer theory, 118-121
Dominant behaviour of DE, 140

Electromagnetic theory
  a DE arising in, 231

[Electromagnetic theory]
  solutions of a BVP relating to, 231-239
Elliptic PDE, 2
Embedding of asymptotic solutions, 164
Emden equation-phase plane analysis of, 301
Emden-Lane-Fowler equation, 54, 60, 61
Equidimensionality, 24
Essential singularities, 423
Euler-Painlevé's equations, 186, 187
Euler-Painlevé transcendents, 187
  connection problems for, 130, 137
  Kamke's collection and, 201-208
  Murphy's collection and, 208-211
  other DE related to, 200
  relation to GBE with damping, 187-188
  series solution for, 189, 196
Existence of solutions of IVP, 8

Factoring of nonlinear DE, 42
Fisher's equation
  IVP for, 304
  phase plane study for travelling waves, 306
  perturbation analysis of, 307
  steady-local analysis of, 135-138
  travelling wave solution of, 304, 305, 306
Fixed singular point, 420

Gas dynamic equations, 2
Generalized Burgers equations (GBE)
  in cylindrical and spherical geometry, 513
  N wave solutions of, 516-519
Generalized Duffing equation, 55
General solution
  of a nonlinear DE, 6, 453
  of an nth order DE, 7
Gravity current releases, global solutions for, 542-545
  self-similar solutions for, 541

Index

Green's function
 for linear BVP, 16-21

Hamiltonian systems
 free end Toda lattice, 411-413
 Henon-Heiles, 394-409
 quartic lattice, 409-411
Henon-Heiles system
 integrability of and chaos in, 394-409
 Painlevé analysis for, 394-403
Homogeneous DE, 3
Hyperbolic PDE, 2

Index of a singularity, 289
Indicial exponent, 72
Inhomogeneous DE, 3
Initial value problem (IVP), 7
Integral equation formulation of BVP, 175
Integrability
 of DE, 357-358
 of a quadratic second order system, 419-420
 of a quadratic third order DE, 420
Internal gravity waves, 324
Inverse scattering transform, 138
Irrational right hand sides
 second order nonlinear DE with, 452-454
 singular solutions of DE with, 453-454
Irregular singular point, 74
Isothermal atmosphere
 nonlinear periodic solutions and, 326-327
 nonlinear travelling waves and, 324
Iteration for nonlinear BVP, 175
Iterative solutions for IVP, 151-153

K-dV equation and PI, 429, 431
Kuramoto model-steady-singularities of and chaos in, 413-418

Laminar boundary layer, BVP for, 244
Laminar convection over a linear heat source, solution of BVP for, 106

Langmuir-Blodgett equation, 55
Langmuir-Boguslavski equation, 55
Langmuir's equation, 62
Laurent series, 140
Levinson's method, 154-158
Lindelöff's majorant, 112-116
Linearly independent solutions, 10
Linearization (exact) of second order autonomous nonlinear DE, 42
Linear second order DE
 relation to nonlinear second order DE, 37
Local analysis of DE, 131-132
 -method of Painlevé, 132
Lorenz system
 integrable cases of, 375-377
 integrability of and chaos in, 365-394
 numerical solutions of and singularities in, 386-394
 Painlevé analysis of, 378-382
 perturbation analysis for periodic solutions of, 368-374
Lotke-Volterra model
 Painlevé property of, 361, 364

Majorants, 108
 for first order DE, 109-110
 Cauchy, 111
 Lindelöff, 112-114
Malmquist theorem, 96
Modified Koreteweg-deVries equation and second Painlevé equation, 139, 428, 432
Movable singular points, 422

Natural convection with viscous dissipation, BVP for, 244
Nature of singular point of two first order linear DE, 256
 center, 258
 nodal star, 259
 node, 256
 saddle, 257
 vortex point, 258
Nonharmonic oscillator
 nonlinear, 100

Nonlinear Bessel equation
  asymptotic solution of, 178-180
  BVP for, 175
  integral equation formulation, 178
  series solutions of, 177
Nonlinear diffusion equation
  BVP solution of, in the phase plane, 314
  similarity solutions of, 308
  study in the phase plane, 309-314
Nonlinear PDE - solutions of - as infinite series, 512
Nonlinear Schroedinger equation and Painlevé equation, 431
Nonsimple singular points for two-dimensional systems, 273-284
Normal quadratic systems
  applications of, 287
  classification of singular points for, 292-301
Nth order system of DE
  critical points of, 246
  unique solution of, 246
N wave solutions, 514
  cylindrical, 517
  planar, 514
  spherical, 518

Order of DE, 3
Ordinary point, 69
Oregonator, 361

Painlevé's α-method, 424
Painlevé's fifth equation
  Briot-Bouquet system and, 494-495
  degenerate cases of, 501
  radial wave equation and, 497
  rational solutions of, 499-501
  relation to PIII, 498, 501
  Riccati equation and, 497
  solution about fixed singularities, 493
  solutions with poles at zero, 495-496
Painlevé's first equation, 460
  and beyond, 472
  asymptotic behavior for, 468-469, 466-467

[Painlevé's first equation]
  BVP, 472
  IVP for, 468
  numerical study of, 469
  qualitative results for, 470-472
  shooting techniques for, 472
  solution with movable poles, 461-468
  solution as quotient of integral functions, 471-472
Painlevé's fourth equation, 482
  Hermite equation and, 492
  rational solutions of, 489, 490-493
  solution at infinity, 488
  solution as ratio of entire functions, 483
  solutions with a simple movable pole, 482
  systems related to, 483-486, 489-490
Painlevé's property, 138-139
  algorithm for testing nth order DE for, 140
  algorithm for testing a system of DE for, 147
  conjecture of Ablowitz, Romani and Segur, 138
Painlevé's second equation
  Airy function and, 476, 480
  asymptotic analysis of, 473, 474
  connection problem for, 475
  dependence of solution on amplitude parameter, 475-478
  numerical study of, 475, 478-482
  relation to K-dV equation, 474-475
  solution with movable simple pole, 473, 475
  Taylor series solution for, 473
Painlevé's second transcendent
  asymptotic form, 184-185
  connection problem for, 181
  integral equation formulation for, 183
  numerical solution of, 183, 185
  symmetrics of, 181
Painlevé's sixth equation, 424
Painlevé's theorem for first order DE, 84, 85-86

Index   577

Painlevé's third equation, 501
    asymptotic representation of solu-
        tions of, 508-510
    expansions of solutions about zero,
        506-507
    perturbation solutions about unity,
        507
    relation to PV, 505-506
    simplified forms of, 502, 503
    solutions to, rational in $z^{1/3}$, 503
    systems equivalent to, 504, 505
Parabolic PDE, 2
Parametric relations between
    solutions of PIII, 448-449
    solutions of PIV, 449-450
    solutions of PV, 450
    solutions of PVI, 450-452
Phase line, singular points on, 248
Phase plane, singular points in, 250
Phase space, 3 dimensional analysis,
    330
Piston problem
    general plane, 537
    spherical and cylindrical, for a
        converging shock, 529
Plasma physics
    a DE arising in, 240
    a 3-dimensional model from, 361, 364
    connection problem in, 240
    solution of a connection problem in,
        240-244
Poisson-Boltzmann equation, 60
Porous media equation, 223
    BVP for, 224
    existence of solutions for BVP for,
        224-230
    similarity forms of, 223
    weak solution for, 224
Psi series, 359
    solution of nonlinear DE, 96

Quadratic systems
    homogenous, 286
    normal form of, 285
Quartic lattice system
    free end Toda lattice, 411-413

{Quartic lattice system]
    integrability of, 409-411

Radius of convergence of series, 127
    computation of, 127-130
    theorems for, 128
Rational solutions of PII, 447
Regular singular point, 72
Relation of
    PII to itself, 446-447
    PII to PXXXIV, 445
Repellor, 354
Resonances in power series expan-
    sions, 141
Riccati equation, 28, 83, 85
    linearization of, 28-29
    and PV, 497
Rikitake two-disk system, Painlevé
    property of, 360, 362-364

Scale invariance, 24
Second order nonlinear DE
    relation to second order linear DE,
        37
    relation to third order linear DE,
        33
Series solution
    computational aspects of, 97
    for nonlinear DE, 97
    radius of convergence of, 97
Shallow water equations
    dam break initial conditions for,
        539-540
    gravity currents releases and, 538
Shock wave, converging, 529
Similarity solutions as intermediate
    asymptoties, 512
Simple pendulum
    nonharmonic nonlinear oscillator,
        100-102
    series solution of IVP for, 103-108
Sine-Gordon equation
    and PIII, 428-429
Singularities
    essential, 423
    fixed, 421

[Singularities]
  movable, 422
Singularities of nonlinear DE, 79
  essential singularities, 82
  isolated singularities, 82
  logarithmic branch points, 81
  poles, 82
Singular points at infinity for a two dimensional linear system, 268-273
Singular points of second order linear DE, 9
Special elementary solutions of
  PIII, 436-437
  PVI, 438-442
Spherical cavity
  collapse of-in a perfect gas, 520
  global solutions for, 521-529
Stability analysis of similarity solutions for arbitrary current releases, 544
Symmetries of DE, 24

Third order linear DE
  relation to second order nonlinear DE, 33
Thomas-Fermi equation, 54
  asymptotic solutions of, 164
  BVP for, 168, 218
  embedding of asymptotic solutions for, 172, 174
  IVP using majorants, 115

Three dimensional linear systems
  canonical forms for, 331
  classification of singular points, 330
  phase space study of, 334-350
Transformations
  of nonlinear nonautonomous second order DE to Lie plane, 58
  of nonlinear nonautonomous second order DE to phase plane and a subsidiary DE, 52
  of nonlinear second order DE to integrable second order DE, 48
  which do not "work", 62, 65
Transformations of Painlevé equations, 442-452
Travelling waves
  for Fisher's equation, 305-306
  in an isothermal atmosphere, 324
Two dimensional linear systems
  canonical forms of, 254
  secular equation of, 253, 255

Unpredictability, 352
Uniqueness of solution of an IVP, 8

Variation of parameters
  for homogeneous linear DE, 13-14
  for inhomogeneous linear DE, 14-15

Well-posed problem, 7
Wronskian, 10